PERSPECTIVES IN CONDENSED MATTER PHYSICS
A Critical Reprint Series

Condensed Matter Physics is certainly one of the scientific disciplines presently characterized by a high rate of growth, both qualitatively and quantitatively. As a matter of fact, being updated on several topics is getting harder and harder, especially for junior scientists. Thus, the requirement of providing the readers with a reliable guide into the forest of printed matter, while recovering in the original form some fundamental papers suggested us to edit critical selections on appealing subjects.

In particular, the present Series is conceived to fill a cultural and professional gap between University graduate studies and current research frontiers. To this end each volume provides the reader with a critical selection of reprinted papers on a specific topic, preceded by an introduction setting the historical view and the state of art. The choice of reprints and the perspective given in the introduction is left to the expert who edits the volume, under the full responsibility of the Editorial Board of the Series. Thus, even though an organic approach to each subject is pursued, some important papers may be omitted just because they lie outside the editor's goal.

The Editorial Board

PERSPECTIVES IN CONDENSED MATTER PHYSICS
A Critical Reprint Series: Volume 4

ELECTRONIC STRUCTURE OF METAL-SEMICONDUCTOR CONTACTS

edited by

Winfried Mönch
Universität Duisburg

KLUWER ACADEMIC PUBLISHERS

Jaca Book

Softcover reprint of the hardcover 1st edition 1990

prima edizione
ottobre 1990

copertina e grafica
Ufficio grafico Jaca Book

Library of Congress Cataloging-in-Publication Data

Electronic structure of metal-semiconductor contacts / edited by W.
Mönch.
p. cm. -- (Perspectives in condensed matter physics ; v. 4)
Includes index.
ISBN-13:978-94-010-6780-5
1. Electric contacts. 2. Semiconductor-metal boundaries.
I. Mönch, W. (Winfried) II. Series.
TK7872.C68E44 1990
621.381'52--dc20 90-40572

ISBN-13:978-94-010-6780-5 e-ISBN-13:978-94-009-0657-0
DOI: 10.1007/978-94-009-0657-0

per informazioni sulle opere pubblicate e in programma
ci si può rivolgere a Editoriale Jaca Book spa
via Rovani, 7, 20123 Milano, telefono 4988927

To Gisela

Preface

Interface and surface science have been important in the development of semiconductor physics right from the beginning on. Modern device concepts are not only based on p–n junctions, which are interfaces between regions containing different types of dopants, but take advantage of the electronic properties of semiconductor-insulator interfaces, heterojunctions between distinct semiconductors, and metal-semiconductor contacts. The latter ones stood almost at the very beginning of semiconductor physics at the end of the last century.

The rectifying properties of metal–semiconductor contacts were first described by Braun in 1874. A physically correct explanation of *unilateral conduction*, as this *deviation from Ohm's law* was called, could not be given at that time. A prerequisite was Wilson's *quantum theory of electronic semi–conductors* which he published in 1931. A few years later, in 1938, Schottky finally explained the rectification at metal–semiconductor contacts by a space–charge layer on the semiconductor side of the junction which is depleted of mobile majority carriers. In appreciation of his fundamental contributions to the understanding of the electronic properties of metal–semiconductor interfaces they are also called *Schottky contacts.* The physical mechanisms, which are responsible for the existence of these depletion layers, have remained under discussion since then.

Almost all of the recent papers contributing to this subject are emphasizing the lack of a comprehensive physical theory of metal–semiconductor contacts. Chemical approaches, on the other hand, often proved to be very successful but are mostly intuitive in nature. A most typical example is the electronegativity concept which had to wait for a quantum mechanical foundation until a few years ago. The electronegativity concept of charge transfer in chemical bonds was finally generalized for solid interfaces and specifically applied to establish chemical trends in electronic properties of metal–semiconductor interfaces. Another hindrance towards a comprehensive model of metal–semiconductor contacts was the search for one single mechanism governing their extremely complex behavior. Recently, however, this path was left and a competition between at least two concurring mechanisms becomes increasingly appreciated.

The Introduction to the articles reprinted is not intended as a review. It rather tries to set a frame for the articles to follow. Many of the papers included in this volume are classical contributions which had a great impact on the development of this field. The selection of other ones will be considered to be a very personal one. Furthermore, not all colleagues will agree with the interpretations of experimental as well as theoretical results given in the Introduction. This, however, seems to be unavoidable in a scientifically active field such as the physics of Schottky barriers.

Many of the interpretations and ideas presented in the Introduction were induced by numerous discussions with colleagues all around the world and with the members of my own research group, and I gratefully acknowledge these stimulations. Furthermore, I should like to thank the Deutsche Forschungsgemeinschaft for the continuous support of my work.

Mülheim a.d. Ruhr, February 1990. Winfried Mönch

Table of Contents

W. Mönch, Electronic Structure of Metal-Semiconductor Contacts

Introduction

1. Preliminary Remarks

In 1874, Braun first reported on the rectifying properties of contacts between metals and metal sulfides[R1]. Power rectifiers, using selenium and cupreous oxide, and crystal rectifiers, made by pressing fine metal wires onto, for example, lead sulfide crystals were fabricated and applied in the fast growing electronic industries already at the beginning of the present century. This technical development is marked by patents issued in 1925[1] and 1904[2], respectively.

While metal–semiconductor diodes were empirically developed a physical understanding of rectification at such contacts had to wait until much later. Already Braun carried out some experiments aiming at a physical explanation of his discovery. He excluded thermoelectric and polarization effects to be responsible and concluded the presence of a thin surface layer of extremely high resistivity. Some 30 years later, more detailed experimental results were available but Pierce[3] still proposed '*to accumulate further data before attempting an explanation of unilateral conduction in crystals*'. A huge step forward was eventually made by Schottky and Deutschmann[R2] in 1929 when they measured the differential capacitance of cupreous oxide rectifiers. They modeled the barrier layer as a parallel capacitor and estimated its thickness as approximately 3×10^{-5} cm. Their conclusion was that in such surface layers a negative space charge exists which considerably penetrates into the semiconductor. This behavior of semiconductors clearly differed from what was known for metals.

After the quantum theory of semiconductors was published by Wilson[4] in 1931 and the positive Hall coefficient of cupreous oxide was correctly determined by Fritsch[5] in 1935 Schottky finally completed his *Halbleitertheorie der Sperrschicht*[R3] (Semiconductor Theory of the Barrier Layer) in 1938. He attributed the rectifying behavior of metal–semiconductor contacts to a depletion of mobile majority carriers on the semiconductor side of the junction causing a space charge which is carried by ionized, static impurities. This is explained schematically on Fig. 1. In the case of cupreous oxide, which is a p–type semiconductor, barrier layers are depleted of holes leaving behind a negative space charge formed by ion ized acceptors. This exactly is what Schottky and Deutschmann[R2] had already concluded from their earlier measurements of the differential capacitance at cupreous oxide rectifiers. The long and "thorny" development of both the experimental data and the theoretical model was described in detail by Schottky himself in one of his papers[6] following the pioneering one[R3].

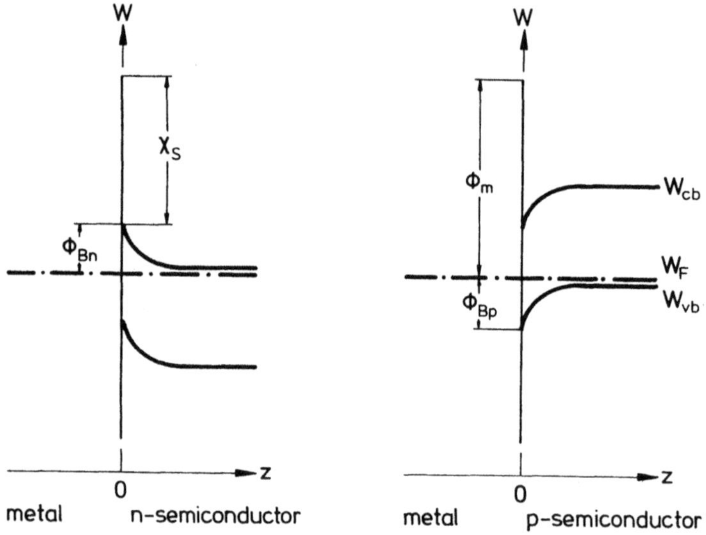

Fig. 1. Schematic band diagrams at abrupt Schottky contacts exhibiting no interface states for semiconductors doped n– and p–type, respectively.

To honor Walter Schottky, depletion layers at interfaces of metal–semiconductor contacts are frequently called *Schottky barriers.* Rectifying metal–semiconductor junctions are also labeled *Schottky contacts* and such devices are also named *Schottky diodes.*

In a paper following his key publication, Schottky[7] derived the current–voltage (I/V) characteristics of such junctions by using his depletion layer model. That characteristics may conveniently be written as (see, for example, reference 8)

$$j = j_s \exp(e_0 V/nk_B T)\,\{1 - \exp(-e_0 V/k_B T)\}\,, \tag{1a}$$

where the saturation–current density is given by

$$j_s = A^* T^2 \exp(-(\Phi_B - \Delta\varphi_{if})/k_B T) \,. \tag{1b}$$

Here, A^* is the effective Richardson constant, $\Delta\varphi_{if}$ accounts for the image–force lowering of the barrier, and n is the ideality factor which is close to the value of unity for ideal junctions at room temperature. The most important parameter is the barrier height Φ_B which Schottky originally called *work function of the majority carriers at the interface.* It measures the energy distance between the Fermi level and the valence–band maximum or the conduction–band minimum at metal interfaces with semiconductors doped p– and n–type, respectively, i.e.

$$\Phi_{Bp} = W_F - W_{vi} \,, \tag{2a}$$

and

$$\Phi_{Bn} = W_{ci} - W_F \,, \tag{2b}$$

respectively. Since Schottky published his basic paper in 1938, i.e. for the past 50 years, the physical mechanisms determining barrier heights in Schottky contacts have remained under discussion. The reasons for that are manifold from both experiment and theory.

The interactions between metals and semiconductors are generally complicated. Abrupt and epitaxial interfaces, which are the most ideal and theoretically the easiest to handle cases, are rarely observed. Mostly, metals evaporated on semiconductors are disrupting surface bonds of the substrate even at room temperature. This may lead to an atom exchange in the top substrate plane or to thicker reacted layers. Such films usually consist of alloys but may also contain compounds between substrate and metal atoms. On the other hand, solid phase reactions and other growth techniques were intentionally used for growing epitaxial films of compounds between metal and semiconductor–substrate atoms. The resulting interfaces are abrupt in most cases. Here, the silicide–silicon contacts have to be mentioned and $NiSi_2/Si$ is the most prominent example among them. Films of nickel disilicide may be grown on Si(111) surfaces in two orientations and, most remarkably, the barrier heights of both types of epitaxial contacts differ by more than two tenths of an eV. This behavior is not yet completely understood.

In general, the knowledge on the chemical composition and the atomic arrangement at metal–semiconductor interfaces is very limited for various reasons. Even if the long–range order were known for a specific epitaxial contact its structural characterization would most certainly be incomplete since interface defects will most probably be present but escape experimental detection. Charged defects, however, will modify the barrier height with respect to what is observed with a perfect junction. Theoretical studies on electronic properties of Schottky contacts thus have to start from simplified models anyway. Such more conceptual approaches have advantages and deficiencies. On the one hand, they are providing trends while even the most detailed investigation of a specific contact is a case study and gives one data point only. On the other hand, such conceptual approaches are mostly considering the influence of one parameter only which then results in a monotonous trend of, for example, the barrier height on that quantity.

Smaller subsets of experimental values of Φ_B were often found to exhibit such simple correlations while the main body of experimental data could not be explained by such *linear models*. The resulting and long–standing discrepancy between experimental and theoretical results has to be attributed to the complicated interactions of metals with semiconductors. Recently, however, two approaches were made to reconcile experimental and theoretical data in this field. One is the application of electrochemical techniques for metal deposition[R25] which aim to preventing major disturbances at the interfaces prepared. Secondly, more than just one of the conceptual models was considered[R28, R35] in explaining the great wealth of experimental data collected. Following both paths, reciprocal results were obtained.

2. Conceptual Models

A metal semiconductor contact may be formed in a *Gedanken experiment* by gradually decreasing the distance d between a metal and a semiconductor. The two samples facing each other will be charged since their work functions generally differ. Naturally, the surface charge Q_m and Q_s per unit area on the metal and the semiconductor, respectively, are equal but of opposite sign and the condition of charge neutrality thus reads

$$Q_m + Q_s = 0 . \tag{3}$$

For a parallel plate capacitor arrangement, the contact potential difference or, with other words, the difference of the work functions and the surface charge density on the plates, which are separated by a vacuum gap of width d, are related as

$$\Delta(d) = \Phi_m - \Phi_s = \frac{e_o}{\epsilon_o} Q_s d . \tag{4}$$

Here, e_o and ϵ_o are the electronic charge and the permittivity of vacuum, respectively. The charge on the semiconductor will be found in a space charge layer, the extension of which is governed by the concentration of the bulk impurities, and possibly also in surface and, when the contact has formed, interface states. A few specific cases will be discussed in the following.

2.1. Schottky–Mott Rule

In the least complicated example, which was first proposed and analyzed by Schottky, the surface charge on the semiconductor resides in a depletion layer only. In this case, the condition of charge neutrality (3) then reads

$$Q_m + Q_{sc} = 0 . \tag{5}$$

In depletion layers, the space charge density is related to the bulk dopant concentration and the band bending $e_o|V_i| = (W_{cb} - W_{ci})$ at the surface as

$$Q_{sc} = +\sqrt{2\epsilon_s \epsilon_o N_d e_o |V_i|} , \tag{6}$$

to better than 10% for $e_o|V_i| \geq 5k_BT$. The semiconductor (dielectric constant ϵ_s) is assumed to be doped n–type with N_d donors per unit volume. For p–type doping, the respective acceptor concentration has to be considered, and the sign on the right side of (6) has to be inverted. The subscript "i" is used to indicate that the same relation also applies to interfaces which are the subject dealt with here. An abrupt metal–semiconductor contact is obtained in the limit of d → 0, and relation (4) gives $\Delta(0) = 0$ or

$$\Phi_{Bn} = \Phi_m - \chi_s , \tag{7}$$

which is the famous Schottky–Mott rule[R4,R5]. Here, χ_s is the electron affinity of the respective semiconductor.

Schweikert was the first to test this rule already in 1939. He evaporated 21 different metals on selenium and measured the initial resistance of the resulting rectifiers which is obtained as $R_o \propto \exp(\Phi_B/k_BT)$ from (1). For Schottky contacts made by using one and the same metal but semiconductor substrates doped p– and n–type, the barrier heights are adding up to the band–gap energy of the semiconductor, i.e.

$$\Phi_{Bp} + \Phi_{Bn} = W_g , \tag{8}$$

as was already noted by Schottky[R5]. Relation (8) was experimentally checked with different semiconductors and was always found to be valid. Selenium is a p–type semiconductor and the initial resistance of selenium rectifiers is thus expected to vary as $\exp(-(\Phi_m - \chi_s)/k_BT)$. Schweikert's data were published by Schottky and are shown in Fig. 2 of reprint R5. Obviously, the barrier heights are varying proportional to the work functions of the metals used but the slope

$$S_\Phi = \frac{d\Phi_B}{d\Phi_m} \tag{9}$$

is much smaller than unity in disagreement with what is *predicted* by the Schottky-Mott rule (7).

2.2. Interface States

Already in 1947, Bardeen[R6] developed the basic ideas to resolve this discrepancy. He considered electronic states localized at the interface and exhibiting levels within the energy gap of the semiconductor in addition to the impurities in the interior of the semiconductor. Such interface states may be either discrete or have a continuous distribution. Interface states may become charged and the charge on the semiconductor side of a Schottky contact is then composed as $Q_s = (Q_{sc} + Q_{is})$. The condition of charge neutrality at the interface (3) now reads

$$Q_m + Q_{sc} + Q_{is} = 0. \tag{10}$$

Since Bardeen published his pioneering paper, i.e. for the past 50 years, the physical nature of the interface states at Schottky contacts has remained under discussion which was very controversial from time to time.

In his basic paper[R6], Bardeen also pointed out that charged interface states are accompanied by a double layer of charge. Part of the total potential drop from the metal to the semiconductor will then occur across that double layer and part across the depletion layer which was only considered in the Schottky–Mott approach. The inclusion of interface states in the *Gedanken experiment* mentioned at the beginning of this paragraph now means that the metal and the semiconductor remain separated by the thickness of that double layer when the intimate contact is formed.

Following Bardeen's ideas, Cowley and Sze[R7] phenomenologically included a continuum of interface states in the *Gedanken experiment* mentioned. They assumed the density D_{is} of interface states to be constant across the energy gap and to have a charge neutrality level at W^0_{is}. This is shown schematically in Fig. 2. Most importantly, the interface states were postulated to be a property of the semiconductor only and thus to be independent of the specific metal used. Within this model the barrier height was found to vary as (see eqation (9) of R7)

$$\Phi_{Bn} = S_\Phi(\Phi_m - \chi_s) + (1 - S_\Phi)\,\Phi^0_{Bn}, \tag{11}$$

where $\Phi^0_{Bn} = (W_{ci} - W^0_{is})$ is the barrier height when the charge neutrality level of the continuum of interface states coincides with the Fermi level at the interface.

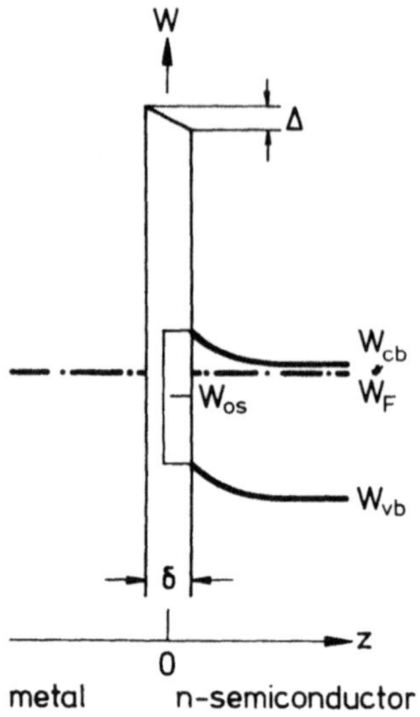

Fig. 2. Schematic band diagram at intimate Schottky contacts exhibiting a continuum of interface states on semiconductors doped n–type.

The slope parameter results as (see equation (7c) of R7)

$$S_{\Phi} = \left(1 + \frac{e_0}{\epsilon_i \epsilon_0} D_{is} \delta_i\right)^{-1}, \tag{12}$$

where ϵ_i is the dielectric constant of the interfacing double layer and δ_i is its width. For a specific semiconductor, relations (11) and (12) clearly give a linear variation of the barrier height as a function of the metal work–function. The main effect of the interface states is to reduce the slope parameter S_{Φ} in comparison with the Schottky–Mott approach which considers no interface states at all and obtains $S_{\Phi} = 1$ in agreement with (12).

2.3. Continuum of Metal–Induced Gap States

At the same time when Cowley and Sze presented their phenomenological analysis on the effect of interface states on the barrier heights of metal–semiconductor contacts, Heine[R8] proposed the first conceptual model on the physical origin of interface states at Schottky contacts. He considered the matching of the wavefunctions at such junctions and concluded that '*for energies in the semiconductor band gap the volume states of the metal all have tails in the semiconductor*' and '*there are actually no surface states*'. The latter statement revised a notion proposed earlier by Bardeen who had indeed assumed the surface states of the initially clean surface to persist under a metal overlayer.

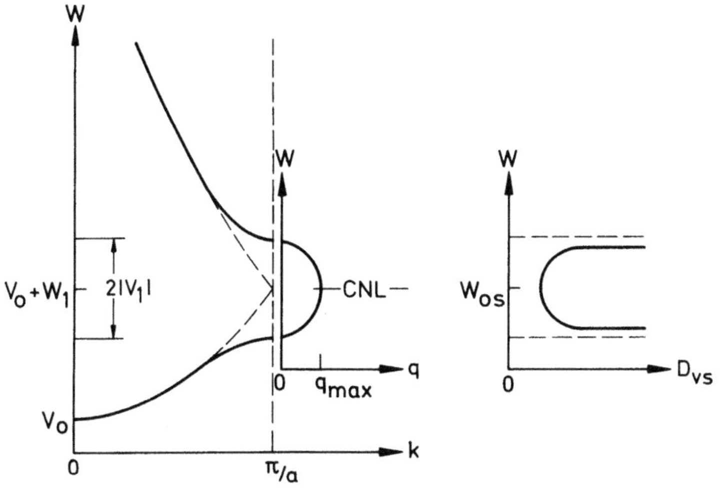

Fig. 3. Band scheme showing virtual gap states and their density of states for a one–dimensional linear lattice.

Heine showed that the exponentially decaying tails of the metal wavefunctions are derived from the continuum of the virtual gap states (ViGS) of the complex semiconductor band structure. These ViGS are solutions of Schrödinger's equation for complex wavevectors at the center and the boundary of the first Brillouin zone. The analysis is most simple for a one–dimensional lattice or linear chain[9,10]. Fig. 3 shows the variation of the imaginary part of the wavevector, the reciprocal of which is a decay length, across the band gap of a linear chain. Considering the respective boundary conditions, the ViG states are not only the source of interface states at junctions between solids but also of surface states at clean and of adsorbate–induced surface states at adsorbate–covered surfaces. For the latter, an example is sketched in Fig. 8b of reprint R8.

With the continuum of metal–induced gap states (MIGS), as they were called later, Heine has presented a physical concept of the interface states at metal–semiconductor contacts. The charge–neutrality level of the MIG states is to be identified as the branch point of the ViG states, where their character changes from predominantly valence–band– or donor–like to mostly conduction–band– or acceptor–like. Although the density of states of the MIGS is by no means constant across the band gap , as shown in Fig. 5 of reprint R8, the phenomenological model worked out by Cowley and Sze may still be used since that density of states is almost flat around its charge–neutrality level.

A rearrangement of (12) shows the slope parameter S_{Φ} to be only determined by the product $D_{is}\delta_i$ as

$$\frac{1}{S_{\Phi}} - 1 = \frac{e_o}{\epsilon_i \epsilon_o} D_{is}\delta_i . \tag{13}$$

For a linear chain, the density of states and the decay length of the virtual gap states vary proportional to the inverse of the width of the band gap. The average band–gap width of a three–dimensional semiconductor is related to the energy of its bulk plasmons and the electronic part of its dielectric constant as

$$\epsilon_{\infty} - 1 = \left[\frac{\hbar\,\omega_p}{<W_g>}\right]^2 . \tag{14}$$

The plasmon energies of the semiconductors considered here differ by 10% only. In consequence, the slope parameters of Schottky contacts are expected to vary as (see equation (20) of reprint R28)

$$\frac{1}{S_{\Phi}} - 1 \propto (\epsilon_{\infty} - 1)^n . \tag{15}$$

Before results of elaborate theoretical approaches and experimental data will be discussed the effect of discrete interface states on the barrier height shall be looked at.

2.4. MIGS–plus–Defects Model

Bardeen[R6] already mentioned that well–defined defects sparsely distributed at the interface would have discrete levels. Depending on the position of the Fermi level, interface defects may become charged. Consequently, they have to be included in the condition of charge neutrality at the interface which instead of (10) then reads

$$Q_{m} + Q_{sc} + Q_{is} + Q_{id} = 0\,. \tag{16}$$

The occupancy of the interface defects is determined by the Fermi–Dirac distribution function $f_{o}(W_{id} - W_{F})$ and, for defects of donor type, one obtains

$$Q_{id} = +e_{o}N_{id}\{1 - f_{o}(W_{id} - W_{F})\}\,. \tag{17}$$

Here, N_{id} and W_{id} are the surface density and the energy level of the interface donors. Considering the now modified charge density $Q_{s} = (Q_{sc} + Q_{is} + Q_{id})$ on the semiconductor side of the interface in relation (4), the barrier heights are found to vary as

$$\Phi_{Bn} = S_{\Phi}(\Phi_{m} - \chi_{s}) + (1 - S_{\Phi})\,\{\Phi^{o}_{Bn} - \frac{N_{id}}{D_{is}}(1 - f_{o})\}\,. \tag{18}$$

Interface donors are thus lowering the barrier height with respect to what is expected for interfaces free of defects. For $\Phi_{Bn} > (W_{ci} - W_{id})$, the interface donors are all charged and the barrier height is then lowered by

$$\delta\Phi^{id}_{Bn} = -(1 - S_{\Phi})\frac{N_{id}}{D_{is}}\,, \tag{19}$$

which is the maximum value for a given density of interface donors. Interface acceptors, on the other hand, will increase the barrier heights.

Here, it shall be mentioned that Zur *et al.*[11] published computational results on the influence of defects on barrier heights in Schottky contacts. They, however, negated the continuum of MIG states. This means that Zur *et al.* added discrete interface states to the Schottky–Mott model. As a consequence, they computed graphs which are most easily obtained from (18) by setting $S_{\Phi} = 1$, which is the Schottky–Mott limit of the slope parameter.

The MIGS–plus–defects model again provides trends. The barrier heights are found to vary linearly as a function of the metal work–function as long as the density of interface defects is either negligible or the same for all metals deposited. If, however, the density of interface defects depends on either the metals used or the technique employed for depositing them onto the semiconductor the relationship between the barrier heights and the metal work–functions is no longer expected to be linear.

2.5. Charge Transfer at Interfaces: MIGS–and–Electronegativity Model

In the phenomenological models discussed so far, the clean metal and semiconductor surfaces, which are joined to form an intimate contact, are characterized by their work functions. It was argued[12] that the work function of a metal may be thought of as composed of two contributions; one is caused by a dipole layer, which varies with surface orientation, and the other one was related to the electronegativity of the metal. The dipole layer was postulated to change in an intimate contact with another solid. Later on, the barrier heights of GaAs Schottky diodes were indeed found to be the same independent of whether the substrates were (110)– or (100)–oriented. The electronegativity of a metal thus seems to be a more relevant property rather than its work function in discussing chemical trends of barrier heights.

In a more conceptual approach, which is favored here, the *charge transfer across interfaces* between solids is thought to be of fundamental importance. Therefore, chemical ideas were employed. Chemical bonds between different atoms are partly ionic. This was learned from, for example, the dipole moments and bond lengths of molecules. Pauling[13] considered diatomic molecules formed by monovalent atoms and estimated the ionicity of such single bonds from the difference of the respective atomic electronegativities. This extremely successful concept was also adopted for explaining sign and size of surface dipoles induced by adatoms on semiconductors[14]. Finally, Mönch[R27] generalized it for modeling chemical trends in the charge transfer at solid–solid and, especially, in metal–semiconductor interfaces. In consequence, barrier heights are plotted versus the electronegativity difference of the semiconductor and the metals in contact.

A MIGS–and–electronegativity model thus *predicts* the barrier heights for such junctions where the electronegativities of the metal and the semiconductor in contact are equal. Across such interfaces no charge transfer is expected and the Fermi level should, therefore, coincide with the charge–neutrality level (CNL) of the continuum of metal–induced gap states, as was first pointed out by Mönch[R27]. The CNL of the MIG states is identical with the branch point of the ViG states of the complex band structure of the semiconductor. This model may also be applied to evaluate the alignment of the band structures across the interfaces in semiconductor heterostructures and semiconductor–insulator junctions[R27]. As long as the electronegativities of such materials in contact are equal the respective branch points of the ViGS will line up at the interface. An additional interface dipole has to be considered when the electronegativities differ[R27]. Similar conclusions were also reached by Tejedor and Flores[15] and, later on, by Tersoff[16].

3. Computational Results

Elaborate theoretical studies on electronic properties of metal–semiconductor contacts were presented by the groups around Flores in Madrid and around Cohen in Berkeley, California. Their detailed calculations[R10-R13] are case studies for Al contacts with silicon as well as various III–V and II–VI compound semiconductors. The results obtained excellently confirm the concepts first proposed by Heine[R8]. Most importantly, the wavefunctions of the metal electrons were indeed found to decay exponentially across the interface into the semiconductor and these tails were also shown to carry electronic charge. This is demonstrated by Figs. 7 of reprint R12 and 8 of reprint R13. Furthermore, Figs. 5 of reprint R11 and 4 of reprint R12 show the density of states of these MIG states to be U–shaped as expected from Heine's concept. Both the magnitude of the density of states and the decay lengths of the MIG states are in excellent agreement with what is estimated from the simple, one–dimensional model.

Following another of Heine's early proposals, which is also found in reprint R8, the effective distance between the electronic charges on both sides of the interface may be approximated by (see equation 19 of R8)

$$\delta_i = \frac{t_m}{\epsilon_m} + \frac{t_s}{\epsilon_s}, \tag{20}$$

where t_m/ϵ_m is the screening length in the metal and t_s/ϵ_s is the tailing length of the wavefunctions of the metal electrons into the semiconductor divided by an appropriate dielectric function. By inserting their computed values for the densities of states and the effective decay lengths of the MIG states into relations (20) and (12), both groups also estimated slope parameters S_X for silicon as well as some of the III–V and II–VI compounds. In the spirit of conceptual arguments given above, they substituted the electronegativities of the metals for their work functions and made use of an empirical linear relation between both quantities.

The influence of charged interface defects on the barrier heights of metal—semiconductor contacts was also examined by Zhang *et al.*[R24] in a detailed theoretical computation. At an Al/Si(111) interface they replaced the first atomic layer of Si closest to the Al boundary by a sheet of positively charged "donors". This "doping" of the interface was found to reduce the barrier height by almost three tenths of an eV compared with an interface free of any such defects. This computational result again confirms conclusions drawn from much simpler, conceptual approaches which were discussed above.

Almost 20 years after Heine had published his pioneering paper, Tersoff[R20] revitalized the discussions on the interconnection between barrier heights and the continuum of metal–induced gap states in Schottky contacts. Tersoff calculated the branch–point energies of the complex band structure for the group IV semiconductors and most of the III–V and II–VI compounds (see, for example, reference 17). His results agree with a smaller set of data evaluated by Tejedor and Flores[15] already some years earlier. Despite the impact of Tersoff's publications, they also added some confusion to the discussions. Tersoff repeatedly stressed[R20] that '*by finding the branch point in the complex energy bands, we immediately have a 'canonical' barrier height for the given semiconductor'*, although he also considered that '*deviations from the canonical barrier heights calculated here may be attributed, at least in part, to the metal–semiconductor electronegativity differences*'. His first statement sets the slope parameters S_Φ to zero but its mitigation would also allow values $S_\Phi > 0$ and, in consequence, the presence of interface dipoles. The latter, however, was rarely realized as demonstrated by many publications only referring to Tersoff's first mentioned statement.

A transfer of electronic charge across Al/Si interfaces into MIG states and thus an interface dipole were already obtained by Louie and Cohen[R12]. Batra and Ciraci[R26] investigated Al/Ge interfaces. Their computed charge–density contours, which are plotted in Fig. 1 of reprint R26, showed the bonding charge in Al–Ge bonds at the interface to be slightly shifted towards the Ge atoms. Pauling's concept and its generalization to interfaces, which were mentioned above, are explaining this finding by the slightly larger electronegativity of germanium in comparison with aluminum.

All the theoretical approaches using different levels of complexity demonstrated the prime importance of metal–induced gap states, which are a property of the semiconductor, in determining the barrier height at a given metal–semiconductor contact. The experimental data, however, did not agree with such a linear model, as will be discussed in the § 4. Therefore, these theoretical results and even the possible lowering of barrier heights by additional interface donors were not fully appreciated by most of the experimentalists until the beginning of 1986.

4. Experimental Data

This paragraph is devoted to the *device approach to metal–semiconductor contacts* which means that barrier heights determined experimentally with Schottky diodes will be considered only. Results on the formation of Schottky barriers during the successive evaporation of metals on semiconductor or, with other words, the *surface science approach to metal–semiconductor interfaces* will be considered in § 5. The following discussion only considers barrier heights determined from I/V characteristics. Most recently, scanning tunneling microscopy was applied for investigations of buried metal–semiconductor interfaces. By using a STM, Kaiser and Bell[R34] successfully mapped variations of the barrier height in Schottky diodes with nanometer spatial resolution. In this new application of scanning tunneling microscopy and spectroscopy, electrons are injected from the STM tip through the metal overlayer into the conduction band of the semiconductor substrate. The principles of this new method, which its inventors named ballistic–electron–emission microscopy (BEEM), is explained in Fig. 1 of reprint R34. By using this technique, the electronic properties of buried metal–semiconductor interface can be directly probed with nanometer spatial resolution. Two examples, near–ideal Au/Si and inhomogeneous Au/GaAs interfaces, are illustrating the potential of BEEM. This technique, however, is not limited to investigations of Schottky diode but may be also used in studies of interfaces in semiconductor heterostructures.

As was frequently mentioned, barrier heights experimentally observed with metal–semiconductor contacts do not meet the linear trends as proposed by the theoretically well–founded MIGS model. Data reported for Si and GaAs Schottky diodes are shown in Figs. 3 of reprint R28 and 1 of reprint R35, respectively. Until the MIGS–plus–defects model was eventually proposed[R28,R35], a number of correlations between smaller or larger subsets of experimental data and different materials parameters were suggested. Some of the proposals which had an impact on the discussions are reprinted in this volume. They will be briefly discussed first and possible implications for the MIGS–and–defects model will also be considered.

4.1. Chemical Trend of the Slope Parameters

Already in 1969, Kurtin *et al.*[R9] correlated slope parameters S_X with the ionicity of the semiconductor and insulator substrates which had been used for fabricating Schottky diodes at that time. They modeled the ionicity of a compound by the electronegativity difference of its anions and cations. Their data plot suggested a covalent–ionic transition for $\Delta X \approx 0.7$ when Pauling's electronegativity units are used. Almost 10 years later, Schlüter[R14] reexamined these experimental data but found no indication of such a correlation as proposed by Kurtin *et al.*[R9]. The MIGS model, on the other hand, predicts the slope parameters to be related to the electronic part ϵ_∞ of the dielectric constant of the semiconductors according to relation (15), as was realized by Mönch[R27]. In Fig. 13 of reprint R27–13 the data as re examined by Schlüter[R14] are plotted according to that relation. A least square fit indeed gives a power law, as expected from the simple analysis of the MIGS model presented in section 2.3, with $n = 2$. This first successful test of the MIGS model was further supported by an additional data point for metal–xenon contacts which was later on provided by Jacob *et al.*[R31]. Their slope parameter $S_{\Phi} = 0.85$ or, what is equivalent, $S_X = 1.52$ completely disagrees with the correlation proposed by Kurtin *et al.* since from their plot one would expect $S_X \lesssim 0.1$ for $\Delta X = 0$. On the other hand, the additional data point perfectly fits the trend which results from the MIGS model and which was established by the respective analysis of experimental data just mentioned (see also Fig. 10 of reference 18).

4.2. Correlations with Chemical Reactivity

The formation of reacted layers and especially of metal–anion compounds at interfaces between metals and II–VI as well as III–V compound semiconductors were carefully studied by Brillson[R15]. His analysis of experimental data suggested two distinctly different barrier heights for contacts exhibiting negative and positive heats of reaction. Remarkably, the reactive metal–semiconductor combinations always showed lowered barrier heights. Previously, Andrews and Phillips[19] had obtained a linear relationship between the barrier heights reported for several silicide–silicon junctions and the heat of formation of the respective silicides. Such correlations did, however, provide no insight in the physical mechanisms which are determining the barrier heights but, nevertheless, they have drawn attention to the complex chemical interactions of metals with semiconductors even at room temperature.

4.3. Effective Work Function Model

An attempt to retrieve the Schottky–Mott rule was made by Freeouf and Woodall[R17]. For that purpose they proposed to substitute an effective work function for the metal work–function in (7). Thus, they negated that interface states are playing a role in determining barrier heights in Schottky contacts. Their model, which they designed for compound semiconductors only, assumes '*microclusters of the one or more interface phases resulting for either oxygen contamination or metal-semiconductor reactions which occur during metallization*'. Then, they focus on excess anions, which are frequently generated by such interface reactions, and suggest to replace the metal work–function by the anion work–function in (7). Till now, no experimental indications were reported for larger concentrations of such anion microclusters. With regard to the MIGS–plus–defects model to be discussed later, however, a possible generation of even a small anion excess, which will escape detection by, for example, photoemission spectroscopy, might be of interest. In the bulk, anti–site defects of type V_{III}, i.e. anions on cation sites, are known as double donors and interface defects of donor–type, on the other hand, are lowering barrier heights in Schottky contacts as was shown in section 2.4.

4.4. MIGS–plus–Defects Model

An evaluation of the experimental data for metal– and silicide–silicon as well as metal–GaAs contacts based on a MIGS–plus–defects model was presented by Mönch[R28,R35]. In Fig. 3 of reprint R28 and Fig. 1 of reprint R35, experimental barrier heights are plotted versus the difference of the electronegativities of the metals and semiconductors in contact, and the branch points of the virtual gap states of the complex semiconductor band–structures as calculated by Tersoff[17,R20] are entered at $\Delta X = 0$ just as suggested by the arguments presented section 2.5. The barrier heights are found either to agree with or to be lower than what is expected from the MIGS model. Part of the Schottky contacts investigated obviously contained almost no interface defects while the deviations from the monotonous trend predicted by the MIGS model may be attributed to interface donors. In GaAs Schottky contacts, donor–type interface defects might be related with As_{Ga} anti–site defects, which are double donors in bulk GaAs. Their formation during metal evaporation onto GaAs substrates was proposed by Spicer *et al.*[R36] in their Advanced Unified Defect Model which will be discussed in § 5. Such As_{Ga} anti–site defects would also represent the smallest–sized anion microclusters which were considered by Freeouf and Woodall[R17] in quite a different context (see section 4.3).

Schottky contacts exhibiting extremely low densities of interface defects may be prepared by electrodeposition of the metal. This technique, which yields interfaces free of contaminants, was first employed by Ludwig *et al.*[R25] to make Ru/GaAs(001) contacts. These diodes exhibited the largest barrier heights of all GaAs Schottky contacts studied. What is most remarkable, however, their data point fits the trend expected from the MIGS model, as Fig. 1 of reprint R35 shows. Meanwhile, this preparational method was also applied for fabricating Pt–, Co–, Rh–, and Ir/GaAs diodes[20,21]. As Fig. 4 shows, these contacts were all found to exhibit barrier heights which are close to the respective values predicted by the MIGS model. These observations demonstrate that electrodeposition of metals on semiconductors may be used for preparing metal–semiconductor interfaces low in defect content. The case of Co/GaAs contacts is of special importance since with diodes prepared by conventional evaporation of the metal the barrier heights were found to be by more than two tenths of an eV smaller in comparison with when electrodeposition was employed. The same dependence on the preparational techniques used was meanwhile also found for Rh/GaAs diodes[21].

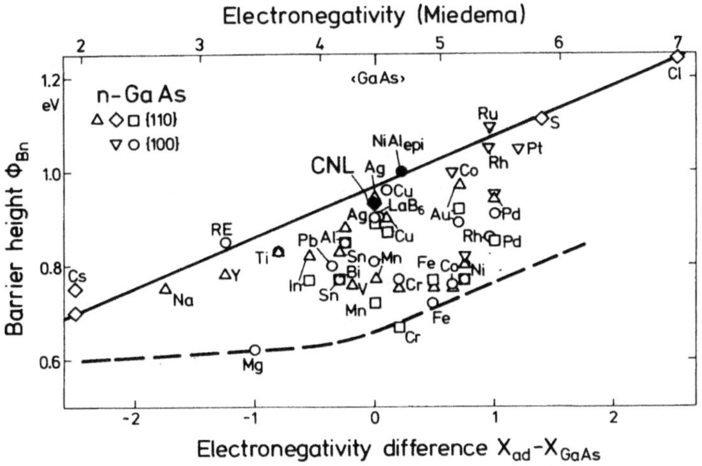

Fig. 4. Barrier heights reported for metal–GaAs contacts versus electronegativity differnce $X_m - X_{GaAs}$. The sources of the data may be found in reference 34.

The most recent data point included in Fig. 4 is the one for *epitaxial* NiAl/GaAs(001) contacts which was provided by Chambers and Loeb[36]. Their results are one of the finest confirmations of the MIGS–and–electronegativity concept[R27, R28] presented in paragraph 2.5. Only then when the epilayer was stoichiometric in composition the epitaxial NiAl/GaAs(001) contacts showed barrier heights of 0.9 to 1.0 eV. The MIGS–and electronegativity model[R35] exactly *predicted* such large barrier heights for abrupt NiAl/GaAs contacts.

Extremely low densities of interface states were also concluded for Cs/GaAs(110) interfaces which are exhibiting the lowest barrier heights observed with GaAs since cesium is the most electropositive element of all. Laubschat *et al.*[R39] investigated the influence of cesium oxide interlayers on the barrier height in Cs/GaAs(110) contacts. Their results are summarized in Fig. 3 of reprint R39. They first deposited two monolayers of Cs on a cleaved GaAs(110) surface kept at approximately 140 K. The barrier height of that abrupt Schottky contact agrees with the values entered in Fig. 1 of reprint R35 and thereby confirms the MIGS model. A subsequent oxidation of the cesium film almost restores flat bands which were initially observed with the clean surfaces. This finding is '*reflecting the fact that the CsO_x/GaAs(110) interface represents some kind of a semiconductor heterojunction*'[R39]. A following deposition of another two monolayers of cesium on top of the oxide film is accompanied by a built–up of depletion layers on the samples doped p– as well as n–type. For Cs/CsO_x/GaAs(110) structures, however, the barrier heights are smaller than observed with the intimate Cs/GaAs(110) interfaces. The most simple explanation for this behavior is a screening of the cesium wavefunctions by the oxide interlayer which tends '*to decrease the density of the MIGS, which may then not be sufficiently high for a complete band bending to a common Fermi–level position for both n– and p–doped substrates*'[R39]. A subsequent oxidation of that cesium overlayer increases the thickness of the cesium oxide film and flattens the GaAs bands again. These experiments demonstrate that interface defects play a minor or even no role in determining barrier heights in nonreactive Cs/GaAs contacts. Otherwise, one would have to postulate that interface defects were first generated during the first deposition of Cs, were then removed during a subsequent oxide formation and were again reinstalled at the now CsO_x/GaAs interface by a Cs deposition on top of the oxide layer. Such assumptions seem to be unreasonable. The experiments by Laubschat *et al.* are, therefore, strongly supporting the MIGS model.

As was already mentioned, $NiSi_2/Si(111)$ contacts are most interesting for several reasons. First, continuous epitaxial films of $NiSi_2$ may be grown on Si(111) surfaces in two different orientations as was first shown by Tung *et al.*[R19]. The orientation of the overlayer is either identical to that of the substrate (type–A interface) or rotated by 180^0 about the surface normal with respect to the substrate (type–B interface). Second, the barrier heights of both types of interfaces differ by 0.14 eV, as was first reported by Tung[R21]. Other observations[22] questioned these findings and seemed to indicate that both epitaxial interfaces exhibit the same, larger barrier height and the lower value is characteristic for sample containing mixtures of type–A and type–B regions. Meanwhile, the controversy was apparently decided in favor of Tung's earlier reports. Most remarkably, Fig. 3 of reprint R28 shows the barrier heights measured with type–B interfaces to be in excellent agreement with what is expected for $NiSi_2/Si$ contacts from the MIGS model. The by 0.14 eV lower barrier heights of type–A interfaces could again be caused by additional interface donors. Third, low frequency capacitances in excess of the capacitance of the space–charge region were also reported for forward–biased $NiSi_2/Si$ diodes and were ascribed to interface states[23]. Werner *et al.*[R32], however, showed such excess capacitances to be related to imperfect back–contacts. These authors concluded that '*previously reported excess capacitances at intimate Schottky contacts were probably caused not by interface states but by imperfect back-contacts*'. Thus, no experimental evidence is available for an increased density of interface states at $NiSi_2/Si$ interfaces of type–A. Furthermore, none of the many theoretical approaches, which were published on that subject so far, presented a convincing explanation of the different barrier heights of those two epitaxial $NiSi_2/Si(111)$ contacts.

An early investigation by Clabes *et al.*[R18] also supports the MIGS–plus–defects model. Clabes *et al.* studied the annealing behavior of V/Si(111) contacts. Their results are shown in Fig. 3 of reprint R18. After the evaporation of vanadium at room temperature, the barrier height is lower than expected from the MIGS trend in Fig. 3 of reprint R28 but annealing treatments at temperatures up to 200 ^{0}C were found to increase the barrier height by 0.1 eV and to bring it exactly on the MIGS line shown in Fig. 3 of R28. In analogy to the recovery of, for example, radiation–damaged samples during heat treatments, this behavior may be attributed to the removal of defects generated during metal evaporation. Up to 200 °C no intermixing was reported to occur at V/Si interfaces but was definitely observed at 350 °C. This process is most likely accompanied by the generation of defects which may then be made responsible for the reported decrease of the barrier height after annealing treatments above 200 °C.

As was shown in section 2.4, interface defects of donor– and acceptor–type are lowering and increasing, respectively, barrier heights with respect to what is expected for ideal Schottky contacts. Waldrop[R23] deliberately doped metal—GaAs(001) interfaces with chalcogens. He found chalcogens to increase the barrier heights of, for example, Ag/GaAs contacts by 0.1 eV. Silver does not react with GaAs and, as Fig. 1 of reprint R35 reveals, the barrier height of an unmodified Ag/GaAs contact equals the value predicted by the MIGS model. Sulfur adsorbed on cleaved GaAs(110) surfaces, on the other hand, was found to induce surface states of acceptor type at approximately 0.34 eV above the valence–band top[24]. The above mentioned experimental results reported by Waldrop are thus supporting the MIGS–plus–defects model.

As was pointed out by Ludeke *et al.*[R37], adatom–induced surface states on semiconductor surfaces broaden when they are buried under metal overlayers. Ludeke *et al.* generalized a model developed by Gadzuk in his studies on the level broadening of metal atoms adsorbed on metal surfaces (see Ref. 19 in reprint R37). Ludeke *et al.* replaced the vacuum in Gadzuk's case by a semiconductor, i.e. a dielectric. This was found to reduce the level broadening considerably.

4.5. Direct Observation of the Continuum of MIG States

The analysis of chemical trends exhibited by the slope parameters and the barrier heights as a function of the electronic part of the dielectric constant and of the difference in electronegativities of metal and semiconductor, respectively, and even the switching of band bending at $Cs/CsO_x/GaAs$ structures with varying oxide thickness are strongly supporting the theoretically well–founded model of metal—induced gap states in Schottky contacts. Direct evidence for the existence of these MIG states, however, came from an investigation[R40] by a research group at the National Institute of Standards and Technology in the US who used a Scanning Tunneling Microscope (STM) in the imaging as well as the spectroscopy mode.

First *et al.*[R40] evaporated iron on GaAs(110) substrates kept at 300 to 450 K and found the deposited atoms to form epitaxial clusters. Particles smaller than ≈ 400 Å show quantum size effects which become appreciable for clusters below ≈ 150 Å^3 in volume while islands with volumes of 1000 Å^3, corresponding to ≈ 85 Fe atoms, were definitely observed to exhibit metallic behavior. In the vicinity of such particles, the metallic states detected on them were observed to overlap in energy the semiconductor band–gap. These gap states were found to decay exponentially as a function of the distance away from the edge of metallic islands. Fig. 4b of reprint R40 shows the decay lengths of these gap states to vary U–shaped across the energy gap with a minimum decay length of 3.4 Å near to mid–gap position. Louie *et al.*[R13] and Tersoff[R20] computed the decay length of MIGS as 2.8 and 3 Å, respectively. These gap states observed around metallic Fe particles on GaAs(110) surfaces exhibit all the features earlier evaluated for the continuum of metal–induced gap states and the MIGS, as they were called for short, were thus experimentally detected for the first time.

5. Surface–Science Approach to Schottky Contacts

The *device approach* to metal–semiconductor contacts presented so far investigates chemical trends of barrier heights which were determined from I/V characteristics measured with Schottky diodes. The *surface–science approach* to Schottky contacts, on the other hand, studies the formation of such interfaces by employing surface–sensitive experimental tools such as, for example, photoemission spectroscopy (PES), electron diffraction (RHEED or LEED), Auger electron spectroscopy (AES), low–energy electron energy–loss spectroscopy (LEELS), Kelvin probes (CPD), and Raman spectroscopy (EFIRS). An introduction to these experimental techniques may be found in references 25 and 26. The use of such techniques provides information on the growth mode of the metal, the surface structure, the chemical composition, and the band bending, all as a function of the *amount* of metal deposited in situ on initially clean and generally well–characterized semiconductor surfaces.

For investigations on the formation of interfaces, the potential of these experimental methods is characterized by their lateral and vertical probing dimensions. A

Kelvin probe, for example, gives information on the work function and its changes, and it thus examines a property of the very surface. The applicability of all the electron spectroscopies for interface studies, on the other hand, is limited by the electron escape–depths at the respective kinetic energies. Finally, Raman spectroscopy[26], which may be used to probe band bending at interfaces, depends on the attenuation of the light used in the overlayer. The areas examined by the different techniques vary from a few μm up to typically 1 mm in diameter. The large–area probes thus provide averaged data unless the property studied is uniform over at least the area examined. Therefore, investigations on the growth mode of nonreactive metals on semiconductors, which should yield abrupt interfaces, became of prime importance.

5.1. Growth Modes of Nonreactive Metals on Semiconductors

As an example, In evaporated on cleaved GaAs(110) surfaces at room (RT) and at liquid nitrogen (LT) temperature was studied by Savage and Lagally[R22] They evaluated angular profiles of high–energy electron diffraction beams and found In to grow at room temperature as three–dimensional islands. *Nominal* coverages of 0.2 and 5 monolayers spread over only 7 and 50%, respectively, of the surface area and the surface between the islands is free of any In atoms. A continuous In film was not yet detected after the evaporation of nominally 150 monolayers. With substrates kept at liquid nitrogen temperature, the formation of three–dimensional islands was found to be delayed for up to coverages of 0.8 monolayers[27]. These growth studies are supplemented by PES observations of Stiles and Kahn[R33] on the development of the hybridized (4d)–bands of silver deposited at room and at reduced temperature on GaAs(110) surfaces. Up to ≈ 0.5 Å of Ag nominally deposited at LT, the Ag(4d) signal exhibits a shape typical for Ag atoms while for larger coverages it develops the form characteristic for bulk and thus metallic silver. Photoemission spectra recorded with nominal submonolayer coverages evaporated on substrates held at RT show Ag(4d) bands as observed with bulk samples. For nominally 5% of a Fe monolayer deposited on GaAs(110) at room temperature, the growth of smaller nonmetallic and larger metallic clusters was also observed by using a STM[R40].

5.2. Surface Band–Bending as Function of Metal Coverage

Investigations of electronic properties of semiconductor surfaces which were successively covered with metals at *low temperatures* provided new insights on the interaction of metal adatoms with semiconductor surfaces and on the formation of Schottky barrier. The research groups around Kahn at Princeton University and around Spicer at Stanford University reported results of their respective investigations at the same conference[R29,R30]. As was already well–known, depletion layers build up almost symmetrically as function of the amount of metal deposited independent of the type of substrate doping when the GaAs samples are held at room temperature during evaporation. With substrates kept at low temperature, however, the position of the Fermi level within the band gap does initially not change with samples doped n–type, while with p–type doping strong depletion layers are forming and after the deposition of $\approx 3\times10^{-2}$ of a monolayer the Fermi level becomes pinned. This different behavior is most clearly revealed by Fig. 2 of reprint R30 and Fig. 4 of reprint R30.

At low submonolayer coverages, isolated adatoms and nonmetallic islands prevail[27], and the observed *asymmetric* built–up of depletion layers on samples doped p–type only indicates the formation of metal–induced surface states of donor type[28]. Fig.1 of reprint R38 shows these initial pinning positions of the Fermi level or, what is the same, the energies of the adatom–induced surface donors to correlate with the ionization energies of the free metal atoms[28]. In reprint R38, Mönch explained this trend by modeling the bonds between ad– and substrate atoms in a most simple tight–binding approach. This interesting topic of semiconductor surface science shall not be further pursued here.

With increasing metal deposition but before *nominal* coverages of a monolayer are reached, i.e. still in the submonolayer regime, depletion layers are also build up on the samples doped n–type and simultaneously the band bending becomes slightly reduced on the p–substrates (see, for example, Fig. 2 in reprint R30). This behavior indicates the onset of a second mechanism. The previously mentioned studies by Savage and Lagally[27,R22] on the growth mode of In on GaAs(110) and by Stiles and Kahn[R33] on the formation of the sd–hybridized band of metallic Ag on GaAs(110) suggest to correlate this second mechanism with the growth of three–dimensional and metallic islands. Under such islands, the metal–semiconductor contact is completely developed as the direct observation of the continuum of metal–induced gap states around Fe particles on GaAs(110) by First *et al.*[R40] has revealed.

The lateral extension of the islands is by many orders of magnitude smaller than the area probed by such large–area techniques as, for example, photoemission spectroscopy while the height of the islands very rapidly exceeds the escape depth of the photoelectrons recorded. Photoemission spectroscopy then probes properties of the area in between the metallic islands. Here, the very details of the growth behavior come into play. Indium on GaAs(110) surfaces, the case studied by Savage and Lagally[R22], apparently follows the Volmer–Weber mode where the surface in between the islands remains free of any adatoms. To mention another example, Stiles and Kahn[R29] reported Sn to grow on GaAs(110) surfaces as islands on a complete monolayer, i.e. in the Stranski–Krastanov mode. Here, systems exhibiting the Volmer–Weber growth mode will be further considered only.

The depletion layer around each of the metallic islands will extend into the semiconductor below but also laterally away from its edge. Underneath an infinitely extended contact, the thickness of the depletion layer is simply obtained by solving Poisson's equation in one dimension. For GaAs with a doping level of 4×10^{18} cm^{-3} and mid–gap position of the Fermi level at the interface, one obtains $d_{\infty} \approx 100$ Å. Around contacts of finite size, the extension of the depletion layer is slightly reduced. For one monolayer of In nominally evaporated on GaAs(110), Savage and Lagally[R22] evaluated the edge separation of the islands as ≈ 150 Å. This distance reduces with increasing amount of In deposited. As the numbers show, the depletion layers around the islands will already overlap at a nominal coverage of one monolayer. This conclusion is supported by observation reported by First *et al.*[R440]. By employing scanning tunneling spectroscopy, they observed the surface band–bending to be homogeneous on GaAs(110) surfaces covered with metallic Fe particles.

5.3. Unified Defect Model

After the deposition of several monolayers of metals such as Al, Ga, and In on GaAs(110) surfaces, two distinctly different positions of the Fermi level were observed with samples doped p– and n–type. Respective experimental data may be found in Figs. 3 of reprint R29, 4 of reprint R29, and 2 of reprint R30. Identical pinning positions of the Fermi level were also observed with oxygen adsorbed on GaAs(110) surfaces. Spicer *et al.*[R16] presented their Unified Defect Model to explain these findings. This model suggests that native defects are generated during the evaporation of the metals and the Fermi level then becomes pinned at the energy levels of these defects. Considering the results of the growth study of In on GaAs(110), the pinning positions of the Fermi level obtained for In deposited at RT have to be attributed to the uncovered GaAs surface regions between metallic In islands. This argument gets further support from the observation reported by Newman *et al.*[29] that Al/GaAs(110) diodes exhibit identical positions of the Fermi level at (0.53 ± 0.02) eV above the valence–band top independent of whether the substrates are doped p– or n–type.

A pinning of the Fermi level by discrete surface states of whatever origin needs a minimum of 10^{12} to 10^{13} surface states per cm^2 depending on the bulk doping level. In case that such defects will persist under an eventually continuous metal overlayer they will only have a marked influence on the barrier height of the then Schottky contact provided their density exceeds 10^{14} cm^{-2}. Otherwise, they will be screened by the MIG states as (18) and (19) reveal when theoretical values of D_{is} and δ_i are inserted.

Allen and Dow[30] proposed anti–site defects to be relevant in this respect and Weber *et al.*[31] realized the energy levels of As_{Ga} anti–sites, which they had determined, to correspond with the pinning positions, which provided the basis for the Unified Defect Model. Since III_V anti–site defects are double donors Mönch[32] pointed out that pinning of the Fermi level at two different positions in the energy gap as observed on samples doped p– and n–type, respectively, can be effected by anti–site defects only when acceptor–type defects are present in addition. In their Advanced Unified Defect Model, Spicer *et al.*[R36] suggested As_{Ga} and Ga_{As} anti–site defects, i.e. double donors and double acceptors, respectively, to be generated during the evaporation of metals on GaAs(110) surfaces. This model needs a fine tuning of the densities of both types of defects. Provided that under a continuous metal overlayer more of the donors than of the acceptors will eventually survive the Advanced Unified Defect Model might also provide an explanation on the nature of the defects which were concluded from an analysis of barrier heights of Schottky diodes by applying the MIGS–plus–defects model.

6. Concluding Remarks

During the past four years, much progress has been achieved towards a generally accepted explanation of the mechanisms which are determining the barrier heights at metal–semiconductor interfaces. Meanwhile, the idea got widely accepted that more than just one mechanism has to be considered. There is mounting experimental evidence that the electronic properties of ideal and then abrupt Schottky contacts are to be understood in terms of a continuum of metal–induced gap states. The most significant support came from

- the chemical trend of the slope parameters S_{Φ},
- the chemical trends of barrier heights reported for metal– and silicide–silicon as well as metal–GaAs contacts, which included the charge–neutrality levels of the respective MIG states as a most essential check point,
- the fabrication–induced variations of barrier heights, where electrochemical metal deposition yielded barrier heights as expected from the MIGS model,
- the switching of interface band–bending at Cs/oxide/GaAs interfaces as a function of the oxide thickness, and, finally but most important,
- the direct observation of a continuum of metal–induced gap states at the edges of metallic islands on GaAs(110) surfaces by using a scanning tunneling microscope.

At this stage, the theoretically well–founded MIGS concept, which was introduced by Heine 25 years ago, should become generally accepted.

The complicated interactions between metals and semiconductors are reflected by the fabrication–induced variations of the barrier heights. With substrates doped n–type, reductions were observed with respect to what is expected from the MIGS model and they were attributed to fabrication–induced interface defects of donor type. Here, the Unified Defect Model and the MIGS model or, with other words, the surface–science and the device approach to metal–semiconductor contacts might meet. However, other physical mechanisms such as, for example, interface strain should also be considered.

Neither this Introduction nor the complete volume are intended to give a status report on the physics and chemistry of Schottky contacts. For that, I should like to refer the reader interested to an earlier review (up to 1982) by Brillson[33] and a most recent one by myself[34] as well as to the proceedings of an Advanced Research Workshop[35].

References

Reprinted Articles

R1. F. Braun, Pogg. Ann. **153**, 556 (1874).

R2. W. Schottky and W. Deutschmann, Physik. Z. **30**, 839 (1929).

R3. W. Schottky, Naturwissenschaften **26**, 843 (1938).

R4. N. F. Mott, Proc. Camb. Phil. Soc. **34**, 568 (1938).

R5. W. Schottky, Physik. Z. **41**, 570 (1940).

R6. J. Bardeen, Phys. Rev. **71**, 717 (1947).

R7. A. M. Cowley and S. M. Sze, J. Appl. Phys. **36**, 3212 (1965).

R8. V. Heine, Phys. Rev **138**, A1689 (1965).

R9. S. Kurtin, T. C. McGill, and C. A. Mead, Phys. Rev. Lett. **22**, 1433 (1969).

R10. F. Yndurain, J. Phys. C: Solid St. Phys. **4**, 2849 (1971).

R11. E. Louis, F. Yndurain, and F. Flores, Phys. Rev. B **13**, 4408 (1976).

R12. S. G. Louie and M. L. Cohen, Phys. Rev. B **13**, 2461 (1976).

R13. S. E. Louie, J. R. Chelikowsky, and M. L. Cohen, Phys. Rev. B **15**, 2154 (1977).

R14. M. Schlüter, Phys. Rev. B **17**, 5044 (1978).

R15. L. J. Brillson, Phys. Rev. Lett. **40**, 260 (1978).

R16. W. E. Spicer, P. W. Chye, P. R. Skeath, C. Y. Su, and I. Lindau, J. Vac. Sci. Technol. **16**, 1422 (1979).

R17. J. L. Freeouf and J. M.Woodall, Appl. Phys. Lett. **39**, 727 (1981).

R18. J. G. Clabes, G. W. Rubloff, B. Reihl, R. J. Purtell, P. S. Ho, A. Zartner, F. J. Himpsel, and D. E. Eastman, J Vac. Sci. Technol. **20**, 684 (1982).

R19. R. T. Tung, J. M. Gibson, and J. M. Poate, Phys. Rev Lett. **50**, 429 (1983).

R20. J. Tersoff, Phys. Rev. Lett. **52**, 465 (1984).

R21. R. T. Tung, Phys. Rev. Lett. **52**, 461 (1984).

R22. D. E. Savage and M. G. Lagally, J. Vac. Sci. Technol. B **4**, 943 (1986).

R23. J. R. Waldrop, Appl. Phys. Lett. **47**, 1301 (1985).

R24. S. B. Zhang, M. L. Cohen, and S. G. Louie, Phys. Rev. B **32**, 3955 (1985).

R25. M. Ludwig, G. Heymann, and P. Janietz, J. Vac. Sci. Technol. B **4**, 485 (1986).

R26. I. P. Batra and S. Ciraci, Phys. Rev. B **33**, 4312 (1986).

R27. W. Mönch, in Festkörperprobleme (Advances in Solid State Physics), Vol. 26, p. 67, ed. by P. Grosse (Vieweg, Braunschweig 1986).

R28. W. Mönch, Phys. Rev. Lett. **58**, 1260 (1987).

R29. K. Stiles, A. Kahn, D. G. Kilday, and G. Margaritondo, J. Vac. Sci. Technol. B **5**, 987 (1987).

R30. R. Cao, K. Miyano, T. Kendelewicz, K. K. Chen, I. Lindau, and W. E. Spicer, J. Vac. Sci. Technol. B **5**, 998 (1987).

R31. W Jacob, E. Bertel, and V. Dose, Europhys. Lett. **4**, 1303 (1987).

R32. J. Werner, A. F. J. Levi, R. T. Tung, M. Anzolowar, and M. Pinto, Phys. Rev. Lett. **60**, 53 (1988).

R33. K. Stiles and A. Kahn, Phys. Rev. Lett. **60**, 440 (1988).

R34. W. J. Kaiser and L. D. Bell, Phys. Rev. Lett. **60**, 1406 (1988).

R35. W. Mönch, Phys. Rev. B **37**, 7129 (1988).

R36. W. E. Spicer, Z. Lilienthal–Weber, E. Weber, N. Newman, T. Kendelewicz, R. Cao, C. McCants, P. Mahowald, K. Miyano, and I. Lindau, J. Vac. Sci. Technol. B **6**, 1245 (1988).

R37. R. Ludeke, G. Jezequel, and A. Taleb–Ibrahimi, J. Vac. Sci. Technol. B **6**, 1277 (1988).

R38. W. Mönch, Europhys. Lett. **7**, 275 (1988).

R39. C. Laubschat, M. Prietsch, M. Domke, E. Weschke, G. Remmers, T. Mandel, J. E. Ortega, and G. Kaindl, Phys. Rev. Lett. **62**, 1306 (1989).

R40. P. N. First, J. A. Stroscio, R. A. Dragoset, D. T. Pierce, and R. J. Celotta, Phys. Rev. Lett. **63**, 1416 (1989).

Additional References

1. L. O. Grondahl, U.S. Patent 164 335 (1925).
2. P. H. Geiger, U.S. Patent 755 840 (1904).
3. G. W. Pierce, Phys. Rev. **25**, 31 (1907).
4. A. H. Wilson, Proc. Roy. Soc. A **133**, 458 (1931) and **134**, 277 (1931).
5. O. Fritsch, Ann. Phys. **22**, 375 (1935).
6. W. Schottky, Z. Physik **113**, 367 (1939).
7. W. Schottky and E. Spenke, Wiss. Veröffentl. a. d. Siemens–Werken **18**, 1 (1939).
8. E. H. Rhoderick and R. H. Williams, *Metal–Semiconductor Contacts*, 2nd Ed. (Clarendon, Oxford 1988).
9. A.–W. Maue, Z. Physik **94**, 94 (1935).
10. E. T. Goodwin, Proc. Cambridge Phil. Soc. **35**, 205 (1939).
11. A. Zur, T. C. McGill, and D. L. Smith, Phys. Rev. B **28**, 2060 (1983).
12. C. A. Mead, Solid–State Electron. **9**, 1023 (1966).
13. L. Pauling, *The Nature of the Chemical Bond*, 3rd Ed. (Cornell University Press, Ithaca, N.Y., 1960).
14. W. Mönch, J. Vac. Sci. Technol. B **4**, 1085 (1986).
15. C. Tejedor and F. Flores, J. Phys. C **11**, L19 (1978).
16. J. Tersoff, Phys. Rev. B **30**, 4874 (1984).
17. J. Tersoff, Surf. Sci. **168**, 275 (1986).
18. W. Mönch, in *Metallization and Metal–Semiconductor Interfaces*, ed. by I. P. Batra (Plenum, New York, N.Y., 1989), p. 11.
19. J. M. Andrews and J. C. Phillips, CRC Crit. Rev. Solid State Sci. **5**, 405 (1975).
20. P. Allongue and E. Souteyrand, J. Vac. Sci. Technol. B **5**, 1644 (1987).
21. G. Heymann, private communication.
22. M. Liehr, P. E. Schmid, F. K. LeGoues, and P. S. Ho, Phys. Rev. Lett. **54**, 2139 (1985).
23. P. S. Ho, E. S. Yang, H. L. Evans, and X. Wu, Phys. Rev. Lett. **56**, 177 (1986).
24. L. Koenders, M. Blömacher, and W. Mönch, J. Vac. Sci. Technol. B **6**, 1416 (1988).

25. D. P. Woodruff and T. A. Delchar, *Modern Techniques of Surface Science* (Cambridge University Press, Cambridge 1986).

26. K. Smit, L. Koenders, and W. Mönch, J. Vac. Sci. Technol. B **7**, 888 (1989) and references cited therein.

27. D. Savage and M. Lagally, J. Vac. Sci. Technol. B **4**, 943 (1986).

28. W. Mönch, J. Vac. Sci. Technol. B **6**, 1270 (1988).

29. N. Newman, W. E. Spicer, and E. R. Weber, J. Vac. Sci. Technol. B **5**, 1020 (1987).

30. R. E. Allen and J. E. Dow, J. Vac. Sci. Technol. **19**, 383 (1981).

31. E. R. Weber, H. Ennen, U. Kaufmann, J. Windscheif, J. Schneider, and T. Woisinski, J. Appl. Phys. **53**, 6140 (1982).

32. W. Mönch, Surf. Sci. **132**, 92 (1983).

33. L. N. Brillson, Surf. Sci. Repts. **2**, 123 (1982).

34. W. Mönch, Rep. Prog. Phys. **53**, in press (1990).

35. I. P. Batra (editor), *Metallization and Metal–Semiconductor Interfaces* (Plenum, New York, N.Y., 1989).

36. S. A. Chambers and V. A. Loeb, J. Vac. Sci. Technol. B **8**, in print (1990).

Reprinted Articles

VI. *Ueber die Stromleitung durch Schwefelmetalle; von* ***Ferdinand Braun.***

Im 9. Hefte dieser Annalen (Bd. 153) befindet sich eine Arbeit von Herwig: „Einige Beobachtungen über das Verhalten von Eisen- und Stahlstäben im galvanischen Strome", wonach diese Körper je nach Richtung, Intensität und Dauer des Stromes demselben verschiedenen Widerstand entgegensetzen. Die Aenderungen schwanken im Allgemeinen zwischen $\frac{1}{3000}$ und $\frac{1}{20000}$ des ganzen Werthes. Diese Arbeit veranlaſst mich, einiges über ähnliche Erfahrungen mitzutheilen, welche ich bei anderen Körpern gemacht habe, welche ich aber noch nicht soweit verfolgen konnte, dafs ich den wahren Grund derselben einfach und präcis auszusprechen im Stande wäre. Ich betrachte daher selbstverständlich die folgende Publication als eine einfache Wiedergabe von Beobachtungen, welche wahrscheinlich noch unter sehr complicirten Bedingungen gewonnen sind, indem ich die genauere experimentelle Analyse mir vorbehalte.

Bei einer grofsen Anzahl natürlicher und künstlicher Schwefelmetalle und sehr verschiedenen Stücken, sowohl Krystallen von so vollkommener Ausbildung, wie ich überhaupt bekommen konnte, als derben Stücken habe ich gefunden, dafs der Widerstand derselben verschieden war mit Richtung, Intensität und Dauer des Stromes. Die Unterschiede betragen bis zu 30 pCt. des ganzen Werthes.

Der Strom von gewöhnlich einem grofsen Bunsen'schen Elemente durchflofs einen zickzackförmigen, 0,6^{mm} dicken Neusilberdraht (von 3,7 S.-E. Widerstand), welcher durch sieben auf demselben vertheilte Quecksilbernäpfchen von Kork hindurchgezogen war. Von diesen Quecksilbernäpfchen wurde Zweigstrom abgenommen, welcher das Schwefelmetall und die strommessende Vorrichtung (eine stark dämpfende Wiedemann'sche Bussole mit meist 0,22 S. E. Widerstand) durchflofs. — Die Schwierigkeit dieser Versuche liegt zunächst in zuverlässigen Contacten. Ich habe benutzt Quecksilbercontact, stark gegen geprefste Kupfer-, Platin- und Silberdrähte und endlich bei einem Stück eine bereits vorhandene Fassung mit dicken Neusilberbügeln, welche durch Schrauben gedrückt waren. Diese letztere Art der Fassung hat Hittorf[1]) als die beste gefunden.

Ich mufs erwähnen, dafs ich keine thermoelektrische Erregung oder Polarisation gefunden habe, welche auch nur entfernt im Stande wäre, die Erscheinungen zu erklären. Ich prüfte dies dadurch, dafs ich eine Wippe rasch umschlug, welche den ersten Strom unterbrach und nur

1) Hittorf. Pogg. Ann. Bd. 84, S. 81.

noch 2 bis 3^{mm} Weg mit ziemlich groſser Geschwindigkeit zurückzulegen hatte, um das Schwefelmetall in den Kreis eines zweiten Multiplicators von passender Empfindlichkeit (Widerstand) einzuschalten.

Bei einer Reihe von natürlichen Schwefelmetallen: Kupferkies, Schwefelkies, Bleiglanz, Fahlerz, bekam ich im Allgemeinen die Erscheinung, daſs die Stromintensität verschieden war, je nach der Stromrichtung, daſs diese Differenz zunahm mit zunehmender Stromintensität und daſs bei Geschlossenhalten des Stromes die Intensität für diejenige Richtung, welche kleineren Widerstand ergab, zunahm, für die entgegengesetzte abnahm. Dabei war stets dafür gesorgt, daſs die Contacte möglichst fest anlagen und waren deshalb mit dem zu untersuchenden Stücke an einem gemeinschaftlichen Brett angebracht. Die Vorrichtungen, welche umgeschaltet werden muſsten, befanden sich auf einem anderen Tische.

Ich verzichte darauf hier mehrere Reihen anzuführen und wähle nur Beispiels halber eine, welche an einem prismatischen Stücke von gegossenem Kupferkies gewonnen wurde. Dasselbe war in Neusilber gefasst und hatte bei ca. 70^{mm} Länge, 20^{mm} Breite nur 15^{mm} Dicke fast 2 S. E. Widerstand. Bei steigender und sinkender Stromintensität (durch veränderte elektromotorische Kraft) wurde es mit metallischem Widerstand verglichen.

Tabelle I.

Elektr. Kraft.	Stromintensität Kupferkies.	Met. Widerstand.
—	15,8	17,0
—	64,2	63,8
—	75,3*)	75,3*)
—	110,7	114,0
—	117 sinkt rasch	126,5
—	119 - -	178
1 Bunsen	159 - -	230

*) Der met. Widerstand wurde so gewählt, daſs diese beiden Stromintensitäten gleich waren.

Die Verschiedenheit der Stromintensität, je nach der Richtung des Stromes habe ich in vielen Versuchen, sowohl mit starken als schwachen elektromotorischen Kräften, mit ersten Ausschlägen als mit constanten Ablenkungen beobachtet und bei sicherer Fassung stets mit demselben qualitativen Resultat: daſs bei kleiner Stromintensität die eine Richtung gröſseren Widerstand bietet, bei wachsender Intensität beide Richtungen sich gleich verhalten und daſs sie dann ihre Rolle vertauschen.

Tabelle II.

Elektr. Kraft.	Stromrichtung I.		Stromrichtung II.	
	Erster Ausschlag.	Constante Ablenkung.	Erster Ausschlag.	Constante Ablenkung.
—	10,0	7,8	9,0	5,7
—	43,0	32,0	47	38,2
—	61	45 fällt bis 39	63	45 steigt bis 49
—	89	59	105	85
—	{ 155	115 - - 106	204	163 - - 167
	{ —	106	—	166
1 Bunsen	—	120	—	230

Es wird die nächste Aufgabe seyn, womöglich einfachere Versuchsbedingungen herzustellen. Aber es erscheint nicht thunlich, und dies hebe ich hervor, die Erscheinungen durch Fehlerquellen zu erklären, welche bei unserem augenblicklichen experimentellen Standpunkt klar erkannt sind. Mangelhafter Contact ist nicht ausreichend zur Erklärung, denn es ist nicht anzunehmen, daſs bei den rasch auf einander folgenden Messungen sich der Contact sollte stets in derselben Periode wie die Stromumkehrung verbessert und verschlechtert haben. In der That, stellt man absichtlich unsichere Contacte her, wie ich dies erreichte, indem ich eine groſse Anzahl von Kupfer- und Zinkblechen in eine Glasröhre zwischen zwei dicke Drähte presste, so erhielt ich bei Stromwechsel gleichfalls Verschiedenheiten, welche aber willkürlich wechselten oder auch so gingen, daſs wenn die Stromintensität eben sank, sie bei plötzlich gewendetem Strom noch mehr fiel.

Nach längerem Liegen (mehrere Monate) zeigte diese Röhre vollständig normales Verhalten, selbst als ich wenig oder viel destillirtes Wasser zwischen die Bleche gab. Ebenso, wenn ich Drähte ganz willkührlich an die verschiedensten metallisch leitenden Gegenstände anlehnen liefs, zeigte sich stets normales Verhalten. Will man die Beobachtungen erklären durch eine Eigenthümlichkeit der Contacte (Uebergangswiderstand), so fehlen uns bis jetzt Untersuchungen hierüber. Allerdings würde ich es nicht für absolut unmöglich halten, dafs sehr dünne Gasschichten die Anomalie bedingen und bei gewöhnlicher Temperatur diese Schichten schon die Verhältnisse der unipolaren Elektricitätsleitung zeigen, welche bei höheren Temperaturen sich in so auffälligem Maafse leicht zeigen lassen. — Durch thermoelektrische Erregung sind die Erscheinungen direct sicher nicht bedingt. Denn einmal müfsten dazu thermoelektrische Kräfte von $\frac{1}{3}$ Bunsen und darüber angenommen werden und aufserdem ist nach der Theorie und der Summe sämmtlicher Erfahrungen die thermoelektrische Kraft proportional der ersten Potenz der Stromintensität, so dafs ein Thermoelement bei geänderter Stromrichtung keine Verschiedenheit des Widerstandes zeigt, vorausgesetzt, dafs nicht erhebliche dauernde Erregungen zurückgeblieben sind. In der That war der scheinbare Widerstand einer 64 gliederigen Thermosäule unabhängig von der Stromrichtung; andererseits gab eine Wippe, welche das Schwefelmetall direct nach dem Umschlagen mit einem Multiplicator verband, wie schon erwähnt, keinen irgend in Betracht kommenden Ausschlag. Endlich fiel beim Oeffnen des Hauptstroms, wobei nun noch der Nebenstrom mit dem Schwefelmetall und dem Multiplicator geschlossen war, der Spiegel bis genau zu demselben Scalentheil, bis zu welchem er im ersten, über die Ruhelage gehenden Ausschlag sank, wenn statt des Kieses ein ihm aequivalenter metallischer Widerstand eingeschaltet war.

Einen Zusammenhang der Richtungen, in welchen verschiedene Leitungsfähigkeit bezw. eine Maximaldifferenz derselben stattfindet, mit den krystallographischen Eigenschaften der Schwefelmetalle habe ich bis jetzt nicht erkennen können, obschon sich häufig Andeutungen zu finden schienen. Ich will nur einen Versuch noch erwähnen, weniger weil ich glaube, derselbe gebe Aufschlufs hierüber, als weil er zeigt, dafs Fehler, wenigstens im äufseren Contact, nicht die Erscheinungen verursachen können. Unter den Krystallen, welche ich untersuchte, fand ich ein Stück glänzendes Fahlerz von ungemein grofsem Widerstande. Die Tetraeder derselben lagen ziemlich frei und besafsen grofse Flächen. Der Strom von 8 Grove'schen Elementen wurde durch den Krystall geleitet und an einer Wiedemann'schen Bussole gemessen, deren Windungen von zusammen 6000 S. E. Widerstand dem Spiegel möglichst genähert waren. Zwei 2^{mm} dicke, an den Enden sorgfältig abgerundete und geglättete Silberdrähte, welche möglichst fest an den Krystall geprefst waren, leiteten den Strom ein und aus. Stand der eine Draht senkrecht auf der horizontal gelegten Tetraederfläche nahe der Spitze, der andere nahe der Grundlinie derselben Fläche, so gab die Bussole unabhängig von der Stromrichtung 27^{sc} Ablenkung. Wurde nun der letztere Draht gegen die Basis des Tetraeders gestemmt, so trat sofort das anomale Verhalten ein. In der einen Richtung gaben die 8 Elemente 17^{sc} Ablenkung, welche allmälig auf 14^{sc} fiel, in der anderen 27^{sc}, welche rasch bis 35 und von da ab langsam auf 40^{sc} stieg. Bei so immensen Widerständen, welche auch blieben, als die Anlegestellen für die Drähte frisch blank geschabt waren, kann einem mangelhaften Contact die Erscheinung nicht mehr zugeschrieben werden. — Andere Fahlerze (z. B. ein Quecksilberfahlerz mit matter Oberfläche) besafsen Leitungsfähigkeiten, welche mindestens mehrere Millionen mal besser sind.

Es wäre möglich, dafs sich die erwähnten Erscheinungen in folgender Weise erklären. Denkt man sich kleine Krystalle, beispielsweise Tetraeder, eingebettet in eine

Grundmasse von anderer Beschaffenheit und seien dieselben so orientirt, dafs im Ganzen nach der einen Seite mehr Spizen liegen als Grundflächen, so wird bei Stromdurchgang Folgendes eintreten: Tritt der Strom aus der Grundmasse ein in die Basen der Tetraeder, so findet an der Basis Abkühlung, an der Spitze Erwärmung statt. Die verschwundenen und die erzeugten Wärmemengen sind einander gleich, vertheilen sich aber auf verschiedene Massen; die Grundfläche wird sich stärker zusammenziehen, als jede der drei anderen Flächen sich ausdehnt. Das Tetraeder bleibt sich nicht mehr ähnlich und weicht vor Allem ab von der Form, welche es bei umgekehrter Stromrichtung annimmt. Bei der ersteren Richtung würde die Spitze schärfer, bei der zweiten flacher werden. Es ist somit denkbar, dafs bei der einen Richtung die Contacte verbessert, d. h. die Anzahl der Berührungspunkte mit der umgebenden Grundmasse vermehrt, bei der anderen Stromrichtung vermindert werden. Diese Aenderung kann im grofsen Ganzen proportional der entstehenden Wärmeausdehnung, d. h. proportional der ersten Potenz der Intensität gesetzt werden.

Ferner werden die Grundmasse und die eingebetteten Krystalle durch den Strom erwärmt; findet dies in den beiden Medien in verschiedener Weise statt, so verschieben sich die einzelnen Theile gegen einander und es findet wieder eine Contactänderung statt, welche dem Quadrate der Stromintensität proportional gesetzt werden kann. Fafst man alles zusammen, so kann der ganze Widerstand gesetzt werden gleich einer Function von der Form

$$w + c \,.\, J + k \,.\, J^2$$

wo w den wahren Widerstand, c und k Constanten, J die Stromintensität bedeutet. Diese Function kann offenbar bei verschiedenem Vorzeichen von c und k kleiner als w, gleich w und gröfser als w werden (vergl. Tab. II).

Ist diese Erklärung richtig, d. h. rühren die Unterschiede her von Contactänderungen, so müssen sich dieselben auch bei längerem Stromdurchgang (wenigstens innerhalb gewisser Grenzen) in derselben Weise zeigen, d. h. für diejenige Stromrichtung, welche gröfseren Widerstand besitzt, mufs der Widerstand steigen, für die entgegengesetzte fallen — vollständig in Uebereinstimmung mit der Erfahrung.

Dafs bei gröfseren Stromintensitäten im Inneren mancher der spröden Schwefelmetalle, z. B. Bleiglanz, Contactänderungen auftreten, ist aufser Zweifel, da sich die Stromintensität dann bisweilen sprungweise ändert. Doch lassen sich solche Mineralien ausschliefsen. Aber auch in festeren, nicht vollständig homogenen Massen halte ich solche Aenderungen wohl für möglich. Ueberraschend erscheint mir nur, dafs auch bei kleinen Intensitäten und ersten Ausschlägen die Anomalien eintreten. Jedenfalls deutet diese Ueberlegung auf die Möglichkeit von Fehlerquellen, welche vorher vollständig eliminirt werden müssen, ehe man sich der interessanteren Auffassung hingeben darf, welche sofort beim Anblick der Erscheinungen entsteht, dafs man es mit einer Art Richtung der leitenden Molecüle und einer gewissen elektrischen Nachwirkung zu thun hat.

Leipzig, den 23. November 1874.

W. Schottky und W. Deutschmann (Berlin-Siemensstadt), **Zum Mechanismus der Richtwirkung in Kupferoxydulgleichrichtern.**

A. Einleitung.

Der Kupferoxydulgleichrichter, in Hunderttausenden von Exemplaren in der Rundfunk- und Signaltechnik, gelegentlich auch schon in der Starkstromtechnik angewandt, ist für den Physiker noch heute ein so gut wie ungelöstes Rätsel. Sein Entdecker, L. O. Grondahl[1]), weist darauf hin, daß die von Schottky[2]) gegebene Theorie der Kristalldetektoren auf ihn kaum anwendbar sei; Grondahl geht davon aus, daß Kupfer eine größere Elektronenaustrittsarbeit besitzt als Cu_2O und glaubt deshalb, daß, unter der Einwirkung einer Schottkyschen „distanzwahrenden Schicht" die Elektronen leichter vom Cu_2O zum Cu übergehen würden als umgekehrt, entgegen dem Befund. Es müsse also eine andere Vorstellung zugrunde gelegt werden; man werde annehmen können, daß, ohne angelegte Spannung, die Elektronen bis zu einem gewissen Grade vom Cu nach dem Cu_2O überquellen und dort eine räumlich verteilte Ladung bilden. Je nachdem das Konzentrationsgefälle der Elektronen in Richtung oder gegen die Richtung des Feldes wirkt, sollen große oder kleine Ströme fließen.

Diese Erklärung ist wenig einleuchtend, und so hat es nicht an weiteren Erklärungsversuchen gefehlt. Es ist auf elektrolytische Deutungen in verschiedener Form zurückgegriffen worden, obwohl Grondahl selbst schon eine Anzahl von Gründen erörtert, die beim Cu_2O-Gleichrichter gegen die Anwendung dieser bei anderen Gleichrichtern wohl fundierten Vorstellung sprechen. Andere Forscher nehmen an, daß der Richteffekt bei den meisten Detektoren und so auch beim Oxydulgleichrichter durch eine Spitzenwirkung zu erklären sein; auch Hypothesen, die die Richtwirkung in das Oxydul selbst verlegen, sind aufgestellt worden und ebenso solche, die mit der Kristallorientierung im Kupfer und Oxydul und dadurch bedingter gegenseitiger Anpassung oder Nichtanpassung der Leitfähigkeiten operieren. Endlich behauptet sich noch die Theorie der verschiedenen Austrittsarbeiten und distanzwahrenden Schicht; geht man nämlich von der lichtelektrischen Grenzwellenlänge des Cu_2O aus, die von V. P. Barton[3]) zu 2570 Å ermittelt wurde, so ergibt sich eine Austrittsarbeit von etwa 4,8 Volt, also eine etwas größere als die des Kupfers, die von A. Goetz[4]) zu etwa

1) Science **36**, 306, 1926; Journ. Am. Inst. El. Eng. **46**, 215, 1927.
2) Zeitschr. f. Phys. **14**, 87ff., 1923.
3) Phys. Rev. **23**, 337, 1924.
4) Zeitschr. f. Phys. **42**, 329 und **43**, 531, 1927.

4,4 V festgestellt wurde. Dann ist aber der Sinn der Detektorwirkung richtig, obgleich sich ihre Größe bei den geringen Unterschieden dieser Austrittsarbeiten schwer verstehen läßt.

Fast allen aufgestellten Erklärungsversuchen ist gemeinsam, daß sie von der Wirkung einer besonderen Sperrschicht von sehr geringer Dicke ausgehen. Daß eine solche vorhanden ist, und zwar praktisch ausschließlich an der Grenze von Mutterkupfer und Oxydul, gibt schon Grondahl an, und wir konnten es für die Sperrichtung durch statische Sondenmessungen bestätigen[1]), bei denen der Potentialverlauf in der Nähe einer begrenzten, auf das Cu_2O aufgebrachten Gegenelektrode untersucht wurde. Im Einklang mit diesen statischen Messungen steht die aus der Praxis bekannte Tatsache, daß sich die Oxydulgleichrichter gegen Wechselstrom etwa wie mit Verlust behaftete Kondensatoren verhalten. Der Zweck der folgenden Messungen sollte sein, die Größe dieser Kapazitäten bei den verschiedensten Gleichstromspannungen und Temperaturen zu bestimmen, wobei wir hofften, auf diese Weise bündige Schlüsse auf die Dicke der wirksamen Sperrschicht ziehen zu können. Gleichzeitig sollten die Abhängigkeiten von Vorspannung und Temperatur dazu dienen, die bisherigen Erklärungsversuche erneut einer Kritik zu unterziehen.

B. Messungen.

Die zu den Untersuchungen benutzte Gleichrichterlamelle bestand aus einer gelochten Kupferscheibe von 1 mm Dicke, deren Außendurchmesser 40 mm und deren Innendurchmesser 15 mm betrug. Sie war auf der einen Seite mit einer Kupferoxydulschicht bedeckt, deren Oberfläche zur Herstellung einer innig berührenden Gegenelektrode mit Zink gespritzt war. Eine ebenfalls zinkgespritzte Kupferscheibe, welche an die erste angepreßt war, diente zur Stromzuführung.

Es wurde zunächst die Gleichstromcharakteristik bei Zimmertemperatur aufgenommen (Fig. 1). Legt man, beispielsweise im Nullpunkt, eine Tangente an diese Kurve, so erhält man aus der Neigung dieser Tangente den Gleichstromwiderstand der Lamelle bei der Vorspannung 0 zu 52 Ohm, was mit den späteren Wechselstrommessungen gut übereinstimmt.

Die Wechselstrommessungen erfolgten bei positiven und negativen Vorspannungen sowie bei Temperaturen zwischen $+20$ und -73^0. Sie erstrecken sich auf einen Frequenzbereich von 800 bis 2500 Hertz und wurden in einer Brückenschaltung (Fig. 2) vorgenommen. Parallel zur Stromquelle lag ein gegen den Brückenscheinwiderstand sehr kleiner Widerstand $W_5 = 0{,}4$ Ohm. Dadurch wird erreicht, daß die an der Brücke liegende Wechselspannung leicht auf dem konstanten Wert 40 mV gehalten werden kann. W_1 und W_4 sind zwei gleiche Widerstände von je

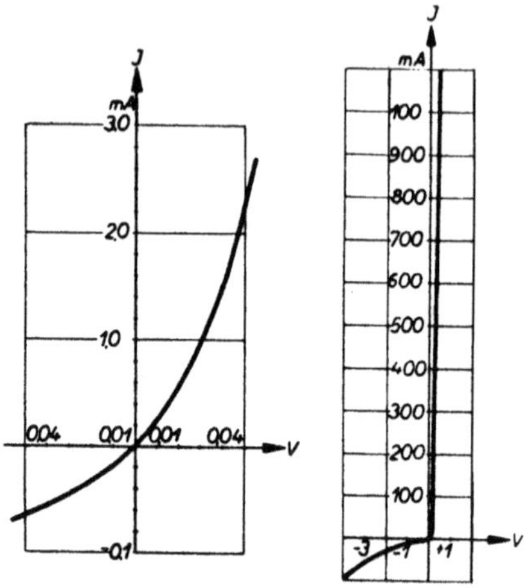

Fig. 1. Gleichstromcharakteristik einer Cu_2O-Lamelle bei Zimmertemperatur.

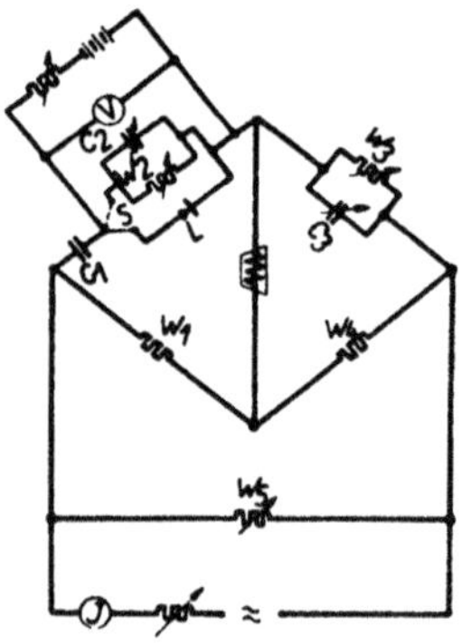

Fig. 2. Meßbrücke zur Scheinwiderstandsbestimmung.

500 Ohm, C_1 ist ein Sperrkondensator von 8 μ F. Mit Hilfe des Schalters S wird zunächst die Lamelle L eingeschaltet und an eine bestimmte, einstellbare Vorspannung V gelegt. Darauf wird mit Hilfe von W_3 und C_3 die Brücke abgeglichen; dann wird in der anderen Stellung des Schalters S die Kombination $W_2 - C_2$ an Stelle der Lamelle L eingeschaltet und W_2 bzw. C_2 solange verändert, bis wieder Gleichgewicht herrscht. Es handelt sich also um eine Substitutionsmethode. Aus den gemessenen Werten von W_2 und C_2 wird der Scheinwiderstand der

1) Ausgeführt von Herrn Dr. Störmer im Forschungslaboratorium des Siemenskonzerns.

Tabelle I. $t = 20^0$.

ω	$V = -5$ *R*	*I*	-2 *R*	*I*	-1 *R*	*I*	$-0{,}5$ *R*	*I*	$-0{,}2$ *R*	*I*	0 *R*	*I*	$+0{,}04$ *R*	*I*	0,06 *R*	*I*
5000	55,9	3,80	106,6	8,3	149,3	20,2	180,0	36,6	179,4	43,2	53,3	4,6	19,6	0,65	10,1	0,18
7000	49,6	2,07	105,1	11,4	146,7	28,1	171,3	47,5	169,7	57,0	52,9	6,3	19,5	0,86	10,1	0,23
10000	49,4	2,85	101,8	15,3	140,1	38,3	159,2	63,4	153,2	73,4	52,2	8,6	22,5	2,56	10,09	0,37
13000	48,7	3,56	98,9	19,2	132,0	45,9	145,1	75,0	135,6	84,0	51,3	10,9	19,4	1,72	10,08	0,46
16000	47,1	4,23	96,8	23,2	125,0	53,4	130,9	83,0	118,6	90,2	50,1	13,0	19,4	2,03	10,17	0,58

Lamelle, getrennt nach reellem und imaginärem Anteil, berechnet. Er gilt dann für die Wechselspannung 20 mV, da die Brücke gleicharmig ist und der Spannungsabfall an dem Sperrkondensator vernachlässigt werden kann. Die erhaltenen Werte wurden in einem rechtwinkligen Koordinatensystem eingetragen, wobei der reelle Anteil auf der Abszissenachse, der imaginäre auf der Ordinatenachse abgegriffen wurde.

Die Meßpunkte, die bei derselben Temperatur und Vorspannung, jedoch bei verschiedenen Frequenzen aufgenommen wurden, liegen durchweg mit sehr guter Übereinstimmung auf Halbkreisen, deren Mittelpunkt auf der reellen Achse liegt.

Tab. I gibt die bei 20^0 gemessenen Scheinwiderstände, getrennt nach reellem und imaginärem Anteil (*R* bzw. *J*), und zwar bei verschiedenen Kreisfrequenzen ω und Gleichstrom-Vorspannungen *V*.

Außerdem wurde ein Meßpunkt bei -10 V Vorspannung und $\omega = 10\,000$ aufgenommen, der den reellen Wert 16,2 und den imaginären Wert $-0{,}16$ ergab. In Fig. 3 sind die durch die Meß-

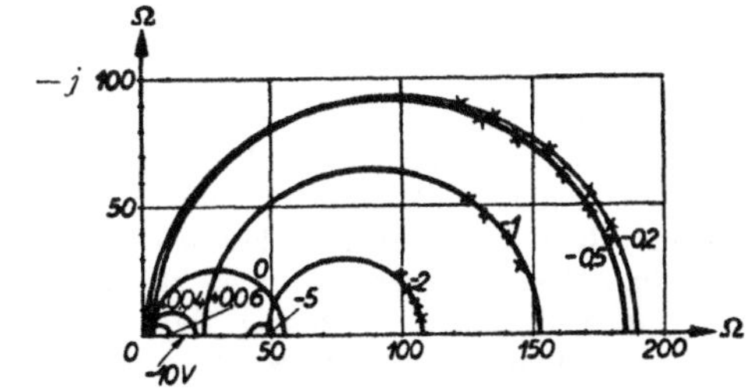

Fig. 3. Scheinwiderstandsdiagramm bei 20^0.

punkte gelegten Kreise eingezeichnet. Bei einigen Kreisen sind die Meßpunkte selber mit eingetragen, um eine Vorstellung von der Genauigkeit zu geben, mit der die Punkte einem Kreise zugeordnet werden können. Die Kurvenpunkte fangen bei negativen Vorspannungen mit kleinen Koordinatenwerten an, wachsen bis zu verhältnismäßig großen Werten bis zu einem Maximum bei etwa $-0{,}3$ Volt und schrumpfen dann sehr stark zusammen; bei $+0{,}08$ Volt Vorspannung sind die Meßwerte bereits so klein, daß ihre Messung Schwierigkeiten bereitet.

Eine zweite Meßreihe wurde bei einer Temperatur von rund -3^0 aufgenommen, und zwar mit der Vorspannung 0 bei der Temperatur $-3{,}6^0$, mit der Vorspannung $-0{,}5$ bei $-2{,}8^0$ und mit der Vorspannung -2 Volt ebenfalls bei $-2{,}8^0$ (vgl. Tab. II und Fig. 4).

Tabelle II. $t = -3^0$

ω	$V = -2$ *R*	*I*	$-0{,}5$ *R*	*I*	0 *R*	*I*
5000	132	14,6	378	200	228	89
7000	132,5	19,6	315	230	203	109
10000	132	27,4	232	240	165	126
13000	129	34,6	172	230	133	130
16000	125,4	41,7	129	210	105	126

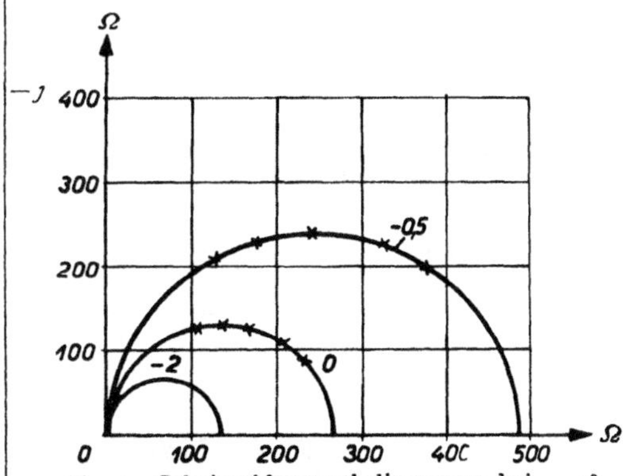

Fig. 4. Scheinwiderstandsdiagramm bei -3^0.

Eine weitere Serie von Messungen wurde bei etwa -18^0 mit den Vorspannungen -2; $-0{,}5$; 0 und $+0{,}1$ Volt aufgenommen (vgl. Tab. III und Fig. 5).

Tabelle III. $t = -18^0$

ω	$V = -2$ *R*	*I*	$-0{,}5$ *R*	*I*	0 *R*	*I*	$+0{,}1$ *R*	*I*
5000	297	72	450	344	320	222	48,3	4,1
7000	280	93	322	365	245	236	48,0	5,6
10000	259	124	217	325	164	223	47,5	7,8
13000	231	144	147	284	114	200	46,7	9,9
16000	202	153	106	249	83	178	45,7	11,7

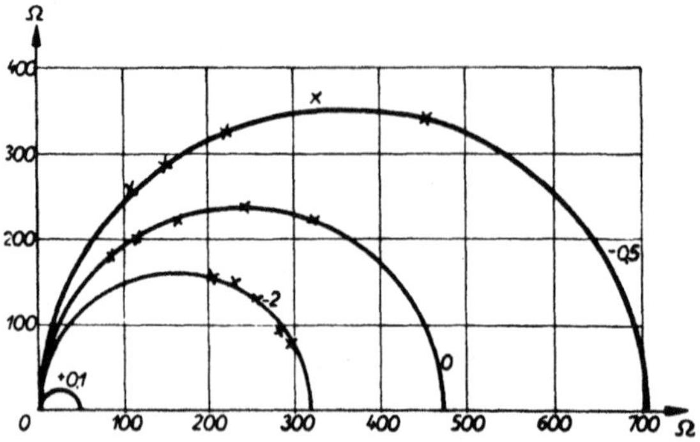

Fig. 5. Scheinwiderstandsdiagramm bei — 18⁰.

Bei — 24⁰ wurde außerdem eine Meßreihe mit o Volt Vorspannung durchgeführt (Tab. IV und Fig. 6).

Tabelle IV. $t = -24^0$

ω	$V = 0$	
	R	I
5000	344	328
7000	245	315
10000	149	269
13000	101	230
16000	69	193

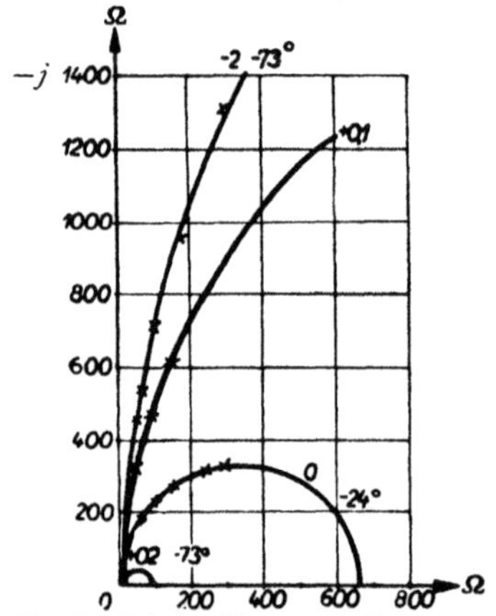

Fig. 6. Scheinwiderstandsdiagramm bei — 24⁰ und — 73⁰.

Schließlich wurde eine größere Meßreihe bei — 73⁰ mit verschiedenen Vorspannungen aufgenommen, deren Ergebnisse sich in Tab. V und ebenfalls in Fig. 6 finden, wo aus Gründen der Übersichtlichkeit nur die Kreise für — 2, + 0,1 und + 0,2 Volt eingezeichnet sind.

Die Messungen bei — 3⁰ zeigen qualitativ ein ähnliches Verhalten wie die bei Zimmertemperatur (20⁰), nur daß die Meßwerte im allgemeinen etwa 3—5 mal höher sind und die Kreise näher dem Anfangspunkt ansetzen. In ähnlicher Weise stellen sich die Messungen bei — 18⁰ dar (Fig. 5). Hier wurde noch bei + 0,1 Volt gemessen.

Bei der tiefsten Temperatur liegen die Meßwerte bereits erheblich höher als bei den anderen Temperaturen, lassen sich aber gleichwohl mit guter Annäherung als Halbkreise darstellen, die sehr nahe dem Anfangspunkt ansetzen. Hier konnte noch bei + 0,2 Volt gemessen werden.

Es fragt sich nun, wie diese Meßresultate zu deuten sind. Für Meßwerte, die auf einem gegen den Anfangspunkt verschobenen Halbkreis liegen, lassen sich zwei vollständig gleichberechtigte Ersatzschaltbilder angeben (Fig. 7). Das eine

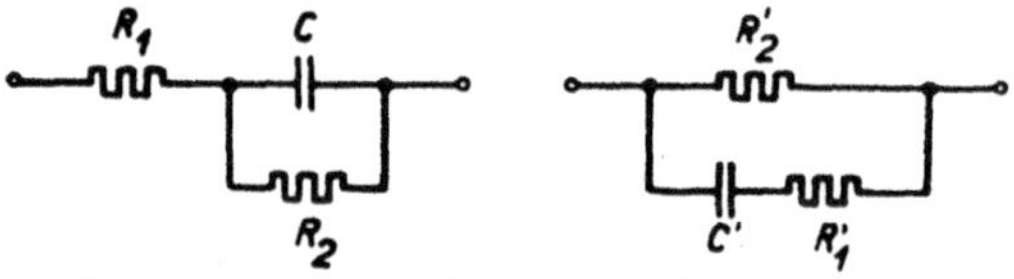

Fig. 7. Ersatzschaltbilder der Cu_2O-Lamelle.

besteht aus einem Widerstand R_1 in Reihe mit einem Kondensator C mit einem Ableitungswiderstand R_2. Das zweite Ersatzschaltbild ist ein Widerstand R_2', dem eine Reihenschaltung von Kondensator C' und Widerstand R_2' parallel liegt[1]). Werten wir die Meßresultate zunächst nach dem ersten Ersatzschaltbild aus, so erhalten wir die Werte der Tab. VI, Spalte 3—5.

Für R_1, also denjenigen Widerstand, der sozusagen einer halbleitenden Schicht entsprechen würde, die der eigentlich wirksamen Schicht vorgeschaltet ist, erhalten wir einen ziemlich merkwürdigen Verlauf. Fig. 8 zeigt den Verlauf von R_1 in Abhängigkeit von der Vorspannung bei Zimmertemperatur. Zunächst ist ein flaches Maximum bei etwa — 3 Volt Vorspannung vor-

Tabelle V. $t = -73^0$.

ω	$V = -2$		$-0,2$		0		$+0,1$		$+0,2$	
	R	I	R	I	R	I	R	I	R	I
5000	293	1300	65,7	817	59,1	708	135,6	604	81,8	12.7
7000	169	954	42,9	587	39,6	509	78,6	443	80,5	16,6
10000	93,5	659	30,9	415	28,7	360	46,5	316	77,8	22,3
13000	61,5	526	23,8	321	22,0	279	32,4	246	73,8	27,8
16000	48,1	429	22,2	263	20,4	228	26,8	201	69,2	31,9

1) Auf die Äquivalenz dieser beiden Ersatzschaltbilder hat zum ersten Male Zobel hingewiesen (Bell. Sy. T. J. **7**, 1928).

Tabelle VI.

t	V	R_1	R_2	C	R_1'	R_2'	C'
$+20^0$	-5	41,9	8,6	4,9	246	50,2	0,142
	-2	48,8	58,4	0,547	89,6	107,2	0,162
	-1	23,5	129	0,253	27,8	152,5	0,181
	$-0,5$	3,0	182	0,223	3,0	185	0,216
	$-0,2$	4,0	186	0,267	4,1	190	0,256
	0	3,8	50	0,364	4,1	53,8	0,314
	$+0,04$	2,8	16,8	0,45	3,3	19,6	0,330
	$+0,06$	2,7	7,4	0,65	3,7	10,1	0,349
-3^0	-2	2,5	132	0,165	2,5	134,5	0,159
	$-0,5$	5	482	0,220	5,1	487	0,216
	0	5	260	0,301	5,1	265	0,293
-18^0	-2	2	316	0,152	2,0	318	0,150
	$-0,5$	2	708	0,213	2,0	710	0,212
	0	3	474	0,294	3,0	477	0,291
	$+0,1$	1,2	476	0,355	1,2	48,8	0,338
-24^0	0	5	660	0,29	5,0	665	0,286
-73^0	-2	17	6200	0,149	17	6200	0,149
	$-0,2$	16	13000	0,244	16	13000	0,244
	0	14	10000	0,281	14	10000	0,281
	$+0,1$	12	3200	0,319	12	3200	0,319
	$+0,2$	1	83	0,376	1	84	0,366

handen, dann eine deutliche Anomalie bei etwa —0,5 Volt Vorspannung. Bei tiefen Temperaturen ist der Wert von R_1 sehr klein und nur sehr ungenau zu bestimmen; er ist daher in Fig. 8 nicht eingezeichnet. Einen wesentlich anderen Verlauf zeigt der Ableitungswiderstand, der in Fig. 9 als Funktion der Vorspannung bei verschiedenen Temperaturen aufgetragen ist; er zeigt bei —0,3 Volt ein deutliches Maximum, das mit der Größe der vorher gezeigten Kreise zusammenhängt, und fällt bei positiven Vorspannungen rapide ab. Schließlich zeigt Fig. 10 die wirksame Kapazität C. Sie strebt bei positiven Vorspannungen sehr großen Werten zu und zeigt bei Zimmertemperatur ein sehr merkwürdiges Minimum bei etwa —6 Volt Vorspannung.

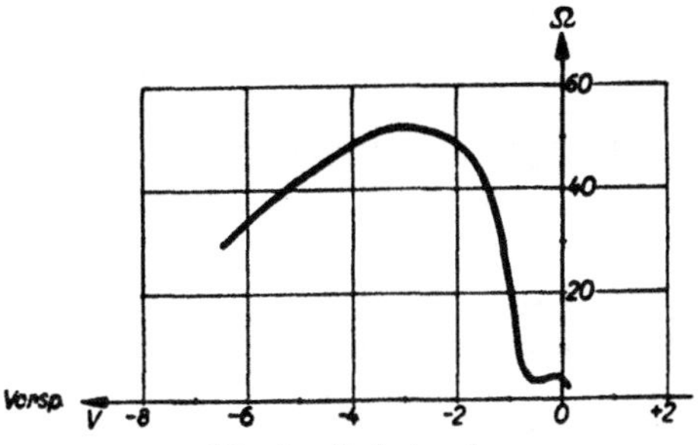

Fig. 8. R_1 bei 20^0.

Es scheint demnach, als ob bei positiven Vorspannungen die wirksame Schicht näher an die Grenzschicht wandert, abgesehen von der besprochenen Anomalie bei Zimmertemperatur.

Obwohl es nahe liegt, das soeben erwähnte Schaltbild zu wählen, also eine Kapazität mit Ableitung und Vorwiderstand anzunehmen, ist es doch von Interesse, das zweite formal gleich-

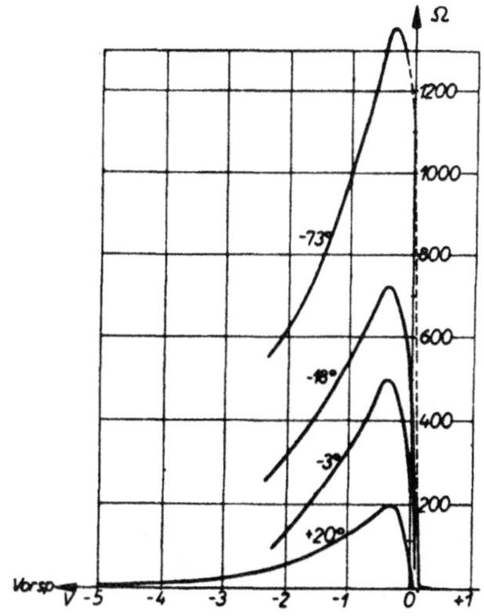

Fig. 9. R_2 bei verschiedenen Temperaturen.

berechtigte Ersatzschaltbild zu betrachten. Die Umrechnung auf die „gestrichenen" Werte finden sich in Spalte 6 bis 8 der Tab. VI. Fig. 11 zeigt den Verlauf von R_1' bei Zimmertemperatur in Abhängigkeit von der Vorspannung. Hier geht R_1' monoton herunter, die Anomalie bei —0,5 Volt bleibt bestehen. Der Verlauf von R_2' ist sehr ähnlich dem von R_2 (Fig. 12). Die Ano-

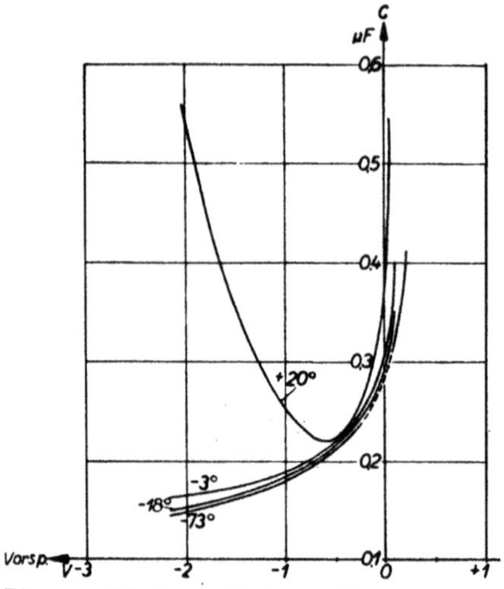

Fig. 10. C bei verschiedenen Temperaturen.

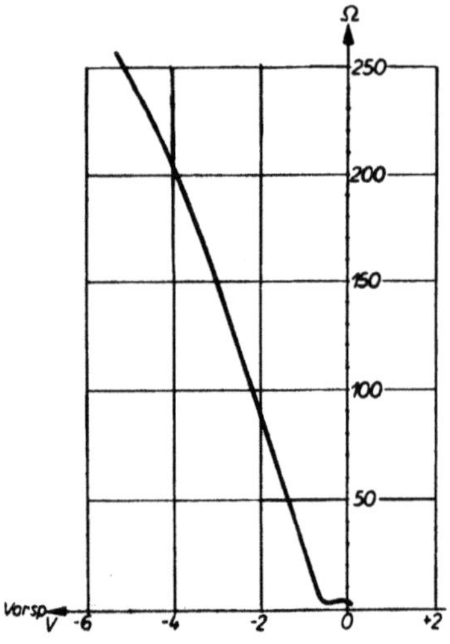

Fig. 11. R_1' bei 20°.

malie für C bei Zimmertemperatur verschwindet jedoch (Fig. 13) vollständig, wenn wir C' betrachten. Wir erhalten dann einen vollständig glatten Verlauf der Kapazität in Abhängigkeit von der Vorspannung mit dem bemerkenswerten Resultat, daß die Kapazität praktisch unabhängig von der Temperatur ist und bei größeren Sperrspannungen einem Grenzwert zustrebt, der auch von der Vorspannung unabhängig ist.

Vor einer weiteren Diskussion dürfte es von Interesse sein, sich zu überlegen, ob ein Dielektrikum mit Nachwirkung für das Gesamtbild der der gefundenen Effekte verantwortlich

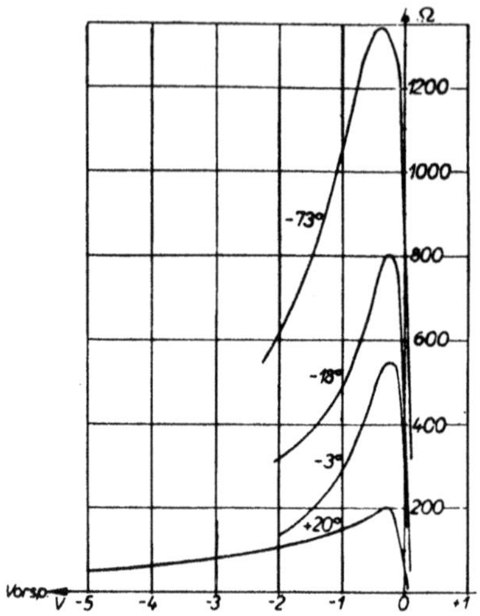

Fig. 12. R_2' bei verschiedenen Temperaturen.

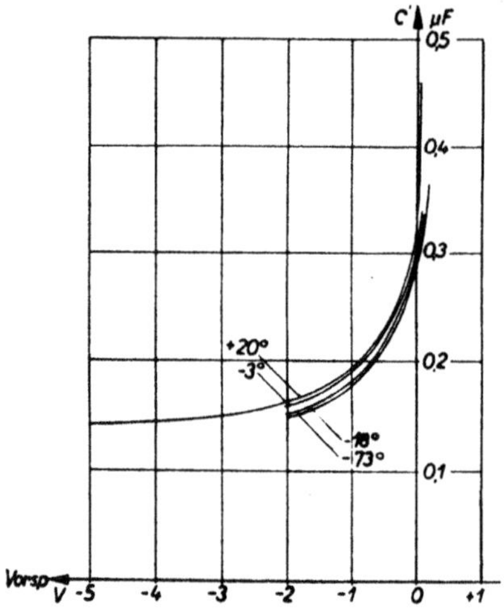

Fig. 13. C' bei verschiedenen Temperaturen.

gemacht werden kann. Ein Kondensator mit reiner Nachwirkung, d. h. mit Wechselstromverlusten, aber ohne Gleichstromverluste, kann als Kondensator C dargestellt werden, dem ein Widerstand $R_1 : \omega$ parallel liegt. Das Diagramm für eine solche Kombination ist eine Gerade durch den Koordinatenanfangspunkt. Hat ein solcher Kondensator außerdem noch eine Gleichstromableitung, liegt ihm also ein Widerstand R_2 parallel, so ergibt sich ein Kreisbogenstück durch den Anfangspunkt, dessen Mittelpunkt unterhalb der Abszissenachse liegt. Wenn unsere Kurven mit großer Genauigkeit Halbkreise sind, so heißt das, daß eine dielektrische Nachwirkung, wenn überhaupt vorhanden, hier nur von ganz unbedeutender Größe sein kann.

C. Diskussion der Meßergebnisse.

Im Anschluß an die eingangs entwickelte Fragestellung besprechen wir zunächst die Absolutgröße der gefundenen Kapazität, die sich ja bei den beiden verwendeten Ersatzschemata von gleicher Größenordnung ergibt, und zwar, einigermaßen unabhängig von der Temperatur, bei der Vorspannung 0 zu etwa 0,3 μF. Bei einer wirksamen Fläche von 11 cm^2 und nach Umrechnung in Zentimeter ergibt das eine Kapazität c von etwa 25 000 cm pro cm^2. Ist D die Dielektrizitätskonstante der Grenzschicht, h der Plattenabstand eines Kondensators von äquivalenter Schichtdicke, so errechnet sich h aus der Beziehung:

$$c = \frac{D \cdot f}{4\pi h}. \tag{1}$$

Hierbei bedeutet f einen Flächenfaktor, der angibt, wie groß die als Sperrschicht in Betracht kommende Fläche im Verhältnis zur mittleren Grenzschichtfläche ist; dieser Faktor kann größer als 1 sein, wenn die Grenzfläche eine erhebliche mikroskopische Rauhigkeit aufweist; er kann aber auch kleiner als 1 sein, wenn die Oxydulschicht nur teilweise fest auf der Kupferfläche aufsitzt. Aus (1) berechnet sich h zu:

$$h = \frac{D \cdot f}{4\pi c} \tag{2}$$

oder, mit Einsetzung des obigen Wertes für c:

$$h = 3{,}3 \cdot 10^{-6} \cdot D \cdot f \,\mathrm{cm}\,. \tag{3}$$

Die Bestimmung der Dielektrizitätskonstanten des Kupferoxyduls steht noch aus; in Analogie zu ähnlichen Verbindungen, die untersucht worden sind, wird man jedoch wohl etwa mit der Größenordnung 10 rechnen können[1]). Hiernach würde sein:

$$h = 3 \cdot 10^{-5} \cdot f \,\mathrm{cm}\,. \tag{4}$$

Nach dem mikroskopischen Befund und aus sonstigen Gründen ist anzunehmen, daß f wesentlich kleiner als 1 ist; leider fehlen bisher noch die nötigen Daten für eine genaue Bestimmung. Läßt man für f das Intervall zwischen 1/10 und 1/1000 zu, so ergeben sich also effektive Abstände von etwa $3 \cdot 10^{-6}$ bis $3 \cdot 10^{-8}$ cm oder von 100 Atomabständen bis zu einem Atomabstand, d. h. man muß annehmen, daß sich bei der Vorspannung Null die Schwerpunkte der überlagerten Wechselstromladungen auf diese Entfernung einander nähern. Bei stark negativer Vorspannung erhöhen sich diese Abstände etwa um das doppelte, bei positiver Vorspannung werden sie erheblich kleiner. Im ganzen kann man sagen, daß der Befund weder sicher dafür noch dagegen spricht, daß wir es hier mit einer erzwungenen Elektronenentladung zu tun haben, bei der eine Zwischenschicht von nur atomarer Dicke vorhanden ist.

Anders ist es aber, wenn wir die sonstigen charakteristischen Züge betrachten, die unsere Meßergebnisse aufweisen. Zunächst steht die außerordentlich starke Temperaturabhängigkeit der gemessenen Größen nicht im Einklang mit dem Mechanismus der erzwungenen Elektronenentladung, von der wir wissen, daß sie nur sehr wenig von der Temperatur beeinflußt wird; zumindest sind bei dieser Entladungsart die Stromerhöhungen, die sich durch Verdoppelung der Feldstärke ergeben, von ganz anderer Größenordnung als die, die sich etwa durch 10 oder 20° Temperaturänderung ergeben. Im Gegensatz hierzu fällt bei den gemessenen Scheinwiderständen, deren Größe im wesentlichen durch R_2 bzw. R_2' gegeben ist, die außerordentlich starke Temperaturabhängigkeit[1]) und, wenigstens in der Sperrichtung, die relativ geringe Feldstärkenabhängigkeit auf; von einer einigermaßen ausgeprägten Grenzfeldstärke, die die vollkommene Überwindung der Austrittsarbeit in der Sperrrichtung anzeigt, ist bei keiner der Kurven Fig. 9 oder 12 etwas zu entdecken.

Dagegen möchten wir auf einen bemerkenswerten Zusammenhang im Temperaturgang hinweisen, der zwischen dem Sperrwiderstand R_2 und dem von Baedeker[2]) bestimmten spezifischen Widerstand des Oxyduls selbst besteht. In beiden Fällen haben wir einen starken Anstieg der Widerstände mit sinkender Temperatur, und wenn auch der Temperaturkoeffizient nicht der gleiche ist — bei R_2 ist er ja von der Vorspannung abhängig —, so ergibt sich doch die gleiche Größenordnung.

---

1) Nach inzwischen im Zentrallabor. der Siemens u. Halske A.-G. durchgeführten Messungen der Herren Dr. Grösser und Lennertz ist der Wert von D bei $T = 180^0$ abs. gleich 12 ± 1. Bei dem bekannten geringen Temperaturgang der Dielektrizitätskonstanten fester Körper wird man diesen Wert auch noch für Zimmertemperatur als gültig ansehen dürfen.

1) Vgl. hierzu auch die Arbeit von H. Kost, Leitfähigkeitsänderungen von Kupferoxydul, Zeitschr. f. Phys. **54**, 367, 1929.

2) K. Baedeker, Ann. d. Phys. **22**, 763, 1907.

Auch die starke Abhängigkeit der Kapazität von der Vorspannung scheint uns nicht im Einklang mit dem Mechanismus der kalten Elektronenentladung zu stehen. Alle diese Befunde scheinen uns vielmehr darauf hinzudeuten, daß wir es bei dem Leitungsmechanismus in der Sperrschicht mit einem Vorgang zu tun haben, der mit den Leitungsvorgängen im massiven Kupferoxydul irgendwie verwandt ist; allerdings müssen in der Grenzschicht gewisse Abänderungen dieses Leitungsmechanismus vorliegen, die für die starke Erhöhung des Widerstandes in der Sperrichtung verantwortlich sind. Man kann sich vielleicht die Vorstellung bilden, daß die Anwesenheit überschüssiger Elektronen in der Grenzschicht die Leitfähigkeit stark herabsetzt, indem gewissermaßen die für die Leitungselektronen verfügbaren Stellen durch statische Elektronenraumladungen besetzt werden und damit der Durchgang für die Leitungselektronen verstopft wird[1]). Die Anlegung eines negativen Potentials an die Gegenelektrode (Zinkelektrode) würde sich hierbei, da an der Grenze Zn/Cu_2O und im Cu_2O selbst fast kein Spannungsabfall eintritt, als Aufspeicherung einer starken negativen Gesamtladung im Kupferoxydul unmittelbar vor der Fläche des Mutterkupfers auswirken; wird dagegen ein positives Potential angelegt, so genügt schon ein Spannungsabfall von etwa 0,2 Volt unmittelbar am Mutterkupfer, um die anscheinend im Nullzustand vorhandene negative Ladung zu kompensieren, und bei hoher Vorspannung ist die Hemmung durch überschüssige Elektronen vollständig beseitigt.

Es ist beabsichtigt, diese Vorstellung durch direktere Versuche nachzuprüfen, über deren Ergebnis an anderer Stelle berichtet werden soll.

Diskussion.

Herr Zenneck (München): 1. Ich würde dankbar sein, wenn ich über die Methode der Sondenmessung Näheres erfahren könnte.

2. In meinem Institut sind seit einiger Zeit ebenfalls Untersuchungen über die Trockengleichrichter im Gange. Dabei hat sich unter anderem herausgestellt, daß der Gleichrichter, der wohl meist die Anordnung $Cu—Cu_2O—Pb$ hat, ungefähr ebenso gute Gleichrichtung ergibt, wenn man die Bleischeibe durch eine Cu-Folie ersetzt. Der Gleichrichter hat dann die scheinbar unsinnige Zusammensetzung $Cu(Cu_2O)Cu$. Die erforderliche Unsymmetrie kommt aber dadurch zustande, daß das eine Cu die Kupferfolie ist, die gegen das Cu_2O nur angepreßt ist, während das andere Cu die Kupferplatte ist, auf der das Cu_2O durch Oxydation gebildet wurde. Für den Stromdurchgang bedeutet dies gerade den wesentlichen Unterschied.

Zu 1. Herr Schottky: Ich möchte Herrn Dr. Störmer, der diese Messungen ausgeführt hat, bitten, selbst die Methode zu erläutern.

Herr Störmer: Auf die Oberfläche des Kupferoxyduls wurde eine scharf begrenzte, gut haftende Elektrode aufgebracht. Zwischen dieser Elektrode und dem Mutterkupfer lag eine Spannung von etwa 2 Volt in der Sperrichtung (obere Elektrode negativ gegen Mutterkupfer). Mit einer nadelartigen Sonde wurde dann elektrometrisch die Spannungsverteilung auf der Oberfläche des Oxyduls in der Nähe der Elektrodengrenze festgestellt. Wäre der Potentialabfall in der Sperrichtung über die ganze Oxydulschicht verteilt gewesen oder an der Grenzfläche zwischen Oxydul und oberer Elektrode aufgetreten, so hätte in einer Entfernung von der Elektrodengrenze, die mit der Oxydulschichtdicke (ca. $^1/_{10}$ mm) vergleichbar ist, bereits praktisch das Potential des Mutterkupfers, also ein Potentialunterschied von 2 Volt gegen die obere Elektrode gemessen werden müssen. Es stellte sich jedoch heraus, daß der Potentialunterschied in $^1/_2$ mm Entfernung erst etwa 0,3 Volt, in 4 mm Entfernung etwa 1 Volt war. Dadurch ist nachgewiesen, daß der Sperrwiderstand praktisch nur an der Grenze von Mutterkupfer und Oxydul liegen kann.

Zu 2. Herr Schottky: Die von Herrn Zenneck für angepreßte Kupferfolien gemachte Feststellung gilt sogar für Kupferüberzüge, die durch Reduktion der Oberseite des Kupferoxyduls gewonnen sind, also Überzüge, die eine denkbar innige Verbindung mit dem Kupferoxydul besitzen. Auf dieser Tatsache sind gewisse im Handel befindliche Oxydulgleichrichter aufgebaut. Der Grund, weshalb die Art des Kontaktes von Kupfer und Kupferoxydul eine so ausschlaggebende Bedeutung für das Auftreten einer Sperrwirkung hat, ist bis jetzt noch nicht mit voller Sicherheit festgestellt. Es bieten sich jedoch verschiedene Erklärungsmöglichkeiten, die mit der versuchsweise von uns eingeführten Hypothese einer Leitfähigkeitsbeeinflussung durch überschüssige Elektronen der Grenzschicht nicht im Widerspruch stehen.

1) Man müßte also annehmen, daß in derartigen elektronischen Halbleitern, im Gegensatz zu den Metallen, eine elektronische Oberflächenladung eine merkliche Eindringtiefe besitzt; eine Vorstellung, die übrigens schon bei dem Grondahlschen Erklärungsversuch eine Rolle spielt.

auftreten, die dem Berührungsradius der Spitze und somit den genannten Eigendimensionen der Randzone entsprechen.

Eine ausführliche Veröffentlichung ist in der Z. Physik beabsichtigt.

Berlin-Siemensstadt, Zentralabteilung der Siemens und Halske AG., den 17. Dezember 1938. W. SCHOTTKY.

Halbleitertheorie der Sperrschicht.

Im Jahre 1929 wurde vom Verf. eine Sperrschichttheorie vorgeschlagen[1], in der die unipolare Leitfähigkeit der Gleichrichter vom Kupferoxydultyp auf Dichteänderungen der Leitungselektronen in einer Randzone an der Grenze Metall-Halbleiter zurückgeführt wurde. Diese Theorie konnte in der Zwischenzeit weiter entwickelt werden unter folgenden Grundvoraussetzungen:

1. An der Grenze Metall-Halbleiter existiert eine thermisch bestimmte Randdichte der freien Elektronen in Überschuß-Halbleitern, der Defektelektronen in Defekt-Halbleitern, die bei nicht zu hohen Feldern und Störstellengehalten von der störstellenbedingten Elektronendichte im Halbleiter-Innern sowie vom hindurchgehenden Strom unabhängig ist. Die für die Randdichte maßgebende Austrittsarbeit Metall-Halbleiter ist sowohl für Elektronen wie für Defektelektronen positiv und so groß, daß die Randdichte im allgemeinen unterhalb der in Störstellenhalbleitern sich einstellenden Innendichte liegt[2].

2. Der Übergang von der Randdichte zur Innendichte vollzieht sich in einer Zone, deren Breite durch das Gleichgewicht von Feldströmen und Diffusionsströmen sowie durch die von der örtlichen Elektronendichte abhängige Raumladungsdichte gegeben ist. Auf diese Weise entstehen feldproportionale Sperrwiderstände, wenn die beweglichen Ladungsträger (negative Elektronen in Überschußhalbleitern, positive Defektelektronen in Defekthalbleitern) durch ein angelegtes Feld von Metall in den Halbleiter hineingetrieben werden. Die Flußrichtung ist diejenige, in der die betreffenden leitenden Teilchen aus dem Halbleiter in das Metall fließen[3]; es wird dann die verarmte Randzone mit leitenden Teilchen überschwemmt, der Sperrwiderstand verschwindet.

3. Bei größeren Spannungen in der Sperrichtung wird die Randdichte durch die an der Metallgrenzfläche auftretenden starken Felder über ihren thermischen Wert vergrößert (Feldemission; auch für Defektelektronen!), so daß der Sperrwiderstand sich wieder verkleinert. Wird auf diese Weise die Randdichte größer als die Innendichte, so tritt die (bei vielen Detektoren zu beobachtende) Umkehr des Gleichrichtungssinnes auf.

Die Durchführung der Theorie führt zu quantitativen Folgerungen, die sowohl mit den Beobachtungen an Flächengleichrichtern wie an Spitzendetektoren (und den diesen meist praktisch gleichartigen „Gleichrichtern mit künstlicher Sperrschicht") in vollem Einklang zu stehen scheinen. Gute Flächengleichrichtung ist hiernach nur zu erwarten, wenn im Halbleiter an der „wirksamen" Metallelektrode eine dünne, chemisch vom Halbleiterinnern abweichende Sperrschicht mit besonders hohem spezifischem Eigenwiderstand existiert, da bei homogen aufgebauten Flächengleichrichtern der Bahnwiderstand der Gesamtschicht den Widerstand der auf 10^{-5} bis 10^{-6} cm konzentrierten Randzone weit überwiegen würde. Spitzengleichrichtung kann dagegen bei allen Halbleitern mit nicht zu hoher Störstellendichte auftreten, da hier wegen der divergenten Ausbreitung des Stromes von selbst nur äquivalente Schichtdicken des Bahnwiderstandes

---

[1] Physik. Z. 30, 839 (1929), insbesondere vorletzter Abschnitt.

[2] *Anm. bei der Korrektur.* Über die Abhängigkeit der Austrittsarbeit Metall-Halbleiter vom Elektrodenmetall finden sich wichtige Aufschlüsse in einer inzwischen erschienenen Veröffentlichung von R. HILSCH und R. W. POHL, Z. Physik 111, 399 (1938). Die dort gezogenen Parallelen zwischen den instationären, dafür aber quasineutralen, Bewegungen von Farbzentrenwolken, und den stationären und raumladungsbedingten Vorgängen in elektronischen Sperrschichten dürften allerdings etwas weniger eng sein, als die Verfasser annehmen.

[3] Vgl. die Diskussionsbemerkung des Verf., Z. techn. Physik 16, 512 (1935): die dort aufgestellte Regel ist inzwischen noch mehrfach bestätigt worden.

NOTE ON THE CONTACT BETWEEN A METAL AND AN INSULATOR OR SEMI-CONDUCTOR

By Mr N. F. MOTT, Gonville and Caius College

[*Received* 1 June, *read* 17 October 1938]

According to quantum mechanics there exists in any non-metallic crystal a band of allowed electronic energy levels which are unoccupied when the crystal is in its state of lowest energy. We call this band the conduction band; the crystal can conduct electricity if electrons are raised into the conduction level from lower levels. According to the theory of semi-conductors given by Wilson, there exist in these substances lattice imperfections at which an electron can exist in a bound stationary state below the conduction band, electrons being raised from these levels into the conduction band by the thermal agitation of the surrounding atoms.

We denote by χ the work necessary to remove an electron from the lowest level of the conduction band to a point outside the crystal. χ is defined for a crystal with clean surfaces and no electrical double layer at the surface.

Suppose now that we place a metal surface near to or in contact with an insulator or semi-conductor. If the two do not influence each other in any way, the work necessary to take an electron from the metal into the conduction band of the insulator is

$$\phi - \chi,$$

where ϕ is the work function of the metal. When, however, a semi-conductor is placed in contact with a metal, an electrical double layer is set up, either between the metal and the semi-conductor, or in the semi-conductor itself. As a result, the energy of the conduction level is displaced*, in such a way that the work necessary to transfer an electron from the metal to the conduction band is about $\frac{1}{2}E$, where E is the energy difference between the energy levels on the imperfections in the semi-conductor and the conduction band. It is the first purpose of this note to investigate the width of this double layer.

For a first treatment we make the three following assumptions: (*a*) that the metal and semi-conductor are not actually in contact, so that the surface atoms of the metal are not distorted in any way; (*b*) that, before the double layer is set up, the filled levels of the insulator or semi-conductor are *below* the surface of the Fermi distribution of the metal (Fig. 1); and, (*c*) that there do not exist on the surface of the semi-conductor any Tamm† levels where an electron can be bound below the level of the conduction band.

* Cf. R. H. Fowler, *Statistical mechanics*, 2nd ed. (Cambridge, 1936), § 11·62.

† Tamm, *Zeit. f. Physik*, 76 (1932), 849.

Electrons can then pass from the occupied band of the semi-conductor into the metal, or from the metal into the conduction band of the salt. There is an excess charge in the salt and an equal and opposite induced charge on the surface of the metal. This gives rise to an electric field; let the potential of an electron in this field, measured from the surface of the Fermi distribution, be $V(x)$, where x denotes distance measured from the surface of the metal. We have then to calculate V.

We assume that in the semi-conductor there are N_2 bound electrons per unit volume which can escape, and N_1 states per unit volume in the conduction band. For simplicity we assume the width of the energy band to be zero. Then, by the Fermi-Dirac distribution law, at any point x the number of electrons per unit volume in the conduction band is

$$N_1/(e^{V/kT}+1)$$

and the number of vacant places in the impurity levels is

$$N_2\left[1-\frac{1}{e^{(V-E)/kT}+1}\right].$$

If kT is small compared with the energies concerned, these become

$$N_1 e^{-V/kT}, \quad N_2 e^{-(E-V)/kT},$$

so that, if e is the charge on the electron, the density of charge per unit volume is

$$e[N_1 e^{-V/kT}-N_2 e^{-(E-V)/kT}].$$

This is connected with V by Poisson's equation; we have

$$\frac{d^2V}{dx^2} = -4\pi e^2[N_1 e^{-V/kT}-N_2 e^{-(E-V)/kT}]. \tag{1}$$

This equation is not soluble in finite terms; we examine two special cases, first, very near, and secondly, far from the metal.

At great distances the charge density tends to zero. The right-hand side of (1) therefore vanishes and we have

$$2V(\infty) = E+kT\log\frac{N_1}{N_2}.$$

This gives the total displacement of the levels. If we put

$$V = V(\infty)+\psi,$$

(1) becomes
$$\frac{d^2\psi}{dx^2} = 4\pi e^2\sqrt{(N_1 N_2)}\sinh\frac{\psi}{kT}e^{-\frac{1}{2}E/kT}.$$

At large distances where $\psi \ll kT$, the solution is

$$\psi = \text{constant}\, e^{-\gamma x},$$

where
$$\gamma^2 = \frac{4\pi e^2}{kT}\sqrt{(N_1 N_2)}\, e^{-\frac{1}{2}E/kT}. \tag{2}$$

37-2

This may be written $\gamma^2 = 4\pi e^2 n_1/kT$,

where n_1 is the number of electrons per unit volume in the conduction band.

At small distances, the form of the solution depends upon whether ϕ or $E-\phi$ is the less. Suppose, for instance, that $V < E-\phi$. Then for sufficiently low temperatures the second term on the right-hand side of (1) is small and the equation becomes

$$\frac{d^2V}{dx^2} = -8\pi e^2 N_1 e^{-V/kT}.$$

This equation has been discussed in connexion with the problem of an atmosphere of electrons in equilibrium with hot electrodes*. In our case it gives the number of electrons in an insulator in equilibrium with a metal, when the number of positive holes is negligible. If we denote by n_0 the number of electrons at the boundary of the insulator, so that

$$n_0 = N_1 e^{-V(0)/kT},$$

and by l the length given by $l^2 = kT/(2\pi n_0 e^2)$, (3)

the solution for the number n of electrons per unit volume near the surface is

$$n = \frac{n_0 l^2}{(x+l)^2}.$$

l is thus the distance in which the electron density falls to one-quarter of its value at the surface.

Since n_0 is necessarily greater than or equal to n_1, we see by comparing (2) and (3) that the electrons are distributed in a layer at least as thick as l. If we take N_1 to be 10^{22}, some values are given below for $T = 300^\circ$ K.

V_0/kT	0	5	10	15	20
n_0	10^{22}	$6{\cdot}7\times10^{19}$	$4{\cdot}5\times10^{17}$	3×10^{15}	2×10^{13}
l (cm. $\times 10^8$)	0·17	2·1	25	310	3800

Let us now apply these ideas to an interface between, say, pure sodium chloride and metallic sodium. For sodium chloride E is about 8 e.V., and so the number of electrons raised from the occupied levels into the conduction band is negligible. In a recent paper† the present author has estimated χ to be $0{\cdot}5_5$ e.V.; ϕ for clean sodium is 2·46 e.V., and so $\phi-\chi$ is 1·9, and l at 700° C. is 4000 A; at lower temperatures the values of l are much larger. Thus in practice the conduction levels are displaced to their equilibrium positions only at a large distance from the interface.

Let us now consider the possibility that there exist on the surface of the insulating crystal Tamm surface levels. It is known from the facts about frictional electricity that charge can reside on the surface of a dielectric, and levels such as

* Cf. R. H. Fowler, *Statistical mechanics*, 2nd ed. (Cambridge, 1936), p. 364.
† *Trans. Faraday Soc.* 34 (1938), 500.

those postulated by Tamm will probably be found necessary to account for these facts.

For an insulator not in contact with a metal the Tamm levels will provide the states of lowest energy for any excess electrons on it, and the electrons will tend to congregate there. At a surface in contact with a metal, however, this may not be the case, since an electron may gain energy by going into the metal.

Let χ' be the work necessary to remove an electron from one of these surface levels to a point outside the crystal, χ' being greater than χ. There are then two possibilities; either $\phi-\chi'$ is positive, in which case the insulator acquires a surface charge of the order
$$ea^{-2}\,e^{-(\phi-\chi')/kT}$$
per unit area, where a is the interatomic distance. If d is the distance between the surface of the insulator and the metal, the potential difference set up is
$$4\pi eda^{-2}\,e^{-(\phi-\chi')/kT}.$$
Taking $a = 3\times10^{-8}$ cm., $d = 10^{-7}$ cm., we see that this is of the order
$$200e^{-(\phi-\chi')/kT}\text{ volts,}$$
so that unless $\phi-\chi'$ is comparable with kT, the potential difference is negligible. Of the total charge on the dielectric in equilibrium with the metal, only a small fraction resides on the surface levels.

If on the other hand $\phi-\chi'$ is negative, the surface levels will collect charge from the metal until they are raised up to or a little above the surface of the Fermi distribution.

The present author* had considered the photoconductivity of crystals of NaCl containing colloidal metal, and has suggested that the low frequency limit ν of the current-frequency curve should be given by
$$h\nu = \phi-\chi. \tag{4}$$
In such crystals the metal must be in actual contact with the salt; from what we have said above, (4) will be valid unless the surface atoms of the metal are actually distorted by the presence of the salt, or by losing electrons to Tamm levels, so that an electrical double layer is set up. Since (4) appears to be in agreement with experiment, we may assume that the effect of any such double layer is inappreciable, the Tamm levels being too high to collect a large charge.

We now turn to the problem of the copper oxide rectifier. The theory has been discussed by Wilson and others; the rectifying contact is made by oxidizing one face of a block of metallic copper to an appreciable depth. The oxidation converts the copper to Cu_2O which is almost pure next to the mother copper but oxygen-rich elsewhere. The oxygen-rich copper oxide is a semi-conductor, while the pure salt is almost an insulator. Wilson discusses the theory of the rectifier by treating

* *Trans. Faraday Soc.* 34 (1938), 500.

the pure copper oxide as introducing a potential barrier between the copper and the copper oxide. It is the origin of this potential barrier that we wish to discuss.

In Cu_2O with excess oxygen there are copper lattice points at which the ion missing. The salt is a defect conductor; at the points where copper ions are missing there are energy levels to which electrons from a normally full band can be raised. These energy levels are only about 0·6 e.V. above the full band; thus they will probably be *below* the surface of the Fermi distribution of the copper. Thus in a copper oxide rectifier the vacant energy levels at missing copper ions nearest to the metal will acquire electrons from the metal. In the pure copper oxide layer a field is set up which brings the vacant levels up to the level of the

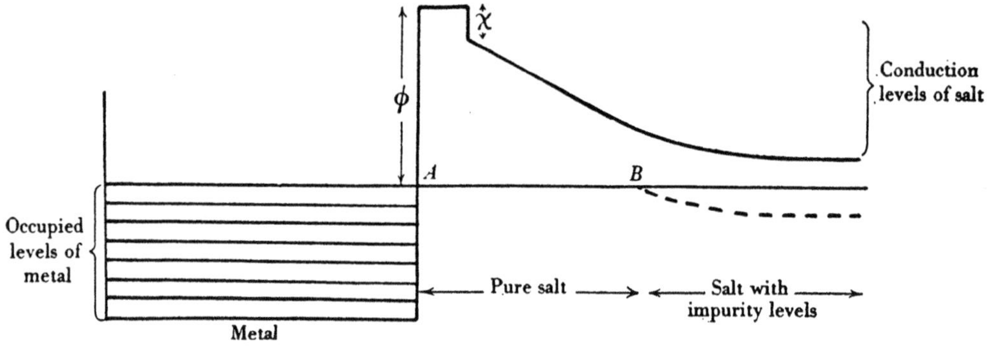

Fig. 1. Energy levels in a metal in contact with a salt. Next to the metal the salt is of stoichiometric composition, but to the right of B there are impurity levels, shown by the broken line. None of these levels can be above AB, since they would then lose their electrons to the metal. The potential barrier separating the conduction levels of the salt from the metal is due to the field caused by excess positive charge to the right of B, and the induced charge on the surface of the metal.

surface of the Fermi distribution of the metal. To the right of this the space charge in copper oxide will bring the levels into equilibrium in a distance which, by (2), will be of the order $(kT/4\pi n_1 e^2)^{\frac{1}{2}}$, where n_1, the number of electrons per cm.^{3} in the conduction band, is for copper-oxide at room temperature, 10^{14}. The distance works out to be $1{\cdot}2\times10^{-5}$ cm.

In Fig. 1 we have illustrated the state of affairs for an electronic semi-conductor; the principles are the same as for a defect conductor, but electronic levels seem to us easier to visualize than defect levels.

The author would like to express his thanks to Mr A. H. Wilson for discussions about this paper.

Abweichungen vom Ohmschen Gesetz in Halbleitern

Von W. Schottky, Berlin-Siemensstadt

(Mitteilung aus dem Wernerwerk Z der Siemens & Halske A.G.)

Mit 4 Abbildungen im Text

Wenn sich elektrisch geladene Teilchen irgendwelcher Art durch ein Medium hindurchbewegen, ist der von ihnen getragene Strom durch das Produkt von Teilchendichte und Teilchengeschwindigkeit gegeben. Das Ohmsche Gesetz ist erfüllt, wenn an der untersuchten Stelle die Teilchendichte von der Größe und Richtung des Stromes unabhängig und ihre Geschwindigkeit der Feldstärke proportional ist.

Im Vakuum und in Gasentladungen ist meist weder die eine noch die andere dieser Voraussetzungen erfüllt, so daß wir uns hier über die Verletzung des Ohmschen Gesetzes nicht wundern. In flüssigen und festen Körpern erleiden aber die bewegten elektrischen Teilchen so viele Zusammenstöße mit ihrer Umgebung, daß in der Regel ihre Fortschreitungsgeschwindigkeit der jeweils vorhandenen Ortsfeldstärke proportional angenommen werden kann. Anders steht es mit der zuerst genannten Voraussetzung des Ohmschen Gesetzes, der Unbeeinflußbarkeit der Teilchendichte durch den Strom. Es hat sich herausgestellt[1]), daß gerade die auffallendste und schon bei kleinsten Spannungen auftretende Abweichung vom Ohmschen Gesetz, nämlich die unipolare Leitfähigkeit gewisser Metall-Halbleiteranordnungen, mit einer durch den Strom bewirkten Erhöhung und Erniedrigung der Elektronendichte in den randnahen Schichten des Halbleiters in Verbindung gebracht werden muß.

Der Grundvorgang ist hierbei nach der seit Ende 1938 entwickelten „Randschichttheorie der Kristallgleichrichter"[2]) folgender. In einer zwischen zwei Metallelektroden angeordneten Halbleiterschicht, von der in Abb. 1 nur die Umgebung der einen

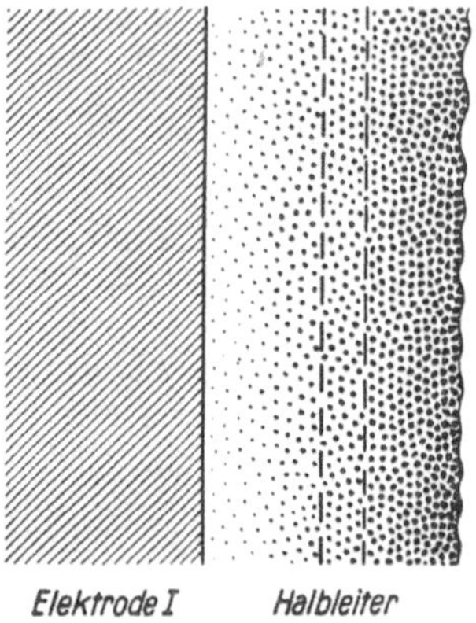

Abb. 1. Konzentrationsgefälle der Halbleiterelektronen in der Nähe einer Metallelektrode

Elektrode dargestellt ist, besteht im stromlosen Zustand aus Gründen, die gleich besprochen werden sollen, ein Unterschied der Elektronendichte zwischen dem Innern des Halbleiters und den unmittelbar an die Elektroden angrenzenden Gebieten. Fließt nun bei der in Abb. 1 dargestellten Elektronenverteilung durch die in der Nähe der Elektrode I eingezeichnete Teilschicht des Halbleiters ein Elektronenstrom von rechts nach links, so erhält die Teilschicht ihre Elektronenzufuhr aus elektronenreicheren Gebieten, sie wird also mit Elektronen überschwemmt. Fließt dagegen der Elektronenstrom von links nach rechts, so erfolgt die Elektronenzufuhr aus einem Verarmungsgebiet, die Teilschicht verarmt deshalb an Elektronen. Damit wird die Elektronendichte von der Größe und Richtung des Stromes abhängig, die Teilschicht leitet „unipolar".

[1]) W. Schottky, Naturwissensch. **26** (1938), 843; B. Davydov, Journ. Phys. USSR **1** (1939), H. 2; N. E. Mott, Proc. Roy. Soc. **171** (1939), 944; W. Schottky, ZS. f. Phys. **113** (1939), 367; W. Schottky u. E. Spenke, Wiss. Veröffentl. a. d. Siemens-Werken **18** (1939), 225.

[2]) Vgl. die unter [1]) genannten Veröffentlichungen des Verfassers.

Ein Dichtegefälle der Elektronen, wie es in Abb. 1 dargestellt ist, kann aus verschiedenen Ursachen entstehen; eine dieser Ursachen ist die örtliche Verschiedenheit des Gehaltes an Elektronen abgebenden Störstellen im Innern oder an der Grenze des Halbleiters. Beim Kupferoxydulgleichrichter ist so etwas, wenn auch nicht als wichtigste, so doch als zusätzliche Ursache der Gleichrichterwirkung nachgewiesen; hier tritt am Mutterkupfer eine Sauerstoffverarmung auf, die als sogenannte „chemische Sperrschicht" die Elektronendichte an der Metallgrenze herabsetzt. Von größerem Interesse ist jedoch der Fall der „physikalischen Sperrschicht" oder „Verarmungsrandschicht", wobei die Elektronenverarmung allein dadurch entsteht, daß das betreffende Elektrodenmetall dem Halbleiter eine wesentlich geringere Elektronendichte aufprägt, als sie, infolge des Störstellengehalts, im Innern des Halbleiters herrscht. Die Vorstellung, daß ein Metall einem angrenzenden Halbleiter eine ganz bestimmte Elektronendichte aufprägt, hat zunächst etwas Ungewöhnliches; wenn man sich aber klar macht, daß die Elektronen, um vom Metall in den Halbleiter zu gelangen, eine gewisse Austrittsarbeit zu überwinden haben, die in erster Näherung als vom Störstellengehalt unabhängig angesehen werden kann, so sieht man, daß die Verhältnisse ganz ähnlich liegen wie auf dem Gebiet der Glühelektronen, wo ebenfalls im stromlosen Zustand unmittelbar über dem glühenden Metall eine Elektronendichte herrscht, die nur durch die Austrittsarbeit des betreffenden Metalls gegeben ist.

Daß es berechtigt ist, diese Vorstellung auch auf den Elektronenübergang Metall-Halbleiter anzuwenden, läßt sich nun aber auch direkt experimentell belegen. Wenn die Elektronendichte in der Verarmungsrandschicht von der Austrittsarbeit Metall-Halbleiter abhängt, so müssen Metalle mit hoher Elektronenaustrittsarbeit eine sehr starke Elektronenverarmung und damit einen hohen Widerstand der betreffenden Gleichrichterkombination hervorrufen, Metalle mit geringer Austrittsarbeit dagegen einen kleinen Widerstand. Hierzu liegt nun eine interessante Untersuchung aus dem Arbeitskreise von Prof. Gudden vor; H. Schweickert hat auf der Gautagung in Erlangen 1939 über Widerstandsmessungen an Selenschichten vorgetragen, die auf der Rückseite an ein möglichst sperrfreies Elektrodenmetall, nämlich Nickel, grenzten, während auf der Vorderseite die verschiedensten Metalle thermisch aufgedampft wurden. Mit Zustimmung von Herrn Schweickert und Herrn Gudden ist es mir möglich, Ihnen eine mir von Herrn Schweickert zugesandte Darstellung seiner Ergebnisse in Abb. 2 vorzuführen. Man sieht, daß in der Tat der gemessene Widerstand der betreffenden Metall-Halbleiterkombination mit der

Vakuum-Austrittsarbeit des Deckmetalls in systematischer Weise variiert, und zwar um über 4 Zehnerpotenzen, wenn die Austrittsarbeit zwischen der des Kaliums mit etwa 2 V bis zu der des Nickels mit etwa 5 V variiert wird.

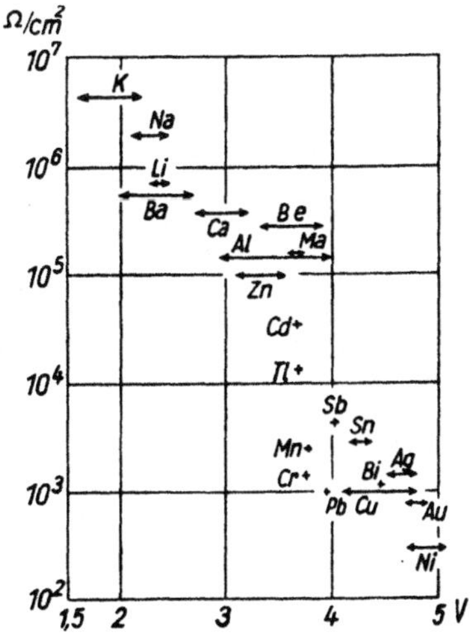

Abb. 2. Austrittsarbeit und Widerstand von Selengleichrichtern nach H. Schweickert 1939

Unerwartet erscheint aber auf den ersten Blick der Sinn dieser Veränderung. Die Stoffe mit der größten Austrittsarbeit ergeben den kleinsten Widerstand, leicht Elektronen abgebende Stoffe, wie die

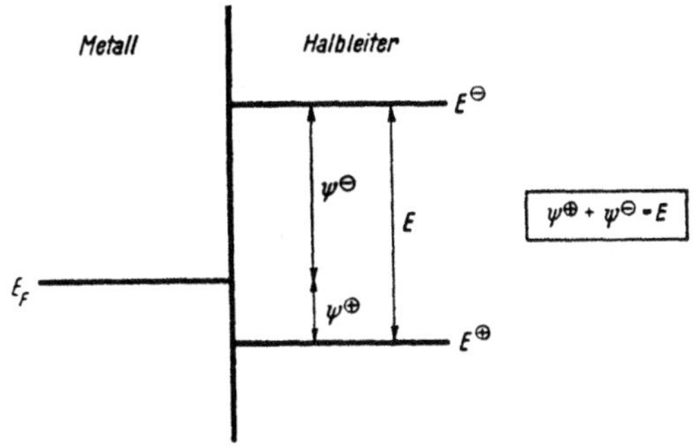

Abb. 3. Berührung Metall – Halbleiter. Lage der für die Austrittsarbeit maßgebenden Energieniveaus

Alkalimetalle, dagegen den größten. Der Widerspruch löst sich, wenn man von der durch verschiedene andere Tatsachen beglaubigten Annahme Gebrauch macht, daß Se kein gewöhnlicher Elektronenhalbleiter, sondern ein Defekthalbleiter ist, daß also in ihm der Ladungstransport durch Elektronenlücken besorgt wird, die sich wie Teilchen mit positiver Ladung und positiver Masse benehmen[3]).

[3]) Vgl. z. B. L. Nordheim, Metallwirtsch. **11** (1931), 121···124 u. 125···128.

Es kommt dann nicht auf die Austrittsarbeit der Überschußelektronen, sondern der Defektelektronen beim Übergang Metall–Halbleiter an, und diese ist, wie die nächste Abb. 3 erläutert, desto kleiner, je größer die Austrittsarbeit für Überschußelektronen ist, weil beide Austrittsarbeiten, $\psi_\ominus$ und $\psi_\oplus$, zusammen, sich zu der Energie E ergänzen müssen, die zur Bildung eines Elektronenpaares, also eines Überschußelektrons und einer Elektronenlücke im ungestörten Kristallgitter notwendig ist (Abb. 3)[4]). Der gegenläufige Gang des Randschichtwiderstandes mit der Elektronenaustrittsarbeit des Elektrodenmetalls ist also in der Tat gerade das, was man bei dem Defektleitungscharakter des Selens erwarten muß.

Zum Schluß möchte ich noch über einige Ergebnisse berichten, die das Wechselstromverhalten der Trockengleichrichter betreffen. Schon auf der Physikertagung in Prag, 1929, konnte darauf hingewiesen werden[5]), daß man durch Kapazitätsmessungen an Kupferoxydulgleichrichtern wichtige Aufschlüsse über die Dicke der — damals noch unbekannten — hochohmigen Schicht gewinnen kann, die für die unipolaren Übergangswiderstände in solchen Gleichrichtern verantwortlich ist. Das wichtigste experimentelle Ergebnis war damals die Feststellung, daß die Kapazitätsschichtdicke sich mit der Vorspannung ändert, und zwar in dem Sinne, daß die Schicht sich in Sperrichtung ausdehnt, in Flußrichtung dagegen auf sehr kleine Abmessungen zusammenzieht.

Zu diesen früheren Ergebnissen am Kupferoxydul ist heute zu sagen, daß sie ganz mit dem übereinstimmen, was die neue Theorie fordert; die vorhin geschilderten Überschwemmungseffekte wirken gerade in dem Sinne, daß die schlechtleitende Verarmungsrandschicht sich in der Sperrichtung ausdehnt, in der Flußrichtung zusammenzieht. Es sieht nun aber so aus, als ob aus Wechselstrommessungen auch zum Verständnis der Selengleichrichter und Selensperrschichtzellen wichtige neue Aufschlüsse zu gewinnen wären. An Selengleichrichtern liegen ältere Messungen vor von Wilfried Meyer mit Alexander Schmidt[6]); Selensperrschichtzellen mit kathodisch aufgestäubten Gold-

[4]) In Abb. 3 bedeutet E_F die Energie der Fermikante im Metall, $E_\ominus$ und $E_\oplus$ sind die potentiellen Energien, die ein im Halbleiter ruhendes Elektron und Defektelektron besitzen würden. Gemäß der Definition der Austrittsarbeiten gilt:

$$\psi_\ominus = E_\ominus - E_F$$
$$\psi_\oplus = E_F - E_\oplus$$

woraus das Komplementaritätsgesetz für die Austrittsarbeiten folgt:

$$\psi_\oplus + \psi_\ominus = E.$$

[5]) W. Schottky u. W. Deutschmann, Phys. ZS. **30** (1929), 839.

[6]) ZS. f. techn. Phys. **14** (1933), 11.

elektroden sind 1936 von F. W. Gundlach[7]), und neuerdings von Frl. Annemarie Schmidt bei Prof. Gudden in Prag untersucht worden, und ich darf Ihnen in Abb. 4 einige Ergebnisse der letztgenannten bisher unveröffentlichten Untersuchungen vorführen. Im unteren Teil der Abbildung ist das Verhältnis der reziproken Kapazitäten $1/C : 1/C^0$ mit und ohne Vorspannung, oder das Verhältnis $l_C : l_C^0$ der kapazitätsmäßig wirksamen Schichtdicken in Abhängigkeit von der Sperrspannung im doppeltlogarithmischen Maßstabe aufgetragen. Die Sperrspannung wächst von rechts nach links. Die ausgezogene schräge Gerade links unten stellt die Asymptote dar, in die nach der Randschichttheorie

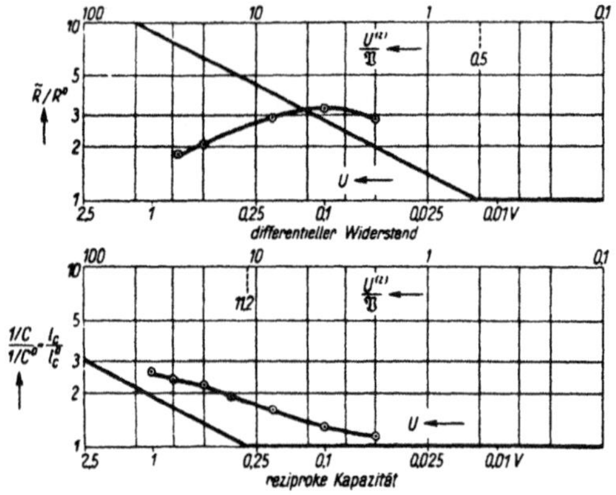

Abb. 4. Relativer Gang der differentiellen Widerstände und der reziproken Kapazitäten von Randschichtgleichrichtern in der Sperrichtung

das Verhältnis l_C/l_C^0 bei großen Sperrspannungen einlaufen müßte. Die Kapazitätsmessungen von Frl. Schmidt sind hiernach in befriedigender Übereinstimmung mit der Randschichttheorie, deren Anwendung auf die exakte Berechnung von Kapazitätswerten neuerdings von meinem Mitarbeiter, Herrn Dr. E. Spenke, durchgeführt worden ist[8]).

Nicht im Einklang mit der idealisierten Randschichttheorie[9]) und anscheinend auch im Widerspruch mit den Kapazitätsmessungen ist jedoch der in Abb. 2 in gleicher Weise dargestellte Verlauf des Widerstandsverhältnisses $\tilde{R}/R^0$ bei wachsender Sperrspannung. ($\tilde{R}$ differentieller Widerstand bei Vorspannung, R^0 Nullwiderstand.) Während nach Abb. 3 zwischen 0,1 und 1 V die Dicke der kapazitativ wirksamen Schicht dauernd wächst, bemerken wir in dem gleichen Spannungsgebiet einen dauernden Abfall von $\tilde{R}/R^0$. Es ist also hier das Auftreten eines neuen Effektes zu vermuten, und zwar eines Effektes, der die mittlere Leitfähigkeit der Randschicht mit wachsender Sperrspannung so stark vergrößert, daß ihr Widerstand trotz wachsender Schichtdicke abnimmt. Nun brauchen wir nach einer Ursache für eine solche Vergrößerung der Randleitfähigkeit nicht weit zu suchen. Die Feldstärken an der Metallelektrode erweisen sich beim Selen schon bei kleinen Sperrspannungen als so groß, daß mit thermischer Feldemission zu rechnen ist; die Austrittsarbeit Metall–Halbleiter wird durch das Feld in der Sperrichtung so stark verkleinert, daß aus diesem Grunde die Randverarmung immer geringer wird und der Widerstand mit wachsender Sperrspannung ab- statt zunimmt.

Beim Kupferoxydulgleichrichter sind die Abweichungen von der idealen Randschichttheorie, wie sich gezeigt hat, in erster Linie auf eine Störstellenverarmung vor der Metallelektrode, eine „chemische Sperrschicht", zurückzuführen. Nicht nur die hier geschilderten Messungen, sondern die Gesamtheit des vorliegenden Beobachtungsmaterials scheint darauf hinzuweisen, daß das beim Selen anders ist; hier sind offenbar die Voraussetzungen der homogenen Randschichttheorie in vielen Fällen verhältnismäßig gut erfüllt, und die Abweichungen von der idealen Theorie sind in ihrem wesentlichsten Teil durch Feldemissionseffekte bedingt[10]).

[7]) Arch. f. Elektrot. **30** (1936), 625.

[8]) Der Schnittpunkt der theoretischen Asymptotengeraden mit der Abszissenachse variiert mit dem Grade der Randverarmung des Halbleiters; die in den beiden Teilen von Abb. 4 angenommenen Lagen des Abszissenschnittpunktes sind noch etwas willkürlich und sollen in einer späteren Mitteilung genauer festgelegt werden.

[9]) Auch für das Widerstandsverhältnis $\tilde{R}/R^0$ fordert die Randschichttheorie in ihrer einfachsten Form bei höheren Sperrspannungen das Einlaufen in eine Gerade mit der Steigung 1 : 2, wie sie im oberen Bild eingezeichnet ist.

[10]) Auch der in Wiss. Veröffentl. a. d. Siemens-Werken **18** (1939), 225, hervorgehobene Unterschied in den Sperrkennlinien des Cu_2O-Gleichrichters (Abb. 21) und Se-Gleichrichters (Abb. 23) scheint viel mehr als dort angenommen durch die beim Selengleichrichter erhöhte Feldemission erklärbar zu sein, die ihrerseits mit Fragen der Austrittsarbeit Metall-Halbleiter zusammenhängt. Vgl. hierzu eine ausführlichere, demnächst erscheinende Darstellung im Schweiz. Arch. f. Wiss. u. Techn.

(Eingegangen am 19. September 1940)

PHYSICAL REVIEW VOLUME 71, NUMBER 10 MAY 15, 1947

Surface States and Rectification at a Metal Semi-Conductor Contact

JOHN BARDEEN
Bell Telephone Laboratories, Murray Hill, New Jersey
(Received February 13, 1947)

Localized states (Tamm levels), having energies distributed in the "forbidden" range between the filled band and the conduction band, may exist at the surface of a semi-conductor. A condition of no net charge on the surface atoms may correspond to a partial filling of these states. If the density of surface levels is sufficiently high, there will be an appreciable double layer at the free surface of a semi-conductor formed from a net charge from electrons in surface states and a space charge of opposite sign, similar to that at a rectifying junction, extending into the semi-conductor. This double layer tends to make the work function independent of the height of the Fermi level in the interior (which in turn depends on impurity content). If contact is made with a metal, the difference in work function between metal and semi-conductor is compensated by surface states charge, rather than by a space charge as is ordinarily assumed, so that the space charge layer is independent of the metal. Rectification characteristics are then independent of the metal. These ideas are used to explain results of Meyerhof and others on the relation between contact potential differences and rectification.

INTRODUCTION

THE generally accepted view[1] of the nature of the rectifying contact between a metal and a semi-conductor is illustrated in Fig. 1 which applies specifically to an excess semi-conductor. Figure 1a shows an energy level diagram of the metal and semi-conductor in equilibrium, but with the contact separated. The Fermi level is the same in both the metal and the semi-conductor. As shown, the work function of the metal, χ_1, is greater than the work function of the semi-conductor, χ_2^0, so that there is a contact difference in potential, $\chi_1-\chi_2^0$. It is assumed that when the metal and semi-conductor are nearly joined, the potential distribution is as shown in Fig. 1b. A double layer is formed such as to give a potential drop, φ_0, from the metal to the interior of the semi-conductor equal to the contact potential difference.

This double layer is assumed to consist of a space charge region in the semi-conductor, extending to a depth of the order of 10^{-6} to 10^{-4} cm,

[1] See, for example, N. F. Mott and R. W. Gurney, *Electronic Processes in Ionic Crystals* (Oxford University Press, London, 1940), Chap. V. The theory of rectification is due in large part to W. Schottky. The most important of his papers is Zeits. f. Physik **118**, 539 (1942).

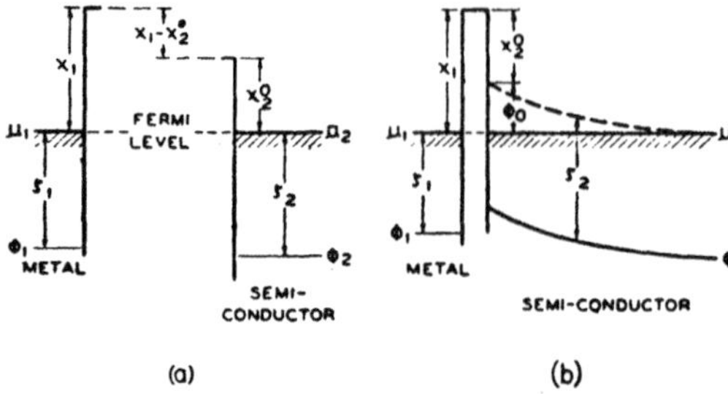

FIG. 1. Energy level diagram for metal and semi-conductor in electrical and thermal equilibrium. μ_1 and μ_2 represent the Fermi levels in metal and semi-conductor, respectively, and ϕ_1 and ϕ_2 the inner potentials. ζ_1 and ζ_2 are the chemical potentials and χ_1 and χ_2^0 the work functions. Figure 1a shows a wide gap with a contact potential difference $\chi_1-\chi_2^0$. In 1b, the gap is very small. There is a space charge region in the semi-conductor which gives an electrostatic potential energy rise at the surface $\varphi_0=\chi_1-\chi_2^0$. This is the usual picture in which no surface states on the semi-conductor are assumed.

and an induced charge on the metal surface. The space charge gives a rise in electrostatic potential energy at the surface of the semi-conductor. Electrons are depleted from the space charge region, giving a layer of high resistivity. If a potential is applied to the junction, most of the drop occurs across this barrier layer. If the potential of the semi-conductor is negative with respect to the metal, the electron energy levels in the semi-conductor are raised, and electrons may flow more easily over the potential hill into the metal. This is the direction of easy flow. On the other hand, if the semi-conductor is positive, the levels are lowered, increasing the height of the hill, and making it more difficult for electrons to travel from the semi-conductor to the metal. This is the direction of high resistance.

According to this view, the equilibrium height of the potential hill, φ_0, and therefore also the degree of rectification, depend on the work function of the metal. For an excess semi-conductor, the larger the work function of the metal, the larger is the potential rise, and the larger is the reverse resistance of the barrier. If the work function of the metal is less than that of the semi-conductor there is a potential drop instead of a potential rise, and no rectification will occur. For a defect semi-conductor, in which the current is carried by holes, just the reverse is true: low metal work function gives high rectification.

A number of investigations have been carried out which have verified these conclusions in some cases, and in other cases have not. H. Schweickert, as quoted by Schottky,[2] has found a correlation between the resistance of selenium rectifiers in the blocking direction and the work function of the metal. Selenium is a defect conductor, and high reverse resistance was found for low work function metals such as K, Na, Li, Ba, and low reverse resistance was found for such high work function metals as Ag, Au, Ni. The metal electrodes were put on by evaporation.

W. H. Brattain,[3] working in this laboratory, has found a good correlation between degree of rectification and work function for metal contacts evaporated on cuprous oxide (a defect conductor) and on both N- and P-type silicon.[4] Metals used, listed in order of decreasing degree of rectification, were Al, Ag, and Pt on cuprous oxide; Pt, Be, Ag, Mg, and Al on N-type silicon; and Mg, Cd, Ag, and Pt on P-type silicon. He found that when contact is made to the semi-conductor by a metal junction in air, the rectification is practically independent of the work function of the metal used. Results somewhat similar to those of Schweickert have been obtained by J. N. Shive[3] who studied the rectification characteristics of a number of contacts made by evaporation of various metals on selenium. Metal contacts, listed in order of decreasing degree of rectification, are Be, Zn, Pb, and Au.

A. V. Joffe[5] has studied contact potential differences and the resistances of the contacts formed from a large number of different semi-conductors and metals. Most gave very poor rectification characteristics. While there was some qualitative correlation of contact potential differences with contact resistance, quantitative agreement with theories of Schottky[1] and Davydov[6] was poor.

---

[2] W. Schottky, Physik Zeits. **41**, 570 (1940).

[3] Unpublished work done at the Bell Telephone Laboratories in 1940 under the general direction of J. A. Becker. The author is indebted to Drs. Brattain and Shive for permission to quote their results.

[4] The designations N- and P-type refer to the direction of rectification. An N-type is an excess semi-conductor, and the direction of easy flow occurs when the semi-conductor is negative relative to the metal. A P-type semi-conductor is a hole conductor, and the direction of easy flow is opposite. In both cases, the direction of easy flow is that in which the carrier moves from the semi-conductor to the metal.

[5] A. V. Joffe, J. Phys. USSR **10**, 49 (1946).

[6] B. Davydov, J. Phys. USSR **4**, 355 (1941).

W. E. Meyerhof[7] has recently been making an extensive study of the relation between contact potential difference and rectification for metal point contacts applied to Si and Ge. He has determined the potential rise, φ_0, from the rectification characteristics at different temperatures and has also measured the ordinary contact potential difference, or Volta potential, by a modified Kelvin bridge method. Preliminary results show little correlation between the two sets of measurements. In fact, he has found that φ_0 is practically independent of the work function of the metal for metal point—Si rectifiers. The same metal may rectify with both P- and N-type silicon, in opposite directions.

These negative results indicate that a closer analysis of the nature of the contact between a metal and semi-conductor is warranted. The main purpose of the present paper is to investigate the effect of electronic states on the surface of the semi-conductor on φ_0. There has been considerable discussion in the literature concerning the possibility of surface states,[8] but little direct evidence as to their existence.

IMPURITY LEVELS AND SURFACE STATES

According to the modern theory of semi-conductors, there is an energy gap between the highest filled band of levels and the lowest state of the conduction band. In the language appropriate to excess semi-conductors, conductivity results from electrons thermally excited to the conduction band from impurity levels. The impurity levels, which are intermediate in energy between the filled band and conduction band, represent states in which the electron is localized around a foreign atom or other defect in the crystal lattice. In normal semi-conductors there may be the order of 10^{-7} to 10^{-3} impurity levels for each atom of the crystal.

In addition to the impurity levels in the interior of the crystal, there may be localized states on the surface with energies in the "forbidden" region between the filled and conduction bands. Shockley[8] and others have investigated the conditions under which surface levels may be expected on an ideal crystal. His analysis, based primarily on a one-dimensional model, indicates that "in a plot of the energy spectrum *versus* interatomic distance the surface levels appear only at lattice constants so small that the boundary curves of the allowed energy bands have crossed." The number of such surface states is equal to the number of surface atoms, and, in a neutral crystal, the surface states are half filled. The conditions for surface states are fulfilled, for example, in a diamond-type lattice.

Surface states may also result from surface imperfections, from foreign atoms on the surface, etc. On general grounds, there is good reason to suppose that the ratio of the number of surface levels to surface atoms may be much higher than the ratio of the number of impurity levels to atoms in the interior.

The energy levels corresponding to the surface states may be discrete, or they may have a continuous distribution with all energies in the gap between the filled and conduction bands. There is little evidence of either an experimental or a theoretical nature on this point. Well-defined impurity states of the same nature, sparsely distributed over the surface, would all have about the same energy and form a discrete level. If the impurity atoms are densely distributed over the surface, so that there is considerable interaction between them, a continuous distribution in energy is to be expected. A continuous distribution is also expected for surface states on clean surfaces of the type discussed by Shockley, Goodwin, and others.[8] In the analysis to follow we assume that the surface levels are continuously distributed in energy. Many of the same general conclusions would follow, however, if discrete levels had been assumed.

We will show that if there is a relatively high density of surface states, there may be a double layer on the free surface of a semi-conductor re-

[7] W. E. Meyerhof, "Contact potential difference in crystal rectifiers," Technical Report No. 5, Univ. of Pennsylvania, BuShips contract NObs-34144, Aug. 10, 1946; see Phys. Rev. 71, 727 (1947).

[8] Surface states with energies in the "forbidden" band were studied first by I. Tamm, Physik. Zeits. Sowjetunion 1, 733 (1932), and are often called Tamm states. Later theoretical work was done by R. H. Fowler, Proc. Roy. Soc. A141, 56 (1933), S. Rijanow, Zeits. f. Physik 89, 806 (1934), A. W. Maue, Zeits. f. Physik 94, 717 (1935), E. T. Goodwin, Proc. Camb. Phil. Soc. 35, 205 (1939), W. G. Pollard, Phys. Rev. 56, 324 (1939), and W. Shockley, Phys. Rev. 56, 317 (1939). The last of these is the most critical, and shows under what conditions surface states are occupied in a normal crystal.

sulting from a surface charge caused by electrons in surface states and a space charge of opposite sign extending into the crystal to a depth of 10^{-6} to 10^{-4} cm. The double layer tends to make the work function independent of the impurities in the interior. If contact is made to a metal, the contact potential difference is compensated largely by a true surface charge rather than by space charge, so that the height of the barrier is largely independent of the metal.

NATURE OF CONTACT

Before discussing the effect of surface states, we will give a brief review of the nature of the contact between metal and semi-conductor.

At a contact between any two electronic conductors there is an electric double layer which adjusts the potential of one relative to the other for equilibrium conditions. This potential difference may be taken as a measure of the strength of the double layer.

Some care must be taken in defining the absolute value of the potential. There is, of course, no way of measuring the potential inside a metal or semi-conductor. The potential just outside the surface can be measured, but the difference between the potential inside and outside the surface depends on the double layer at the surface. Only differences in potential between different parts of a conductor can be observed directly.

Nevertheless, it is desirable to consider the potential in the interior. The true time-average electrostatic potential is periodic, with the period of the lattice. The macroscopic potential, Φ, may be defined as the space average of the actual potential over a region large compared with atomic dimensions. This definition is somewhat arbitrary; another definition might yield a different value for the strength of the double layer at a contact or at a surface. However, only differences in double layers have direct significance, and these are independent of the way the potential in the interior is defined.

The general theory of equilibria[9] shows that when two electronic conductors are in contact and in thermal equilibrium, the electrochemical potentials, μ_1 and μ_2, must be the same in both:

$$\mu_1 = \mu_2 \text{ in equilibrium.} \tag{1}$$

In the usual picture of an electronic conductor, μ is just the Fermi level which determines the probability, p, that a state of energy E is occupied.

$$p = 1/(1+\exp[(E-\mu)/kT]). \tag{2}$$

The electrochemical potential in a region may be defined thermodynamically, in a way which is independent of any particular model, by the equation:

$$\mu = \frac{\partial}{\partial n}(U-TS),$$

where n is the number of electrons in the region, and U and S are the energy and entropy. The partial differentiation corresponds to a reversible change in which the volume and temperature are held constant. The internal energy U includes an electrostatic potential energy, $-e\Phi$, per electron. The choice of the zero or reference level from which this potential energy is measured is arbitrary, and there is the same arbitrary choice in the definition of μ.

It is convenient to introduce a quantity which depends on the chemical constitution of the material, and on the electron density, but not on the electrostatic potential. This quantity, ζ, which is called the chemical potential, to distinguish it from the electrochemical potential, is defined by

$$\zeta = \mu - \varphi, \tag{3}$$

where

$$\varphi = -e\Phi$$

is the electrostatic energy per electron.

Let φ_e be the potential energy of an electron just outside the surface of a conductor. This potential is to be evaluated at a point where the image potential is negligible, but at a distance from the surface small compared with macroscopic dimensions. The work function of the surface is then:

$$\chi_s = \varphi_e - \mu. \tag{4}$$

This is the energy required to take an electron from the conductor and place it at rest at a point, defined as above, just outside the surface.

[9] An excellent discussion is given by R. H. Fowler and E. A. Guggenheim, *Statistical Thermodynamics* (Cambridge University Press, England, 1939), Chap. XI. The author is indebted to Dr. C. Herring for the method of presentation used above.

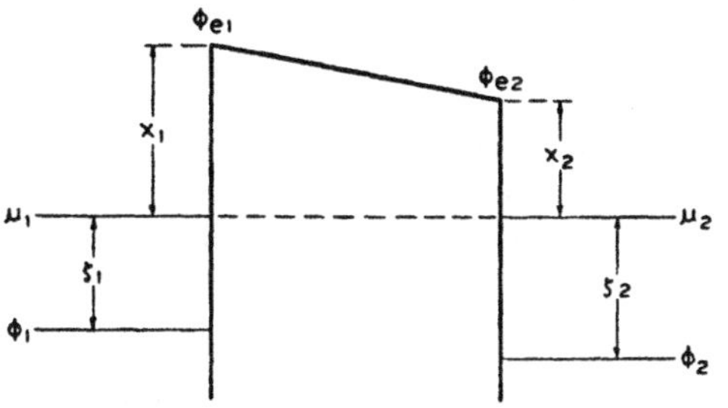

FIG. 2. Diagram to illustrate double layers at a contact between two conductors in equilibrium. There is a double layer of strength $\varphi_{e1}-\varphi_0$ at the surface of conductor 1, a double layer of strength $\varphi_{e2}-\varphi_2$, in the opposite direction at the surface of conductor 2, and a double layer of strength $\varphi_{e2}-\varphi_{e1}=-(\chi_1-\chi_2)$ formed by surface charges on the two conductors. The total strength, $\varphi_2-\varphi_1$, is independent of the surface double layers.

The work function depends on the double layer at the surface. By replacing μ by $\zeta+\varphi$, according to Eq. (3), the work function may be expressed in the form:

$$\chi_s=\varphi_e-\varphi-\zeta, \tag{5}$$

which shows explicitly that the work function is the sum of the energy required to take an electron through the surface double layer, $\varphi_e-\varphi$, and a body term, $-\zeta$, independent of the surface.

It should be emphasized that the separation of the work function into a surface term and a body term depends on the precise way the potential energy, φ, in the interior of the conductor is defined. We have suggested above that the electrostatic potential Φ be defined as space average of the actual potential. Another definition, might yield, for example, a smaller value for $\varphi_e-\varphi$ and a correspondingly larger value for $-\zeta$. Differences in double layers, such as might yield different work functions for different crystal faces of the same material, are not affected by the arbitrary definition of the inner potential. Since the double layer involves a potential difference, the value is independent of the reference level, or zero of potential, which is also arbitrary.

If two electronic conductors are in contact, and in thermal equilibrium, the electrochemical potentials must be the same in both. Thus,

$$\zeta_1+\varphi_1=\zeta_2+\varphi_2,$$

or

$$\varphi_2-\varphi_1=\zeta_1-\zeta_2. \tag{6}$$

The strength of the double layer at the contact, $\varphi_1-\varphi_2$, is equal to the difference between the chemical potentials. Since the chemical potentials depend only on the internal constitutions of the two conductors, the double layer is independent of the work functions of the two surfaces before they are brought into contact.

The situation when there is a small gap between the two conductors in equilibrium is illustrated in Fig. 2. There is a double layer of strength $\varphi_{e1}-\varphi_1$ at the surface of conductor 1, a double layer of strength $\varphi_{e2}-\varphi_2$, in the opposite direction, at the surface of conductor 2, and a double layer of strength $\varphi_{e2}-\varphi_{e1}=-(\chi_1-\chi_2)$, formed by surface charges on the two conductors. The total strength of the double layer is:

$$(\varphi_{e1}-\varphi_1)+(\varphi_{e2}-\varphi_{e1})-(\varphi_{e2}-\varphi_2)=\varphi_2-\varphi_1. \tag{7}$$

The surface double layers drop out, as they should according to Eq. (6).

If the surfaces are in very intimate contact, so that the electron space charges of the two surfaces overlap, it is not possible to divide the total double layer into different parts which depend on the work functions of the individual surfaces. All that can be said is that the strength of the total double layer is $\varphi_2-\varphi_1$. The work functions of the surfaces before they are brought into contact play no role.[10]

USUAL PICTURE OF RECTIFYING CONTACT

If one of the conductors is a semi-conductor, and the other is a metal, it has generally been assumed that as they are brought into contact, no surface charge forms on the semi-conductor.[11] Instead, there is a space charge near the surface which may extend to a depth of the order 10^{-6} to 10^{-4} cm. If the gap is small compared this distance, but is still large enough so that there is

[10] H. Y. Fan, Phys. Rev. 61, 365 (1942); 62, 388 (1942), has attempted to calculate the electron space charge distribution in the double layer at the contact between two metals, and at the contact between a metal and a semi-conductor. Although his calculations are based on a highly idealized model, they may serve to give a rough picture of the charge density in the contact region. He does not consider the effect of surface states on the semi-conductor.

[11] N. F. Mott, Proc. Camb. Phil. Soc. 34, 568 (1938) gives a discussion of the contact between a metal and an insulator or a semi-conductor. A brief analysis is given of the effect of a surface charge resulting from electrons in surface states on the contact between a metal and an insulator. Mott assumes a discrete energy level for the surface states.

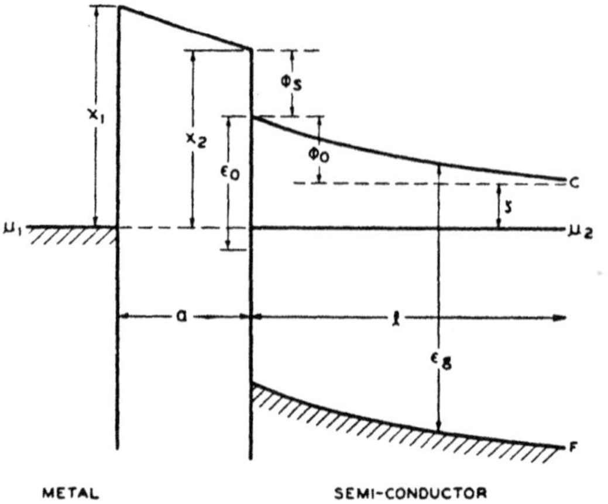

FIG. 3. Energy level diagram for metal semi-conductor contact illustrating notation used in text. The Fermi levels are μ_1 and μ_2 and the work functions χ_1 and χ_2. The lowest state of the conduction band is denoted by C and the highest level of the filled band by F. If the surface states are filled to an energy ϵ_0 below the conduction band there is no net charge on the surface atoms.

no appreciable overlap, most of the energy drop $\chi_1-\chi_2^0$ will occur in the semi-conductor rather than in the gap. This situation is illustrated in Fig. 1b. The space charge raises the energy at the surface of the semi-conductor by an amount $\chi_1-\chi_2^0$. This is the usual picture which indicates that the height of the potential barrier should be equal to the difference in work functions, and is that which is discussed in the introduction.

This picture is deficient in two respects:

(a) The contact may be so intimate that a division of the double layer into one at the surface of the metal, one at the surface of the semi-conductor, and one caused by space charge may not be possible. There will be a double layer at the immediate interface, and another due to space charge, but it is not possible to say how the total strength is divided between them. A cruder way of stating this is that the work functions of the two surfaces are modified by the contact. However, if the contact is intimate, the separate work functions have no meaning.

(b) There may be electronic states localized on the surface of the semi-conductor, so that the surface atoms can become charged. An appreciable field may then exist in the gap. Part of the total drop from metal to semi-conductor will occur across the gap and part across the space charge region. If the density of surface states is sufficiently high, part of the double layer at the *free* surface of a semi-conductor may be formed by a surface states charge compensated by a space charge. The height of the potential barrier of a rectifier formed from the material will then be determined in part by the normal space charge of the free semi-conductor and in part by the work function of the metal.

FREE SURFACE OF SEMI-CONDUCTOR

We will first consider the free surface of a semi-conductor, and then discuss the rectifying contact. The notation to be used is illustrated in the energy level diagram of Fig. 3. The lowest state of the conduction band and the highest state of the filled band of the semi-conductor are indicated, with an energy gap ϵ_g. It is assumed that the distribution of surface states is such that the surface states give no net charge if the states are filled to an energy ϵ_0 below the conduction band.[12] Since the Fermi level cuts the surface above the level determined by ϵ_0, the surface as shown will be negatively charged, this charge resulting from electrons in states between ϵ_0 and the Fermi level. The picture applies to an excess semi-conductor.[13] In the body of the

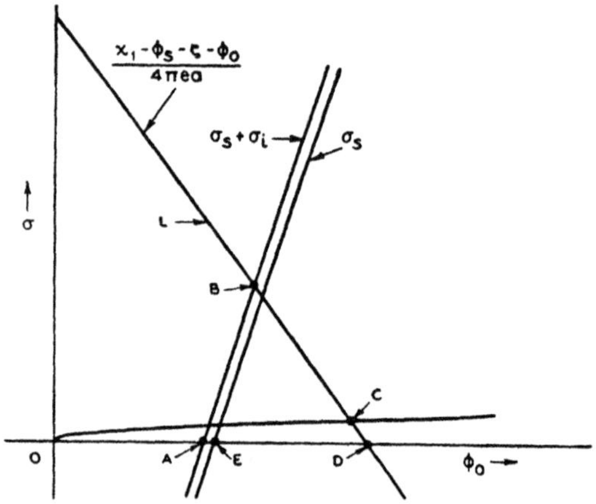

FIG. 4. Schematic diagram showing how the potential rise ϕ_0 is determined from the density of surface levels and the contact potential difference (see text).

[12] We refer to the net charge of the surface states as "surface states" charge to distinguish it from the total charge, including space charge, in the surface layer, which we call "surface charge."

[13] The case of an excess semi-conductor seems to be easier to visualize than that of a defect semi-conductor. All results derived for one case, of course, apply to the other with obvious changes in signs of the charges.

semi-conductor, the Fermi level is an energy ζ below the conduction band. The space charge region extends for an approximate distance l into the semi-conductor, giving a potential energy rise φ_0 at the surface. An energy φ_s is required to remove an electron from the lowest state of the conduction band near the surface to a point just outside the semi-conductor. The work function, χ_2, depends on φ_s as well as on φ_0:

$$\chi_2 = \varphi_s + \varphi_0 + \zeta. \tag{8}$$

The amount and extent of the space charge inside the free surface is determined by the density of surface levels. For zero external field, the positive space charge raises the potential at the surface by an amount just sufficient to give a compensating negative surface charge. The larger the positive space charge region, the larger is φ_0, and the smaller is the negative surface states charge. For some φ_0 the two will be equal in magnitude. This is the equilibrium value.

These relations are indicated in a schematic way in Fig. 4, which shows the variation of σ_i, the total space charge per unit area, of σ_s, the surface states charge, and of the surface charge, $\sigma_s + \sigma_i$, with φ_0. The other curves on the diagram will be referred to later in connection with the discussion of a metal contact. For a uniform space charge in the boundary layer, σ_i is proportional to the square root of φ_0 (see the appendix). The plot of σ_s has a steep slope corresponding to a fairly high density of surface states. It passes through zero (point E in Fig. 4) when

$$\varphi_0 = \epsilon_0 - \zeta, \tag{9}$$

because, according to the definition of ϵ_0 the surface states will then be filled up to the level corresponding to zero surface states charge. The point A represents a value of φ_0 for which $\sigma_s + \sigma_i = 0$, which corresponds to a neutral surface. It can be seen that if the density of surface states is high, this value of φ_0 will be close to that given by Eq. (9).

For the limiting case of a very high density of surface states, the line σ_s becomes vertical, points A and E coincide, and Eq. (9) must be satisfied; hence, according to Eq. (8):

$$\chi_2 = \varphi_s + \epsilon_0. \tag{10}$$

The work function is determined entirely by the

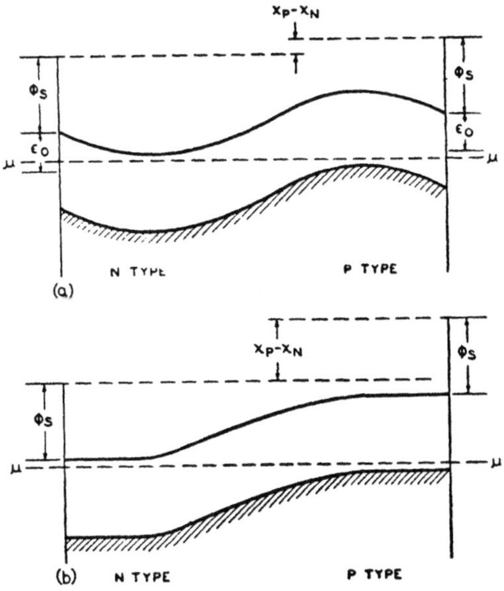

FIG. 5. Energy level diagram showing contact potential difference between two ends of a semi-conductor which changes from N-type to P-type along its length. (a) High density of surface states, small contact potential difference. (b) No surface states, large contact potential difference.

surface, and is independent of the position of the Fermi level in the interior.

On the other hand, if the density of surface states is small, $\sigma_s + \sigma_i$ is approximately equal to σ_i and the condition for a neutral surface leads to a small value for φ_0. In the limiting case of vanishing surface states charge, $\varphi_0 = 0$, and the work function is

$$\chi_2 = \varphi_s + \zeta, \tag{11}$$

which of course *does* depend on the position of the Fermi level in the interior of the semi-conductor.

Some semi-conductors can be made either excess or defect, or N- or P-type,[4] depending on the nature and concentration of impurities. The distribution of impurities in a single sample may be such as to make one end N-type and the other end P-type. Silicon and germanium are examples of materials which may behave in this way. For an N-type conductor, ζ is small; for a P-type conductor ζ is almost equal to the energy gap ϵ_g, which is the order of one electron volt for these materials.

If the density of surface states is low, a large difference in work function between N- and

P-type is to be expected. The difference will be almost equal to the energy gap ϵ_g. For a high density of surface states, the difference in work function will be small. Figures 5a and b show schematic energy level diagrams of these two limiting cases. Both show the variation in potential along, and the work functions at the two ends of a sample which changes from N- to P-type from left to right along its length. Figure 5a is that corresponding to a high density of surface states. There is a space charge region at each end. The potential at the left is raised so as to bring the energy level corresponding to a neutral surface close to the Fermi level. The potential at the right is similarly lowered. The difference in work functions, $\chi_P - \chi_N$, is small. Figure 5b is the case corresponding to a small density of surface states. There are no space charge regions at the ends, and the difference in work functions is large.

A difference in work function can be detected experimentally as a difference in contact potential. Meyerhof[7] has in this way measured the differences in work functions of various samples of N- and P-type silicon and has investigated the effect of surface treatment on these differences. He finds that the difference between N- and P-type samples is about 0.25 ev, which is considerably smaller than the energy gap of about 1.1 ev. It is quite possible that there is a high density of surface states, either on the pure material or resulting from surface films or impurities, which is sufficient to account for this discrepancy.

We have so far discussed the density of surface states in a purely qualitative way. It is of interest to make an estimate of the density required to produce an appreciable space charge layer at the free surface of a semi-conductor. The density depends on the density of charged centers, or donors, in the boundary layer. If there are N such centers per unit volume, and if the thickness of the boundary layer (assumed uniform) is l, we have for the total space charge per unit area:[1]

$$\sigma_i = eNl. \tag{12}$$

The potential energy rise at the surface is

$$\varphi_0 = (2\pi/\kappa)e^2Nl^2, \tag{13}$$

where κ is the dielectric constant.

Let the number of surface states in the energy interval $d\epsilon$ be $nd\epsilon/\epsilon_0$. We assume, for simplicity that n is a constant, which is of the order of the total number of surface levels per unit area with energies in the gap. The change in surface states charge density corresponding to a potential energy rise φ_0 at the surface is then

$$\Delta\sigma_s = en\varphi_0/\epsilon_0. \tag{14}$$

The condition for an appreciable space charge layer is that φ_0 be of the order of magnitude of ϵ_0 when $\Delta\sigma_s = \sigma_i$. This requires that n be the order of Nl. Taking, for example, $\kappa \sim 15$, and $N \sim 10^{17}$, Eq. (13) gives $l \sim 10^{-5}$, so that

$$n \sim Nl \sim 10^{12} \text{ states/cm}^2. \tag{15}$$

This density corresponds to about one surface state per thousand surface atoms. Other things being fixed, the limiting density of surface states is proportional to the square root of the density of charged centers in the boundary layer.

METAL SEMI-CONDUCTOR CONTACT

As a metal surface approaches that of a semi-conductor, the electrostatic field in the gap increases. There is a surface charge σ_M on the metal surface, and a charge of equal magnitude and opposite sign divided between a surface charge, σ_s, and a space charge, σ_i (the latter has been defined as the total space charge per unit area of surface). Thus

$$\sigma_M = -(\sigma_s + \sigma_i). \tag{16}$$

In the case of the metal, the surface charge causes only a very slight change in work function. However, the charge on the semi-conductor may cause appreciable changes in its work function, so that its value χ_2 will differ appreciably from the zero field value χ_2^0. These changes can be analyzed with the aid of Figs. 3 and 4.

According to Fig. 3, the field in the gap between metal and semi-conductor is $(\chi_1 - \chi_2)/ea$. This is produced by the surface charge which gives a field $4\pi(\sigma_s + \sigma_i)$. Equation (8) expresses χ_2 as a function of φ_0. Equating the two expressions for the field leads to:

$$\frac{\chi_1 - \zeta - \varphi_s - \varphi_0}{4\pi ea} = \sigma_s(\varphi_0) + \sigma_i(\varphi_0). \tag{17}$$

The left-hand side of the equation is plotted in Fig. 4 as the line "L," and is a straight line of negative slope. The solution to (17) is represented by the point B, which is the intersection of "L" with the line representing $\sigma_s+\sigma_i$. The line "L" crosses the axis at the point D, which corresponds to $\chi_2=\chi_1$, or

$$\varphi_0=\chi_1-\varphi_s-\zeta. \qquad (18)$$

It can be seen that if the density of surface levels is high, so that σ_s and $\sigma_s+\sigma_i$ have steep slopes, the point B will lie close to the point A which gives the value of φ_0 for the free surface. In this case,

$$\varphi_0\sim\epsilon_0-\zeta, \qquad (19)$$

and is practically independent of the work function of the metal. This is a possible explanation of the results of Meyerhof, who found that the value of φ_0 for metal points on silicon does not depend very much on the metal used.

In the limiting case of vanishing surface states charge density, $\sigma_s=0$, and the solution of (17) is given by the point C of Fig. 4. The value of φ_0 is then close to that given by the intersection of the line L with the horizontal axis (point D). In this case,

$$\varphi_0\sim\chi_1-\varphi_s-\zeta, \qquad (20)$$

and the usual picture of the contact between metal and semi-conductor applies.

As the metal is brought closer to the semi-conductor, the slope of the line "L" increases with decreasing "a" and one might expect a transition from a solution near A to one near D with a resulting value of φ_0 given by Eq. (20). However, even if there is only one surface state per hundred surface atoms, the slope of the line σ_s is so great that the solution still lies near point A even when a is reduced to atomic distances, say 3A. This may be verified by making a rough calculation based on Fig. 4. For the value of φ_0 to be largely determined by the semi-conductor, the value of φ_0 for the solution B must lie closer to A then to D. The necessary and sufficient condition for this is that the slope of the line σ_s be much greater, in absolute value, than the slope of "L." We have

$$\text{slope of } \sigma_s=m_s=en/\epsilon_0,$$
$$\text{slope of } L=-m_L=-1/4\pi ea. \qquad (21)$$

The condition that m_s/m_L be large is

$$m_s/m_L=4\pi e^2an/\epsilon_0\gg1, \qquad (22)$$

or

$$n\gg\epsilon_0/4\pi e^2a. \qquad (23)$$

Setting for example, $\epsilon_0\sim10^{-12}$ erg and $a\sim3\times10^{-8}$ cm, this condition requires that

$$n\gg10^{13}. \qquad (24)$$

Thus if there is appreciably more than about one surface state per hundred surface atoms, the metal work function will have little effect on φ_0. This is about an order of magnitude larger than the surface density required for the existence of an appreciable boundary layer at the free surface.[14]

The theory is worked out in detail for the special case of a uniform Schottky exhaustion layer[1] in the appendix.

CONCLUSIONS

If the density of surface levels with energies in the "forbidden" band is sufficiently high ($>\sim10^{12}/\text{cm}^2$), there will be a double layer at the free surface of a semi-conductor formed from a surface states charge and a space charge of opposite sign. The space charge region is similar to that which exists at a rectifying contact. This double layer tends to make the work function independent of the height of the Fermi level in the interior, and so independent of the impurity content.

The total strength of the double layer at a rectifying junction between a metal and semi-conductor is fixed by the difference in chemical potentials, and so depends on the body properties of the metal and semi-conductor, and is independent of the work functions of the surfaces before they are brought into contact. The double layer consists of the following parts:

[14] S. Benzer, Phys. Rev. 71, 141 (1947), has recently reported on the current-voltage characteristic observed when contact is made between two pieces of the same (homogeneous) germanium crystal. He states that "the characteristic observed for both polarities is more the order of the back resistance observed when either piece of the crystal is contacted with a metal; in both directions the negative resistance at high voltage appears, which is typical of the back characteristics of metal-Ge contacts using these alloys." The presence of a space charge layer at the surface of each piece which is little modified by contact would result in a characteristic similar to that of two rectifiers in series opposition. Such boundary layers are to be expected if the density of surface states is the order of that given by Eq. (24) and if the contact is not too intimate. Benzer's results are thus indirect evidence for surface states on germanium crystals.

(1) A double layer of atomic dimensions at the metal surface.

(2) A double layer of atomic dimensions at the semi-conductor surface.

(3) A double layer formed from surface charges on the metal and semi-conductor, both of atomic dimensions.

(4) A double layer formed from a surface charge of atomic dimensions and a space charge extending to a depth of 10^{-6} to 10^{-4} cm into the semi-conductor.

The strengths of the double layers may be estimated in different cases as follows:

(a) If the density of surface levels is sufficiently high ($>\sim 10^{13}/\mathrm{cm}^2$) the double layer (4) will be independent of the metal, and will be the same as that for the free surface of the semi-conductor. The rectification properties will then be largely independent of the work function of the metal. The difference in contact potentials is compensated by (3).

(b) If the density of surface levels is small ($<\sim 10^{13}/\mathrm{cm}^2$), the double layer (3) will be small, and (4) will be determined approximately by the difference in work functions.

(c) If the contact between the metal and semi-conductor is very intimate, it may not be possible to distinguish between the double layers (1), (2), and (3). The metal will tend to broaden the surface levels, but if the broadening is small compared with the energy gap, conclusion (a) will still be valid.

(d) If the broadening of the surface levels by the metal is large, no conclusions about the space charge can be drawn from measurements of contact potential differences.

It is believed that all of these cases may be realized.

The author is indebted to various members of the technical staff of the Bell Telephone Laboratories, particularly to W. Shockley, C. Herring, and W. H. Brattain, for the benefit of numerous discussions concerning the subject matter of this paper, and for helpful comments on the preparation of the manuscript.

APPENDIX

Theory for Schottky Exhaustion Layer

The detailed calculation of φ_0 requires a knowledge of the distribution of space charge in the semi-conductor and its dependence on φ_0. The dependence of surface charge on φ_0 is also required. We will carry through the calculation explicitly only for the case of a Schottky exhaustion layer[1] with a uniform density of charge. This case is particularly simple and brings out the essential features of the problem. For simplicity we also assume that the surface states are uniformly distributed in energy.

Let eN be the positive charge density in the barrier region, assumed constant. Let l be the thickness of the charged layer. Let $nd\epsilon/\epsilon_0$ be the number of surface levels per unit area with energies in the range $d\epsilon$. We assume that n is a constant, independent of energy. For other notation see Fig. 3.

We have the following relations:

(a) The total space charge per unit area is:

$$\sigma_i = Nel. \tag{12}$$

(b) The potential energy at the surface of the semi-conductor is

$$\varphi_0 = \frac{2\pi e^2 N}{\kappa} l^2, \tag{13}$$

where κ is the dielectric constant.

(c) The surface states charge is

$$\sigma_s = -en(\epsilon_0 - \varphi_0 - \zeta)/\epsilon_0. \tag{15}$$

(d) The total surface charge is

$$\sigma_i + \sigma_s = Nel - en(\epsilon_0 - \varphi_0 - \zeta)/\epsilon_0. \tag{25}$$

(e) The work function of the semi-conductor is:

$$\chi_2 = \varphi_s + \zeta + \frac{2\pi e^2 N}{\kappa} l^2. \tag{26}$$

Equating $\sigma_i + \sigma_s$ with $(\chi_1 - \chi_2)/4\pi e a$, we get Eq. (17) from which φ_0 or the thickness of the barrier layer, l, can be determined. With the notation:

$$l_0 = \frac{n}{N}(1 - \zeta/\epsilon_0), \tag{27a}$$

$$l_1 = \kappa\epsilon_0/2\pi e^2 n, \tag{27b}$$

$$l_2^2 = \kappa(\chi_1 - \varphi_s - \zeta)/2\pi e^2 N, \tag{27c}$$

Eq. (17) reduces to:

$$l^2/l_1 + l - l_0 = (l_2^2 - l^2)/2\kappa a, \tag{28}$$

which is a simple quadratic equation for l.

Whether the semi-conductor or the metal predominates in determining l, and thus the barrier height, depends on the relative magnitudes of the terms on the left and right sides of Eq. (28). The terms on the right side will be negligible if

$$l_1 \ll 2\kappa a \tag{29}$$

and

$$l_0 \gg l_2^2/2\kappa a. \tag{30}$$

The first condition requires that

$$n \gg \epsilon_0/2\pi e^2 a, \tag{31}$$

and the second that

$$n \gg (\epsilon_0/4\pi e^2 a)(\chi_1 - \varphi_s - \zeta)/(\epsilon_0 - \zeta). \tag{31a}$$

These conditions are essentially equivalent to that given by Eq. (23), and lead to the conclusion that $n \gg 10^{13}$ in order that the metal have little influence on the space charge region. In this case of high density of surface states, the equation for the layer thickness reduces to that for the free surface of the semi-conductor:

$$l^2/l_1 + l - l_0 = 0. \tag{32}$$

Equation (32) may, of course, be used to estimate the thickness of the space change region at the free surface of the semi-conductor, regardless of the density of surface states. The limiting case of high density corresponds to

$$l_0 \gg l_1, \tag{33a}$$

or

$$n^2 \gg N\kappa\epsilon_0^2/2\pi e^2(\epsilon_0 - \zeta). \tag{33b}$$

This is essentially the requirement, stated above Eq. (15), that n be large compared with Nl. The approximate solution of Eq. (32) is then

$$l^2 \approx l_0 l_1.$$

The product $l_0 l_1$ is independent of n, and is just the square of the thickness of the Schottky layer for a barrier height

$$\varphi_0 = \epsilon_0 - \zeta. \tag{34}$$

This is the condition that the Fermi level cross the surface near the energy corresponding to zero surface states charge.

In the limiting case of a vanishingly small density of surface states, the thickness of the barrier layer is determined by setting the right-hand side of Eq. (28) to zero. This gives

$$\varphi_0 = \chi_1 - \varphi_s - \zeta. \tag{35}$$

The height of the barrier is equal to the difference in work functions.

Equation (28) may be used for intermediate cases.

Surface States and Barrier Height of Metal–Semiconductor Systems*

A. M. COWLEY
Electrical Engineering Department, University of California, Santa Barbara, California
AND
S. M. SZE
Bell Telephone Laboratories, Incorporated, Murray Hill, New Jersey

(Received 30 November 1964; in final form 1 June 1965)

The dependence of the barrier height of metal–semiconductor systems upon the metal work function is derived based on the following assumptions: (1) the contact between the metal and the semiconductor has an interfacial layer of the order of atomic dimensions; it is further assumed that this layer is transparent to electrons with energy greater than the potential barrier but can withstand potential across it. (2) The surface state density (per unit area per electron volt) at the interface is a property only of the semiconductor surface and is independent of the metal. The barrier height φ_{Bn} is defined here as the energy needed by an electron at the Fermi level in the metal to enter the conduction band of the semiconductor.

With the above assumptions, the barrier height for n-type semiconductor–metal contacts is found to be a linear combination of the metal work function φ_m and a quantity φ_0 which is defined as the energy below which the surface states must be filled for charge neutrality at the semiconductor surface. The energy φ_0 is measured from the edge of the valence band. For constant surface state density the theoretical expression obtained is

$$\varphi_{Bn}=\gamma(\varphi_m-\chi)+(1-\gamma)(E_g-\varphi_0)-\Delta\varphi_n,$$

where χ and E_g are electron affinity and the band gap of the semiconductor, respectively, $\Delta\varphi_n$ is the image force barrier lowering, and γ is a weighting factor which depends mainly on the surface state density and the thickness of the interfacial layer.

The theoretical expression is compared to the presently available φ_{Bn} vs φ_m data for Si, GaP, GaAs, and CdS, by fitting the data to straight lines using the method of least squares. The best straight-line fit was obtained for the GaP data, with probable error limits on the slope and intercept of ± 0.03 and ± 0.13 eV, respectively.

The parameter γ in the theoretical expression is found to range from 0.07 for GaAs to almost unity for the CdS data reported by Goodman indicating weak and strong dependence of the surface barrier height on the metal work function, respectively.

The value of φ_0 is roughly a third of the respective band gap energies for Si, GaP, and GaAs, and the surface state density for these semiconductors is found to be in the range 10^{13}–10^{14} states/cm^2/eV, for the experiments cited.

Excessive scatter in the data points for the CdS data of Mead and Spitzer casts doubt on the significance of a straight-line fit for this case. The data of Goodman for CdS obey the Schottky theory for a metal–semiconductor barrier, but this agreement requires a value of the electron affinity χ which is different from the vacuum-photothreshold value measured by other authors.

I. INTRODUCTION

IN 1947 Bardeen[1] proposed a theory to explain the rectification characteristic of a metal–semiconductor contact. He showed that if the density of localized surface states having energies distributed in the semiconductor energy gap is sufficiently high, a double layer at the free surface of a semiconductor is formed from a net charge of electrons in surface states and a space charge of opposite sign. He concluded that this double layer will tend to make the work function independent of the height of the Fermi level in the interior of the semiconductor, and the rectification characteristics of a metal–semiconductor contact are then practically independent of the metal. It is well known that such surface states and surface barriers do exist.[2] The surface states can occur either from the termination of the periodic structure of the semiconductor crystal at the surface or from the presence of adsorbed foreign atoms on the surface.

The work of Archer and Atalla[3] on metal contacts on cleaved silicon surfaces seems to indicate that for gold on cleaved silicon, there is good agreement between experiment and the simple theory proposed by Schottky[4] that the barrier height depends only on the work functions of the metal and the semiconductor and is independent of the semiconductor doping. Recently, however, Mead and Spitzer[5] have studied the barrier height of metal–semiconductor systems for 14 elemental

* Work performed at W. W. Hansen Laboratories of Physics, Stanford, California, supported in part by the U. S. Electronics Command Laboratory, Ft. Monmouth, New Jersey, under Contract DA36(039) SC-85387 and by the U. S. Navy under contract Nonr-225(24).

[1] J. Bardeen, Phys. Rev. 71, 717 (1947).

[2] R. H. Kingston, *Semiconductor Surface Physics* (University of Pennsylvania Press, Philadelphia, 1957).

[3] R. J. Archer and M. M. Atalla, Am. Acad. Sci. N. Y. **101**, 697 (1963).

[4] H. K. Henisch, *Rectifying Semiconductor Contacts* (Oxford University Press, New York, 1957), Chap. VII.

[5] C. A. Mead and W. G. Spitzer, Phys. Rev. **134**, A713 (1964).

and III–V semiconductors using cleaved semiconductor surfaces. In all cases, they report that the position of the Fermi level at the surface is essentially independent of the metal work function and is very close to one-third of the band gap from the valence-band edge.

The results of the experiments cited above are evidently quite different; the Schottky model seems to hold for Si–metal systems, but the results for the Mead and Spitzer experiments are in direct conflict with the Schottky theory.

Crowell, Sze, and Spitzer[6] have observed that the temperature dependences of the gold n-type silicon surface barrier height and the silicon energy gap are the same. They also showed that the Fermi level at the metal–semiconductor interface is pinned in relation to the valence–band edge, independent of lattice temperature.

In this paper the dependence of the barrier height on metal work function, surface states, and the thickness of the interfacial layer is derived. An attempt will be made to explain published experimental results in accordance with the derived theoretical expression; additional data on metal–GaP systems are given in Sec. III to confirm the validity of the expression. The temperature dependence of the barrier height is also explained. A discussion of the basic assumptions involved in this treatment is given in Sec. IV.

II. THEORY OF METAL–SEMICONDUCTOR SURFACE BARRIER IN THE PRESENCE OF SURFACE STATES

The energy band diagram of a metal–n-type-semiconductor contact is shown in Fig. 1. The various quantities used in the derivation which follows are defined in this figure. The first quantity which is of interest to us is the energy φ_0; this energy is measured from the valence–band edge at the semiconductor surface and specifies the level below which all surface states must be filled for charge neutrality at the semiconductor surface. The second quantity is φ_{Bn}, the barrier height of the metal–semiconductor contact; φ_{Bn} is the energy needed by an electron at the Fermi level in the metal to enter the conduction band of the semiconductor, and includes the effect of the image force lowering $\Delta\varphi_n$. The interfacial layer will be assumed to have a thickness of a few angstroms and will be assumed transparent to electrons whose energy is greater than the potential barrier.

We first consider a semiconductor with acceptor surface states whose density is D_s states/cm^2/eV, and assume, as did Bardeen,[1] that D_s is a constant over the energy range from φ_0 to the Fermi level. The case of a nonuniform density of surface states in the forbidden gap will be considered in Sec. IV.

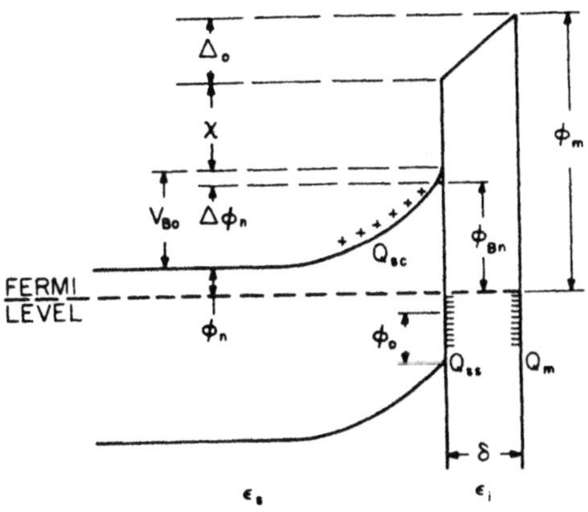

FIG. 1. Energy band diagram of a metal–n-type-semiconductor contact with an interfacial layer. ϕ_m = work function of metal, ϕ_{Bn} = barrier height of metal–semiconductor surface barrier, ϕ_0 = energy level at surface defined in Sec. II, $\Delta\phi_n$ = image force barrier lowering, ϕ_n = energy difference between conduction band and E_F in bulk semiconductor, Δ_0 = potential across interfacial layer, χ = electron affinity of semiconductor, V_{BO} = diffusion potential, ϵ_s = dielectric constant of semiconductor, ϵ_i = dielectric constant of interfacial layer, δ = thickness of interfacial layer, Q_{sc} = space-charge density in semiconductor, Q_{ss} = surface state charge density on semiconductor, and Q_m = surface charge density on metal.

For a uniform distribution the surface state charge density on the semiconductor Q_{ss} is given by

$$Q_{ss} = -eD_s(E_g - \varphi_0 - \varphi_{Bn} - \Delta\varphi_n)\ \mathrm{C\ cm^2}, \tag{1}$$

where $\Delta\varphi_n$ is the image force barrier lowering[7] and e is the electronic charge. The quantity in parentheses is simply the difference between the Fermi level at the surface and φ_0. D_s times this quantity yields the number of surface states above φ_0 which are full.

The space charge which forms in the depletion layer of the semiconductor can be expressed as an equivalent surface charge density, which is the net charge/cm^2 looking into the bulk semiconductor from a point just inside the semiconductor surface. The charge is obtained by solving Poisson's equation for the depletion layer of the semiconductor and can be written as

$$Q_{sc} = [2e\epsilon_s N_D(\varphi_{Bn} + \Delta\varphi_n - \varphi_n - kT/e)]^{\frac{1}{2}}\ \mathrm{C/cm^2}, \tag{2}$$

where N_D is the donor density of the bulk semiconductor. Equation (2) is valid only if there is no inversion layer in the semiconductor; the metal–semiconductor systems considered in this paper all meet this restriction.

The total equivalent surface charge density on the semiconductor surface is given by the sum of Eqs. (1) and (2). In the absence of any space charge effects in the interfacial layer, an exactly equal and opposite charge Q_m develops on the metal surface. For thin interfacial layers, such effects are negligible, and Q_m can

[6] C. R. Crowell, S. M. Sze, and W. G. Spitzer, Appl. Phys. Letters 4, 91 (1964).

[7] S. M. Sze, C. R. Crowell, and D. Kahng, J. Appl. Phys. 35, 2534 (1964).

be written as

$$Q_m = -(Q_{ss}+Q_{sc})$$
$$= eD_s(E_g-\varphi_0-\varphi_{Bn}-\Delta\varphi_n) - [2e\epsilon_s N_D(\varphi_{Bn}+\Delta\varphi_n-\varphi_n-kT/e)]^{\frac{1}{2}}. \quad (3)$$

The potential Δ_0 across the interfacial layer with no voltage applied to the junction can be obtained by the application of Gauss' law to the surface charge on the metal and semiconductor:

$$\Delta_0 = -\delta(Q_m/\epsilon_i), \quad (4)$$

where ϵ_i is the dielectric constant of the interfacial layer and δ its thickness. Another relation for Δ_0 can be obtained by inspection of the energy band diagram of Fig. 1:

$$\Delta_0 = \varphi_m - (\chi+\varphi_{Bn}+\Delta\varphi_n). \quad (5)$$

This results from the fact that the Fermi level must be constant throughout the metal–interfacial-layer–semiconductor system at equilibrium.

If Δ_0 is eliminated from Eqs. (4) and (5), and Eq. (3) is used to substitute for Q_m, we obtain

$$(\varphi_m-\chi)-(\varphi_{Bn}+\Delta\varphi_n) = \left[\frac{2e\epsilon_s N_D\delta^2}{\epsilon_i^2}(\varphi_{Bn}+\Delta\varphi_n-\varphi_n-kT/e)\right]^{\frac{1}{2}} - \frac{eD_s\delta}{\epsilon_i}(E_g-\varphi_0-\varphi_{Bn}-\varphi_n). \quad (6)$$

Equation (6) can now be solved for φ_{Bn}. Introducing the quantities V_1, α, and γ:

$$V_1 = 2e\epsilon_s N_D\delta^2/\epsilon_i^2, \quad (7a)$$

$$\alpha = eD_s\delta/\epsilon_i, \quad (7b)$$

$$\gamma = 1/(1+\alpha) = \epsilon_i/(\epsilon_i+e\delta D_s), \quad (7c)$$

we can write the solution to (6) as

$$\varphi_{Bn} = [\gamma(\varphi_m-\chi)+(1-\gamma)(E_g-\varphi_0)-\Delta\varphi_n] + \{\gamma^2V_1/2-\gamma^{\frac{3}{2}}[V_1(\varphi_m-\chi) + (1-\gamma)(E_g-\varphi_0)V_1/\gamma - V_1(\varphi_n+kT/e)/\gamma+\gamma V_1^2/4]^{\frac{1}{2}}\}. \quad (8)$$

Equation (7a) can be used to calculate V_1 if values of δ and ϵ_i are estimated: For vacuum-cleaved or well-cleaned semiconductor substrates the interfacial layer will have a thickness of atomic dimensions, i.e., 4 or 5 Å. The dielectric constant of such a thin layer can be well approximated by the free-space value, and since this approximation represents a lower limit for ϵ_i, it leads to an overestimation of V_1. For $\epsilon_s \cong 10\epsilon_i$ and $N_D \leq 10^{18}$ cm^{-3}, V_1 is small, of the order of 0.01 eV, and the { } term in Eq. (8) is estimated to be less than 0.04 eV. Neglect of the { } term in Eq. (8) reduces the equation to

$$\varphi_{Bn} = \gamma(\varphi_m-\chi)+(1-\gamma)(E_g-\varphi_0)-\Delta\varphi_n \equiv \gamma\varphi_m+b. \quad (9)$$

If γ and b can be determined experimentally and if χ is known, then using the measured values we can determine φ_0 from the relation

$$\varphi_0 = E_g - (\gamma\chi+b+\Delta\varphi_n)/(1-\gamma), \quad (10)$$

and from (7c) a relation for determining D_s is

$$D_s = (1-\gamma)\epsilon_i/\gamma\delta e. \quad (11)$$

Using the previous assumptions for δ and ϵ_i, Eq. (11) becomes

$$D_s \cong 10^{13}(1-\gamma)/\gamma \text{ states/cm}^2\text{/eV}. \quad (11a)$$

Three special cases of Eq. (9) are of interest:

(1) When $\gamma\to 0$, then

$$\varphi_{Bn} \cong (E_g-\varphi_0)-\Delta\varphi_n. \quad (12)$$

In this case the Fermi level at the interface is "pinned" by the surface states at the value φ_0 above the valence band. The barrier height is independent of the metal work function, and is determined entirely by the doping and surface properties of the semiconductor.

(2) When $\gamma\to 1$, then

$$\varphi_{Bn} \cong (\varphi_m-\chi)-\Delta\varphi_n. \quad (13)$$

This is recognized as the familiar expression (except for the $\Delta\varphi_n$ term) for the barrier height of a simple Schottky barrier where surface effects are neglected.

The barrier height φ_{Bn} of a metal–n-type-semiconductor contact is the threshold for photoemission of electrons from the metal into the semiconductor. The corresponding quantity for a metal–p-type semiconductor is φ_{Bp}, which is the threshold for photoemission of holes from the metal into the p-type semiconductor. If φ_0 is assumed to be independent of doping, then for a given semiconductor φ_{Bn} and φ_{Bp} are related by

$$\varphi_{Bn}+\varphi_{Bp}+\Delta\varphi_n+\Delta\varphi_p = E_g, \quad (14)$$

where $\Delta\varphi_p$ is the image force barrier lowering for the metal–p-type-semiconductor contact. In checking the validity of (14) for cases where both φ_{Bn} and φ_{Bp} have been measured for a particular semiconductor, the image effects should not be neglected as some authors have done.[8,9] In experiments to be cited later in this paper, we find that $\Delta\varphi_n+\Delta\varphi_p$ can exceed 0.1 eV.

(3) The temperature effect on the barrier height: If we assume that the surface state density and the thickness of the interfacial layer are slowly varying functions of the lattice temperature, then γ and φ_0 are essentially independent of temperature. If the temperature variation of the image force barrier lowering can be neglected, we obtain from Eq. (9)

$$\partial\varphi_{Bn}/\partial T = \gamma\partial\varphi_m/\partial T - \gamma\partial\chi/\partial T + (1-\gamma)\partial E_g/\partial T. \quad (15)$$

[8] C. A. Mead and W. G. Spitzer, Phys. Rev. Letters **10**, 471 (1963).

[9] W. G. Spitzer and C. A. Mead, J. Appl. Phys. **34**, 3061 (1963).

If the temperature dependence of $(\varphi_m - \chi)$ and the band gap are approximately equal or if γ is small compared to unity, then from Eq. (15)

$$\partial\varphi_{Bn}/\partial T \cong \partial E_g/\partial T. \tag{16}$$

III. EXPERIMENTAL DETERMINATION OF φ_0 AND SURFACE STATES DENSITY

The Si–metal system studied by Archer and Atalla[3] will be considered first. An attempt is made here to fit the data of the experiments to the theory of Sec. II. The GaAs system studied by Mead and Spitzer[9] and the CdS–metal systems studied by Mead and Spitzer[9] and by Goodman[10] are then investigated.

Also included are the experimental results for metal–GaP systems. The details of these experiments have been previously reported.[11,12]

1. Si–Metal Systems

Archer and Atalla attempted to verify the relation between the barrier height of a metal–silicon junction and the metal–semiconductor work function by using one metal and varying the doping density in the silicon. They proposed the simple Schottky model for the junction which yields Eq. (13). This expression can be rewritten in terms of φ_n and the diffusion potential V_{BO} by using the fact that $\varphi_{Bn} + \Delta\varphi_n = V_{BO} + \varphi_n$ (Fig. 1):

$$V_{BO} = \varphi_m - \chi - \varphi_n. \tag{17}$$

This relation corresponds to the case where the electronic charge due to surface states is entirely negligible. However, the second part of Archer and Atalla's experiment, in which different metals were used in forming the junctions, shows that the metal–semiconductor barrier height measured from the Fermi level in the metal is quite insensitive to the work function of the metal used, indicating that surface states charge is not negligible. The data of Crowell *et al.*[13] for Au, Ag, Cu, and Pd support this observation. Equation (13) should therefore be replaced by Eq. (9), which is rewritten here in terms of φ_n and V_{BO}:

$$V_{BO} = \gamma(\varphi_m - \chi) + (1-\gamma)(E_g - \varphi_0) - \varphi_n. \tag{18}$$

Archer and Atalla obtained the following expression for V_{BO} vs φ_n from a least-squares fit of a straight line to their experimental data:

$$V_{BO} = 0.846 - 1.2\varphi_n. \tag{19}$$

They were not able to explain the fact that Eq. (19) has a slope of 1.2 rather than unity as predicted by Eqs. (17) and (18). They reanalyzed the data by fitting it to a line with slope of unity and minimizing the

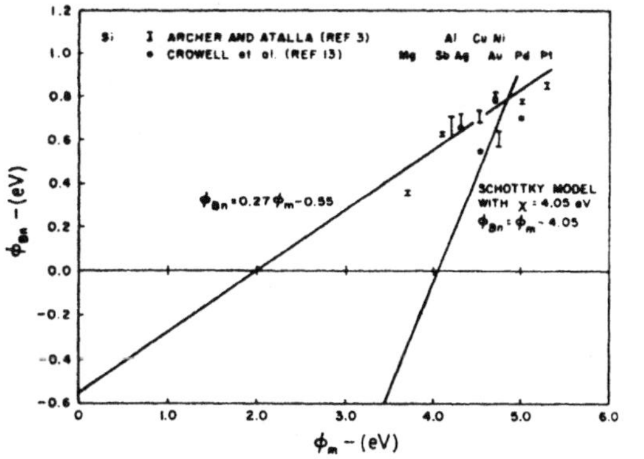

FIG. 2. Least-squares straight-line fit to the silicon φ_{Bn} vs φ_m data calculated from the results of Archer and Atalla (Ref. 3); vertical lines indicate range of φ_{Bn} values for each metal. The photothreshold measurements of Crowell *et al.* (Ref. 13) are shown for comparison.

average deviation of the data from the line, obtaining the relation

$$V_{BO} = 0.810 - \varphi_n. \tag{20}$$

They compare the intercept of this equation to the intercept of Eq. (17) obtaining

$$\varphi_m - \chi = 0.810. \tag{21}$$

The metal used was gold, with $\varphi_m = 4.70$ eV,[14] and χ for silicon is 4.05 eV,[15] so that

$$\varphi_m - \chi = 0.65 \tag{22}$$

for these materials. The agreement between Eqs. (21) and (22) is fair.

Use of Eq. (18) to interpret the experimentally derived Eq. (21) would seem to be more reasonable than the procedure of Archer and Atalla, in light of the fact that surface states are known to be nonnegligible in their experiments. From a comparison of Eqs. (20) and (18) we find

$$\gamma(\varphi_m - \chi) + (1-\gamma)(E_g - \varphi_0) = 0.810. \tag{23}$$

The values for γ and φ_0 can be obtained as follows: Archer and Atalla measured V_{BO} for various metals on *n*- and *p*-type silicon. Their results were presented as a plot of V_{BO} vs $\varphi_m - (\chi + \varphi_n)$, and for each experimental point, the metal used to form the contact was specified. We have calculated the value of φ_{Bn} corresponding to each of Archer and Atalla's experimental points, and the results of these calculations are plotted in Fig. 2. The photothreshold data of Crowell *et al.*[13] are shown for comparison.[16] A least-square straight line fit to Archer

[10] A. M. Goodman, J. Appl. Phys. **35**, 573 (1964).
[11] M. Cowley and H. Heffner, J. Appl. Phys. **35**, 255 (1964).
[12] S. M. Sze, J. L. Moll, and T. Sugano, Solid State Electron. **7**, 509 (1964).
[13] C. R. Crowell, W. G. Spitzer, L. E. Howarth, and E. E. LaBate, Phys. Rev. **127**, 2006 (1962).

[14] J. C. Reviere, Proc. Phys. Soc. (London) **870**, 676 (1957).
[15] F. G. Allen and G. W. Gobeli, Phys. Rev. **127**, 150 (1962).
[16] The data of Crowell *et al.* were obtained from samples prepared by cleaving Si in vacuum, and subsequently evaporating metals onto the cleaved surface. Using the same cleavage apparatus,

TABLE I. Experimental values for metal–GaP barrier heights.

Metal	V_0 (intercept of $1/C^2$ vs V plot)	φ_{Bn} (capacitance)	φ_{Bn} (photoresponse)	φ_{Bn} (capacitance)– φ_{Bn} (photoresponse)
Copper	1.3	1.34	1.20±0.02	0.14
Aluminum	1.10	1.14	1.05±0.005	0.09
Gold	1.30	1.34	1.28±0.02	0.06
Platinum	1.48	1.52	1.45±0.03	0.07
Magnesium	1.05	1.09	1.04±0.005	0.05
Silver	...	...	1.20±0.02	...

and Atalla's data yields

$$\varphi_{Bn}=0.27\varphi_m=0.55 \qquad (24)$$

with probable error limits on the slope and intercept of ±0.05 and ±0.22 eV, respectively. Comparing this expression to Eq. (9), and using Eqs. (10) and (11a), we obtain

$$\gamma=0.27\pm0.05,$$
$$\varphi_0=0.30\pm0.36 \text{ eV},$$
$$D_s=(2.7\pm0.7)\times10^{13} \text{ states/cm}^2\text{/eV},$$

where the error limits in φ_0 and D_s are calculated on the basis of the error limits in the slope and intercept of (24). Using the above values for γ and φ_0, and the values of φ_m, χ, and E_g for Au and Si, we find

$$\gamma(\varphi_m-\chi)+(1-\gamma)(E_g-\varphi_0)=0.76\pm0.26. \qquad (25)$$

Equation (25) is in satisfactory agreement with (23) in view of the uncertainties in γ and φ_0.

It has been shown experimentally[6] that the height of a gold-*n*-type-silicon barrier has the same temperature dependence as the silicon energy gap. Therefore, the Fermi level at the metal–semiconductor interface is pinned with respect to the valence-band edge (in the temperature range 100° to 400°K); in other words, the relation $\partial\varphi_{Bn}/\partial T=\partial E_g/\partial T$ holds. In order to explain this equality we have to consider the temperature dependence of E_g, φ_m, and χ. In the above temperature range, $\partial E_g/\partial T$ for Si has been measured by Macfarlane *et al.*[17] to be approximately -2.4×10^{-4} eV/°K. $\partial\chi/\partial T$ is determined by Allen[18] to be of the order of $+10^{-4}$ eV/°K. The temperature dependence of the metal work function has been discussed by Herring and Nichols[19] to be of the order of 10^{-5} to 10^{-4} eV/°K. $\partial\varphi_m/\partial T$ for freshly evaporated films of silver has been determined by Crowell and Armstrong[20] to be -1.3×10^{-4} eV/°K. Since gold is a monovalent metal similar to silver, $\partial\varphi_m/\partial T$ for gold films can be assumed to be about -10^{-4} eV/°K. Thus the sum of $\partial\varphi_m/\partial T$ and $-\partial\chi/\partial t$ is very close to $\partial E_g/\partial T$, and we obtain the relation as shown by Eq. (16).

Archer and Atalla cleaved the Si in a stream of evaporating metal. Since both experiments were done in a vacuum of 10^{-6} Torr, it would seem that the method of Archer and Atalla produced diodes with less interfacial contamination. For this reason, the data of Archer and Atalla are used in this paper.

[17] G. G. Macfarlane, T. P. Mclean, J. E. Quarrington, and V. Roberts, Phys. Rev. **108**, 1377 (1957).

[18] F. G. Allen, J. Phys. Chem. Solids **8**, 119 (1959).

[19] C. Herring and M. H. Nichols, Rev. Mod. Phys. **21**, 185 (1949).

[20] C. R. Crowell and R. A. Armstrong, Phys. Rev. **114**, 1500 (1959).

2. GaP–Metal Systems

The GaP–gold barrier height has been determined by both differential capacitance and photothreshold measurements and the details of these experiments have been presented.[11,12] Using similar techniques, the barrier heights for GaP–aluminum, platinum, copper, and magnesium diodes have been measured. The experimental results for all five metals are presented in Table I (with an additional photothreshold value for silver).

The last column in Table I shows the difference in the values of φ_{Bn} determined by the capacitance and photoresponse methods. The value for φ_{Bn} (capacitance) was calculated using the relation

$$\varphi_{Bn}(\text{capacitance})=V_0+\varphi_n+kT/e-\Delta\varphi_n, \qquad (26)$$

where the kT/e term is a correction due to the reserve layer in the semiconductor.[4] The presence of this term in the capacitance relation (26) and in Eq. (2) is discussed by Goodman.[21] The value for φ_n was determined to be 0.06 eV, from a calculation using the known donor density in the GaP samples and taking the density-of-states effective mass m^* to be 0.09 m.[22]

There is a consistent discrepancy between the values of φ_{Bn} determined from the capacitance and photoemission measurements: φ_{Bn} (capacitance) is always larger than φ_{Bn} (photoresponse) by an amount of the order of 100 MeV. We offer the following explanation for this phenomenon: Goodman[21] finds that in the presence of an interfacial layer between the metal and semiconductor, two correction terms arise in the equation relating the intercept of the $1/C^2$ plot and the diffusion potential V_{BO}. In terms of our notation it can be verified that Goodman's result takes the form

$$\varphi_{Bn}(\text{capactiance})=V_0+\varphi_n+kT/e-\Delta\varphi_n-V_1/4 -[V_1V_{BO}]^{\frac{1}{2}}. \qquad (27)$$

[21] A. M. Goodman, J. Appl. Phys. **34**, 329 (1963).

[22] L. L. Chang and G. L. Pearson, J. Phys. Chem. Solids **25**, 23 (1964).

Since the diffusion potential V_{BO} is approximately 1 V for all the GaP diodes, it is evident that a value of about 0.01 eV for V_1 explains the discrepancy between φ_{Bn}(capacitance) and φ_{Bn}(photoresponse). This is in good agreement with the estimate made for V_1 in Sec. II.

A straight-line least-squares fit to the photoresponse data of Table I has been made in Fig. 3. The resulting line obeys the equation

$$\varphi_{Bn}=0.27\varphi_m-0.01. \tag{28}$$

The probable errors in the slope and intercept of Eq. (28) are ± 0.05 and ± 0.13 eV, respectively. By comparing Eq. (9) and (27), using the known value for E_g and estimating χ(GaP) to be 4.0 eV, we obtain

$$\gamma=0.27\pm 0.05$$
$$\varphi_0=0.66\pm 0.2 \text{ eV},$$
$$D_s\simeq(2.7\pm 0.4)\times 10^{13} \text{ states/cm}^2\text{/eV}.$$

Equation (14) can be checked for GaP–Au diodes by use of the value of φ_{Bp} obtained by Crowell *et al.*[23] They obtained a value of 0.715 ± 0.035 eV by measuring the threshold for photomission of holes in *p*-type GaP–Au diodes. $\Delta\varphi_p$ for their diodes was determined[7] to be 0.069 eV, and $\Delta\varphi_n$ for the diodes used in the present work is 0.078 eV. From these numbers and Eq. (4), $E_g=2.14\pm 0.04$ eV, compared to 2.24 eV for the room-temperature bandgap energy of GaP.

The Schottky relation, Eq. (13), for GaP is shown in Fig. 3 for comparison with the fitted line, showing clearly that the GaP–metal system is not described by this relation.

3. GaAs–Metal and CdS–Metal Systems

Mead and Spitzer[9] have obtained photothreshold data for CdS and GaAs diodes prepared using vacuum-cleaved semiconductor substrates. In Figs. 4 and 5, we show their data fitted to straight lines, in the same way as for Si and GaP, with the following results: For GaAs,

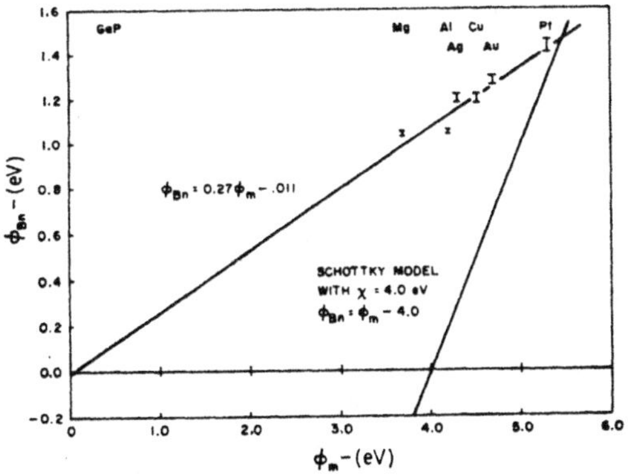

FIG. 3. Least-squares straight-line fit to GaP φ_{Bn} vs φ_m data obtained in the present work.

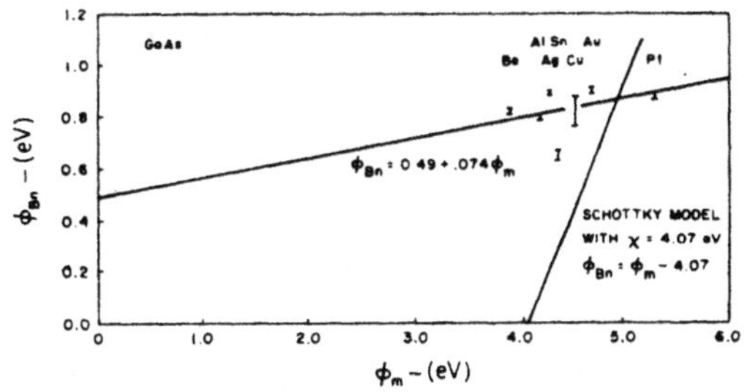

FIG. 4. Least-squares straight-line fit to the GaAs φ_{Bn} vs φ_m data of Mead and Spitzer (Ref. 9, Table III).

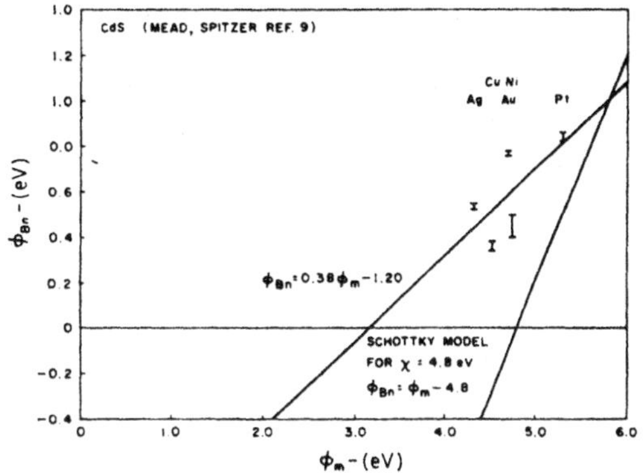

FIG. 5. Least-squares straight-line fit to the CdS φ_{Bn} vs φ_m data of Mead and Spitzer (Ref. 9, Table I).

$$\varphi_{Bn}=0.074\varphi_m+0.49 \tag{29}$$

with probable error limits in slope and intercept of ± 0.05 and ± 0.24, respectively. Taking the electron affinity to be 4.07 eV for (110) GaAs,[24] we obtain

$$\gamma=0.074\pm 0.05,$$
$$\varphi_0=0.53\pm 0.33 \text{ eV},$$
$$D_s=(1.25\pm 1.0)\times 10^{14} \text{ states/cm}^2\text{/eV}.$$

Using the most probable value of φ_0 to calculate the quantity $E_g-\varphi_0$ for GaAs, we obtain $E_g-\varphi_0\simeq 0.91$ eV. This is approximately equal to the barrier height measured by the photoelectric method for most of the metals used, indicating that the GaAs–metal system is closely described by the limiting case of $\gamma\rightarrow 0$ as given by Eq. (12). The relatively weak dependence of φ_{Bn} on φ_m for GaAs is also apparent from inspection of Fig. 4.

[23] C. R. Crowell, W. G. Spitzer, and H. G. White, Appl. Phys. Letters, 1, 3 (1962).

[24] G. W. Gobeli and F. G. Allen, "Photoelectric Threshold and Work Function of III–V Semiconductors," to be published in *Physics of III–V Compounds*, edited by R. K. Willardson and A. C. Beer. Also Phys. Rev. 137, A245 (1965).

TABLE II. Summary of barrier height data and calculations for Si, GaP, GaAs, and CdS. Entries in the last column are calculated using most probable value of φ_0.

Semiconductor	γ	b(eV)	χ(eV)	$D_s\times10^{-13}$ (eV^{-1} cm^{-2})	φ_0(eV)	φ_0/E_g
Si	0.27±0.05	−0.55±0.22	4.05	2.7±0.7	0.30±0.36	0.27
GaP	0.27±0.03	−0.01±0.13	4.0 (est.)	2.7±0.4	0.66±0.2	0.294
GaAs	0.07±0.05	+0.49±0.24	4.07	12.5±10.0	0.53±0.33	0.38
CdS (Mead, Spitzer)	0.38±0.16	−1.20±0.77	4.8	1.6 ±1.1	1.5 ±1.5	0.6
CdS (Goodman)	0.84±0.05	−3.3 ±0.23	4.8	0.2 ±0.07	−2.1 ±1.5	...

Equation (14) can be checked for GaAs–Au and GaAs–Al diodes, using the photothreshold data of Spitzer and Mead.[9] For GaAs–Au, $\varphi_{Bn}+\varphi_{Bp}+\Delta\varphi_n+\Delta\varphi_p$ is found to be 1.42 eV at 300°K compared to the accepted value of 1.40 eV for the 300°K GaAs band gap.[25] Taking averages of the φ_{Bn} and φ_{Bp} data (at 300°K) from Mead and Spitzer's experiments on GaAs–Al also gives $E_g=1.42$ eV.

For the φ_{Bn} vs φ_m data of Mead and Spitzer for CdS we obtain from the least-squares analysis

$$\varphi_{Bn}=0.38\varphi_m-1.20, \qquad (30)$$

with error limits of ±0.16 and ±0.77 eV in slope and intercept, respectively. The data and the fitted line are shown in Fig. 5. Taking χ for CdS to be 4.8 eV,[26] we find

$$\gamma=0.38\pm0.16,$$
$$\varphi_0=1.5\pm1.5 \text{ eV},$$
$$D_s=(1.6\pm1.1)\times10^{13} \text{ states/cm}^2\text{/eV}.$$

Goodman[10] has determined φ_{Bn} for contacts prepared by evaporating various metals on chemically cleaned CdS single crystals. He finds that the φ_{Bn} vs φ_m variation in this case is well described by the Schottky relation [Eq. (13) without the $\Delta\varphi_n$ term] with χ taken to be 4.0 eV. This is not consistent, however, with the

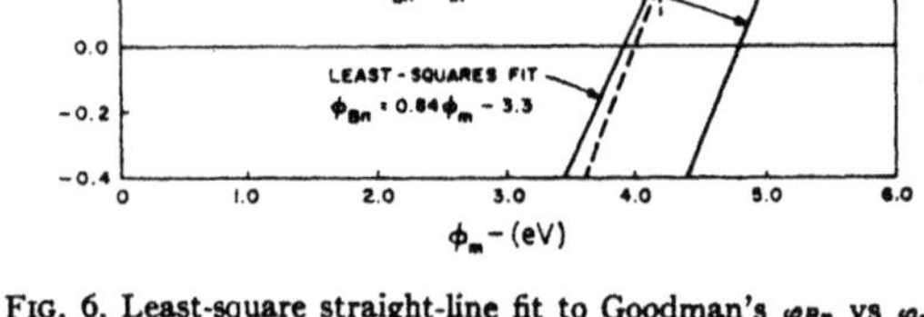

FIG. 6. Least-square straight-line fit to Goodman's φ_{Bn} vs φ_m data for chemically cleaned CdS (Ref. 10, Table VI). The Schottky relations for $\chi=4.0$ eV and $\chi=4.8$ eV are shown for comparison.

[25] C. Hilsum and A. C. Rose-Innes, *Semiconducting III–V Compounds* (Pergamon Press, New York, 1961), p. 59.

[26] N. Kindig and W. E. Spicer, Bull. Am. Phys. Soc. 9, 573 (1964).

measured value of χ(CdS) = 4.8 eV reported by Kindig and Spicer.[26] Figure 6 shows the data fitted to a straight line by the least-squares method; this line does not differ significantly from the Schottky relation line shown for comparison with $\chi=4.0$ eV. The Schottky relation for $\chi=4.8$ eV is also shown.

The experimental and calculated data for the Si, GaP, GaAs, and CdS–metal systems are summarized in Table II. The ratio φ_0/E_g has been calculated in each case on the basis of the most probable value for φ_0.

IV. DISCUSSION

It has been shown in Sec. III that Eq. (9) provides a reasonable explanation for the φ_{Bn} vs φ_m data for metals on Si, GaP, and GaAs. It is also found in the case of Si that Eq. (9) is compatible with the V_{BO} vs φ_n data obtained by Archer and Atalla. The significance of the calculated value of φ_0 and D_s for Si, GaAs, and CdS is questionable due to the large probable error in these quantities (Table II). The GaP data seems to give the best "fit" to a straight line, and the calculations of φ_0 and D_s can probably be considered reliable for this case. The model chosen for the derivation of Eq. (9) is a highly idealized one, and the fitting of experimental φ_{Bn} vs φ_m data to this equation is a procedure which merits the further discussion which follows.

As pointed out in Archer and Atalla's paper,[3] the vacuum values of χ and φ_m may not be the appropriate values to use in Eq. (9) to calculate $\varphi_m-\chi$ for an intimate metal–semiconductor contact, due to dipole interactions between the semiconductor and the different metals. Since the dipole interactions may differ from metal to metal, there may be a nonsystematic variation between the metal and the semiconductor work functions for the vacuum and intimate contact cases. This effect might partially account for the scatter in the experimental points for Si, GaAs, and for the CdS data of Mead and Spitzer. It is interesting in this connection to note that the greatest amount of scatter in the φ_{Bn} vs φ_m data is observed for the cases where the semiconductor surface was cleaved in vacuum, i.e., the Si–metal and GaAs–metal systems, and the CdS–metal system studied by Mead and Spitzer. In contrast, the GaP–metal system and the CdS–metal system studied by Goodman have a relatively small amount of scatter. This difference might be explained on the basis that the presence of a somewhat thicker interfacial layer on the chemically cleaned semiconductor surfaces decreases the importance of dipole interactions between metal and semiconductor.

The assumption of a uniform (constant) density of surface states in the forbidden gap is difficult to justify, and in fact may not be valid. Experimental evidence does exist, however, in support of a uniform or near-uniform distribution of surface states in germanium and silicon.[27–29] For a nonuniform distribution of surface states in the forbidden gap, Eq. (1) should be replaced by

$$Q_{ss}=-e\int_{\varphi_0}^{\infty}\frac{1}{1+\exp[(E-E_F)/kT]}D_s dE, \quad (31)$$

where E_F, the Fermi energy, is given according to Fig. 1 by

$$E_F=E_g-\varphi_{Bn}-\Delta\varphi_n. \quad (32)$$

An adequate approximation to Q_{ss} can be obtained by taking the limit of Eq. (31) as $T\rightarrow 0$:

$$Q_{ss}\approx -e\int_{\varphi_0}^{E_g-\varphi_{Bn}-\Delta\varphi_n} D_s dE. \quad (33)$$

The development leading to Eq. (6) can now be repeated in terms of Eq. (33) to give

$$\frac{\epsilon_i}{\delta}(\chi+\varphi_{Bn}+\Delta\varphi_n-\varphi_m)=e\int_{\varphi_0}^{E_g-\varphi_{Bn}-\Delta\varphi_n} D_s dE -[2e\epsilon_s N_D(\varphi_{Bn}+\Delta\varphi_n-\varphi_n-kT/e)]^{\frac{1}{2}}. \quad (34)$$

Taking the derivative of Eq. (34) with respect to φ_m yields

$$\frac{\epsilon_i}{\delta}\left(1-\frac{\partial\varphi_{Bn}}{\partial\varphi_m}\right)=e\frac{\partial\varphi_{Bn}}{\partial\varphi_m}D_s(E_g-\varphi_{Bn}-\Delta\varphi_n) +\frac{1}{2}\frac{\partial\varphi_{Bn}}{\partial\varphi_m}\left[\frac{2e\epsilon_s N_D}{(\varphi_{Bn}+\Delta\varphi_n-\varphi_n-kT/e)}\right]^{\frac{1}{2}} \quad (35)$$

or

$$D_s \text{ (at the Fermi level)}=\frac{1-\partial\varphi_{Bn}/\partial\varphi_m}{\partial\varphi_{Bn}/\partial\varphi_m}\frac{\epsilon_i}{e\delta}-\frac{1}{2e}\times\left[\frac{2e\epsilon_s N_D}{(\varphi_{Bn}+\Delta\varphi_n-\varphi_n-kT/e)}\right]^{\frac{1}{2}}. \quad (36)$$

Based on the same assumptions for ϵ_i, δ, and ϵ_s as given in Sec. II, and for $N_D\leq 10^{18}$, the second term on the right-hand side of Eq. (36) is of the order of 2×10^{12} states/cm^2/eV or less. For $\partial\varphi_{Bn}/\partial\varphi_m$ equal to 0.5 or smaller the first term of Eq. (36) is always the dominant term, and if $\partial\varphi_{Bn}/\partial\varphi_m$ does not vary greatly, D_s will then be essentially a constant.

[27] W. Shockley and G. L. Pearson, Phys. Rev. 74, 232 (1948).
[28] W. H. Brattain and W. Shockley, Phys. Rev. 72, 365 (1947).
[29] H. C. Montgomery and W. L. Brown, Phys. Rev. 103, 865 (1956).

It is perhaps just possible to decide that a better fit to the silicon data of Fig. 2 is provided by a curved line whose slope is about the same as the line shown on the graph near the Pt and Pd points, becoming less steep from Ni to Al, and becoming steeper in the region between Sb and Mg. If this is the case, then according to Eq. (36) the surface states density D_s in the forbidden gap of silicon is somewhat greater in the vicinity of 0.35 eV above the valence band edge. This could correspond to the band of surface state levels 0.4 to 0.6 eV above the valence band edge postulated by Allen and Gobeli[15] as an explanation for their photoemission data. The same authors point out that a discrete surface state level 0.45 eV above the valence band edge is also consistent with their data.

The interfacial layer thickness δ between metal and semiconductor atoms is dependent on both the exposure time of the semiconductor surface to the residual gases and the manner in which the first layer of metal atoms "sits" on the semiconductor surface. It is reasonable to assume that the value for δ varies $\pm 50\%$ about some mean value for cleaved surfaces; such a variation will be averaged out to some extent in the least-squares analysis of the data, and in any case the qualitative features of the model based on constant δ will still be observed. Since the residual pressure in the vacuum system is about 2×10^{-7} to 10^{-6} Torr in the previously cited experiments,[3,5] a monolayer can form on the cleaved surface in less than one second to a few seconds depending on the sticking coefficient of the residual gases.[30] For etched and protected surfaces such as those used in the GaP experiments, it is likely that a monolayer or more of adsorbed residual gas forms the interfacial layer.

The potential Δ across the interfacial layer varies from 0.01 to about 1 V, depending mainly on the metal work function; this results in a field of 10^5 to 10^7 V/cm in the layer. It is well known that SiO_2 films can withstand fields exceeding 10^7 V/cm, but due to the different nature of the interfacial layer for an intimate contact, higher fields may exist in the layer.

The most probable values for the surface-state neutrality level φ_0 as derived from the least-squares analysis for GaP, GaAs, and Si are roughly a third of the respective band gaps. This seems to be consistent with the findings of Mead and Spitzer,[5] since if D_s has a high value, as is apparently the case for GaAs, the Fermi level tends to become pinned at the surface φ_0 eV above the valence band, producing a barrier height of $E_g-\varphi_0$ or about $\frac{2}{3}E_g$. Allen and Gobeli,[15] who propose a surface state distribution different from ours for Si, find the surface state neutrality for cleaved (111) Si "free" surfaces to be about 0.3 eV above the valence band edge. This value has been verified by Kawaji

[30] J. H. de Boer, *The Dynamical Character of Adsorption* (Clarendon Press, Oxford, England, 1953), Chap. 2.

and Takashima,[31] and by Handler.[32] More recent work by Gobeli and Allen[24] indicates that the surface neutrality level for the cleaved (110) GaAs surface is about 0.76 eV, which does not correspond to the position of the Fermi level in metal–GaAs contacts. Fischer[33] has performed preliminary photoemission measurements on cleaved GaP (110) surfaces following the procedures described in Ref. 24. Tentative results of this work indicate an electron affinity of approximately 3.45 eV and a surface state neutrality level of about 0.94 eV for the (110) GaP surface. It is not clear that these results can be applied in the present work, however, since the GaP diodes described here were prepared using chemically cleaned (111) GaP surfaces.

The results of Mead and Spitzer[9] and of Goodman[10] for CdS present some difficulties when compared to the theory. In view of the large amount of scatter in Mead and Spitzer's experimental points for CdS, and the concomitant probable errors in the slope and intercept of the fitted line, it is doubtful that any significant conclusions can be drawn regarding the effect of surface states on the potential barrier in this case. The results presented by Goodman appear to obey the Schottky relation, indicating little or no dependence of surface barrier height on surface states, but the value of χ required for agreement with the Schottky relation is in poor agreement with the reported value determined by another method.[26] We have no explanation for this behavior. It seems clear, however, that the chief reason for the difference in results of the experiments of Goodman and of Mead and Spitzer for CdS is the different method of semiconductor surface preparation, and it could be conjectured that the additional interfacial layer thickness for Goodman's diodes had a "smoothing" effect on the φ_{Bn} vs φ_m data, due to the partial elimination of dipole interactions.

It should be clear at this point that the conclusions regarding the surface states density and distribution drawn on the basis of the preceding work are valid only for the specific methods of surface treatment used in the experiments cited. For cleaved surfaces, it is possible that the surface states can be attributed to "dangling bonds"; modification of this scheme by adsorbed gas molecules is expected in the case of chemically cleaned substrates.

The available data for GaP and GaAs provide excellent verification of Eq. (14) if the image corrections are included. The values for φ_{Bn} and φ_{Bp} were taken to be the appropriate photothresholds, since these provide the most direct determinations of φ_{Bn} and φ_{Bp}; the capacitance values of φ_{Bn} and φ_{Bp} are uncertain to the extent that the magnitude of the correction terms of Eq. (27) are not known accurately.

Using the values of γ and V_1 obtained from the data, we have found that for any reasonable value of $E_g - \varphi_0$, the terms involving V_1 in Eq. (8) are unimportant, thus justifying the use of the approximate form, Eq. (9).

ACKNOWLEDGMENTS

The authors wish to thank Dr. C. R. Crowell, Dr. J. Godfrey, Dr. H. K. Gummel, and Dr. H. Heffner for helpful comments and encouragement during this work. Critical reading of the manuscript by Dr. G. W. Gobeli and Dr. F. G. Allen is also gratefully acknowledged.

[31] S. Kawaji and Y. Takashima, Surface Sci. 1, 119 (1964).
[32] P. Handler, Appl. Phys. Letters 3, 96 (1963).
[33] T. E. Fischer, Bell Telephone Laboratories, Murray Hill, New Jersey (private communication).

Theory of Surface States

VOLKER HEINE*
Bell Telephone Laboratories, Murray Hill, New Jersey
(Received 22 January 1965)

The properties of metal-to-semiconductor junctions and of free semiconductor surfaces are usually explained on the basis of surface states. The theory of the metal contacts is discussed critically, because strictly speaking localized surface states cannot exist in such junctions. However, it is shown that virtual or resonance surface states can exist which behave for practical purposes in the same way. They are really the tails of the metal wave functions rather than separate states. In the past, the length of this tail has often been ignored. Some estimates of its length are made and its consequences pointed out. A semiquantitative discussion is given of various recent data, including the effect of an oxide layer on barrier height, the variation of barrier height with the metal, the work function of a free surface at high doping, and the effect of a cesium layer on the work function.

I. INTRODUCTION

THIS paper is in the nature of a commentary and elaboration of the theory of metal-to-semiconductor junctions and of free semiconductor surfaces, with a discussion of some of the more recent data. As is well known, when a contact is made between a metal and a semiconductor, there is a potential barrier Φ. The Fermi level E_F at the surface of a semiconductor lies in the energy gap at a point,[1]

$$\xi = E_g - \Phi \qquad (1)$$

above the top of the valence band [Fig. 2(a)] where E_g is the gap width. The remarkable thing is that ξ is constant to within about 0.2 eV independent of:

(a) the degree of doping[2,3] which swings the bands in the bulk material by something of the order of 1 eV;

(b) the metal that is used,[4–8] some differing from the semiconductor by several volts in the work function φ;

(c) the crystallographic orientation of the surface[8,9];

(d) whether there is a monolayer or so of oxygen or oxide between the metal and semiconductor[3,8];

(e) any bias voltage applied across the junction[4,8];

(f) whether one dispenses with the metal completely and has a free surface, at least in the case of silicon, again either atomically clean or with an oxide layer.[10,8]

Furthermore, ξ is roughly a constant fraction ~0.3 of the energy gap in a whole range of group IV and III–V semiconductors.[6,7] Of course there are exceptions to all these statements, but by and large this is the story.

All these facts receive a natural explanation in terms of Bardeen's theory,[1] namely, that there are surface states on the semiconductor in the band gap. A density of even a fraction of a surface state per surface atom per eV is quite sufficient to absorb any required extra charge into the band of surface states without the Fermi level in this band moving very much. Thus E_F at the surface of the semiconductor is effectively pegged to some fixed value of ξ. A few of the data are shown in Fig. 1.

At the time Bardeen's theory was first advanced, the semiconductor-metal contacts were made by pressing the metal against the semiconductor, each coated with an oxide film. Now, however, many of the junctions are made by cleaving the semiconductor in a stream of the metal vapor used to deposit the metal. The effective partial pressure of the condensing metal is ten or more times the residual partial pressure of gas,[7,8] so that there is intimate atomic contact between the metal and semiconductor without any intervening oxygen or other impurity layer. Under such conditions it appears at first

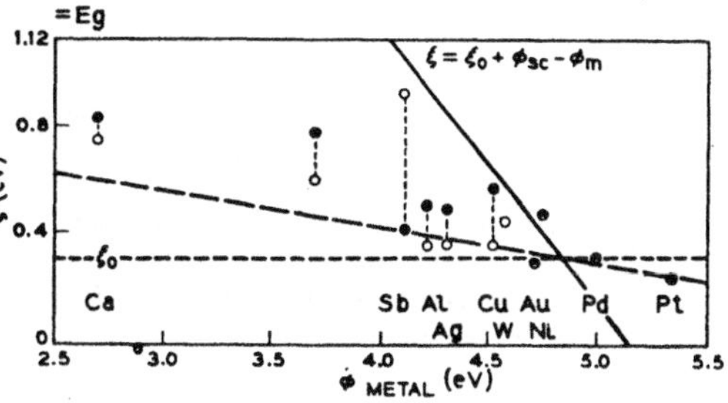

FIG. 1. Position of Fermi level at the surface (ξ) above the valence-band edge for metal-to-silicon contacts. ξ_0 is the value for a free-silicon surface. ξ is given by $E_g - \Phi$, where Φ is the measured barrier height. White circles refer to intimate contact between metal and semiconductor, black circles to the case of a thin intervening oxide layer. Circles half-black are used where the two values coincide within experimental error. The data points second from the left are for magnesium. (After Ref. 8. Data from Refs. 2–8.)

* Permanent address: Cavendish Laboratory, Free School Lane, Cambridge, England.
[1] J. Bardeen, Phys. Rev. **71**, 717 (1947).
[2] C. R. Crowell, S. M. Sze, and W. G. Spitzer, Appl. Phys. Letters **4**, 91 (1964).
[3] R. J. Archer and M. M. Atalla, Ann. N. Y. Acad. Sci. **101**, 697 (1963).
[4] C. R. Crowell, J. C. Sarace, and S. M. Sze, Trans. AIME (to be published).
[5] C. R. Crowell, W. G. Spitzer, L. E. Howarth, and E. E. La Bate, Phys. Rev. **127**, 2006 (1962).
[6] C. A. Mead and W. G. Spitzer, Phys. Rev. **134**, A713 (1964); Phys. Rev. Letters **10**, 471 (1963).
[7] W. G. Spitzer and C. A. Mead, J. Appl. Phys. **34**, 3061 (1963).
[8] C. R. Crowell (private communication and to be published); D. Kahng, Bell System Tech. J. **43**, 215 (1964); S. M. Sze, C. R. Crowell, and D. Kahng, J. Appl. Phys. **35**, 2534 (1964).
[9] F. G. Allen and A. B. Fowler, J. Phys. Chem. Solids **3**, 107 (1957). See also F. G. Allen and G. W. Gobeli, J. Appl. Phys. **35**, 597 (1964); Y. Takeishi and H. D. Hagstrum (to be published).
[10] F. G. Allen and G. W. Gobeli, Phys. Rev. **127**, 150 (1962).

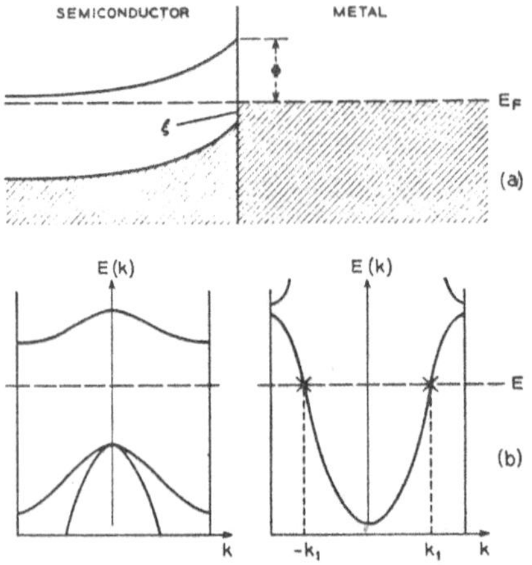

FIG. 2. (a) Definition of Φ and ξ at semiconductor-metal junction. (b) An energy E of a surface state in the band gap of the semiconductor corresponds to two propagating Bloch functions k_1, $-k_1$ in the metal.

sight unreasonable to talk about surface states at all. For an energy E below E_F in the gap of the semiconductor, the solutions of the Schrödinger equation will decay exponentially in the semiconductor but propagate as Bloch states on the metal side of the junction to form the ordinary volume states of the metal.[11,12] This follows from simple considerations of matching the wave functions at the boundary as shown in Fig. 2. Let us assume the x axis perpendicular to the surface. For some value of $\mathbf{k}=\mathbf{k}_{\parallel}$ parallel to the surface, e.g., $k_y=k_z=0$, we have the bands shown in Fig. 2(b). At energy E, the exponential solution in the semiconductor can always be joined onto the two Bloch states with wave vector $k_\perp=k_1$, $-k_1$ in the metal, the existence of the two k's ensuring that both ψ and its derivative can be matched at the boundary. Thus for energies in the semiconductor band gap the volume states of the metal all have tails in the semiconductor; there are actually no surface states.

Our first task is therefore to discuss in Sec. II how the Bardeen explanation in terms of surface states may be rescued. Although the exact form of the band of surface states is not known, there are some semiquantitative statements one can make. For example, the form of the exponential tail in the semiconductor is determined by the ordinary volume band structure. In Sec. III we make some estimates of its length and discuss its relation to some of the experimental data.

II. MATCHING THE WAVE FUNCTION

Let us first assume that $\mathbf{k}_{\parallel}$, the component of $\mathbf{k}$ parallel to the boundary plane, is conserved as an electron crosses the junction. In reality there is sufficient disorder at the boundary for diffuse scattering to be probably a better approximation than the specular assumption made here. But the effect of scattering is to scramble states of the same energy E, leaving invariant any total property such as the density of states or the charge distribution $\psi^*\psi$ summed over all states in some energy interval dE. Thus the device of considering states of definite $\mathbf{k}_{\parallel}$ leads to correct results for total or average properties.

As a first step we treat the valence states of the semiconductor and the conduction electrons of the metal as free electron gases, which is a good approximation in view of the small band gap.[11,13,14] Because of the higher density of four electrons per atom in the semiconductor than the metal, the potential and Fermi surfaces in the two regions are as shown in Fig. 3. The important point is that in this simple model all semiconductor states with $\mathbf{k}_{\parallel}$ greater than $A'A''$ do not penetrate into the metal. Their $\mathbf{k}_{\parallel}$ is sufficiently large that their perpendicular kinetic energy

$$E_\perp = E - \hbar^2 k_{\parallel}^2/(2m) \quad (2)$$

is negative in the metal, i.e., they decay exponentially on the metal side of the boundary. The value of the Fermi momentum k_F for silicon is compared with that of some typical metals in Table I. We have not counted the d electrons in the case of transition metals because they have such different wave functions with small value at the cell edge that one does not expect them to match at all well onto the silicon wave functions. Let us concentrate for the present on the top part of the table. We note that a large fraction f of the high-energy states of the silicon, given by

$$f = 1 - k_{F,\mathrm{m}}/k_{F,\mathrm{sc}} \quad (3)$$

do not propagate into the metal. The same applies to any surface states in the band gap formed when the

TABLE I. Fermi momenta k_F of metals (m) compared with silicon (sc).

	$k_{F,\mathrm{m}}/k_{F,\mathrm{sc}}$
sodium	0.50
palladium,[a] platinum[a]	0.55
calcium, nickel[a]	0.61
silver, gold	0.66
magnesium, copper	0.75
antimony	0.93
aluminum	0.97

[a] Assuming half an s-p electron per atom.

[11] V. Heine, Surface Sci. **2**, 1 (1964).

[12] V. Heine, Proc. Phys. Soc. (London) **81**, 300 (1962); P. J. Price, *Report of the International Conference on the Physics of Semiconductors* (Institute of Physics and the Physical Society, London, 1962), p. 99.

[13] We append a selection of general articles which contain more detailed documentation. V. Heine, in Proceedings of the Ninth Conference on Low Temperature Physics, 1964 (to be published); J. M. Ziman, Advan. Phys. **13**, 89 (1964); L. Kleinman and J. C. Phillips, Phys. Rev. **125**, 819 (1962); M. H. Cohen and V. Heine, *ibid.* **122**, 1821 (1961).

[14] D. Brust, Phys. Rev. **134**, A1337 (1964).

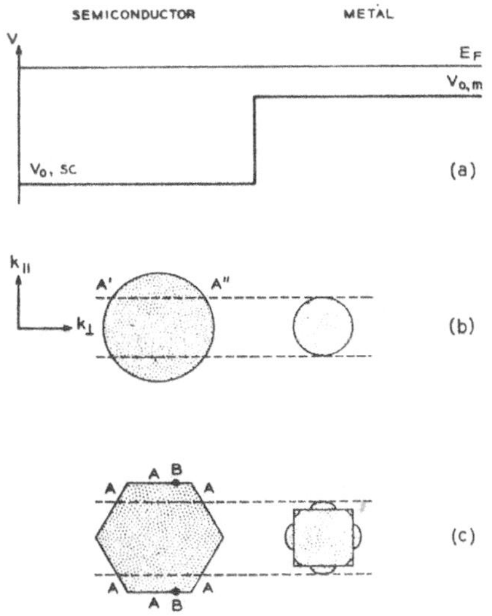

FIG. 3. (a) The potential, (b) the Fermi spheres, and (c) the electron distributions in the Jones zones, in the nearly free-electron model.

Fermi distribution is enclosed in the Jones zone. (We reserve the word Brillouin zone for the reduced zone, and use Jones zone for any zone constructed in extended **k** space out of prominent reflection planes.) A surface state at B, if one exists, is made out of plane waves at B and B', with a small imaginary component in $k_\perp$ which does not affect the behavior parallel to the boundary. Thus along the regions AAA of the zone [Fig. 3(c)] genuinely localized surface states are possible in this simple approximation. The remainder would propagate into the metal as discussed in connection with Fig. 2.

This argument can be taken a step further by considering the actual band structure of the semiconductor. The crystal face most commonly chosen for making junctions is the (111) cleavage plane. The surface state is related to the band edges of a section $E(k_\perp)$ of the volume band structure for fixed $\mathbf{k}_{\parallel}$, these edges in almost all cases occurring at $k_\perp=0$. The wave function of the surface state is therefore a mixture of the volume states at $\mathbf{k}=\mathbf{k}_{\parallel}$, $k_\perp=0$ with of course the slow exponential decay into the semiconductor superimposed. The plane of the vector $\mathbf{k}_{\parallel}$ is defined by the three directions $(10\bar{1})$, $(1\bar{1}0)$, $(01\bar{1})$ at 120°, and a typical $\mathbf{k}_{\parallel}$ about halfway between the center and the edge of the Brillouin zone is $(2\pi/a)(\frac{1}{3}, 0, -\frac{1}{3})$. The Bloch states at the gap are linear combinations of plane waves, in the plane-wave pseudopotential representation of band structures,[11,13] with **k** differing from $(2\pi/a)(\frac{1}{3}, 0, -\frac{1}{3})$ by various reciprocal lattice vectors. The lowest ones relevant at the gap are

$$\frac{2\pi}{a}(-5/3, 0, -\tfrac{1}{3}) \quad \text{and} \quad \frac{2\pi}{a}(\tfrac{1}{3}, 0, 5/3). \tag{4}$$

These have $k_{\parallel}=0.79k_F$. In addition, the wave functions contain an admixture of

$$\frac{2\pi}{a}(-\tfrac{2}{3}, \pm 1, \tfrac{2}{3}), \quad \frac{2\pi}{a}(\tfrac{4}{3}, \pm 1, \tfrac{2}{3}), \quad \frac{2\pi}{a}(-\tfrac{2}{3}, \pm 1, -\tfrac{4}{3}), \tag{5}$$

with $k_{\parallel}/k_F=0.79$, 0.30, 1.08 in pairs. Moreover, they have quite a high $k_\perp$, varying between $0.37k_F$ and $1.10k_F$. Thus for all the top group of metals in Table I, we have to a very good approximation surface states in the semiconductor which decay exponentially on the metal side.

Strictly speaking, however, there is some coupling to the states in the metal as envisaged in the discussion in Sec. I. It comes through having some admixture of the wave with $k_{\parallel}=0.30k_F$ in (5) and to a lesser extent through the admixture of higher waves in the metal by Bragg reflections. One ought to discuss the quantity

$$\rho(E)=\int \psi^*\psi\, n(E)\, d\tau, \tag{6}$$

where $n(E)$ is the density of states in the metal and the volume integration is over all the semiconductor. This is the total charge residing in the semiconductor from the tails of the metal wave functions per unit range dE. For a true surface state of energy E_s, it is just a delta function at E_s containing one electron. In our situation, since the coupling between semiconductor and metal is weak, one has the conventional Breit-Wigner situation,[15] and $\rho(E)$ goes through a resonance peak around E_s, the total again being one electron. Following the customary nomenclature, we shall term these "resonance" or "virtual" surface states.

We return now to the opposite case such as aluminum in the bottom part of Table I. Here k_F in the metal is comparable with that in the semiconductor and greater than $k_{\parallel}$ for a typical wave function in the band gap, e.g., the waves (4). For an energy in the gap, the $E_\perp$ [Eq. (2)] in the metal is positive and we can only talk about the density of charge $\rho(E)$ in the semiconductor tails of the metal wave functions. For a particular $\mathbf{k}_{\parallel}$ the question of matching the tails to the metal volume states becomes a one-dimensional problem of the idealized form shown in Fig. 4. We choose as our model a weak one-dimensional potential with lattice constant b, and consider the first band gap in a nearly free-electron band structure. The algebra of matching wave functions in the metal of type $\sin(kx-\beta)$ to those in the semiconductor which have the form

$$e^{qx}\sin(\pi x/b-\gamma) \tag{7}$$

for all energies in the band gap, has been performed. The calculation involves three steps: (a) determining $q(E)$ and $\gamma(E)$ in the gap using a two-plane-wave

[15] L. D. Landau and E. M. Lifshitz, *Quantum Mechanics* (Pergamon Press, Inc., London, 1958), p. 440.

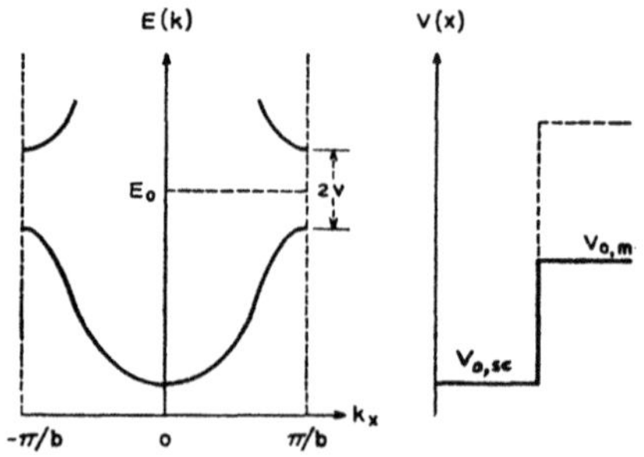

FIG. 4. Band structure of semiconductor, and potential $V(x)$ in the semiconductor (sc) and metal (m), shown on the same energy scale. The value of $V_{0,m}$ shown corresponds to positive $E_\perp$ in the metal, the dashed value to negative $E_\perp$. The latter corresponds roughly to a free surface.

approximation, (b) matching the wave functions at the boundary, and (c) integrating $|\psi|^2$ over the tail in the semiconductor to obtain the total charge contained in it. The following approximations were made:

$$v = \text{half the band gap},$$

$$\ll E_0 - V_{0,m} = \hbar^2 k_m^2/2m \approx E_\perp, \text{ in the metal}, \tag{8}$$

$$\ll E_0 - V_{0,sc} = \hbar^2\pi^2/(2b^2m) \approx E_\perp, \text{ in the semiconductor},$$

where E_0, $V_{0,m}$, and $V_{0,sc}$ are defined in Figs. 3 and 4. v is chosen positive corresponding to "crossed bands" in the sense of Shockley,[16] and the boundary between metal and semiconductor was taken at $x=\frac{1}{2}b$ halfway between the "atoms." The result is that $\rho(E)$ is proportional to

$$\frac{1}{(1-\epsilon^2)^{1/2}} \frac{1}{(\alpha^2+1)+(\alpha^2-1)\epsilon}, \tag{9}$$

where

$$\epsilon = (E-E_0)/v, \quad \alpha^2 = (E_0 - V_{0,sc})/(E_0 - V_{0,m}). \tag{10}$$

It is plotted in Fig. 5, and the total charge for energy integrated over the gap is exactly one electron. Although (9) has a mild infinity at each band edge, this is a

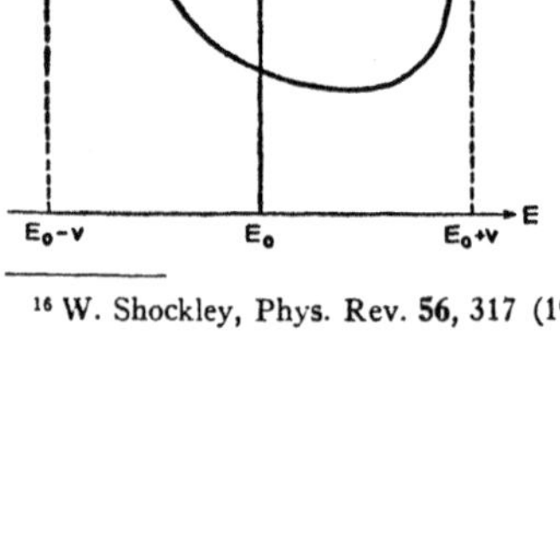

FIG. 5. Density of charge per unit energy in the semiconductor tails of the metal wave functions.

[16] W. Shockley, Phys. Rev. 56, 317 (1939).

general feature of one-dimensional band structures which disappears on integrating over $\mathbf{k}_{\parallel}$ and the integrated charge involved is intrinsically small. It is the second factor in (9) which controls where the bulk of the contribution to the total charge comes from. In our case it is a maximum at the bottom of the band gap. It is precisely here, at an energy

$$E\text{ (surface state)} = E_0 - v + 2v/\alpha^2 \tag{11}$$

just above the bottom of the gap, that we get a surface state if we raise $V_{0,m}$ as high above E_0 (shown dotted in Fig. 4) as we have so far taken it below E_0. Incidentally, if we take the boundary at the point $x=\frac{1}{4}b$, which perhaps corresponds a bit more closely to the spacing of atomic layers in the (111) direction because of the two atoms per unit cell, both the maximum of $\rho(E)$ and the surface state come nearly at the center of the gap: It is difficult to say whether the model is too crude for this to have any physical significance. We conclude therefore that the tails of the wave functions behave rather like a localized surface state, both in the total charge and approximately in its distribution in energy in the gap. That is, the resonance has become considerably broader but is still recognizable, as seen near the bottom of the gap in Fig. 5.

We have therefore demonstrated how it can be possible for the system to be relatively stable against the application of different metals with or without an oxide layer, i.e., against different boundary conditions. Since it is the most important property, it is perhaps worth commenting how it may be looked at from various points of view. Firstly, in the limit of small band gaps, q in (7) becomes small and the length of the tail of the wave function long compared with the Fermi wavelength $\lambda/(2\pi)$. Under such conditions we might expect a theorem to hold for $\rho(E)$ analogous to that for black-body radiation, namely that the energy density per unit frequency range is constant except within λ of the walls, independent of boundary conditions. A second factor is the Shockley theorem[16] and the rapid variation of ψ'/ψ with energy on the semiconductor side so that the surface-state energy is insensitive[11] to relatively small changes in boundary condition ψ'/ψ on the other side. Thirdly, a comment about "dangling bonds." In accordance with the Shockley theorem, the existence of surface states has everything to do with the formation of covalent bonds in the sense of the crossing of bands, i.e., the fact that the potential tends to heap up charge between the atoms rather than at the atoms. But the broken orbital on the actual surface atom is not very important, at least so the experimental evidence indicates. A model which depended sensitively on what happens to the dangling orbital would not have the insensitive properties mentioned in (a) to (f) in Sec. I, particularly as regards surface orientation. Similar evidence comes from the electrical behavior of dislocations which does not correlate at all with the presence or absence of dangling bonds and can be explained quite well without them.[17,18] This also makes sense from a theoretical point of view. A wave function cannot be localized more than a Wannier function, and this is spread over several atomic distances in a semiconductor because of the small band gaps; i.e., q in (7) is small. This means that of any localized or surface state, only a small probability $\psi^*\psi$ is concentrated on any one atom or layer of atoms. Consequently that one atom or layer cannot influence the energy very much as we see in the shallow donor and acceptor levels. Of course deep traps do exist too, but they need enormous perturbations like doubly charged zinc or negative copper,[19] and the hybridization energy of a broken bond is not like that. Dangling bonds might tend to form shallow states, but they would be near the valence- and conduction-band edges for each particular $k_{\parallel}$, which means almost all of them would be well hidden in the total valence and conduction bands. Finally surfaces tend to reconstruct atomically so as to join up into conjugated systems, thus eliminating the dangling bonds.

III. DISCUSSION OF THE DATA

We turn first to Fig. 1 and the effect of a thin oxide layer between the semiconductor and metal. If $k_{F,m}$ is small so that the resonance surface state is sharp and only slightly coupled to the metal, the intervention of the oxide layer would just reduce the coupling further, leaving the energy of the resonance and thus ξ unchanged. If $k_{F,m}$ is large and the coupling therefore strong, the tails of the wave functions will suffer a reduction in amplitude in the oxide, by a larger factor than in the same thickness of semiconductor because the band gap in the oxide is larger. Thus the amplitude of the tails in the semiconductor is smaller and E_F has to rise in the band gap to achieve the same total charge in the tails needed for electrical neutrality. The magnitude of the increase in ξ, i.e., the amount that the black circles lie above the white ones in Fig. 1, correlates well with the position of the metal in Table I in accordance with our explanation. Only antimony in Fig. 1 behaves completely anomalously, and we note it behaves normally when there is an intervening oxide layer, but not when in intimate contact with the semiconductor, a point to which we return later.

Having discussed the differences in Fig. 1 between with and without an oxide layer, we turn now to the absolute value of ξ for the intimate contacts (white circles in Fig. 1) and its deviation from ξ_0, the value at a free surface. The most elementary explanation would run as follows: As we move to the left in Fig. 1, the dipole

$$D=\varphi_{sc}-\varphi_m \tag{12}$$

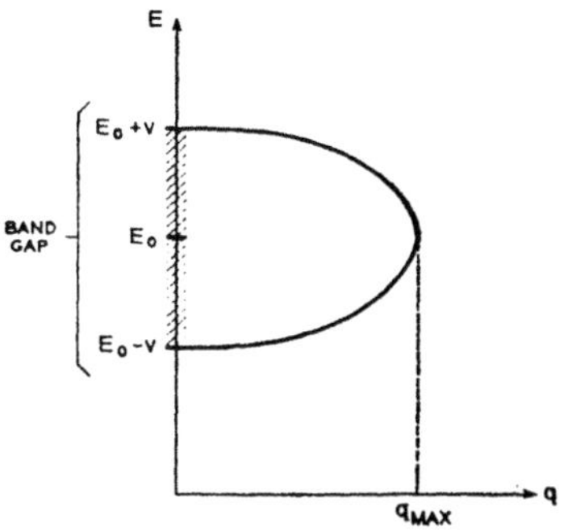

FIG. 6. Variation of q in the band gap.

that has to exist between the surface states and the metal increases, i.e., more electrons flow into the surface states to set up the dipole, and E_F in the surface states has to rise. We have

$$D=4\pi\sigma t/\epsilon\,, \tag{13}$$

where $t\approx 1/q$ is the mean separation between the negative charge in the surface states and the positive charge on the metal, ϵ is the dielectric constant, and σ is the charge density per unit area. Further

$$\delta E_F\equiv\xi-\xi_0=\sigma/\rho(E)\,, \tag{14}$$

where $\rho(E)$ is also expressed per unit area. Thus

$$\xi=\xi_0+\left[\frac{1}{4\pi\rho(E)(t/\epsilon)}\right](\varphi_{sc}-\varphi_m). \tag{15}$$

Incidentally, in (12) and (15) it is not quite the real work functions φ_{sc} and φ_m that we want, but the volume contributions to them without the surface dipoles on the free surfaces. It is this which represents the difference in binding of an electron in the two materials. However, since the surface dipoles are small and anyway unknown, we use the measured φ's.

In order to calculate ξ from (15) we must estimate t/ϵ and $\rho(E)$. We start with t/ϵ, discussing both theoretical and experimental estimates. Theoretically q in (7) varies in the band gap as shown in Fig. 6, with q_{max} being given approximately by[20,21]

$$q_{max}\approx k_{at\ gap}(v/2E_0)\,, \tag{16}$$

where E_0 is the energy at the gap $\hbar^2k_{gap}^2/(2m)$ measured from the bottom of the band. Taking again the representative point defined by Eqs. (4) and (5), and a mean vertical band gap $2v$ of $4\frac{1}{2}$ eV to the first two conduction bands,[14] we obtain[22]

$$1/q_{max}=6.0\ \text{Å}. \tag{17}$$

A calculation following Kane[23] for the center of the Brillouin zone gives 5.5 Å. Since $|\psi|^2$ in the surface

[17] R. K. Mueller, J. Appl. Phys. 32, 640 (1961); and (private communication).
[18] V. Heine (to be published).
[19] N. B. Hannay, *Semiconductors* (Reinhold Publishing Corporation, New York, 1959), p. 342.
[20] W. Kohn, Phys. Rev. 115, 809 (1959).
[21] E. T. Goodwin, Proc. Cambridge Phil. Soc. 35, 205 (1939).
[22] The estimate below Eq. (5) of Ref. 11 was somewhat too large.
[23] E. O. Kane, J. Appl. Phys. 32, 83 (1961).

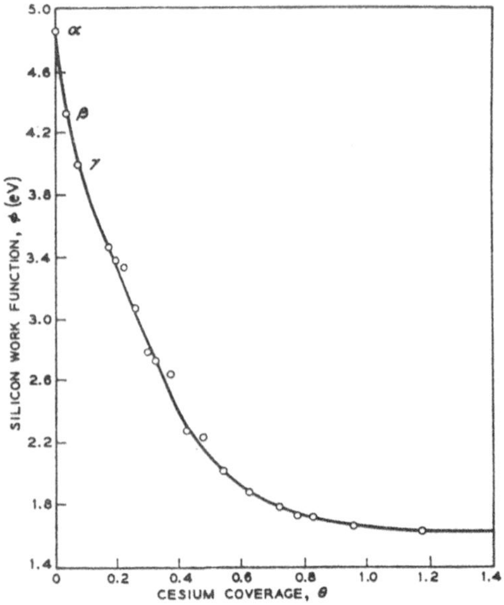

FIG. 7. Variation of silicon work function φ with cesium coverage. (After Ref. 24.)

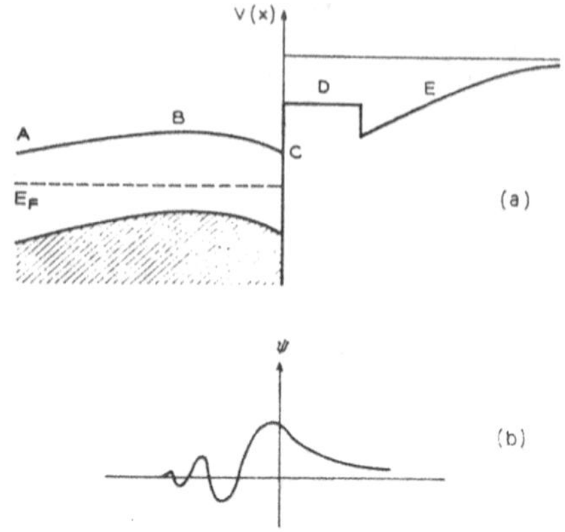

FIG. 8. Potential around Cs^+ ion on the silicon surface. AB are the bands bent by the space-charge layer. The bending in BC represents the Coulomb potential from the Cs^+ ion. DE is the pseudopotential of the ion [V. Heine and I. Abarenkov, Phil. Mag. 9, 451 (1964)].

state varies as $\exp(-2qx)$, the center of gravity of the charge is a distance $1/(2q)$ from the surface. We note that our estimates give a *minimum* length of the tail, applicable when the surface state has an energy at the center of the direct gap.

Experimentally the measurement most directly involving t is the change of work function φ of silicon when a fraction θ of a monolayer of cesium is deposited (Fig. 7).[24] Let us consider the limit of low coverage. The ionization energy of a cesium atom is 3.9 eV, and if we bring it up to the silicon surface the proximity of the dielectric reduces this further. The work function of silicon is 4.8 eV, so that the electron prefers by 0.9 eV to go into surface states at E_F in the semiconductor. The potential and wave function are shown schematically in Fig. 8, and the dipole moment contributed by each cesium atom gives a measure of the length of the tail. We have

$$\varphi(0)-\varphi(\theta)=4\pi Q n_0[(t/\epsilon)_1+(t/\epsilon)_2]\theta, \qquad (18)$$

where Q is the effective charge of the dipole and suffixes 1 and 2 refer to the part of the separation t inside and outside the semiconductor, respectively. Since the diameter 3.2 Å of a cesium ion is comparable with the penetration in the semiconductor, a fraction of the electron will be situated on the ion, and we take $Q\sim\frac{3}{4}e$. The number of sites n_0 is 8×10^{14} cm^{-2}. Turning to the data, what is "low coverage" for our purposes? At the point β the φ has already dropped 0.5 eV so that the dipole potential produced at one ion by its neighbors is of this order of magnitude. That is already a considerable perturbation compared with the energy differences depicted in Fig. 8, and indeed the experimental curve appears to be still increasing in slope between α and β. The straight line slopes from β to γ and α to β are, respectively, 9 and 15 eV from which we guess a slope at $\theta=0$ of 25 eV. It would be interesting to have more precise data in this region. (18) now gives

$$(t/\epsilon)_1+(t/\epsilon)_2=2.2 \text{ Å}. \qquad (19)$$

For $(t/\epsilon)_2$ we take $t=1.6$ Å the radius of the core, and $\epsilon=1$ because the core is tightly enough bound. Using $\epsilon(\text{Si})=11.6$ we obtain for the mean length of the tail in the silicon

$$t_1=8 \text{ Å}. \qquad (20)$$

This figure compares satisfactorily with our theoretical estimate $1/(2q)\geqslant3$ Å and a figure $t/\epsilon\leqslant1$ Å (i.e., $t\leqslant11$ Å) deduced by Crowell[8] from the lack of variation of barrier height Φ with applied bias voltage in some junctions.

It is more difficult to estimate the density of states $\rho(E)$. As regards the total number of surface states, there is roughly speaking only one number which can be invariant under all the circumstances outlined in Sec. I, and that is 2 electrons (counting both spin directions) per surface atom. One might argue that it should be half this because of counting per unit cell, but, on the other hand, there are at least two bands at the valence- and conduction-band edges likely to give rise to surface states. The question is how spread out in energy they are. Scaling in energy the results of Pugh[25] for diamond would give a very high value, 6.4×10^{15} electrons cm^{-2} eV^{-1}. However, looking at Fig. 5, one might be more inclined to go to the opposite extreme and assume

[24] F. G. Allen and G. W. Gobeli, *Report of the International Conference on the Physics of Semiconductors* (Institute of Physics and the Physical Society, London, 1962), p. 818.

[25] D. Pugh, Phys. Rev. Letters **12**, 390 (1964).

uniform spreading of the "states" in the band gap, at least for the junctions with metals with high k_F. Since these do not behave radically different from other junctions or the free surfaces, we might infer the same true there. In any case such an assumption gives a reasonable low-estimate $\rho_{l.e.}(E)$. Adopting a mean band gap of $4\frac{1}{2}$ eV again,[14] we obtain

$$\rho_{l.e.}(E)=3.5\times10^{14} \text{ electrons cm}^{-2} \text{ eV}^{-1}. \quad (21)$$

We shall adopt (21) in subsequent calculations, because as we shall see a value much larger would imply very small values of $\xi-\xi_0$. We note that (21) is only twice the "straight-line" estimate of Allen and Gobeli.[10] We reject their detailed model with two very narrow bands for the same reason as above, namely it seems too fine a structure to be retained in junctions with metals with high k_F (though one could also argue that their two bands are placed symmetrically above and below the Fermi level so that any broadening would also be symmetrical and retain the same E_F). Such very dense bands might also be expected to show up in photoemission.[26] At dopings less than 10^{19} cm^{-3}, (21) is quite consistent with their data, predicting a curve which bends up 0.02 eV at the p-type end and down 0.03 eV at the n-type end. However, the experimental variation of φ at and beyond 10^{19} cm^{-3} does not fit. At 10^{19} cm^{-3}, the thickness of the space-charge region is 50 Å (p-type) and 90 Å (n-type) which is becoming comparable with the length of the surface-state tails. Thus, the bands bend over the length (20) by about 0.12 eV. It seems reasonable that under conditions of such steep band bending there might be deviations of $\sim$0.03 eV from the simple theory as would be required to fit the data. (We note that at even higher dopings the change of φ reverses sign.) At least we suggest this as an alternative interpretation of the data. Measurements on diodes at a lower temperature should give a clearer indication of the density of states applicable to the central range of doping. We shall later come across other deviations from the simple theory under conditions of extreme band bending.

We return at last to Eq. (15) and the data of Fig. 1. We put

$$\frac{t}{\epsilon}=\frac{8.0}{11.6}+\frac{0.5}{1}=1.2 \text{ Å}, \quad (22)$$

the first term being taken from (20) and the second term representing the screening distance in the metal.[27] With the use of (21) and (22), Eq. (15) becomes

$$\xi=\xi_0+0.13(\varphi_{sc}-\varphi_m), \quad (23)$$

which is shown on Fig. 1 as the dashed line. It reproduces the general trend of the data as well as any straight line can. If we accept t as correct to 50%, then $\rho(E)$ cannot be much less than half of (21) without producing a line that is too steep to represent the data. On the other hand, if filling up the surface states is indeed the explanation of the trend to higher ξ values at lower φ_m, then $\rho(E)$ cannot be much larger than (21) either. However we shall later mention a possible alternative explanation of the high values of ξ in Fig. 1.

We first discuss two other consequences of (21), (22), and filling up the surface states. One is the variation of barrier height with applied bias voltage. Since the surface states are in good electrical contact with the metal, we assume there is no extra dipole field between them. All the charge $\sigma=\epsilon E/4\pi$ on which the lines of force from the applied field E terminate therefore reside in the surface band, raising its Fermi level by $\sigma/\rho(E)$: i.e., Φ would decrease by this amount to keep E_F constant. Experimentally[8] a bias of 10^5 V/cm produces no measurable change in Φ, i.e., $\delta\Phi\leq2\times10^{-3}$ eV. The $\delta\Phi$ predicted from (21) for this field is 2×10^{-3} eV, just on the limit of error. Again we find $\rho(E)$ cannot be much lower than (21) without leading to contradictions. A similar effect to bias is the space-charge field produced by doping. High n-type doping produces a strong field which on the present model would lead to a reduction in Φ by 0.01 eV for a 0.2-Ω-cm sample ($N_i\sim9\times10^{17}$cm^{-3}). However, an *increase* of 0.03 eV in Φ is found,[2,8] which cannot in any way be reconciled with the mechanism of the present model.

This discrepancy leads us to point out that we have in fact been using the idea of a Fermi level in the band of surface states in a very cavalier fashion. The very fact that the charge in these states is spread out over some distance t means that there will be very severe bending of the bands over this distance, as shown in Fig. 9. Indeed the concepts of band bending must not be taken too seriously over such distances since the band states themselves cannot be localized to smaller than the size of the Wannier function which is of the same order[20] as t. From (22), over half of the dipole potential difference between a metal and a semiconductor in the junction occurs in the semiconductor. Thus if φ_m is less than φ_{sc} by more than 1 eV, the bands are bent so much that the conduction-band edge dips below the Fermi level [Fig. 9(c)]. This might be the explanation of the good surface conductivity found on silicon when covered with a monolayer of metallic cesium.[26] The same should apply for a calcium monolayer. Certainly such bending of the conduction band is compatible with the existence of a barrier. Calcium produces a barrier (Fig. 1), and it would be interesting to know if cesium does too. We come now to an alternative explanation of the trend of ξ in Fig. 1. Basically a higher value of $\rho(E)$, i.e., a less sloping line, would not be incompatible with the data for metals with high $\varphi_m\geq4$ eV. That leaves calcium and magnesium anomalously high. This might be related to the extreme bending of the bands in the sense that in Fig. 9(c) the length of the tails would be seriously curtailed, at least in the lower half of the gap, necessi-

[26] F. G. Allen (private communication).

[27] N. F. Mott and H. Jones, *The Theory of the Properties of Metals and Alloys* (Clarendon Press, Oxford, 1936), p. 88.

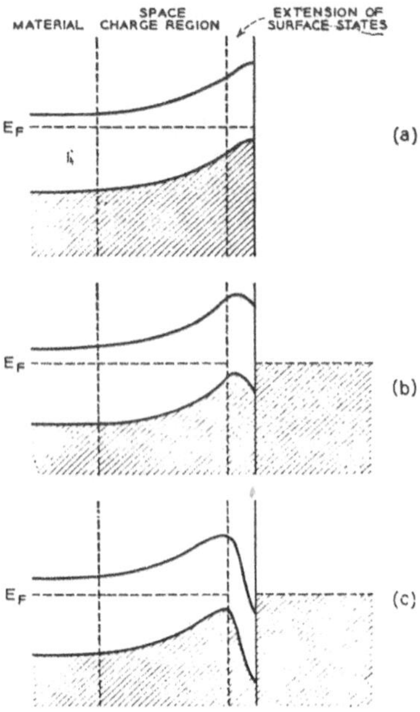

FIG. 9. Surface-state band bending (a) on a free surface, (b) in a metal junction when $\varphi_m < \varphi_{sc}$, (c) in a junction when $\varphi_m \ll \varphi_{sc}$. On the left is the semiconductor with bent valence- and conduction-band edges shown, on the right the metal [cases (b) and (c)].

tating a higher ξ and lower barrier to make up the required total charge. Another effect would be as follows: The charge density $|\psi|^2$ is so high in the tails that the bands bend downward already at a depth considerably greater than t which represents the center of gravity of the charge, which would also tend to lower the barrier. We conclude by reiterating that this surface-state band bending is an inescapable consequence of the surface-state charge being spread out in a tail, but the exact consequences of it for barrier heights are far from clear. In particular we have not been able to use it to provide a convincing explanation for the sign of the change of barrier height with doping.

Our present models say nothing specific about the case of antimony in Fig. 1. Here ξ lies close to the solid sloping line, which represents the ξ that would be observed if the whole of the dipole potential $\varphi_{sc} - \varphi_m$ appeared across the space-charge region rather than between the surface states and the metal. This suggests that the surface layer of the semiconductor is highly conducting and acts as one conducting system with the metal so that no dipole field can exist between them. If the first layer of antimony bonded tetrahedrally to the silicon and acted as donor centers, it might well produce such a conducting n-type layer: Its thickness would only have to be comparable with (20) for it to be effective.

Finally a comment about the work function of silicon with a monolayer of cesium or oxygen: Since the thickness of the surface states is as large as (20), the charge that has to be transferred per atom to produce the required dipole $|\varphi_{sc} - \varphi_m|$ is only about one-tenth of an electron per atom. Consequently, the electronic condition in a monolayer is very nearly the same as in bulk material. Thus the work function with a monolayer will be close to that of pure cesium or "oxygen metal," respectively. By the latter term we mean that the oxygen is likely to be bonded to the silicon as a close-packed monolayer rather than existing as diatomic molecules. In that case it would be metallic, and we can estimate crudely its work function as 8 eV, the mean of the ionization energy (13.6 eV) and the electron affinity (2.3 eV) of atomic oxygen. As a smaller second-order effect, the existence of the dipole leaves the cesium and the oxygen, respectively, with a small net positive and negative charge, thus increasing and decreasing φ, respectively. We can estimate the order of magnitude by taking one-tenth of the difference between the ionization potential and the electron affinity of the atom, since that represents the Coulomb energy for one extra electron, obtaining 0.3 eV for cesium and 1 eV for oxygen. Thus φ for cesium on silicon might be expected to drop $\sim$0.3 eV below the value for bulk cesium at about one monolayer coverage, and then increase again to the latter for thicker layers, of course. Something like that happens for cesium on metals,[28] and there appears to be some experimental evidence for it on semiconductors.[29] The φ for a monolayer of oxygen on silicon becomes $8-1=7$ eV, compared with the experimental value[26] of 6 eV. For thicker layers φ returns to 4.6 eV, very near the value for a clean surface, which presumably indicates the conversion of the oxygen layer to an electrically inert oxide layer.[26]

ACKNOWLEDGMENTS

It is a pleasure to acknowledge the kind hospitality of Bell Telephone Laboratories during the summer of 1964, where this work was done. Several useful conversations with Dr. J. Bardeen, Dr. F. G. Allen, Dr. C. R. Crowell, Dr. E. O. Kane, and Dr. C. A. Mead have contributed significantly to the present work, and their help is gratefully acknowledged.

[28] J. M. Houston and H. F. Webster, Advan. Electron. Electron Phys. 17, 125 (1962).
[29] J. Bardeen (private communication).

FUNDAMENTAL TRANSITION IN THE ELECTRONIC NATURE OF SOLIDS*

S. Kurtin, T. C. McGill, and C. A. Mead
California Institute of Technology, Pasadena, California 91109
(Received 6 March 1969)

Striking evidence for a fundamental covalent-ionic transition in the electronic nature of solids is presented.

The fundamental electronic properties of nonmetallic, inorganic, crystalline solids depend in a natural manner on the character of the chemical bond which is developed between constituent atoms. For example, the familiar Group-IV semiconductors are totally covalent and exhibit characteristics which are qualitatively different from highly ionic materials, such as the alkali halides. However, many materials of fundamental and practical importance are intermediate between these two extremes; it is not immediately evident how their properties may be treated. We will demonstrate here that if crystalline solids are ordered by some measure of their bond ionicity, then one finds striking evidence for a universal and surprisingly abrupt transition in many electronic properties associated with the quantum mechanical valence state. This transition divides crystalline solids into two well-defined classes: "covalent" and "ionic." Within each class there is a unifying character to the observed properties.

The first systematic treatment of the partial ionic character of chemical bonds was given by Pauling.[1] Over the range of materials we discuss, the difference in electronegativity (X) of the two constituents of a binary compound gives a crude but monotonic measure of the ionicity of the compound. Although improved methods for characterizing the ionicity of compounds have been proposed,[2] the nature of the transition under discussion here is not materially dependent upon which ionicity scale is used and for convenience we will use the electronegativity difference ΔX.

It has been generally recognized for some years that semiconductors can be classified into two broad groups based on the properties of the semiconductor-metal interface.[3,4] "Ionic" materials display little or no Fermi-level stabilization at the interface; "covalent" materials display virtually complete stabilization. The classic experiment consists of measuring the barrier energy at the metal-semiconductor interface as a function of the electronegativity X_M of the metal employed. Typical results[5,6] are shown in Fig. 1 for SiO_2 and vacuum-cleaved Si. The slope $S=(d\Phi_{Bn}/dX_M)$ of a straight line fitted to the data is inversely correlated to the extent of Fermi-level stabilization. This slope is found to be $\simeq 1$ (no Fermi-level stabilization) for "ionic" materials and $\simeq 0.1$ (nearly complete Fermi-level stabilization) for "covalent" materials. Recent data[7] obtained for the layer compound GaSe are also shown in Fig. 1; this material displays partial Fermi-level stabilization, $S\simeq 0.6$. Experiments on the isostructural layer compounds GaS and GaTe reveal[8] that this result is not a consequence of the layer structure, but appears to be related to a continuous transition from the "ionic" properties of GaS to the "covalent" properties of GaTe. A similar, although less well-defined, transition has been observed in the CdS_xSe_{1-x} mixed-crystal system.[4] These results suggest that the covalent-ionic transition may be elucidated by plotting slope S as a function of ΔX. Such a plot is shown as Fig. 2(a). There is evidently a well-defined transition between the interface properties of "ionic" materials and those of "covalent" materials.

Lattice termination is a severe disruption of

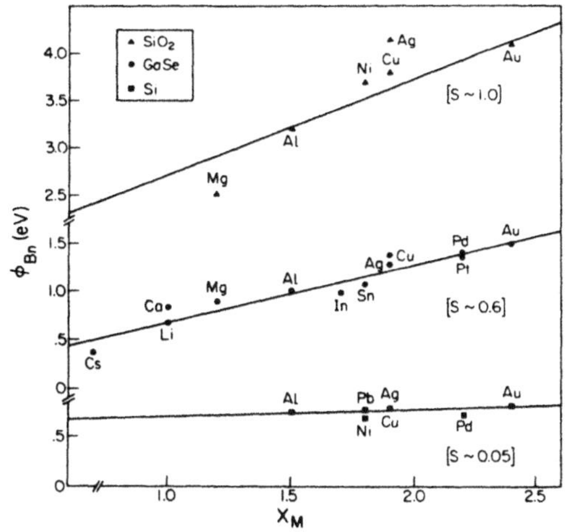

FIG. 1. Barrier energies of various metals on n-type semiconductors plotted versus electronegativity X_M of the metal. For each material, slope S of the reference line is inversely proportional to the extent of Fermi-level stabilization at the semiconductor-metal interface.

crystalline perfection and hence one would expect a large perturbation of valence-state energy at the surface of a solid. However, the data presented above indicate that this large perturbation occurs only for covalent materials. A simple Madelung calculation for highly ionic solids reveals that the electrostatic contribution to the binding energy for an atom at the surface is virtually identical to that for an atom in the bulk.[9] This result emphasizes the nonuniqueness of the surface in an ionic solid (i.e., there is virtually no energy differential for an atom at the surface; a lack of states at a given energy in the bulk would thus imply a lack of states of the same energy at the surface). On the basis of such elementary considerations we are led to the possibility that in "ionic" crystals effects attributable to long-range order may not dominate the observed properties (see also McGill, Kurtin, and Mead[10]).

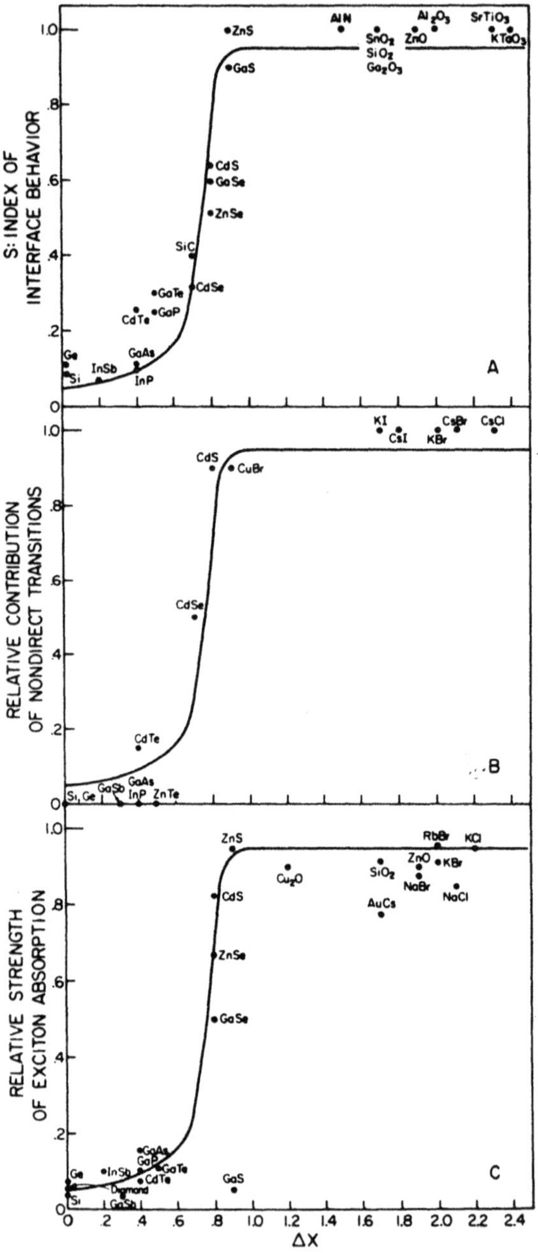

FIG. 2. Collected data representing several independent experiments plotted as a function of lattice electronegativity difference ΔX. (a) The slope S obtained from plots similar to Fig. 1; S is inversely proportional to the Fermi-level stabilization at the semiconductor-metal interface. (b) The relative contribution of nondirect transition in photon induced electronic excitations, as measured by photoemission studies. (c) The relative strength of exciton absorption as determined from optical spectra. These three independent experiments all illustrate the fundamental covalent-ionic transition in solids. (Details of the compilation of these data although straightforward are too lengthy to appear here. They will be discussed in detail in Ref. 10.) The same arbitrary curve is shown in all three sections of this figure for illustrative purposes.

If the transition displayed in Fig. 2(a) is fundamental to the nature of electron states in solids, it should be evident not only in phenomena involving lattice termination, but also in totally independent experiments, as for example in photoemission studies. Spicer et al.[11] have used energy analysis of photoemitted electrons to study photon-induced electronic transitions in a wide variety of solids. These experiments allow an assessment of the relative importance in a given solid of the different types of optical transitions: "direct" (single-particle crystal momentum k is conserved, "nondirect" (single-particle k conservation is not an important selection rule), and "indirect" (single-particle k is conserved with phonon participation). Of course, in all cases total momentum is conserved; however, in many materials nondirect transitions dominate the observed optical excitations. In Fig. 2(b) we plot the relative importance of nondirect transitions as a function of ΔX for many solids.[12] The similarity between Figs. 2(a) and 2(b) strongly suggests that the covalent-ionic transition is directly related to conservation of crystal momentum in optical excitations.

To summarize, the electronic valence states

of covalent materials are appropriately described in terms of an independent-particle k and lattice termination leads to a significant perturbation of electron energy near the surface (a large density of surface states). These observations are consistent with the simplified but historically significant treatment of surface states by Shockley.[13] In contrast, ionic materials ($\Delta X \geq 0.9$) are characterized by the absence of k conservation in optical transitions and by a small density of surface states ($<10^{11}/\mathrm{cm}^2$). Lack of single-particle k conservation, which occurs solely in the more ionic crystals, may be viewed as a result of localization of valence electronic wave functions; i.e., the spatial coordinates $\vec{r}$ become a more appropriate basis for describing electronic valence states than momentum coordinates $\vec{k}$.

The fundamental nature of the covalent-ionic transition in solids may be further illustrated by examining its influence on optical spectra in the neighborhood of the absorption edge. "Ionic" materials exhibit large absorption peaks near the edge, a consequence of localized excitons. Even at extremely low temperature, "covalent" materials show a smooth onset of absorption with only a very small peak. The extent to which the exciton peak dominates the absorption edge can be estimated from the ratio of the area of the peak itself to the total area of the peak plus general background under the peak. The relative strength of the exciton peak was determined in this manner from published optical spectra[14] on a number of materials and is plotted versus ΔX in Fig. 2(c). Although these data are the result of an experiment quite different from those previously discussed the covalent-ionic transition is once again very much in evidence.

Several independent types of experiments have been discussed. These experiments were chosen because they bear directly on the nature of the electronic valence state, and because they have been performed on a wide variety of materials and crystal structures. Strong evidence of the fundamental transition is also seen in many other phenomena involving electronic properties associated with valence states in solids.[12]

From a theoretical point of view, the overwhelming complexity of a realistic solution to the many-body problem in solids has necessitated the development of single-particle-like approximations. Since the procedure for obtaining independent-particle approximations is not unique, physically meaningful approximations are best motivated by experimental evidence. In the covalent limit a band approximation based on extended Bloch-like valence states has met with great success. In principle, "ionic" materials may be treated by starting with the band approximation and including correlation effects as a perturbation. However, the experimentally observed properties of "ionic" crystals indicate that an independent-particle approximation based on localized valence states may be more appropriate for these materials. In fact, the very abruptness of this transition implies that any attempt to describe "ionic" solids by the inclusion of electron-electron correlation as a perturbation upon delocalized valence states will be extremely involved since it is necessary to change the character of the initially assumed states. We believe that the experimental evidence indicates that localized valence states provide a more appropriate (and conceptually correct) basis from which refined treatments of "ionic" materials may be developed.

The authors would like to thank R. P. Feynman, J. M. Ziman, K. K. Thornber, W. A. Goddard, III, and J. Bardeen for much helpful discussion, and J. C. Phillips for supplying preprints of his recent work. The cooperation of W. E. Spicer, without which this paper would have suffered immeasurably, is deeply appreciated.

---

*Work supported in part by the Office of Naval Research.

[1]See, for example, L. Pauling, The Nature of the Chemical Bond (Cornell University Press, New York, 1967), 3rd ed.

[2]J. C. Phillips, Phys. Rev. Letters 20, 550 (1968), and Bull. Am. Phys. Soc. 14, 366 (1969).

[3]C. A. Mead, Solid-State Electron. 9, 1023 (1966).

[4]C. A. Mead, Appl. Phys. Letters 6, 103 (1965).

[5]B. E. Deal, E. H. Snow, and C. A. Mead, J. Phys. Chem. Solids 27, 1873 (1966).

[6]M. J. Turner and E. H. Rhoderick, Solid-State Electron. 11, 291 (1968).

[7]S. Kurtin and C. A. Mead, J. Phys. Chem. Solids 29, 1865 (1968).

[8]S. Kurtin and C. A. Mead, to be published.

[9]$\alpha_b \equiv \sum_j(\pm 1/P_{ij}) = 1.748$ for an atom in the bulk of an fcc structure; $\alpha_s \equiv \sum_j(\pm 1/P_{ij}) = 1.682$ for an atom at the surface of a terminated 3-D array with the same structure. Therefore, on the basis of this model there is an energy shift at the surface of less than 4%.

[10]T. C. McGill, S. Kurtin, and C. A. Mead, to be published.

[11]See, for example: K. K. Thornber, Sci. Progr. (Gt. Brit.) 57, 149 (1969); W. E. Spicer and R. E. Simon, Phys. Rev. Letters 9, 385 (1962); W. E. Spicer, Phys. Rev. 154, 385 (1967), and 161, 526 (1967), and Phys.

Letters 20, 325 (1966); C. N. Berglund and W. E. Spicer, Phys. Rev. 136, A1030 (1964); J. L. Shay, W. E. Spicer, and F. Herman, Phys. Rev. Letters 18, 649 (1967); J. L. Shay and W. E. Spicer, Phys. Rev. 161, 799 (1967); N. B. Kindig and W. E. Spicer, Phys. Rev. 138, A561 (1965); and others.

[12]The authors are deeply indebted to W. Spicer for providing much unpublished data and for assessing the relative importance of nondirect transitions in these materials.

[13]W. Shockley, Phys. Rev. 56, 317 (1939).

[14]See, for example, F. Harbeke, Z. Naturforsch. 19a, 548 (1964); M. D. Sturge, Phys. Rev. 127, 768 (1962); W. W. Piper, P. D. Johnson, and D. T. F. Marple, J. Phys. Chem. Solids 8, 457 (1959); D. G. Thomas, J. Phys. Chem. Solids 15, 86 (1960); and many others.

J. Phys. C: Solid St. Phys., 1971. Vol. 4. Printed in Great Britain

Density of states and barrier height of metal–Si contacts†

F. YNDURAIN
Centro Coordinado de Física (CSIC–UAM). Departamento de Física. Universidad Autónoma de Madrid. Spain
MS. received 29*th June* 1971

Abstract. Using the surface Green function method of García–Moliner and Rubio the extra density of states of metal–Si(110) contacts is studied in detail. This is then used to calculate the barrier height for a wide range of Sommerfeld metals. The surface position of the Fermi level relative to the band edges of the semiconductor depends on both the work function and the Fermi energy of the metal. Some arguments are put forward to the effect that a similar behaviour is expected for crystallographic orientations of the Si surface other than the (110). The calculations are actually in fair agreement with experimental data on (111) surfaces.

1. Introduction

We consider the standard situation described in figure 1. which also displays the notation used in this paper. The two extreme models are well known. Schottky (1931) attributed all the barrier height to the difference in work functions. that is

$$\phi_{\mathrm{Bn}} = \phi_{\mathrm{m}} - \chi \tag{1.1}$$

whereas Bardeen (1947) assumed a high density of surface states on the semiconductor. distributed over the energy gap. For densities larger than 10^{13} states $\mathrm{cm}^{-2}\ \mathrm{eV}^{-1}$ one expects a pegging of the Fermi level. with

$$\phi_{\mathrm{Bn}} = E_{\mathrm{g}} - \phi_0. \tag{1.2}$$

Modern technological developments have made it possible to obtain nearly ideal contacts. Barrier height measurements using different procedures (I–V, C–V, photoelectric, etc) yield equal values for the same sample (eg Sze 1969). These results are in open disagreement with both equations (1.1) and (1.2). Cowley and Sze (1965) proposed a hybrid approach in which

$$\phi_{\mathrm{Bn}} = \gamma(\phi_{\mathrm{m}} - \chi) + (1 - \gamma)(E_{\mathrm{g}} - \phi_0). \tag{1.3}$$

where γ is a weighting factor ($0 \leqslant \gamma \leqslant 1$). somehow related to the density of states in the semiconductor gap. and $\Delta\phi_{\mathrm{n}}$ for an n type semiconductor is the barrier lowering due to the image force. The weakest point in this analysis was the assumption that the said density of states is a property of the semiconductor alone. irrespective of the metal. Likewise Turner and Rhoderick (1968) assumed that this was just the density of states of the free semiconductor surface.

The philosophy of this paper is very different. As Heine (1965) pointed out one cannot have surface states in the semiconductor gap if this energy range is inside the metallic band. The analysis of García-Moliner and Rubio (1971) characterizes this as resonance like solutions of the secular determinant obtained from the appropriate surface Green

† Work partially supported by the General Electric Company (through General Eléctrica Española) and by the 'Fundación Pedro Barrié de la Maza. Conde de Fenosa'. The financial support of both these institutions is gratefully acknowledged.

function $\mathscr{G}_s$, corroborating some of Heine's conclusions and carrying the study a step further. This paper follows along this line. A detailed study of the extra density of states at a metal–semiconductor interface is described in § 2. The main point is that it *does* depend on the metal. Furthermore, the in-band density of states for lower energies is also distorted and this in turn is related to the concept of 'point of charge neutrality' usually somewhat vaguely defined and here discussed as part of the information one obtains from $\mathscr{G}_s$. We also emphasize that the usual reliance of one dimensional models is inherently unwarranted for the interface problem and make instead a three dimensional analysis.

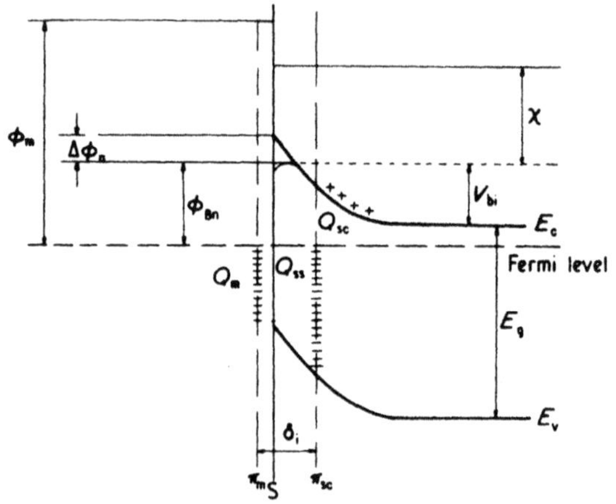

Figure 1. Energy band diagram for a metal–n type semiconductor intimate junction showing the different quantities used in this work. ϕ_m is the metal work function: ϕ_{Bn} the barrier height: $\Delta\phi_n$ the image force barrier lowering: χ the electron affinity of the semiconductor: V_{bi} the diffusion potential: Q_{sc} the space charge density in the semiconductor; Q_{ss} the surface density of charge in the semiconductor; Q_m the surface charge density on the metal: δ_i the mean separation between Q_{ss} and Q_m. S is the surface plane: π_m, π_{sc} are the positions, with respect to the surface plane, of the centre of gravity of Q_m and Q_{ss} respectively.

The calculation thus proceeds in two stages:

(i) The extra density of states is studied as a property of the spectrum of a metal semiconductor interface without allowing charge transfer.

(ii) The electrons are then allowed to distribute themselves on the new energy spectrum and to flow across the boundary, describing the resulting equilibrium in terms of a Poisson equation. We thus estimate ϕ_0 (see below) and the relative position of the Fermi level at the surface of separation for a wide range of metals both in intimate and oxide coated contact with Si.

The work described here is totally one-electron, the emphasis being on the fact that the part of the effects due to interface states is, for a given semiconductor, a specific property which depends not drastically, but by all means to an extent calculable and measurable, on the metal involved. Nevertheless some attempt is made, though admittedly very crude, to estimate the magnitude of the many body effects involved.

2. Extra density of states in the metal–semiconductor interface

The formula for the density of states, including surface states, has been given by García-Moliner and Rubio (1971), who have also discussed the meaning of symmetrization about a surface plane. In this case the extra density of states is

$$\Delta\rho_s(E) = -\pi^{-1}\frac{d}{dE}\arg\det\{\mathscr{G}_1(E) + \mathscr{G}_2(E)\}. \tag{2.1}$$

In our problem, 1 means the metal (described in a Sommerfeld model) and 2 the (two band) semiconductor.

For the one dimensional model the analysis of García-Moliner and Rubio (1971) yields (*a*) the extra density of states (higher in the lower half of the semiconductor gap) first derived by Heine (1965) for a 'bonding' gap from an estimate of the charge 'in the tails of the metallic wavefunctions' and (*b*) a similar formula, symmetric about the midgap, for the case of an 'antibonding' gap, which had been customarily assumed to contain no extra states whatever. In both cases the integral of $\Delta\rho_s$ over the gap (including spin degeneracy) amounts to exactly one electronic state, this being one of the most direct results obtainable from equation (2.1).

Now, in the case of Si we shall deal with a (110) surface, and we have to study the gap associated with the corresponding face of the Jones zone (Heine and Jones 1969). When remapped onto the first Brillouin zone this gap lies inside the zone boundaries. The origin of a gap of this type is the crossing of two bands of the same symmetry which correspond to different irreducible representations: hence the wavefunctions of the band edges must have the two appropriate (different) Fourier components parallel to the surface. This need not make the results very different for different surface orientations (in fact we shall presently contend that for Si the differences are unimportant), but it does constitute a feature of a three dimensional analysis (Forstmann 1970) which is absent from one dimensional models. Following up its consequences we find that the one dimensional concepts of 'bonding' and 'antibonding' gaps cease to be meaningful and distinct.

The full analysis requires the complete evaluation of the three dimensional Green function, which is rather lengthy and will not be reproduced here. Suffice it to say that application of equation (2.1) yields

$$\Delta\rho_s(\mathscr{E}) = \pi^{-1}\frac{2\alpha(1+\alpha^2)V_g^2}{(V_g^2-\mathscr{E}^2)^{1/2}\{(1+\alpha^2)^2(V_g^2-\mathscr{E}^2)+4\alpha^2\mathscr{E}^2\}}. \tag{2.2}$$

Here $\mathscr{E}$ is the energy referred to the midgap and α^2 is $(W_{sc} + |V_g|)/(W_m - \mathscr{E})$ while $V_g = V_{eff}$ (220) (Heine and Jones 1969). $W_{sc/m}$ is the semiconductor/metal valence band width. Although not explicitly labelled, α and V_g depend on $\boldsymbol{\kappa}$ the component of $\boldsymbol{k}$ parallel to the surface for the state of energy $\mathscr{E}$ under study. It follows from equation (2.2) that, for a fixed $\boldsymbol{\kappa}$, (i) $\Delta\rho_s$ depends only on $|V_g|$ and is symmetrical about the midgap (ie, so to speak, the 'bonding' and 'antibonding' contributions of a one dimensional model are merged together) and (ii) the integral of $\Delta\rho_s$ over the gap (including spin degeneracy) is two electronic states. This is shown in figure 2 which exhibits the origin of $\Delta\rho_s$ in the lower half of the gap as 'coming from' an absence of states (one in total) in the valence band. Furthermore, we must also consider $\Delta\rho_s$ for energies below the semiconductor gap. This is related to the concept of surface charge neutrality. Discarding the possible role of impurities in the earlier experiments one expects in contemporary measurements to deal essentially with intrinsic states. How can one then have charge neutrality at the surface if there is an extra charge in that part of $\Delta\rho_s$ whose occupation is permitted by the selfconsistent position of the Fermi level? The answer comes from the study of arg det $\mathscr{G}_s$ for energies in the semiconductor valence band. It turns out that this produces there a negative $\Delta\rho_s$ whose integral exactly cancels that of $\Delta\rho_s$ from the top of the valence band to the midgap. Since those states missing from the valence band would always be occupied, we shall say no

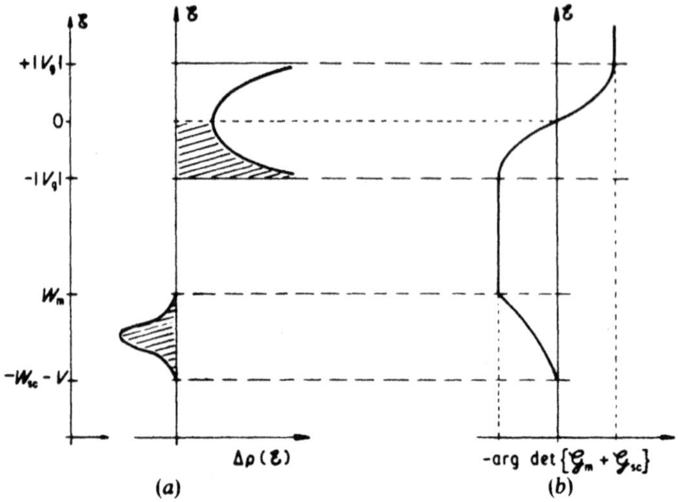

Figure 2. (*a*) Extra density of states in the semiconductor energy spectrum against energy for a three dimensional model. The shaded areas are identical. (*b*) Behaviour of arc det$\{\mathcal{G}_{sc} + \mathcal{G}_m\}$ along the energy spectrum.

more about them. but they will be taken into account as extra positive charge to determine the point of charge neutrality.

Finally. for a complete three dimensional analysis we must integrate equation (2.2) for all values of $\boldsymbol{\kappa}$ in the corresponding Jones zone. This leads us from optical gap to thermal gap (figure 3). In this process the details of the bulk band structure (Cohen and

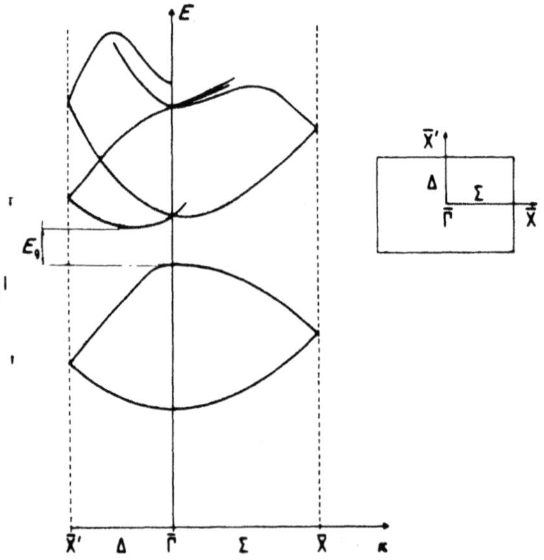

Figure 3. Energy band structure of silicon parallel to the (110) surface (after Cohen and Bergstresser 1966). plotted along two symmetry directions of the two dimensional Brillouin zone. which is also drawn.

Bergstresser 1966, Heine and Jones 1969) become important because for different κ the optical gap width and its position with respect to the bottom of the metallic band are different and so is equation (2.2) for different values of κ.

Table 1. Fermi energies and work functions

Metal	E_F(eV)	ϕ_m(eV)	D_s(states $cm^{-2}eV^{-1}$)
K	2·1	2·2	$\sim 5{\cdot}9 \times 10^{13}$
Na	3·1	2·3	$\sim 6{\cdot}6 \times 10^{13}$
Pt, Pd	3·6	5·3, 4·9	$\sim 7{\cdot}0 \times 10^{13}$
Ca, Ni	4·5	2·7, 4·5	$\sim 7{\cdot}0 \times 10^{13}$
Ag, Au	5·5	4·2, 4·75	$\sim 7{\cdot}4 \times 10^{13}$
Mg, Cu	7·0	3·6, 4·5	$\sim 8{\cdot}0 \times 10^{13}$
Sb	10·3	4·1	$\sim 8{\cdot}0 \times 10^{13}$
Al	11·7	4·1	$\sim 8{\cdot}0 \times 10^{13}$
Si*	12·04	4·05*	

Fermi energy after Kittel (1956) and work function after Szydlo (1970) corresponding to metals (for the silicon the affinity is given) and density of states at the semiconductor gap of the Si(110)–metal junction.

For typical metals (Ag, Au, Cu, Ni, Al, etc) we find, upon integrating, a nearly constant extra density of states (henceforth denoted $D_s(\mathscr{E})$) in the thermal gap. D_s is symmetrical about the midgap, where it has a shallow minimum. Since $\Delta\rho_s$ depends on W_m, D_s is different for different metals (table 1), in disagreement with the assumption often made without evidence in support. For metals with a small Fermi energy (eg Na, K) it is possible to have, for some fixed κ, genuine surface states, in the corresponding optical gap, because $\mathscr{G}_1$ and $\mathscr{G}_2$ have a common energy range in which both are real. More physically, the bottom of the metallic band may be inside the semiconductor gap. Of course this will only happen for certain values of κ and their total effect must be included in the global calculation. This analysis for the Si–Na junction is in progress, but we shall ignore such a situation in the remainder of this article.

3. Origin and calculation of the barrier height

Following common practice we shall refer locally, for each z (the coordinate normal to the interface) the energy levels to the midgap. The Fermi level, actually constant throughout, becomes then the z dependent variable $\mathscr{E}_F(z)$ and its bending describes in fact the opposite of the band bending. The surface value of the distance in energy from $\mathscr{E}_F$ to the bottom of the conduction band is called the barrier height ϕ_{Bn} (for an n type semiconductor).

Having obtained the new energy spectrum we shall carry out the second stage indicated in § 1 in two steps. (i) We study the selfconsistent distribution of electrons as a consequence of the occupation of these states, strictly governed by the position of the Fermi level but without allowing charge transfer from the metal to the semiconductor and (ii) we switch on the electric contact; the electrons are allowed to flow and the final equilibrium of the whole system is reached.

(i) For a semiconductor with an energy spectrum which includes $\Delta\rho_s$ the charge density $\rho(z)$ is related to the electrostatic potential $\psi(z)$ by

$$\nabla^2\psi(z) = -\frac{4\pi}{\epsilon_{sc}}\rho(z). \qquad (3.1)$$

We define $u(z) = \mathscr{E}_F(z)/k_BT$, so that

$$\mathscr{E}_F(z) - e\psi(z) = \mu = \text{constant} \qquad (3.2)$$

whence equation (3.1) becomes

$$\nabla^2 u(z) = -\frac{4\pi e}{\epsilon_{sc} k_B T} \rho(u). \tag{3.3}$$

Gauss' theorem yields the surface charge

$$Q_{ss} = -2en_i L_D F(u_s, u_b) \tag{3.4}$$

of electrons (of charge $-e$) accumulated in the semiconductor gap. L_D, n_i and F are as in Frankl's book (1967), s meaning surface and b meaning bulk.

But, having calculated $\Delta\rho_s$ we also have

$$Q_{ss} = e\int_{-\infty}^{\infty} \Delta\rho_s(\mathscr{E})\, f(\mathscr{E} - \mathscr{E}_F)\, d\mathscr{E}. \tag{3.5}$$

On account of equations (2.1), (2.2), (3.4) and (3.5) we have, measuring V_g in units of $k_B T = 0{\cdot}026$ eV,

$$\pi^{-1} \tan^{-1}\left\{\frac{2\alpha}{1+\alpha^2} \frac{u_s}{(V_g^2 - u_s^2)^{1/2}}\right\} \simeq \frac{n_i L_D}{5\times 10^{14}} F(u_s, u_b). \tag{3.6}$$

From tabulated values of $n_i L_D$ and $F(u_s, u_b)$ (Frankl 1967) we find that $|u_s|$ is small and and decreases for decreasing $|u_b|$; for instance for $|u_b| = 10$, equation (3.6) yields $|u_s| \leqslant 10^{-1}$. In particular equation (3.6) gives $u_s = 0$ for $u_b = 0$. As indicated before, after integration over $\boldsymbol{\kappa}$ on the (110) face of the Jones zone, it turns out that the energy levels below the thermal gap are the ones filled and, therefore, $\phi_0 \simeq E_g/2 \simeq 0{\cdot}55$ eV (E_g being the silicon thermal gap) in going from p to n type bulk doping. This shows how naturally the flat band condition comes from equation (3.6).

(ii) We now allow electrical contact. A surface dipole D is then set up which in first approximation can be taken as the difference between the work functions. This dipole is compensated by the charge in the gap.

$$D = \frac{\delta_i}{\epsilon_i} Q_{ss} \tag{3.8}$$

where δ_i is the distance between Q_{ss} and Q_m, and ϵ_i the dielectric constant of the medium lying between Q_{ss} and Q_m. Q_{ss} is given by

$$Q_{ss} = -\int_{\phi_0}^{\infty} f(\mathscr{E}'_F - \mathscr{E}')\, \Delta N(\mathscr{E}')\, d\mathscr{E}' \tag{3.9}$$

$\mathscr{E}'$ being the energy referred to the thermal midgap and $\Delta N(\mathscr{E}')$ the extra density of states (2.6) integrated over $\boldsymbol{\kappa}$.

Equations (3.8) and (3.9) give the position of the Fermi level at the metal-semiconductor interface when the equilibrium is reached, that is

$$\frac{\epsilon_i}{\delta_i} D = -\int_{\phi_0}^{\infty} f(\mathscr{E}'_F - \mathscr{E}')\, \Delta N(\mathscr{E}')\, d\mathscr{E}' \tag{3.10}$$

which is, in first approximation, equation (1.3) of Cowley and Sze (1965), only that $\Delta N(\mathscr{E}')$ is here calculated *a priori*.

If the dipole is taken as the difference in work functions, then (figure 1),

$$D = \phi_m - (\chi + \phi_{Bn} + \Delta\phi_n). \tag{3.11}$$

Two special cases of equation (3.10) are of interest:

(i) When there is no extra density of states at the interface ($\Delta N(\mathscr{E}') = 0$), equation (3.10) takes the form (1.1) given by Schottky (except for the $\Delta\phi_n$ term).

$$\phi_{Bn} = \phi_m - \chi - \Delta\phi_n. \tag{3.12}$$

(ii) When there is a very high extra density of states in the semiconductor gap ($\Delta N(\mathscr{E}') \rightarrow \infty$), equation (3.10) takes the form (1.2) given by Bardeen (except for the $\Delta\phi_n$ term).

$$\phi_{Bn} = E_g - \phi_0 - \Delta\phi_n. \tag{3.13}$$

As stated above ϕ_0 (ie. the position of the Fermi level at the interface) is nearly independent of the degree of bulk doping of the semiconductor. Hence, in agreement with experimental evidence (see eg Mead 1966):

$$\Delta\phi_n + \Delta\phi_p + \phi_{Bn} + \phi_{Bp} = E_g \tag{3.14}$$

where ϕ_{Bp} is the barrier height for a metal–semiconductor p type contact and is defined as the energy threshold needed to transfer a hole from the metal to the semiconductor. $\Delta\phi_p$ represents the barrier lowering due to image force for a p type semiconductor.

In order to calculate ϕ_{Bn} we must know the different quantities in (3.10).

For D we would take, following common practice,

$$D = \phi_{sc} - \phi_m. \tag{3.15}$$

The metallic work function (for a semiconductor it ought to be similar) is governed by electrostatic and many-body forces. When an electron is at the metal surface, these forces are strongly affected by the polarizability of the medium in contact with the metal. Therefore (3.15) cannot be correct because ϕ_{sc} and ϕ_m are measured with free surfaces.

A way of getting information about the effect of many-body effects on the barrier of the metal–Si junction is to study behaviour of the barrier height of the metal–ionic semi-conductor contact for different metals. The density of states in the gap for an ionic semi-conductor is so low that a Schottky-like barrier behaviour is expected. We are concerned here with the metal–CdS junction. The experimentally observed behaviour of such junction is actually Schottky-like, so that ϕ_{Bn} against ϕ_m is a straight line of 45° slope (Geppert *et al.* 1966). Nevertheless, equation (1.1) does not give the experimental values at ϕ_{Bn}. In order to get them we must replace equation (1.1) by,

$$\phi_{Bn} = \phi_m - \chi_{CdS} + \Delta. \tag{3.16}$$

with $\Delta \simeq 0{\cdot}8$ eV and χ_{CdS} being the electron affinity of CdS.

The origin of Δ is the dressing of the electron in going from the conduction band of the semiconductor, where the electron density is low, to the metal, where the electron density is high and many-body effects become very important.

We expect a similar behaviour for an electron on both Si and CdS conduction bands. Hence, instead of equation (3.15) we venture to write

$$D = \phi_{sc} - \phi_m - \Delta. \tag{3.17}$$

Very crudely, we assume Δ to be the same for all metals, and take this as an elementary attempt to introduce some plausible correction into equation (3.15) which is obviously incorrect as it stands.

We shall assume that the charge accumulated on the extra density of states is concentrated at its centre of probability. Therefore, in order to know δ_i/ϵ_i, we do not assume any kind of interface layer between the metal and the semiconductor.

The local density of states is given by

$$\rho(\boldsymbol{r}, E) = -\frac{1}{\pi} I_m G_s(\boldsymbol{r}, \boldsymbol{r}; E^+). \tag{3.18}$$

In our problem equation (3.18) has the following form (García-Moliner and Rubio 1971):

$$\rho(\mathbf{r}, E) = -\frac{1}{\pi} I_m \{ G_{sc}(\mathbf{r}, \mathbf{r}) + \langle \mathbf{r} | G_{sc} \mathcal{G}_{sc}^{-1} G_{sc} | \mathbf{r} \rangle - 2 \langle \mathbf{r} | G_{sc} (\mathcal{G}_{sc} + \mathcal{G}_{m})^{-1} G_{n} | \mathbf{r} \rangle \}. \quad (3.19)$$

Using this formula one can estimate (E. Louis, private communication) that the charge centre of probability lies at approximately 2·5 Å from the surface of the semiconductor; as this distance is smaller than the semiconductor interatomic distance, it cannot be polarized and its dielectric constant is 1 (in vacuum units).

On the other hand,

$$\frac{\delta_i}{\epsilon_i} = \frac{\delta_{sc}}{\epsilon_{sc}} + \frac{\delta_m}{\epsilon_m}. \quad (3.20)$$

Taking for δ_m/ϵ_m the typical metal screening distance, we have

$$\frac{\delta_i}{\epsilon_i} \simeq 2{\cdot}5 + 0{\cdot}5 = 3{\cdot}0 \text{ Å}. \quad (3.21)$$

Using equations (3.17) and (3.21) in equation (3.10) and taking into account $\Delta N(\mathscr{E}')$ we have calculated the barrier height for different metals (see table 2 and figure 4).

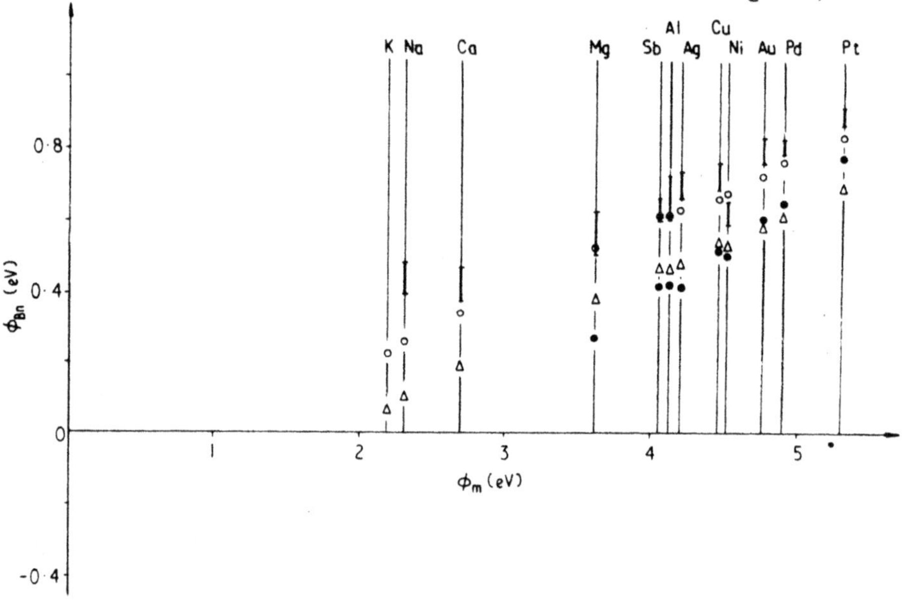

Figure 4. Barrier height for the metal–Si junction against metal work function showing experimental (after Szydo 1970: denoted by error bars) and the theoretical results calculated in this paper: ○ intimate contact (Δ = 0·8 eV); △ intimate contact (Δ = 0); ● oxide coated interface.

We have neglected $\Delta\phi_n$ altogether because (Szydlo *et al.* 1970)

$$\Delta\phi_n \simeq \left(\frac{eE_m}{\epsilon_{sc}}\right)^{1/2} \quad (3.22)$$

where E_m is the maximum electric field between the metal and the semiconductor, and $\Delta\phi_n \ll \phi_{Bn}$ in our problem.

The results for the barrier height are in good agreement with experimental data. The behaviour of the barrier is intermediate between Schottky and Bardeen-like. The extra

Table 2. Values of barrier heights (in eV) for different metals and for different interface parameters of the Si(110)–metal junction.

Metal	$\phi_{Bn}(\delta/\epsilon_i = 3Å)$	(oxid)	$(\delta_i/\epsilon_i = 1Å)$	$(\delta_i/\epsilon_i = 2{\cdot}2Å)$	$(\delta_i/\epsilon_i = 5Å)$
K	0·231		−0·132	0·111	0·321
Na	0·260		−0·102	0·162	0·351
Ca	0·342		0·091	0·280	0·412
Mg	0·515	0·270	0·479	0·505	0·528
Sb	0·602		0·657	0·693	0·583
Al	0·602	0·425	0·657	0·693	0·583
Ag	0·626	0·426	0·699	0·642	0·641
Cu	0·661	0·522	0·800	0·704	0·628
Ni	0·675	0·518	0·784	0·721	0·637
Au	0·726	0·596	0·902	0·770	0·662
Pd	0·759	0·646	0·938	0·819	0·687
Pt	0·828	0·774	1·092	0·917	0·737

density of states in the semiconductor gap region is neither sufficient to peg the Fermi level at the interface nor negligible.

When D is positive, the semiconductor must be negatively charged, hence there is a flow of electrons to the metal until the equilibrium is reached and therefore $\phi_{Bn} < E_g - \phi_0$. When D is negative, $\phi_{Bn} > E_g - \phi_0$.

So far, it has usually been assumed that the behaviour of the barrier height is linear in D (ie, in ϕ_m). When D increases in absolute value, more electrons are needed to build up the dipole and therefore a nearly linear relation between ϕ_{Bn} and ϕ_m is expected. Nevertheless $\Delta N(\mathscr{E}')$ does not depend on the metal work function but on its bandwidth. Hence, ϕ_{Bn} also depends on the metal bandwidth, and there is no reason to have a strictly linear relation between ϕ_{Bn} and ϕ_m. A clear example is the case of Ni and Cu, both with the same work function but giving different barrier heights as shown in figure 4. $\phi_{Bn}(Cu) > \phi_{Bn}(Ni)$. As the bandwidth for Ni is smaller than for Cu, D_s is smaller for the Si–Ni junction than for Si–Cu, and in order to build up the same dipole D in both junctions, the Fermi level must be higher for Ni–Si than for Cu–Si. Hence $\phi_{Bn}(Ni) < \phi_{Bn}(Cu)$. (This argument has been made with $\Delta = 0$ because this parameter is not known in detail and masks the results.)

In table 2 we have listed the barrier heights calculated for different values of δ_i/ϵ_i. As we are working at very short distances, a small change in δ_i/ϵ_i induces a large change in the charge needed to set up the dipole, and $\mathscr{E}_F$ at the semiconductor gap changes quite appreciably.

4. Barrier height for an oxide coated interface

According to Turner and Rhoderick (1968), when the semiconductor surface is chemically prepared one can assume that there is a thin oxide film between the metal and the semiconductor. For lack of a better model for the oxides we shall make an estimate in the manner of Yndurain and Rubio (1971).

The extra density of states in this case is

$$\Delta\rho_s(E) = -\frac{1}{\pi}\frac{d}{dE}\arg\det\left\{\mathscr{G}_{sc}\left(1+\frac{\mathscr{G}_m}{\mathscr{G}_{ox}}\tanh\kappa_{ox}L\right)+\mathscr{G}_m\left(1+\frac{\mathscr{G}_{sc}}{\mathscr{G}_{ox}}\tanh\kappa_{ox}L\right)\right\} \quad (4.1)$$

where m means metal, sc semiconductor, ox oxide, L is the film width and κ_{ox} is a wave-vector parameter characteristic of the oxide.

It is easy to see that, when $L = 0$, equation (4.1) becomes equation (2.1). When L increases the resonance is broadened because the oxide gap is larger than the semiconductor one. Therefore, the density of states in the thermal gap of the semiconductor decreases when L increases.

We have estimated a density of states in the thermal gap of the semiconductor of roughly $\lesssim 2 \times 10^{13}$ states cm^{-2} eV^{-1} for a 10 Å oxide layer. This density of states is symmetrical about the midgap. Using this and the formula (3.10) we have calculated the barrier height. In this case we take $D = \phi_{sc} - \phi_m$ because the Δ parameter is meaningless here. The results, listed in table 2 and shown in figure 4, are in rough agreement with Turner and Rhoderick's measurements.

5. Concluding remarks

We interpret the extra density of states at the metal–semiconductor interface in terms of resonance-like states using the Green function formulation of García-Moliner and Rubio (1969, 1971) which turns out to be very appropriate for this particular problem. We have used a realistic three dimensional model for the semiconductor in order to calculate the corresponding Green function and stressed that one dimensional models are inherently unsuitable for this problem (as for many others, in spite of their frequent use).

With the extra density of states and the condition of surface charge neutrality which arises from it, we have calculated the barrier height following arguments similar to those of Cowley and Sze (1965). Our aim has not been to make a detailed calculation of the barrier height (which emphatically requires a many body treatment of the interface) but rather to estimate in a plausible way the differences found for different metals, both for intimate and for oxide coated contact. We find that these differences are in fact reasonably well explained by a calculation *ab initio* (except for the insertion of Δ).

Our calculation is for the (110) silicon surface, because it is the easiest to handle, whereas the experiments are usually performed for the (111) surface; however, we do not expect a critically different density of states for different orientations of the surface plane because (*a*) the surface states (for a free silicon surface) for different orientations of the surface plane are very similar (Yndurain, to be published) and (*b*) it can be proved using true three dimensional Green functions for both metal and semiconductor that, for an arbitrary orientation of the surface plane the extra density of states integrated across the optical gap is, for fixed $\boldsymbol{\kappa}$, two states (taking into account spin degeneracy). Both statements suggest that the barrier height ought not to be very sensitive to the surface orientation and, hence, we can compare our results to experimental values.

Acknowledgments

I am indebted to F. García-Moliner and J. Rubio for much helpful discussion and for reading the manuscript.

References

BARDEEN, J., 1947, *Phys. Rev.*, **71**, 717–27.
COHEN, M. L. and BERGSTRESSER, T. K., 1966, *Phys. Rev.*, **141**, 789–96.
COWLEY, A. M., and SZE, S. M., 1965, *J. appl. Phys.*, **36**, 3212–20.
FORSTMANN, F., 1970, *Z. Phys.*, **235**, 69–74.
FRANKL, D. R., 1967, *Electrical properties of semiconductor surfaces*, (Oxford: Pergamon).
GARCIA-MOLINER, F., and RUBIO, J., 1969, *J. Phys. C: Solid St. Phys.*, **2**, 1789–96.
1971, *Proc. R. Soc.* A, **324**, 257–73.
GEPPERT, D. V., COWLEY, A. M., and DORE, B. V., 1966, *J. appl. Phys.*, **37**, 2458–67.
HEINE, V., 1965, *Phys. Rev.*, **138**, A1689–96.
HEINE, V., and JONES, R. O., 1969, *J. Phys. C: Solid St. Phys.*, **2**, 719–32.
KITTEL, C., 1956, *Introduction to Solid State Physics*, (New York and London: Wiley).
MEAD, C. A., 1966, *Solid St. Electron*, **9**, 1023–33.
SCHOTTKY, W., 1931, *Phys. Z.*, **32**, 833–8.
SZE, S. M., 1969, *Physics of semiconductor devices*, (New York and London: Wiley).
SZYDLO, N., 1970, *Thèse*, Faculté des Sciences, Université de Dijon.
SZYDLO, N., POIRIER, R., and KLEEFSTRA, M., 1970, *Appl. Phys. Lett.*, **17**, 477–8.
TURNER, M. J., and RHODERICK, E. H., 1968, *Solid. St. Electron.*, **11**, 291–300.
YNDURAIN, F., and RUBIO, J., 1971, *Phys. Rev. Lett.*, **26**, 138–40.

Metal-semiconductor junction for (110) surfaces of zinc-blende compounds

E. Louis, F. Yndurain,* and F. Flores

Centro Coordinado de Física, C-12 6', Universidad Autónoma, Canto Blanco, Madrid 34, Spain

(Received 24 February 1975)

The behavior of the metal-semiconductor junction is studied for some III–V and II–VI zinc-blende compounds, within the framework of a one-electron theory. Using a pseudopotential description of the semiconductor crystals, the density of interface states is analyzed performing the matching of the wave functions at the interface. The characteristics of the different types of interface states are discussed in detail, and particular examples are given for contacts with metals of high and low electronic densities. As a by-product, the surface states at the semiconductor vacuum interface are also calculated. Finally, the barrier potential of the junction is obtained as a function of the metal work function (for metals of high density). Although very good agreement has been found with the experimental data for the compounds of the covalent group, large discrepancies arise for those of the ionic group. Some comments on the reasons for these discrepancies are made.

I. INTRODUCTION

The metal-semiconductor (M-S) interface is of prime interest not only because of its technical applications,[1] but also owing to other theoretical and experimental reasons; for instance, the study of the properties of the M-S junction could help one attain a better understanding of the characteristics of the junction components. In particular, the M-S interface has recently been used[2] to study the ionic character of dielectric crystals. It turns out, after the analysis of the experimental data carried out by Kurtin *et al.*,[2] that the dielectric crystals could be classified into two main groups, ionic and covalent, according to the degree of stability of the Fermi level when the crystal is placed in intimate contact with different metals. The more ionic the crystal, the weaker the pegging of the Fermi level, and the junction has a more Schottky-like[1] behavior. On the other hand, in the extreme covalent case the Fermi level is pegged by a high density of interface states[3] that confers a Bardeen-like behavior[1] to the contact. Moreover, this transition from Schottky to Bardeen-like behavior seems to occur abruptly. This striking transition is explained by Kurtin *et al.*[2] in terms of the electronic differences between the valence-band states of those two groups; namely, in the covalent crystals the valence states have an extended character, while in the ionic solids this character is localized.

In a previous paper[4] we reported the preliminary results of a study of the behavior of some zinc-blende solids at the M-S interface; we studied just the (110) surface of those crystals. At this stage our purpose is to describe and extend the model and calculations we performed in I. We shall investigate the role played by the interface states in the M-S junction of (110) surfaces of zinc-blende crystals and whether the transition outlined above can be explained in terms of those interface states.

Our model is mainly based upon a pseudopotential description of the semiconductor bulk band structure. Although we shall perform the calculation using the abrupt potential model (APM), we shall discuss, by means of a model previously applied to the semiconductor-vacuum system,[5] the relevance of a smooth interface potential at the M-S junction. The density of interface states is analyzed by performing the matching of the wave functions at the interface; in the present work we analyze two points of high symmetry of the two-dimensional Brillouin zone (BZ). The characteristics of the different types of interface states[3] which can appear at the M-S junction are discussed in detail and particular examples are given for contacts with metals of high and low electronic densities. Detailed calculations are performed for some zinc-blende crystals joined to a metal of high density (Al); our results confirm the conclusions of I (that calculation was carried out considering a single point of the BZ) in the sense that while our theoretical calculation agrees with the available experimental data for solids of the covalent group, the differences between the theoretical and the experimental results for crystals of the ionic group are remarkable. At the end of this paper we shall discuss these discrepancies.

As the most important by-product of our analysis, we shall study the semiconductor-vacuum interface; this study will be a way of checking our description of the semiconductor wave functions by comparing our results for the surface-state energies with those obtained by means of more elaborate calculations.[6] Our analysis allows us to write secular equations for the surface-state energies which consider in a clear way the heteropolar character of the (110) surface of the zinc-blende structure.

This paper is organized as follows: In Sec. II we shall present the calculation of the evanescent wave functions at two points of the BZ, which we

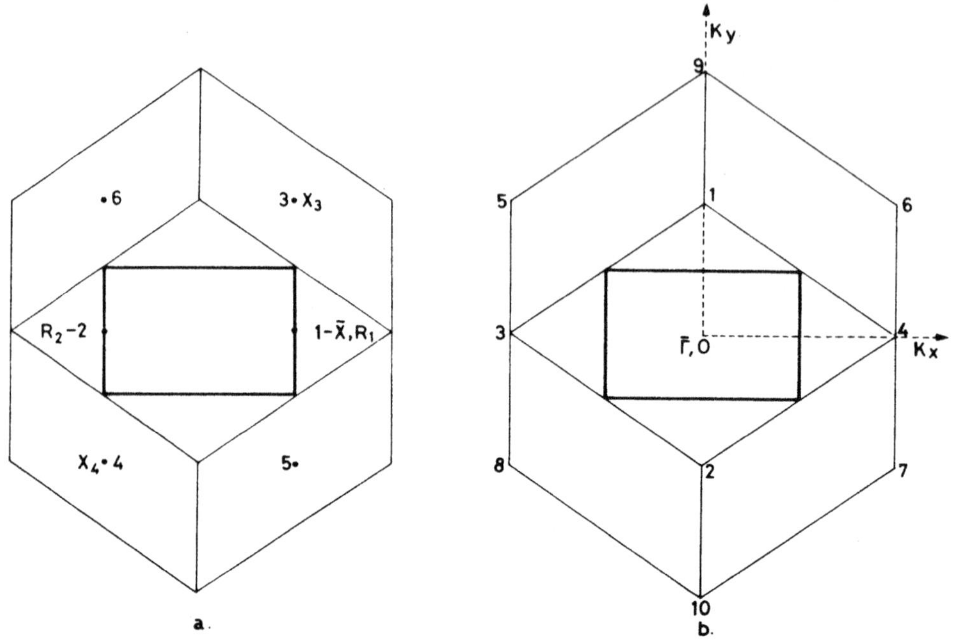

FIG. 1. Two-dimensional BZ (inner rectangle) and projected Jones zone for the (110) face of a fcc lattice. The parallel components of the crystal momentum which were included in the analysis of the $\bar{\Gamma}$ and $\bar{X}$ points of the two-dimensional BZ are shown in (a) and (b), respectively.

have carried out in an analytical way, with a description of the approximations involved in our analysis; in Sec. II we also calculate the surface-state energies at the free surface of seven zinc-blende crystals. In Sec. III the characteristics of the interface states at the *M-S* junction are studied for the two points of the BZ analyzed in Sec. II. Although the detailed calculation of interface states is restricted to metals of high electronic density, the junctions with a metal of low density (Na) are also discussed and some details of the interface states which can be found, are given. The relevance of the interface potential in the calculation of interface states is also discussed. In Sec. IV we apply the analysis of Sec. III to the quantitative study of the barrier height at the *M-S* interface for the (110) surface of seven zinc-blende crystals; our results are discussed and compared with the available experimental data.[2] Finally, in Sec. V we make some concluding remarks.

II. EVANESCENT WAVE FUNCTIONS AT (110) ZINC-BLENDE SURFACES: SURFACE STATES

In this section we shall study analytically the semiconductor wave function at the forbidden energy region. We shall base our analysis on the pseudopotential scheme; in particular, we shall take the pseudopotential form factors from the empirical pseudopotential method.[7]

In principle, the exact study of the matching process at the interface would require including an infinite set of evanescent waves in correspondence to the infinite number of parallel components of the crystal momentum associated with a given $\vec{\kappa}$ of the first two-dimensional BZ.[8] Unfortunately, since we cannot get an infinite set of waves to be matched, we must take a finite number of evanescent functions. As discussed elsewhere,[9] an essential requirement for an approximation to give physically sensible results is that the number of real lines must equal the number of parallel reciprocal-lattice vectors used in the calculation. In order to satisfy the above requirement we shall use here the method employed by Elices *et al.*[9] As discussed by these authors, the evanescent waves at the (111) surface of silicon can be described with enough accuracy by including in the analysis of a particular $\vec{\kappa}$ of the BZ just those parallel components lying inside the (111) projection of the Jones zone.[10,11] In this section we shall use that criterion for studying the evanescent waves at two points of high symmetry of the BZ ($\bar{\Gamma}$ and $\bar{R}$,[12] see Fig. 1).

A. Point $\bar{R}$

Consider the point $R_1(\frac{1}{2}, \frac{3}{2}, 0)$ of the three-dimensional bulk band structure. As one moves along the [110] direction, one comes across three other points of this symmetry, (0, 1, 0), $(\bar{\frac{1}{2}}, \frac{1}{2}, 0)$, and $(\bar{1}, 0, 0)$, respectively, equivalent to $X_4(1, 0, \bar{1})$, $R_2(\frac{3}{2}, \frac{1}{2}, 0)$, and $X_3(0, 1, 1)$ [Fig. 1(a)]. As sketched in Fig. 1(a), at the point R_1 we have the following six parallel components of the crystal momentum lying inside the (110) projection of the Jones zone[13]:

$$|\vec{\kappa}_1\rangle = (\tfrac{1}{2}\sqrt{2}, 0)\ , \quad |\vec{\kappa}_4\rangle = (\tfrac{1}{2}\sqrt{\bar{2}}, \bar{1})\ ,$$
$$|\vec{\kappa}_2\rangle = (\tfrac{1}{2}\sqrt{\bar{2}}, 0)\ , \quad |\vec{\kappa}_5\rangle = (\tfrac{1}{2}\sqrt{2}, \bar{1})\ , \tag{1}$$
$$|\vec{\kappa}_3\rangle = (\tfrac{1}{2}\sqrt{2}, 1)\ , \quad |\vec{\kappa}_6\rangle = (\tfrac{1}{2}\sqrt{\bar{2}}, 1)\ .$$

We shall describe the bulk band structure at X_3 by means of the six plane waves

$$\begin{matrix} (0,1,1) & (0,\bar{1},1) & (0,\bar{1},\bar{1}) & (0,1,\bar{1}) & (1,0,0) & (\bar{1},0,0) \\ |\vec{\kappa}_3\rangle & |\vec{\kappa}_6\rangle & |\vec{\kappa}_4\rangle & |\vec{\kappa}_5\rangle & |\vec{\kappa}_2\rangle & |\vec{\kappa}_1\rangle \end{matrix} \tag{2}$$

which have the parallel components indicated in (2). These waves are only those included by Heine and Jones[14] in their study of the bulk band structure of diamond crystals. In order to find the real lines and evanescent wave functions at this point, we add $\lambda(110) = \lambda\vec{h}$ to each one of the plane waves of (2).

For energies in the principal forbidden gap we find two loops joining through the complex $\vec{\kappa}$ plane the valence-band level E_5 with the conduction levels E_1 and E_3. In diamond crystals[15,16] E_1 and E_3 are degenerate, giving rise to a single degenerate loop. The following wave function is associated to the first loop [in components of the plane waves in (2)]:

$$\phi_1 = e^{-q_1 z}[(i, 1, 0, 0, \tfrac{1}{2}(1+i)\alpha_4, \tfrac{1}{2}(1-i)\alpha_4) + a_1(0, 0, -i, 1, \tfrac{1}{2}(1+i\alpha_4, \tfrac{1}{2}(1-i)\alpha_4)]\ , \tag{3}$$

where

$$q_1 = (1/2h)[(E_1 - E)(E - E_5)]^{1/2}\ ,$$
$$a_1 = \frac{1 - i[(E_1 - E)/(E - E_5)]^{1/2}}{1 + i[(E_1 - E)/(E - E_5)]^{1/2}}\ , \tag{4}$$
$$\alpha_4 = \frac{\sqrt{2}(V_3^S + V_4^A)}{E_1 - V_4^A - \frac{1}{2}h^2}\ .$$

The wave function through the second loop is given by

$$\phi_2 = e^{-q_2 z}[(-i, 1, 0, 0, \tfrac{1}{2}(1-i)\alpha_3, \tfrac{1}{2}(1+i)\alpha_3) + a_2(0, 0, i, 1, \tfrac{1}{2}(1-i)\alpha_3, \tfrac{1}{2}(1+i)\alpha_3)]\ , \tag{5}$$

where

$$q_2 = (1/2h)[(E_3 - E)(E - E_5)]^{1/2}\ ,$$
$$a_2 = \frac{1 - i[(E_3 - E)/(E - E_5)]^{1/2}}{1 + i[(E_3 - E)/(E - E_5)]^{1/2}}\ , \tag{6}$$
$$\alpha_3 = \frac{\sqrt{2}(V_3^S - V_3^A)}{E_3 + V_4^A - \frac{1}{2}h^2}\ .$$

In (3) and (5) V_3^S, V_3^A, and V_4^A are the symmetric and antisymmetric pseudopotential form factors,[7] and z is the direction perpendicular to the surface. As stated above, in diamond crystals the conduction band is degenerate, giving rise to a degenerate loop $a_1 \equiv a_2$; in this case the wave functions along the loop are usually written as $\frac{1}{2}(\phi_1 + \phi_2)$ and $(1/2i)(\phi_1 - \phi_2)$, ϕ_1 and ϕ_2 being the waves in (3) and (5), respectively.

The X_4 point is analyzed in an analogous way. Two wave functions (ϕ_3 and ϕ_4) are calculated after the analysis of the complex band structure at X_4; as they are similar to those given in (3) and (5), we shall omit details.

The bulk band structure at the R_1 point will be described by means of the plane waves

$$\begin{matrix} (\frac{1}{2}, \frac{3}{2}, 0) & (\bar{\frac{3}{2}}, \bar{\frac{1}{2}}, 0) & (\bar{\frac{1}{2}}, \frac{1}{2}, 1) & (\bar{\frac{1}{2}}, \frac{1}{2}, \bar{1}) & (\frac{1}{2}, \bar{\frac{1}{2}}, 0) \\ |\vec{\kappa}_1\rangle & |\vec{\kappa}_1\rangle & |\vec{\kappa}_3\rangle & |\vec{\kappa}_5\rangle & |\vec{\kappa}_2\rangle \end{matrix}\ . \tag{7}$$

The analysis of the complex band structure leads to a loop joining the valence band (E_2) to the conduction band (E_4). The wave function along this loop is given by

$$\phi_5 = e^{-q_5 z}[(0, 1, \tfrac{1}{2}i\gamma_1, \tfrac{1}{2}i\gamma_2, 0) + a_3(\bar{1}, 0, \tfrac{1}{2}i\gamma_1, \tfrac{1}{2}i\gamma_2, 0)]\ , \tag{8}$$

where

$$q_5 = (1/2h)[(E_4 - E)(E - E_2)]^{1/2}\ ,$$
$$a_3 = \frac{1 - i[(E_4 - E)/(E - E_2)]^{1/2}}{1 + i[(E_4 - E)/(E - E_2)]^{1/2}}\ ,$$

and (9)

$$\gamma_1 = \frac{\sqrt{2}(V_3^A + iV_3^S)}{E_4 - \frac{3}{4}h^2}\ ,$$
$$\gamma_2 = \frac{\sqrt{2}(V_3^A - iV_3^S)}{E_2 - \frac{3}{4}h^2}\ .$$

The analysis of the real lines at R_2 gives rise to a single wave function ϕ_6 similar to that in (8). This function completes the set of six evanescent waves needed to perform the matching correctly.

We shall apply our analysis to the study of surface states at the semiconductor-vacuum interface. For the sake of simplicity we shall use the APM throughout. As stated elsewhere[5] this is not a reliable model as far as the calculation of surface-state energies is concerned; nevertheless, as we are interested in (a) checking our description of the evanescent wave functions by comparing our results with calculations which use that model (APM) and (b) understanding the role played by the ionic potential in the surface-state calculation, we shall use the APM everywhere in this section. In performing the surface-state calculation, the six wave functions described above (ϕ_i, $i = 1, \ldots, 6$) should be matched to six plane waves at the vacuum side.

A standard analysis gives rise to a secular equation which can be factored into the determinants

$$\det[D \pm (V_3^A/V_3^S)Z] = 0 , \quad (10a)$$

where D (diamond) and Z (zinc-blende) are the following matrices:

$$D = \begin{vmatrix} d_1 + id_2 & -d_1 + id_2 & d_5 \\ d_3 - id_4 & 0 & d_6 \\ 0 & d_3 + id_4 & -d_6 \end{vmatrix} \quad (10b)$$

and

$$Z = \begin{vmatrix} d_1 + id_2 & d_1 - id_2 & 0 \\ 0 & 0 & d_6 \\ 0 & 0 & d_6 \end{vmatrix} , \quad (10c)$$

where d_i, $i = 1, \ldots, 6$, are given by the following expressions:

$$\begin{aligned} d_1 &= [-\sqrt{2}\, V_3^S/2(E_1 - \tfrac{1}{2}h^2)](1 + a_1)(1 + \tfrac{1}{2}ih\mathfrak{L}_1) , \\ d_2 &= [\sqrt{2}\, V_3^S/2(E_1 - \tfrac{1}{2}h^2)](1 + a_1)(1 - \tfrac{1}{2}ih\mathfrak{L}_1) , \\ d_3 &= 1 - \tfrac{1}{2}ih\mathfrak{L}_3 , \\ d_4 &= -a_1(1 + \tfrac{1}{2}ih\mathfrak{L}_3) , \\ d_5 &= (1 - a_3) - ih\mathfrak{L}_1(1 + a_3) , \\ d_6 &= [\sqrt{2}\, V_3^S/(E_4 - \tfrac{3}{4}h^2)](1 + a_3) , \end{aligned} \quad (11)$$

with $\mathfrak{L}_i = (W_0 + \kappa_i^2 - E)^{-1/2}$, $i = 1, 3$ (W_0 fixes the vacuum level). In obtaining Eqs. (10) we have performed the following approximations: First, we have supposed the small splitting of the conduction band at the X point to be negligible, that is, $E_1 \sim E_3$ and $a_1 = a_2$ in expressions (3)–(6), and second, we have taken $V_4^A \sim 0$.

We want to stress that the secular equation (10a) gives rise to a degenerate surface state for covalent solids (diamond crystals), since in this case $V_3^A = 0$ and Eq. (10a) then becomes $[\det(D)]^2 = 0$. This agrees with previous theoretical analysis.[15,16] The ionic component V_3^A leads to an interaction between both surface states and then to the splitting of the two surface states, degenerate in the covalent limit. This shows, therefore, that Eq. (10a) includes clearly the heteropolar character of the (110) surface of zinc-blende solids.[17]

It is worthwhile studying the secular equation (10a) within the lowest approximation (two-band limit). This approximation consists in neglecting the (111) Fourier component of the crystal pseudopotential. In this case $d_1 = d_2 = d_6 = 0$, and Eq. (10a) can be factored in the following secular equations:

$$a_3 = (1 - ih\mathfrak{L}_1)/(1 + ih\mathfrak{L}_1) , \quad (12a)$$

$$a_1 = i(1 - \tfrac{1}{2}ih\mathfrak{L}_3)/(1 + \tfrac{1}{2}ih\mathfrak{L}_3) , \quad (12b)$$

$$a_1 = -i(1 - \tfrac{1}{2}ih\mathfrak{L}_3)/(1 + \tfrac{1}{2}ih\mathfrak{L}_3) , \quad (12c)$$

which correspond, respectively, to three one-dimensional models.[18] The meaning of Eqs. (12) can be clearly understood. Equation (12a) is related to the loop at the R_1 point and therefore corresponds to a one-dimensional model in which the bulk band gap at R_1 is described by means of an effective pseudopotential Fourier component $V_{eff}(220)$; in this approximation we retain the energy levels calculated by means of more elaborate models,[7] in the spirit that our approach describes the essential features of the wave functions and thus of the matching process. Equations (12b) and (12c) are typical for a two-dimensional system[18,19] and are related to the two loops of the complex band structure at X. In this approximation we have decoupled the points X and R in such a way that a surface state could appear at one of these points for energies corresponding to the allowed energy region at the other point; in such a case the interaction between both points, introduced by the (111) pseudopotential Fourier component, would change the surface state into a resonance. Finally, we stress that this approximation does not take into account the heteropolar character of the solid as the $V(220)$ form factor is purely covalent and all information on the ionic character of the pseudopotential is contained in the (111) Fourier component neglected here.

B. Point $\bar{\Gamma}$

Two points of high symmetry of the bulk band structure project into $\bar{\Gamma}$ [Fig. 1(b)], $\Gamma(000)$ and $X(110)$. A standard calculation gives the lines of real energy and the wave functions along them. Figure 2 shows the real lines in the region of the fundamental gap. Omitting details, we shall restrict ourselves to the discussion of the secular equations for surface-state energies in the lowest approximation (two-band model). As discussed in Sec. II A, this approximation consists in neglecting the terms involving the (111) form factors. We shall describe just the Σ_1 symmetry which is the only one giving surface states.[6] For this symmetry the surface states are given by

$$b_1 = (1 - ih\mathfrak{L}_0')/(1 + ih\mathfrak{L}_0') , \quad (13a)$$

$$b_2 = (1 - ih\mathfrak{L}_1')/(1 + ih\mathfrak{L}_1') , \quad (13b)$$

where

$$b_1 = \frac{1 - i[(E_2' - E)/(E - E_3')]^{1/2}}{1 + i[(E_2' - E)/(E - E_3')]^{1/2}} , \quad (14a)$$

$$b_2 = \frac{1 - 2i[(E_5' - E)/(E - E_4')]^{1/2}\{[\tfrac{1}{2}(E_5' + E_6') - E]/(E_6' - E)\}^{1/2}}{1 + 2i[(E_5' - E)/(E - E_4')]^{1/2}\{[\tfrac{1}{2}(E_5' + E_6') - E]/(E_6' - E)\}^{1/2}} , \quad (14b)$$

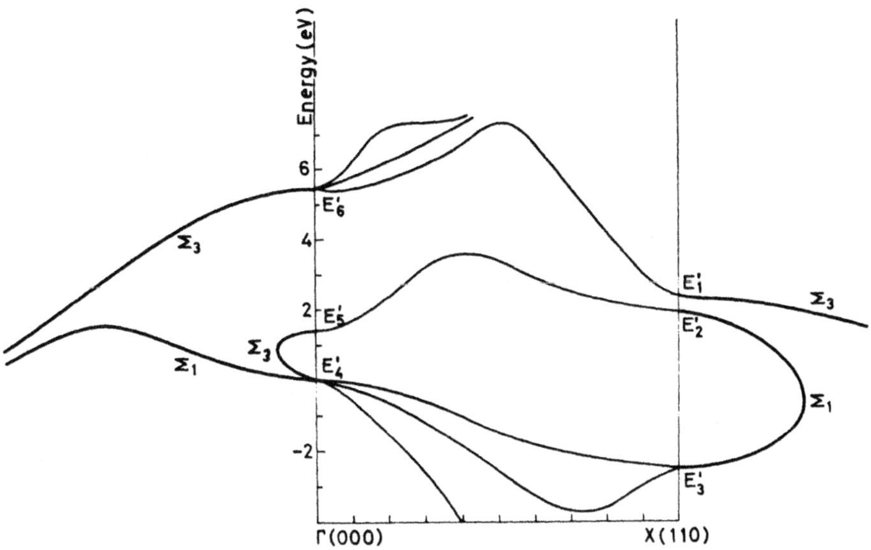

FIG. 2. Real lines of GaAs, along the Γ(000)–X(110) direction, included in the analysis of the $\bar{\Gamma}$ point. The symmetries attached to the lines of real energy are those corresponding to the diamond lattice (see text).

while the $\mathcal{L}_i'$ are

$$\mathcal{L}_0' = (W_0 - E)^{-1/2} \quad \text{and} \quad \mathcal{L}_1' = (W_0 + \kappa_1'^2 - E)^{1/2} ,$$

$|\kappa_1'\rangle = (0, 1)$. The energy levels $E_1', \ldots, E_6'$ (X_5, X_1, X_3, Γ_{15}, Γ_1, and Γ_{15} in the usual terminology) are specified in Fig. 2. In the covalent limit Eq. (13a) corresponds to the Σ_1 symmetry[20] whereas Eq. (13b) belongs to the symmetry Σ_3. The meaning of these equations is clear. The first equation is associated to the loop at the X point (Fig. 2), while Eq. (13b) is associated to the loop at Γ (Fig. 2). This can be seen in a clearer way by noting that the level E_6' is, in ionic crystals, at much higher energies than E_4' and E_5', in such a way that

$$b_2 \simeq \frac{1 - \sqrt{2}\, i[(E_5' - E)/(E - E_4')]}{1 + \sqrt{2}\, i[(E_5' - E)/(E - E_4')]} .$$

Finally, we stress that the fact that the lowest approximation decouples the loops at X and Γ implies that we could obtain surface states below the fundamental gap,[15] in the energy region $E' \leq E \leq E_4'$; however, in higher approximations such a surface state would emerge into a resonance state.

C. Surface-state energies and discussion of the approximations

In order to check the approximations we have performed in describing the semiconductor evanescent wave functions, we have calculated surface-state energies for seven zinc-blende compounds; this calculation allows us to compare our results with those obtained by means of more elaborate approaches, in particular the calculation of Ball and Morgan.[8] In Table I we report our results (points $\bar{\Gamma}$ and $\bar{X}$) and those obtained by Ball and Morgan; the agreement between both theoretical calculations is greater for the $\bar{\Gamma}$ point. This seems reasonable since in this point we have described the evanescent wave functions by means of more parallel components $|\vec{\kappa}\rangle$ than in the case of $\bar{X}$. Note that at the $\bar{\Gamma}$ point of GaAs we have found a single surface state, whereas Ball and Morgan find two surface states; the reason is that those authors obtain the surface state close to the top of the valence band, while in our case the surface state is just appearing.

Finally, it is worth noting that the point $\bar{X}$ could be a measure of the ionic character of the solid; as discussed previously in this section, Eq. (10a) gives the surface states which are degenerate in the covalent limit and split in the ionic limit. The results obtained with that equation are reported in Table I and display an increasing gap on going from less to more ionic crystals; in other words, as the ratio V_3^A/V_3^S increases.

Our results for surface-state energies indicate that our description of the semiconductor evanescent wave functions is good enough for the particular case we shall study in Sec. III, namely, the continuum density of states characteristic of junctions with metals of high density (Al). This point will be cleared up in Sec. IV.

III. ELECTRONIC STATES AT METAL-SEMICONDUCTOR INTERFACE

In this section we shall discuss the characteristics of the interface states at the M-S junction.

TABLE I. Surface-state energies (eV) at the points $\bar{\Gamma}$ and $\bar{X}$ of the two-dimensional BZ of the (110) zinc-blende surface for some crystals. In the cases of GaAs and InP, (a) corresponds to Ref. 6 and (b) to the present calculation; the results for all of the other solids are those calculated in this work. The energies are referred to the top of the valence band for each semiconductor.

	GaAs		InP						
	(a)	(b)	(a)	(b)	GaP	InSb	CdTe	ZnSe	ZnS
$\bar{\Gamma}$	0.272	···	0.272	0.05	0.05	0.35	1.05	0.85	0.8
	1.088	0.8	1.224	0.9	1.3	0.4	1.2	1.6	2.2
$\bar{X}$	0.544	1.0	0.408	0.9	0.92	0.9	−0.06	−0.2	−0.1
	1.088	1.48	1.088	1.44	43	1.37	2.02	2.4	2.55

The metal will be described by the Sommerfeld model. Depending on the width of the metallic band we shall find different types of interface states; namely, virtual states[3] for metals of high density, Breit-Wigner resonances[3] for narrow-band metals, and even true surface states for metals of very narrow band. All of these cases will be discussed by means of particular examples for the points $\bar{X}$ and $\bar{\Gamma}$ and for junctions with different types of metals. At the end of this section we shall discuss the relevance of the interface potential for the density of interface states at the M-S junction.

In order to investigate the density of interface states we shall use the surface-Green's-function method.[21] In this scheme the density of interface states minus the density of states associated with the hard cores of both media is directly given by (for a given $\vec{\kappa}$)

$$\rho(E,\vec{\kappa}) = -\frac{1}{\pi}\frac{d}{dE}\arg\det \mathcal{G}_s^{-1}(E^+) , \qquad (15)$$

where $\mathcal{G}_s^{-1}$ is the inverse matrix of the surface Green's function (see Ref. 21 for details). There are many ways to calculate $\mathcal{G}_s$; once we have analyzed the semiconductor evanescent wave functions and the matching at the interface, the most straightforward way to calculate $\mathcal{G}_s$ is to get it from the secular equation obtained through the matching at the interface.[22] Nevertheless, to avoid details we shall derive the determinant of $\mathcal{G}_s^{-1}$ in an heuristic way.

A. Point $\bar{X}$

The surface-state energies were calculated by means of Eq. (10a). Keeping in mind the definition of $\rho(E)$ in (15), the density of states is

$$\rho(E,\bar{X}) = -\frac{1}{\pi}\frac{d}{dE}\arg\det\left(D \pm \frac{V_3^A}{V_3^S} Z\right) + \frac{1}{\pi}\frac{d}{dE}\arg\det\left(D^0 \pm \frac{V_3^A}{V_3^S} Z^0\right) , \qquad (16)$$

where D^0 and Z^0 are the matrices D and Z calculated by setting $\mathcal{L}_i = 0$ in Eq. (11), that is, the semiconductor joined to an infinite potential barrier (hard core). It is worth noting that the right-hand side of (16) includes contributions coming from possible surface states associated with the infinite-barrier case (hard-core surface states); these contributions should be neglected as they cancel with the other terms involved in the definition of $\rho(E)$,[21,23] which we have omitted.

In this case we shall also analyze in detail the lowest approximation (two-band limit), since, in this limit, Eq. (16) factors into three terms,

$$\rho(E,\bar{X}) = \rho_1 + \rho_2 + \rho_3 , \qquad (17)$$

where

$$\rho_1(E) = -\frac{1}{\pi}\frac{d}{dE}\arg\frac{a_3 - (1 - ih\mathcal{L}_1)/(1 + ih\mathcal{L}_1)}{a_3 - 1} , \qquad (18a)$$

$$\rho_2(E) = -\frac{1}{\pi}\frac{d}{dE}\arg\frac{a_1 - i(1 - \frac{1}{2}ih\mathcal{L}_3)/(1 + \frac{1}{2}ih\mathcal{L}_3)}{a_1 - i} , \qquad (18b)$$

$$\rho_3(E) = -\frac{1}{\pi}\frac{d}{dE}\arg\frac{a_1 + i(1 - \frac{1}{2}ih\mathcal{L}_3)/(1 + \frac{1}{2}ih\mathcal{L}_3)}{a_1 + i} . \qquad (18c)$$

In the case of semiconductor-vacuum interfaces, it is clear that Eqs. (17) give the density associated with the surface states calculated by means of Eq. (10a) minus that quantity associated with the infinite-barrier case (hard cores).

Equations (17) allow a straightforward discussion of the different types of interface states one can find at the M-S junction. In those equations, the magnitudes $\mathcal{L}_1$ and $\mathcal{L}_3$ can be either imaginary or real quantities, telling us whether the wave propagates or does not propagate, respectively, at the metal side. If the metallic band is wide enough, both quantities will be imaginary, giving a continuous density of virtual states along the forbidden gap (Fig. 3) and displaying no structure (this is the case usually analyzed[3,24,25]); the total density of states is $\frac{1}{2}$ for each loop. If the metal has a narrow band, it could happen that $\mathcal{L}_3$ would

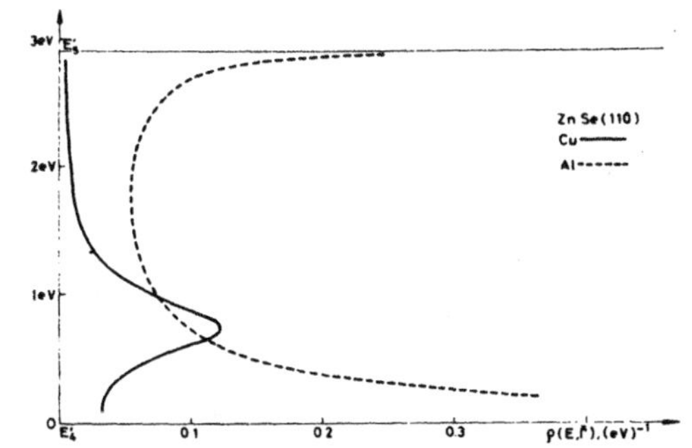

FIG. 3. Broken line: density of virtual states for the Al-ZnSe(110) junction. Solid-line: Breit-Wigner resonance characteristic of the junction Cu-ZnSe(110). The results correspond to the σ_3 symmetry of the $\overline{\Gamma}$ point.

be real and $\mathcal{L}_1$ imaginary. In this case we would find "true surface states" for the loops associated with the X points and a density of virtual states for the R points; then, if we introduce the interaction between X and R, we find a resonance state of the Breit-Wigner type (Fig. 3). In the last case, the total density of states is one-half of the virtual state and one true surface state. In Fig. 4 we plot, for the junction ZnS-Pt, two distinct energy regions at the $\overline{X}$ point; in region II we can find resonance surface states and in region III true surface states. These features of the density of states at the M-S junction indicate that the characteristics of the contact depend on the particular metal.

B. Point $\overline{\Gamma}$

The lowest approximation at this point allows us to write the density of interface states, as in $\overline{X}$, in the following form:

$$\rho(E,\overline{\Gamma}) = \rho_1 + \rho_2 + \rho_3 \ , \tag{19}$$

where in this case the terms ρ_i, $i = 1, 2, 3$, are given by

$$\rho_1(E) = -\frac{1}{\pi}\frac{d}{dE}\arg\frac{b_1 - (1 - ih\mathcal{L}_0')/(1 + ih\mathcal{L}_0')}{b_1 - 1} \ , \tag{20a}$$

$$\rho_2(E) = -\frac{1}{\pi}\frac{d}{dE}\arg\frac{b_2 + (1 - ih\mathcal{L}_1')/(1 + ih\mathcal{L}_1')}{b_2 + 1} \ , \tag{20b}$$

$$\rho_3(E) = -\frac{1}{\pi}\frac{d}{dE}\arg\frac{b_3 - (1 - ih\mathcal{L}_1')/(1 + ih\mathcal{L}_1')}{b_3 - 1} \ . \tag{20c}$$

In this case we can also have the types of interface states discussed in Sec. III A. The virtual states appear for the loop associated with the parallel component of the crystal momentum of lowest energy (that is, b_1), while for the loops b_2 and b_3 we can find true surface states. As in $\overline{X}$, the interaction between the loops b_1, b_2, and b_3 lead to a resonance of the Breit-Wigner type (Fig. 3). In Fig. 4 we have sketched the situation for the junction ZnS-Pt; in this case we have at $\overline{\Gamma}$ an energy region where one can find virtual states (region I of Fig. 4) and a much larger region (II of Fig. 4) where resonance states can be obtained. These features of the density of interface states indicate, as in $\overline{X}$, that the characteristics of the contact can depend on the particular metal.

As we are interested in the role played by the ionic character of the semiconductor at the M-S junction we shall restrict ourselves to junctions with metals of high density, for which the analysis of the density of interface states is much simpler. In Table II we have reported our results for the density of interface states for junctions of several zinc-blende crystals (InSb, GaAs, InP, GaP, CdTe, ZnSe, and ZnS) with Al; the calculation has been done for an energy which is $\epsilon_0 = \frac{1}{3}E_g$, where the energy refers to the top of the valence band and E_g is the thermal gap for each semiconductor (this energy is where the Fermi level is supposed to be placed for most of the junctions,[1] as will be discussed in Sec. IV). Although we have performed our calculation by means of the highest approximation we have studied, it is worth pointing out that the density of states does not vary very much (20%)

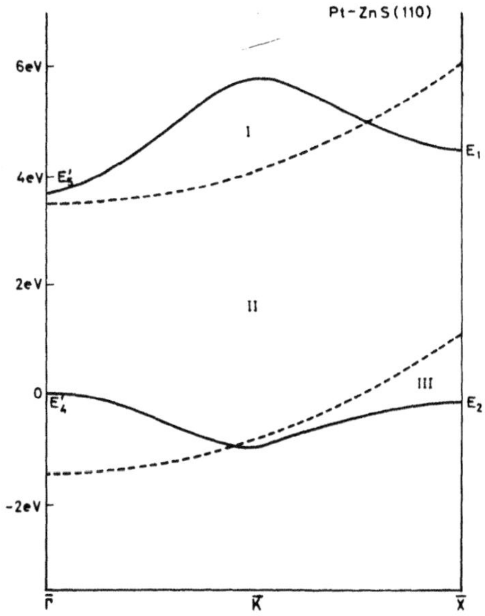

FIG. 4. Bulk band structure of ZnS projected on the (110) face (solid line) and the two free electron parabolas (see text) of lowest energy for Pt (broken line). Different regions are to be noted: in region I only virtual states can be found, whereas in regions II and III resonances or true surface states, respectively, can be obtained.

if we use the two-band model. This is in contrast to the case of the semiconductor-vacuum system, where both approximations differ qualitatively [at $\bar{X}$ the lowest approximation gives degenerate surface states, while the highest approximation splits those states by an amount proportional to the ionic character of the crystal; see Eq. (10a)]. As a consequence, the density of interface states is not directly a function of the ionic potential but only of the band effective masses and of the magnitude of the forbidden gap (both quantities are functions of the strength of the atomic potential). Although the calculated density of states decreases as the ionic character of the solid increases, the differences between the densities corresponding to GaAs and ZnS (Table II) are not as large as one could expect from the experimental data.[1,26] One feature to be noted is the slight difference between the densities calculated at the $\bar{\Gamma}$ point and at the $\bar{X}$ point; this independence of $\rho(E, \vec{\kappa})$ from the particular $\vec{\kappa}$ has also been found for the Si(111)-Al junction.[22] As a final remark, we stress that these features are also valid for all junctions with metals of high density; the density of states being, in this case, almost independent of the particular metal.

C. Interface potential

So far we have assumed the APM; the question now is the quality of this model for the M-S junction. We know that for the study of the free surface of semiconductors the APM is a poor model,[5] at least for the electronic surface-state energies, which are very sensitive to the profile of the potential at the surface.

In order to elucidate this point we have studied a model with the following features: (a) As we shall study only the case of metals with a wide Sommerfeld band, the model displays a density of "virtual states"; (b) we shall use a one-dimensional model; this point is justified by the reliability of the lowest approximation, as discussed earlier; and (c) we shall use a linear barrier as a model of the interface potential.[27,28]

TABLE II. Density of interface states (eV^{-1}) at the metal (al)–(110-zinc-blende junction for some semiconductors. The density of states has been calculated for an energy equal to $\frac{1}{3} E_g$, referred to the top of the valence band, where E_g is the thermal gap for each semiconductor. Results are given for the points $\bar{\Gamma}$ and $\bar{X}$ of the two-dimensional BZ.

	$\bar{\Gamma}$	$\bar{X}$
GaP	0.224	0.43
GaAs	0.348	0.49
InSb	0.88	0.62
InP	0.38	0.46
CdTe	0.28	0.40
ZnSe	0.185	0.33
ZnS	0.117	0.29

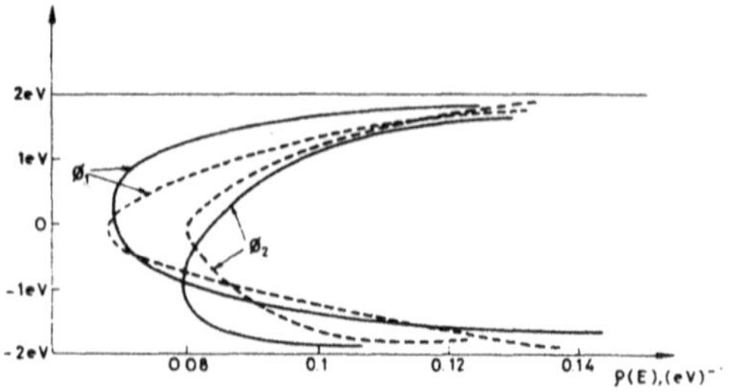

FIG. 5. Density of virtual states at the M-S junction for a one-dimensional model, calculated using either the APM (broken line) or the linear-barrier model (solid line). Two cases are shown: $\phi_1 = 180°$ and $\phi_2 = 0°$, that is, the periodic potential is finished at a minimum (bonding case) or a maximum (antibonding case), respectively (see Ref. 5).

We have studied the density of virtual states for a semiconductor like GaAs and a metal of a Sommerfeld band of 5.0 eV; we then varied the slope of the linear potential and thus the width of the interface region. The results for two extreme cases are drawn in Fig. 5. We have used an interface region 8 Å wide; this is clearly unphysical but it amounts to a test which is even stricter than physical reality. The results show that the densities of virtual states of the Fermi level corresponding to that model and to the APM do not differ by more than 5%. Moreover, the total density of states at the semiconductor gap is nearly independent of the interface potential, the differences being negligible. Our conclusion, then, is that the density of interface states is not very sensitive to interface potential, justifying in this way the calculations we have performed in this section.

IV. CALCULATION OF THE BARRIER HEIGHT: COMPARISON WITH EXPERIMENTAL DATA

In order to obtain the relation between the barrier height at the junction and the metal work function, we shall follow the work of Cowley and Sze.[26] First, we consider the experimental data reported by Mead,[29] according to which the Fermi level for several dielectrics in contact with Au is placed, as stated in Sec. III, at $E_F = \frac{1}{3}E_g$. (A theoretical calculation of the Fermi level at the junction would require a complete analysis of the charge redistribution throughout the whole spectrum of occupied states.[23,24]) The fact that the density of interface states, D, is almost independent of the metal (for metals of high density), and that the distance δ between these (virtual) interface states and the opposed charge placed at the metal surface is also independent of the particular metal, as will be discussed later, allow us to write the barrier height

TABLE III. Calculated values of the slope S and the electronic density of resonance surface states for different III-V and II-VI compounds. The density of resonance surface states has been calculated for an energy referred to the top of the valence band equal to $\frac{1}{3}E_g$, where E_g is the thermal gap for each semiconductor.

	InSb	GaAs	InP	GaP	CdTe	ZnSe	ZnS
$10^{-14}D$ (eV^{+1} cm^{+2})	5.05	3.72	3.45	3.12	2.34	2.28	1.96
S	0.05	0.1	0.09	0.12	0.14	0.155	0.18

ϕ_{Bn} (for type n semiconductors) at the junction as a linear function of the metal workfunction[28] ϕ_M:

$$\phi_{Bn} = S\phi_M + b \ , \tag{21}$$

where

$$S = (1 + 4\pi D\delta e^2)^{-1} \tag{21a}$$

b is a constant which can be determined through experimental measurements[29] and e is the electron charge.

In Sec. III we discussed the calculation of the density of interface states for seven zinc-blende solids joined to a metal of high density (Al). As shown in Table I, the density of interface states does not depend very much on the particular $\vec{\kappa}$. This allows us to calculate the total density of states for a given energy in the following way:

$$D(E_F) = \frac{1}{N}\sum_{\vec{\kappa}} \rho(\vec{\kappa}, E_F) \ , \tag{22}$$

where, in the present case, the sum over $\vec{\kappa}$ is restricted to $\overline{\Gamma}$ and $\overline{X}$, and where N is the number of terms included in the sum (in this case $N = 2$).

In Table III we have reported our results for the total density of states. As pointed out in Sec. III, this density decreases as the ionic character of the solid increases; nevertheless, the difference between the densities corresponding to the covalent solid (GaAs) and those found for the ionic crystals (ZnS) are not as large as expected from experimental measurements.[26]

We shall now discuss how to calculate the effective distance between the virtual states and the opposed charge at the metal surface, which will complete the information needed to calculate the slope S of Eq. (21). In order to estimate δ, let us first consider the screening of the resonance charge placed at the semiconductor half-space. We assume that this screening can be described by a bulk electronic dielectric function; in particular we shall use that proposed by Inkson[30] (which is a simplified version of that given by Penn[31])

$$\epsilon_S = 1 + \frac{\epsilon_\infty - 1}{1 + (\epsilon_\infty - 1)k^2/k_s^2} \ , \tag{23}$$

where k_s and ϵ_∞ are the Fermi-Thomas wave vectors and the static dielectric constant, respectively. Bearing in mind that the resonance charge decays as $\exp(-2q|z|)$, where q is directly obtained from the equations of the loops discussed in Sec. II, and symmetrizing into the whole space,[32] the effective distance from the charge to the semiconductor surface turns out to be approximately $1/k_s + 1/2\epsilon_\infty q$. We may suppose that the screening of the induced charge in the metallic side is describe by the usual dielectric function

$$\epsilon_m = 1 + k_m^2/k^2 \ , \tag{24}$$

where k_m is the Fermi-Thomas wave vector of the metal. Finally, we obtain the total effective distance to be inserted into (21),

$$\delta = \frac{1}{2\epsilon_\infty q} + \frac{1}{k_s} + \frac{1}{k_m} \ . \tag{25}$$

This expression indicates that the effective distance is a sum of a distance $(k_s^{-1} + k_m^{-1})$ within which there is no screening of the charge plus the distance $1/2\epsilon_\infty q$ which is nothing but the effective distance where the charge, completely screened by the crystal, is located.

This result allows us to include the effect of the ionic screening. Since the effective distance related to the ionic screening [this distance is characterized by the distance between equivalent (110) planes in the solid] can be expected to be larger than k_s^{-1}, we can write δ as follows:

$$\delta \approx \frac{1}{2\epsilon_0 q} + \frac{1}{k_s} + \frac{1}{k_m} \ , \tag{26}$$

where ϵ_0 includes the ionic screening in the long-wavelength limit.

The value of $k_s^{-1} + k_m^{-1}$ is very approximately 1 Å and is largely independent of the particular metal, as was assumed at the beginning of this section. The term $1/2\epsilon_0 q$ can be calculated as follows: As discussed in Sec. III, the density of virtual states can be calculated by considering the different loops, included in the analysis independently; this allows us to attach to each loop an effective distance related to the imaginary part of the component of the crystal momentum perpendicular to the surface, q. The value of δ does not depend very much on the particular semiconductor since (a) the value of $k_s^{-1} + k_m^{-1}$ dominates and (b), on the other hand,

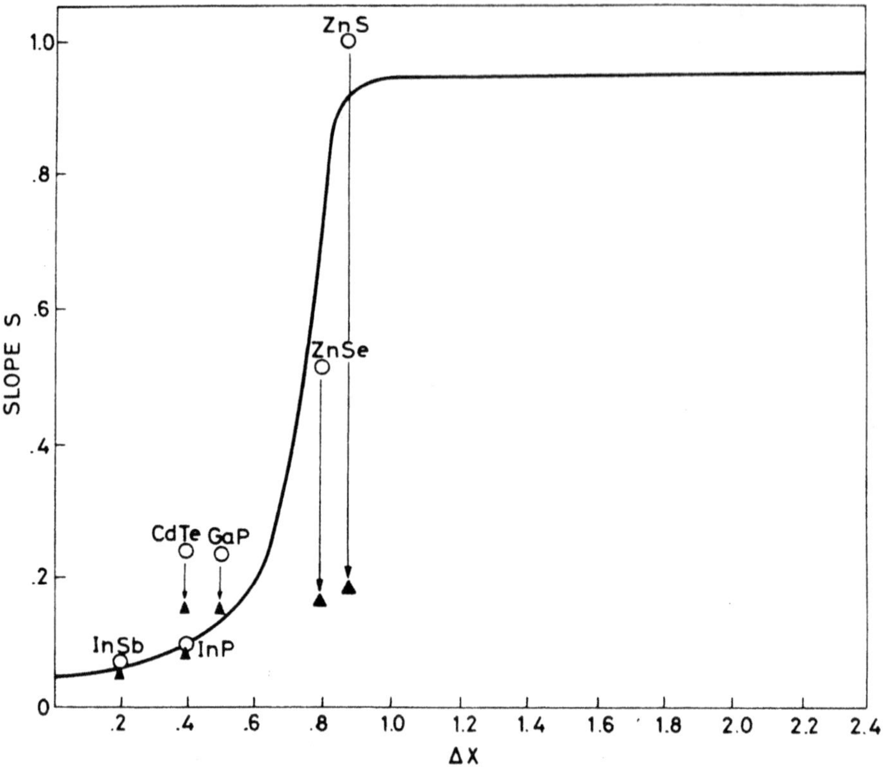

FIG. 6. Theoretical values (▲) and experimental data (○) for the slope S plotted as a function of the lattice electronegative difference ΔX. The curve is taken from Ref. 2.

$(\epsilon_0 q)^{-1}$ is almost constant because as the ionic character of the crystal increases, ϵ_0 decreases and q increases, keeping the value of $E_F q$ almost constant.

Now we are able to calculate the magnitude $D\delta$ as the sum of the charges associated with each loop included in the particular $\vec{\kappa}$ under study, multiplied by the effective distance associated to that loop. Thus $D\delta$ for the energy E_F is just

$$D\delta(E_F) = \frac{1}{N}\sum_{\vec{\kappa}, i}^{N} \rho(\vec{\kappa}, i; E_F)\delta(\vec{\kappa}, i; E_F) \ , \qquad (27)$$

where N is the number of $\vec{\kappa}$ considered and i runs over the loops included in the analysis of each $\vec{\kappa}$. In our case $N = 2$, and the $\vec{\kappa}$ are $\overline{X}$ and $\overline{\Gamma}$.

Our results for the slope S are given in Table III. The theoretical values of the slope are plotted in Fig. 6, together with the experimental data collected by Kurtin *et al.*[2] The present results differ slightly from those reported in I, because in that case we included just one point in the analysis ($\overline{\Gamma}$). The inclusion of $\overline{X}$ has increased the values of the total density of interface states and then lowered the values of S; nevertheless, the discrepancies are not considerable. As in I, there is an excellent agreement between theory and experiment for crystals of the covalent group (InSb, GaAs, and InP) and the discrepancy between theory and experiment increases with increasing ionicity (ZnS).

V. CONCLUDING REMARKS

In this paper we have presented a study of the properties of the (110) zinc-blende surfaces and of their junction with different metals. Although our main purpose was the analysis of the covalent-ionic transition in the M-S interface, the study of the free surface which we have carried out has allowed us to analyze clearly the heteropolar character of the (110) surface. Regarding the interface density of states for junctions with metals of high density, we have proved that the lowest approximation gives quite accurate results; this conclusion shows that since this approximation does not involve the (111) pseudopotential form factors, the ionic character of the crystal is not of prime relevance in calculating densities of virtual states.

As for the covalent-ionic transition, we have found that our model, based on a one-electron scheme within the pseudopotential framework, is unable to explain the data reported by Kurtin *et al.*[2]; our model fails in treating highly ionic compounds. Since the slope S calculated in this paper is nearly independent of the ionic character of the solid, one can ask whether the agreement between theory and experiment for covalent solids is fortuitous, the one-electron model being basically inadequate for the study of metal-semiconductor junctions. Nevertheless, it should be noted that the covalent-ionic transition has been observed in different kinds of experimental properties, as discussed by Kurtin *et al.*[2] Their results regarding optical absorption (excitons and nondirect transitions) tend to show that the one-electron approach is not a proper scheme for the study of properties of ionic solids. We think that the experimental data reported by Kurtin *et al.*,[2] the model discussed in this paper, and other theoretical works about many-body effects in ionic solids[33] point in the same direction; namely, they show that the one-electron scheme is an adequate framework for the treatment of some properties of covalent solids, but cannot explain most of the characteristics of the ionic solids (the correlation effects are greatly enhanced as the ionicity of the solids, and then the localization of the electrons, increases). Hence we think that the metal-ionic junctions should be studied by including many electron effects and, of course, the entire charge of the system, that is, the valence electrons[34] and the resonance surface states.

ACKNOWLEDGMENTS

We are indebted to Dr. F. García-Moliner and Dr. J. Rubio for many useful discussions and for the critical reading of the manuscript.

---

*Present address: Department of Physics, University of California, Berkeley, Calif. 94720.

[1]S. M. Sze, *Principles of Semiconductor Devices* (Wiley, New York, 1969).
[2]S. Kurtin, T. C. McGill, and C. A. Mead, Phys. Rev. Lett. 22, 1433 (1969).
[3]V. Heine, Phys. Rev. 138, A1689 (1965).
[4]F. Flores, E. Louis, and F. Yndurain, J. Phys. C 6, L465 (1973), hereafter referred as I.
[5]F. Flores, E. Louis, and J. Rubio, J. Phys. C 5, 3469 (1972).
[6]G. Ball and D. J. Morgan, Phys. Status Solidi B 50, 199 (1972).
[7]M. L. Cohen and T. K. Bergstresser, Phys. Rev. 141, A789 (1966).
[8]V. Heine, Proc. Phys. Soc. Lond. 81, 300 (1963).
[9]M. Elices, F. Flores, E. Louis, and J. Rubio, J. Phys. C 7, 3020 (1974).
[10]This has been proved to work also for the (111) surface of zinc-blende crystals (see Ref. 11). Nevertheless, we are aware of the important differences between that surface and the (110), which could be of primary importance in choosing the approximations to describe the evanescent waves.
[11]E. Louis and M. Elices, Phys. Rev. B 12, 618 (1975).
[12]Points of two-dimensional symmetry will be denoted by capital letters with bars above them.
[13]Hereafter we shall write the plane waves in units of $2\pi/a$, a being the lattice constant.
[14]V. Heine and R. O. Jones, J. Phys. C 2, 719 (1969).
[15]R. O. Jones, Phys. Rev. Lett. 20, 992 (1968).
[16]F. Yndurain and M. Elices, Surf. Sci. 29, 540 (1972).
[17]The heteropolar character of the (110) surface has been previously introduced, although in a rather simplified fashion, by E. Louis and F. Yndurain (see Ref. 25).
[18]F. Flores and J. Rubio, J. Phys. C 6, L258 (1972).
[19]F. Forstmann, Z. Phys. 235, 69 (1970).
[20]R. O. Jones, Ph.D. thesis (University of Cambridge, 1966) (unpublished).
[21]F. García-Moliner and J. Rubio, J. Phys. C 2, 1789 (1969).
[22]J. Rubio, F. Flores, and E. Louis (unpublished).
[23]C. Tejedor, E. Louis, and F. Flores, Solid State Commun. 15, 587 (1974).
[24]F. Yndurain, J. Phys. C 4, 2849 (1971).
[25]E. Louis and F. Yndurain, Phys. Status Solidi B 57, 175 (1973).
[26]A. M. Cowley and S. M. Sze, J. Appl. Phys. 36, 3212 (1966).
[27]The use of the linear barrier to describe the interface potential has been justified for the semiconductor-vacuum system (see Ref. 28).
[28]E. Larrea and F. Flores, Surf. Sci. 49, 339 (1975).
[29]C. A. Mead, Solid State Electron. 9, 1023 (1966).
[30]J. C. Inkson, J. Phys. C 5, 2599 (1972).
[31]D. R. Penn, Phys. Rev. 128, 2093 (1962).
[32]F. Flores, Nuovo Cimento B 14, 1 (1973).
[33]S. T. Pantelides, D. J. Mickish, and A. B. Kunz, Phys. Rev. B 10, 2602 (1974).
[34]A. J. Bennett and C. B. Duke, Phys. Rev. 162, 578 (1967).

Electronic structure of a metal-semiconductor interface*

Steven G. Louie† and Marvin L. Cohen

Department of Physics, University of California, and Inorganic Materials Research Division, Lawrence Berkeley Laboratory, Berkeley, California 94720

(Received 25 August 1975)

The electronic structure of a jellium-Si interface is calculated using a jellium density corresponding to Al and self-consistent Si pseudopotentials. Local densities of states and charge densities are used to study states near the interface. At the Si surface, a high density of extra states is found in the energy range of the Si fundamental gap. These states are bulklike in jellium and decay into Si with a high concentration of charge in the dangling-bond free-surface-like Si state. Truly localized interface states are also found but at lower energies. The calculated barrier height is in excellent agreement with recent experimental results.

I. INTRODUCTION

In this paper we present self-consistent pseudopotential calculations on the electronic structure of a metal-semiconductor interface. The calculations model an Al-Si interface with a jellium potential representing the aluminum-ion potential in contact with the Si (111) surface. This model describes an ideal interface; i.e., there is no oxide layer between the two materials. A local density of states (LDOS) which displays the density of states as a function of distance away from the interface has been calculated for this Al-Si junction. Various states which exist near the interface are identified and discussed in terms of the LDOS and their charge densities. Our calculated interface barrier height is found to be in excellent agreement with recent experimental results.[1] To our knowledge, this is the first realistic self-consistent calculation for a metal-semiconductor interface.

Metal-semiconductor interfaces are of great importance because of their rectifying properties, which are crucial to the operation of many electronic devices. Many experimental efforts have been devoted to the study of their properties. With the advent of recent ultrahigh-vacuum techniques, ideal interfaces can now be fabricated and studied systematically,[1-5] and the detailed electronic structure at the interface can be probed using modern photoemission techniques.[3-5] On the theoretical side, metal-semiconductor interfaces have been the subject of much discussion and speculation.[6-15] Many models have been proposed to explain the interface properties. However, regrettably, past theoretical investigations into their electronic structure have been mostly qualitative or semiquantitative. A clear picture of the electronic structure at a metal-semiconductor interface has yet to emerge.

Experimentally, the electrical barrier height ϕ_B (Schottky barrier) at a metal-semiconductor interface can be accurately determined using many different methods (I-V, C-V, photoelectric, etc.). To avoid confusion over n- and p-type semiconductors, we measure here the barrier height from the Fermi level E_F to the semiconductor conduction band. For covalent semiconductors such as Si and Ge, the barrier height is found to be virtually independent of the metal contact and of the doping in the semiconductor.[1,15,16] Bardeen[6] attributed this behavior of the barrier height to a high density of surface states in the semiconductor band gap; i.e., the filling or emptying of these surface states pins the Fermi level to a nearly constant value. Heine,[7] on the other hand, pointed out that semiconductor surface states cannot exist in the semiconductor gap if this energy range is inside the metallic band. He suggested that the pinning of the Fermi level is due to states of a different type in the semiconductor gap. These states are composed of the states from the tails of the metallic wave functions decaying into the semiconductor side.

Theories[8,10-12] which do not explicitly involve extra states in the semiconductor gap have also been proposed to explain the barrier-height behavior. Inkson,[8,10] using a model-dielectric-function formulation, proposed that the pinning of the Fermi level is due to the narrowing of the semiconductor gap at the interface. According to Inkson, the screening of the valence and conduction bands of the semiconductor is different near the interface. This causes the valence band to bend up and the conduction band to bend down and eventually the bands merge together at the interface for a covalent semiconductor. In addition, Phillips[11,12] claimed that polarizability effects play the dominant role at the metal-semiconductor interface. He suggested that it is the elementary excitations[11] and chemical bonding[12] at the interface which determine the behavior of the Schottky barrier.

The purpose of the present work is to study the electronic structure of a metal-semiconductor interface in detail using the Al-Si junction as a prototype and to gain some insights into the nature of metal-semiconductor Schottky barriers. The model and methods of calculation presented here can be applied to general metal-semiconductor contacts. The main features of this calculation which are absent in previous work are as follows: (i) A realistic interface is constructed through a jellium-semiconductor model and (ii) the calculation is self-consistent. It is noted that, as in all previously existing self-consistent surface calculations, self-consistency in the present context means self-consistency in the electronic responses to a given structural model.

The remainder of the paper is organized as follows: In Sec. II the model for the interface and the steps in the self-consistent calculations are discussed in detail. In Sec. III the results for the electronic structure of the Al-Si interface are presented. And in the final section, IV, some discussion and conclusions are presented.

II. CALCULATIONS

Our model for an ideal metal-semiconductor interface consists of jellium in contact with a semiconductor described in the pseudopotential formalism. Present experimental and theoretical methods do not allow a detailed determination of the geometry at the metal-semiconductor interface; however, we believe that the important properties of the interface are dominated by the properties of the free electrons residing next to the semiconductor surface. The present model is expected to contain all of the essential features of a metal-semiconductor interface.

The method we have employed to calculate the electronic structure of the Al-Si interface is similar to the method which we have used previously in surface and molecular calculations.[17-20] The main difficulties in calculating the electronic structure of solid interfaces are the following:

(i) Periodicity along the direction perpendicular to the interface is absent. Therefore the established methods for bulk calculations which depend on the periodicity of crystalline solids cannot be used.

(ii) Self-consistency is essential in obtaining realistic solutions. It is necessary to allow the electrons to react to the boundary conditions imposed by the interface and the resulting readjustment and screening is a fundamental part of the problem.

The essence of our method is to retain (artificial) periodicity perpendicular to the interface and thus allow the use of well-established tools in pseudopotential crystal calculations to calculate the interface electronic structure. In addition, the method goes beyond the usual pseudopotential approach through the requirement of self-consistency. This method has been successfully applied to surfaces[18-20] (both for metal and semiconductor surfaces) and molecular[17] calculations.

For the present calculation, we consider a unit cell consisting of a slab of Si with the (111) surfaces exposed to a jellium of Al density on both sides. This cell is then repeated and the electronic structure of the system is calculated self-consistently. The basic idea consists of considering periodic interfaces which are separated by large distances, and then obtaining the essential features of a single interface by calculating the electronic structure of this periodic system. The unit cell used consists of 12 layers of Si plus an equivalent distance of jellium. It is spanned in two dimensions by the shortest lattice vectors parallel to the Si (111) surface, i.e., hexagonal lattice vectors with length 7.26 a.u., and by a long c axis of length $c = 71.1$ a.u. The volume of the cell is equal to 3241 a.u.[3]

With the above geometry, the jellium edge is one-half of a Si-Si bond length away from the atoms of the Si (111) surface. This is a physically reasonable choice since the length of an Al-Si bond is approximately the same as a Si-Si bond. To simulate noninteracting interfaces, the Si and the Al slab sizes have to be chosen such that (a) the bulk properties of the materials are adequately reproduced and (b) the surfaces from opposite sides of the same slab do not interact appreciably. Calculations on the Si (111) surface[18,19] and various test calculations on jellium slabs of Al density showed that the assumed slab thickness which is equivalent to 12 layers of Si satisfies the above requirements well.

The electronic structure of this "periodic" system can now be determined in a self-consistent manner using pseudopotentials. The steps leading to a self-consistent solution are shown in Fig. 1. We expand the electron wave functions in plane waves with reciprocal-lattice vectors $\vec{G}$:

$$\psi_{n\vec{k}}(\vec{r}) = \sum_{\vec{G}} a^{n}_{\vec{k}}(\vec{G})\, e^{i(\vec{k}+\vec{G})\cdot\vec{r}} . \tag{1}$$

This leads to a matrix eigenvalue equation of the usual kind,

$$\sum_{\vec{G}'} (H_{\vec{G},\vec{G}'} - E\delta_{\vec{G},\vec{G}'})\, a_{\vec{k}}(\vec{G}') = 0 , \tag{2}$$

which is solved by standard methods.[21] Here, the Hamiltonian matrix elements are of the form

$$H_{\vec{G},\vec{G}'} = (\hbar^2/2m)|\vec{k}+\vec{G}|^2 \delta_{\vec{G},\vec{G}'} + V_{ps}(\vec{G},\vec{G}') , \tag{3}$$

where $V_{ps}(\vec{G},\vec{G}')$ are the pseudopotential matrix elements. In general the pseudopotentials are nonlocal and energy dependent.[21] However, for bulk Si and Si surfaces, local pseudopotentials are known to yield satisfactory results. Therefore

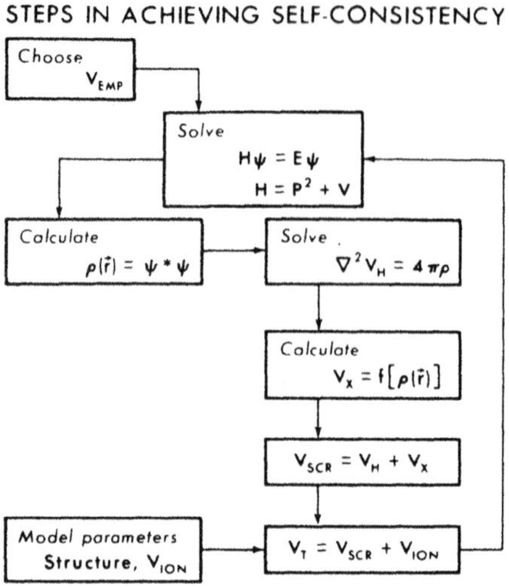

FIG. 1. Self-consistent loop in calculating the electronic structure of a metal-semiconductor interface.

local pseudopotentials will be used throughout for the present calculation.

The self-consistent cycle is initiated by the following potential:

$$V_{\text{start}}(\vec{G}) = S(\vec{G})\, V^{\text{Si}}_{\text{emp}}(|\vec{G}|) + V^{\text{Al}}_{\text{start}}(\vec{G}) \, . \tag{4}$$

The first term is the starting potential for the Si slab and the second term is the starting potential for the Al slab. The Si structure factor

$$S(\vec{G}) = \frac{1}{M}\sum_{\tau_i}^{M} e^{-i\vec{G}\cdot\vec{\tau}_i} \tag{5}$$

describes the positions of the Si atoms in the unit cell. $V^{\text{Si}}_{\text{emp}}(|\vec{G}|)$ are Si atomic pseudopotential form factors derived from empirical bulk calculations.[22] Since empirical form factors are only known for discrete $\vec{G}$ vectors and the $\vec{G}$ vectors are different for different crystal structures, a continuous extrapolation is performed to obtain the form factors corresponding to the new $\vec{G}$ vectors in the interface problem. We fitted a curve of the form

$$V(q) = \frac{a_1(q^2 - a_2)}{e^{a_3(q^2 - a_4)} + 1} \tag{6}$$

to the three form factors for bulk Si, $V(111) = -0.2241$ Ry, $V(200) = 0.0551$ Ry, $V(311) = 0.0724$ Ry, and renormalized it for the different unit-cell volume. The four parameters a_i in Eq. (6) are given in Table I. The potential is normalized to an atomic volume of 270 a.u.[3] and the units are in Ry if q is entered in a.u.

A starting potential for the Al slab is less obvious. We assumed that in zeroth order, the Al electronic charge is uniform and confined completely inside the Al slab. Then the Hartree part of the electron screening will cancel the positive jellium background and the starting potential for the Al slab can be taken to contain only an exchange term

$$V^{\text{Al}}_{\text{start}}(\vec{G}) = -\alpha(3/2\pi)(3\pi^2)^{1/3} e^2 \rho^{1/3}_{\text{jel}}(\vec{G}) \, , \tag{7}$$

where $\alpha = 0.79$ and $\rho^{1/3}_{\text{jel}}(\vec{G})$ are the Fourier components of the jellium density to the one-third power. Here we have replaced the nonlocal Hartree-Fock exchange potential $V_x(r, r')$ by the statistical exchange model of Slater.[23] The choice of $\alpha = 0.79$ will be discussed later. In principle, for a self-consistent calculation, the starting potentials should be unimportant. However, in practice, a good starting potential reduces the number of iterations needed enormously.

From Eq. (2) we obtain the band structure $E_n(\vec{k})$ and the pseudo-wave-functions $\psi_{n\vec{k}}(\vec{r})$. To perform the next step in the self-consistent loop, the total valence charge density

$$\rho(\vec{r}) = 2\sum_n \sum_{\vec{k}} \psi^*_{n\vec{k}}(\vec{r})\, \psi_{n\vec{k}}(\vec{r}), \quad E_n(\vec{k}) \le E_F \tag{8}$$

has to be accurately determined. This requires good convergence in the electron wave functions and a precise location of the Fermi level. To assure good convergence, the electronic wave functions were expanded in a basis set consisting of approximately 270 plane waves. This expansion corresponds to a kinetic-energy cutoff[21] $E_1 = |\vec{G}_{\max}|^2 \approx 2.7$ Ry. In addition, another 300 plane waves were included via Löwdin's perturbation scheme.[21] The total valence charge density was evaluated at 21 $\vec{k}$ points in the irreducible part ($\frac{1}{12}$) of the *two*-dimensional hexagonal Brillouin zone, with the Fermi level determined by demanding charge neutrality in the unit cell. That is, the Fermi level is determined by filling the eigenlevels in the Brillouin zone until the number of occupied levels corresponds to the number of electrons in the unit cell required by charge neutrality.

We note that, for our "periodic" system, we should in principle evaluate the total charge over the whole three-dimensional Brillouin zone. However, for a large elongated cell, as in the present case, the energies and wave functions are quite independent of the component of the $\vec{k}$ vectors along the c direction. As we shall see later, the final

TABLE I. Parameters entering Eqs. (6) and (16) to define the empirical and ionic Si pseudopotentials.

	$V^{\text{Si}}_{\text{emp}}$	$V^{\text{Si}}_{\text{ion}}$
a_1	0.174 59	−0.573 15
a_2	2.221 44	0.790 65
a_3	0.863 34	−0.352 01
a_4	1.534 57	−0.018 07

charge density away from the interface is in accord with the bulk calculations, thus indicating that our sampling in $\vec{k}$ space is sufficiently fine and the wave functions are converged.

Once the valence charge density $\rho(\vec{r})$ is known in terms of its Fourier components $\rho(\vec{G})$, the Hartree-Fock-type screening potentials V_H and V_x can be evaluated easily. V_H, the so-called Hartree screening potential, is the repulsive Coulomb potential seen by an electron and is generated by all the valence electrons. It is related to the valence charge density by Poisson's equation

$$\nabla^2 V_H(\vec{r}) = -4\pi e^2 \rho(\vec{r}) \tag{9}$$

and can be written as a Fourier series

$$V_H(\vec{r}) = \sum_{\vec{G}} V_H(\vec{G})\, e^{i\vec{G}\cdot\vec{r}} , \tag{10}$$

with

$$V_H(\vec{G}) = 4\pi e^2 \rho(\vec{G}) / |\vec{G}|^2 . \tag{11}$$

Physically, over-all charge neutrality in the solid requires that $V_H(\vec{G}=0) = -V_{\text{ion}}(\vec{G}=0)$, where V_{ion} is the ionic potential generated by the positive Si^{+4} ion cores and by the positive jellium slab. Therefore, for the present calculations, we can arbitrarily set $V_H(\vec{G}=0) = V_{\text{ion}}(\vec{G}=0) = 0$. Numerically, however, the divergent character of $V_H(\vec{G})$ and $V_{\text{ion}}(\vec{G})$ for small $\vec{G}$ values causes some problems with the stability of the self-consistency process. This is discussed in detail in Ref. 19. The Hartree-Fock exchange potential was approximated using the Slater exchange model, as we did for the Al starting potential. In $\vec{G}$ space, the exchange potential then has the form

$$V_x(\vec{G}) = -\alpha(3/2\pi)(3\pi^2)^{1/3} e^2 \rho^{1/3}(\vec{G}) , \tag{12}$$

where $\alpha = 0.79$ and $\rho^{1/3}(\vec{G})$ are Fourier components of $\rho^{1/3}(\vec{r})$. The justification for the present value for α is that this choice of α will bring Slater's exchange in an approximate agreement with Wigner's[24] interpolation formula at the average valence charge density of Al and Si. Thus, from the total charge density, the electronic screening potential

$$V_{\text{SCR}}(\vec{r}) = \sum_{\vec{G}} [V_H(\vec{G}) + V_x(\vec{G})]\, e^{i\vec{G}\cdot\vec{r}} \tag{13}$$

is obtained at each iteration in the self-consistent loop.

After the screening potential is determined, the self-consistent process is continued by adding V_{SCR} to an ionic potential V_{ion} to form a potential for the next iteration. The ionic potential consists of two terms,

$$V_{\text{ion}}(\vec{G}) = S(\vec{G})\, V_{\text{ion}}^{\text{Si}}(\vec{G}) + V_{\text{ion}}^{\text{Al}}(\vec{G}) , \tag{14}$$

where the first term is generated by the Si^{+4} ionic cores and the second term is generated by the Al slab. $S(\vec{G})$ is the Si structure factor as defined in Eq. (5).

First let us discuss $V_{\text{ion}}^{\text{Al}}$. This is just the Coulombic potential generated by repeated slabs of uniform positive charge. For an origin at the center of a metallic slab, $V_{\text{ion}}^{\text{Al}}$ has the form

$$V_{\text{ion}}^{\text{Al}}(\vec{G}) = \frac{-8\pi e^2 n_+ \sin G_z a/2}{C G_z^3}\, \delta_{G_x 0}\, \delta_{G_y 0} , \tag{15}$$

where a is the width of the jellium slab, C is the length of the unit cell along the direction (z) perpendicular to the interface, and n_+ is the positive background density.

For the Si-ion core potential, we have used an atomic model potential which was fitted to atomic term values by Abarenkov and Heine.[25] The repulsive cores of the ionic model potentials as given by Abarenkov and Heine are nonlocal (i.e., l dependent). For the present calculation, a local, "on the Fermi sphere" approximation[21] has been made and the Fourier transform of the resulting local potential was fitted to a four-parameter curve,

$$V_{\text{ion}}^{\text{Si}}(q) = (a_1/q^2)[\cos(a_2 q) + a_3]\, e^{a_4 q^4} . \tag{16}$$

The values of the a_i's are given in Table I. The normalization and the units for Eq. (16) are the same as those for Eq. (6). Using the parameters given in Table I, this ionic core potential has proven to yield excellent results in bulk and surface self-consistent calculations.[19]

With the above V_{ion}, the first two cycles of the self-consistent loop were performed using

$$\begin{aligned} V_{\text{IN}}^{(1)}(\vec{r}) &= V_{\text{start}}(\vec{r}) , \\ V_{\text{IN}}^{(2)}(\vec{r}) &= V_{\text{ion}}(\vec{r}) + V_{\text{SCR}}^{(1)}(\vec{r}) . \end{aligned} \tag{17}$$

However, owing to the divergent character of V_H and V_{ion} mentioned earlier, an input potential V_{IN} which deviates from the truly self-consistent potential will lead to an output potential which "overshoots" and is further away from the true potential. Therefore further iterations based on a straightforward extension of Eq. (17) would not yield a converging potential. This unstable behavior of the screening potential, especially for very small $\vec{G}$ vectors, is commonly found in surface calculations.[18–20,26,27] The procedure to deal with these instabilities is to compute adjusted input potentials $V_{\text{IN}}^{(n)}(\vec{G})$ for $n > 2$ from preceding input and output potentials. This can be done by obtaining the input potential of the nth iteration from a linear combination of input and output potentials of the $(n-1)$th iteration or from inspecting V_{OUT}-versus-V_{IN} graphs separately for each small $\vec{G}$. A detailed discussion of this problem and the procedures to overcome it are given in Ref. 19. The criterion for self-consistency is now the stability of the adjusted input screening potential as compared to the output screening potential calculated from Eq. (13). In the present calculation, the final self-consistent potential is stable to within 0.01 Ry.

After self-consistency has been reached, the electronic structure of the interface can then be analyzed in terms of charge densities. For this purpose, charge densities have been calculated as a function of different energy intervals and different $\vec{k}$ points in the Brillouin zone. In addition, we performed a local density of states (LDOS) calculation for the Al-Si interface. This LDOS, which displays the density of states in real space, facilitates the identification and illustrates the characteristics of the various kinds of states at the interface. Analogous to the projected density of states in tight-binding calculations, the LDOS for a given region in real space is given by

$$N_i(E) = \sum_{\vec{k}_\parallel, n} \int_{\Omega_i} |\psi_{\vec{k}_\parallel, n}(\vec{r})|^2 d^3r \, \delta(E - E_n(\vec{k}_\parallel)) , \qquad (18)$$

where $\vec{k}_\parallel$ is the wave vector parallel to the interface, n is the band index, $\psi_{\vec{k}_\parallel, n}$ is the electronic wave function, and Ω_i is the volume of the chosen region. Physically $N_i(E)$ can be interpreted as the probability that an electron with energy E is found in the region i.

III. RESULTS

In this section our results for the Al-Si interface which have been briefly reported recently[28] are discussed. We find that four different types of states can exist near the Al-Si interface. Aside from the usual states which are bulklike in both materials, there are states with energy below the Al conduction band which are bulklike in the Si side but decay rapidly in the Al side. Also, in the two-dimensional Brillouin zone, we find extra "gap" states in the semiconductor energy gaps whenever the range of the gap is inside the metallic band. They are somewhat similar to the states suggested by Heine; i.e., they are bulklike in Al and decay rapidly in Si. However, at the Si surface, these gap states retain the characteristics of the "free-surface" Si *surface states* which existed in the absence of the metal. It is these states which pin the Fermi level and dominate the properties of the Al-Si junction. In addition, we find truly localized interface states which decay in both directions away from the interface. These appear in the Si energy gaps in the energy range below the Al conduction band.

First let us examine the total, self-consistent valence charge density. The total charge density is a good indicator of the quality of the present work. For the present calculations to adequately represent noninteracting interfaces, the charge densities away from the interface should resemble the bulk densities of the two materials. Figure 2 displays the total valence charge density in a (110) plane along with the function $\bar{\rho}_{\text{total}}(z)$, which is the total charge density averaged parallel to the interface with z being the direction perpendicular to the interface. For the purpose of discussing the charge densities and the local density of states, we have also divided the unit cell into 12 equal regions (slices) as shown partially in Fig. 2(b). The jellium edge is indicated by the double-dashed line. Only the charge within a few angstroms from the interface is significantly perturbed from the bulk values. The charge densities in regions I and II and regions V and VI are in accord with bulk densities.[22,29] The slight differences between the present Si charges away from the interface and those calculated in Refs. 22 and 29 are due to the difference in the cutoff energy E_1.

From Fig. 2(b) one sees the well-known Friedel oscillations in the Al charge density and there is a net transfer of charge from aluminum to silicon. On the Al side, regions I and II each contain 7.9% of the total charge in the unit cell, whereas region III contains only 7.6%. On the Si side, regions V and VI each contain 8.8% of the total charge but region IV contains 9.1%. Thus approximately 0.3% of the total charge in the unit cell has been transferred from region III to region IV. A dipole potential with an electric field pointing toward the Si side is hence set up at the interface. This is a consequence of equalizing the Fermi levels in the two materials. From Fig. 2(a), the Al charge appears to be spilling into the empty "channels" in the Si charge density and into the dangling-bond

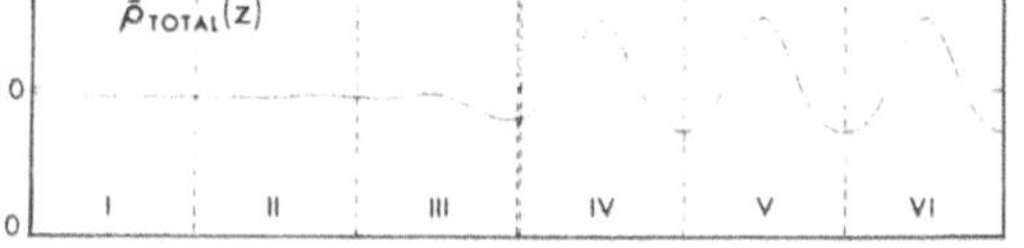

FIG. 2. (a) Total valence-charge-density contours in a (110) plane. The Si atoms are indicated by dots. (b) Total valence charge density averaged parallel to the interface and plotted along the direction perpendicular to the interface. The charge densities are normalized to 1 electron per unit cell.

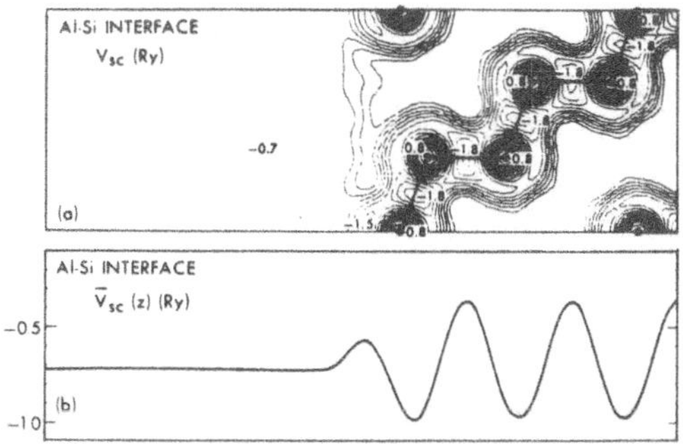

FIG. 3. (a) Contour plot of the final self-consistent potential V_{sc} in a (110) plane. (b) Final self-consistent potential averaged parallel to the interface and plotted along the direction perpendicular to the interface. The potential values are in rydbergs.

sites. The charge density at the dangling-bond sites in the present case is significantly higher than a sum of the jellium electron charge density and the Si charge density from the free-surface calculations. This indicates the formation of a metallic-covalent-like bond between Si and a jellium of Al density.

Figure 3 displays the self-consistent pseudopotential V_{sc} in a (110) plane along with $\bar{V}_{sc}(z)$, which is V_{sc} averaged parallel to the interface. The total charge density discussed earlier is the self-consistent response to this potential. The potential on the Al side is flat and does not show pronounced Friedel oscillations. Similar properties have been found in self-consistent calculations on the Al surface using the jellium model.[26] In the course of self-consistency, the Si potentials on the first two layers are made slightly deeper than the Si potentials further away from the interface. As expected, the perturbation to the Si potentials due to the presence of the metal appears to be much less than the perturbation due to the free surface.[19,30]

Now let us discuss the local density of states (LDOS) as defined in Eq. (18). We have calculated the LDOS for the six regions indicated in Fig. 2 by using 21 points in the irreducible part of the two-dimensional zone. The histograms of the LDOS for the six regions are shown in Fig. 4. To facilitate comparisons, the density of states of bulk Si from Ref. 22 is superimposed on the LDOS of regions IV–VI, and a free-electron density of states [i.e., $N(E) \sim E^{1/2}$] is superimposed on the LDOS of regions I–III. The Fermi level is indicated by the dashed line. Most of the interesting features appear in the LDOS of region IV. To investigate the energy positions of the extra states and their origins, we subtracted the LDOS of region VI from the LDOS of region IV to obtain a difference local density of states (DLDOS). The result is presented in Fig. 5. The positive portion of the histogram indicates an addition of states in that particular energy range in region IV and the negative portion of the histogram shows a depreciation of states.

The LDOS reveals much information about the electronic structure of the interface. From the position of the Fermi level and the position of the conduction-band edge of the semiconductor, one

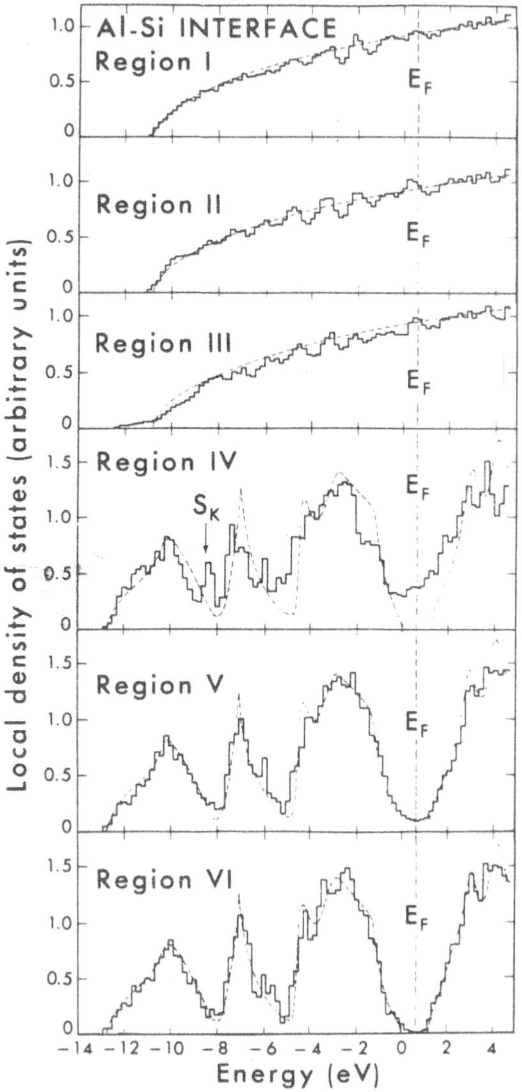

FIG. 4. Local density of states in arbitrary units as defined by Eq. (18). The regions are as shown in Fig. 2(b).

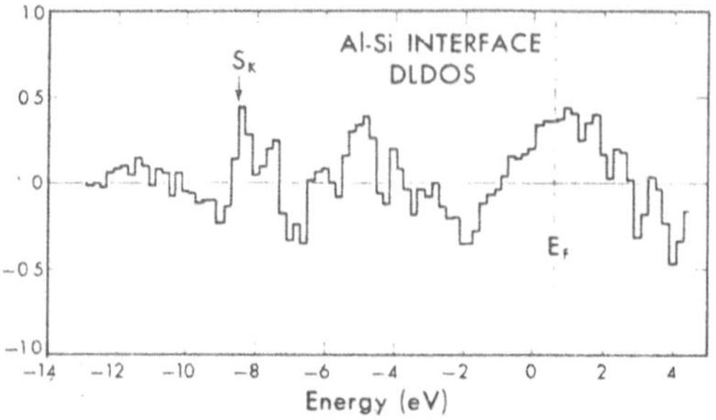

FIG. 5. Difference local density of states (DLDOS) obtained by subtracting the LDOS of region VI from that of region IV. The units are the same as in Fig. 4.

can calculate the barrier height at the interface. We obtained a barrier height of 0.6 ± 0.1 eV for the Al-Si interface, which is in excellent agreement with the recent experimental result[1] of 0.61 eV. There are other experimental values for the Al-Si barrier height, ranging from ~0.55 to ~0.70 eV. (See, for example, Ref. 15.) However, we believe that the value from Ref. 1 is the best for an ideal Al-Si interface because of the ultrahigh-vacuum conditions used in this particular experiment.

The various types of states which appear near the interface can be seen from the LDOS. States with energy below −11.1 eV (i.e., below the onset of the Al conduction band) are bulklike in Si and do not penetrate into the bulk of Al. Of course there are states with higher energy which can behave similarly. For example, at the $\vec{k}$ point K, states with energy up to −6.5 eV are below the Al conduction band. To illustrate this type of state, the charge density for all states with energy below −11.5 eV is presented in Fig. 6. On the Si side, the charge-density contours strongly resemble the charge-density contours for the bottom band of bulk[22] Si whereas the charge on the Al side is completely zero.

From the LDOS of region IV (Fig. 4) or the DLDOS (Fig. 5), we see that the dips in the bulk

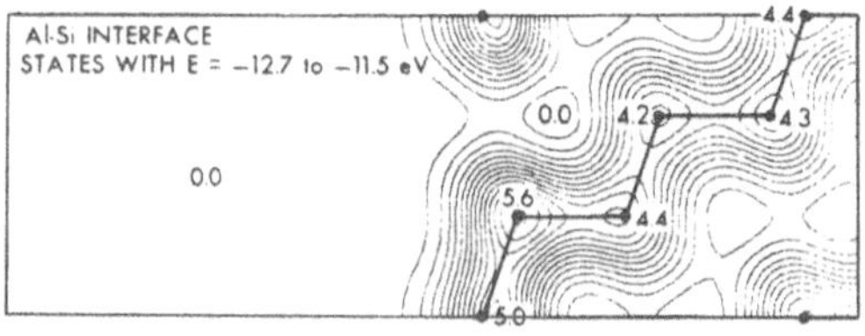

FIG. 6. Charge-density contours for states with energy below −11.5 eV in the same plane and normalization as in Fig. 1(a).

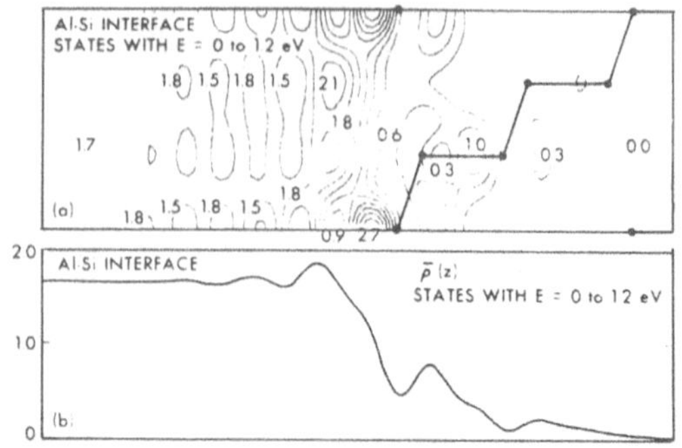

FIG. 7. (a) Charge-density contours for "gap" states with energy between 0 and 1.2 eV in the same plane and normalization as in Fig. 1(a). (b) Charge density in (a) averaged parallel to the interface and plotted along the direction perpendicular to the interface.

Si density of states which are due to gaps in the Si band structure are being filled up by either interface states or "gap" states at the interface. The extra states centered at ~ −8.2 eV are partially interface states and partially gap states, whereas the states centered at ~ −5.0 eV and states in the optical gap are gap states.

The gap states in the optical gap are of particular importance because the density of these states sensitively influences the position of the Fermi level with respect to the semiconductor band edges. These states have a charge density which is metallic in the Al slab, becomes dangling-bond-like at the Si surface, and decays rapidly to zero in the Si slab. The charge density for these states in the thermal gap, i.e., states with energy between 0 and 1.2 eV, is plotted in Fig. 7 along with $\bar{\rho}(z)$, which is the same charge density averaged parallel to the interface. The dangling-bond surface states which exist at these energies for the free-surface case have been matched to the continuum of metallic states. Thus, as seen from Fig. 7(a), the charge is quite uniform in the Al slab but retains the dangling-bond character at the Si surface. We note that the charge density displayed in Fig. 7 is for all states with energy in the thermal gap. The decaying rates are different for states at different energies. The charge for states near midgap decays most rapidly into the Si side.

An examination of the LDOS of region IV from −1.0 to 2.0 eV indicates that there is an apparent asymmetry in the distribution of extra states about the optical gap. A plausible physical explanation is the following: The states in the optical gap are derived from the valence band and the conduction band. Note the large depreciation of states near −1.8 eV and near +4.0 eV. (See Fig. 5.) Since

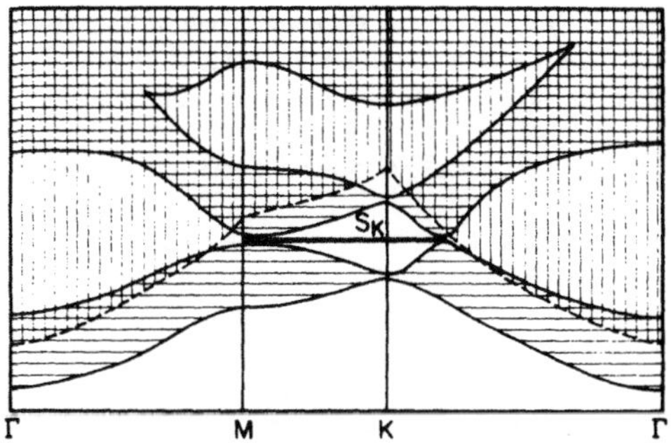

FIG. 8. Schematic diagram of the bottom two bands of the Si band structure (horizontally hatched) projected to the two-dimensional Brillouin zone. Superimposed on it is the projected Al conduction band (vertically hatched). S_K denotes the interface states discussed in the text.

these gap states are dangling-bond-like (i.e., p_z-like) in region IV and the top of the Si valence band is p-like, whereas the bottom of the conduction band is s-like, bulk states from the top of the valence band will be "robbed" to form the gap states while only states higher in the conduction band will be strongly affected by the formation of the gap states. Therefore the depreciation of bulk state densities will be larger at the top of the Si valence band than at the bottom of the conduction band. This results in the apparent asymmetry.

The interface states centered at −8.5 eV, labeled S_K in Fig. 4, appear near the point K in the two-dimensional hexagonal Brillouin zone. At first sight, localized states should not appear because there are aluminum states in this energy range. This appearance of interface states is a band-structure effect. Near the point K in $\vec{k}$ space, the Si two-dimensional band structure has a gap between −7.2 and −9.5 eV which is below the Al conduction band. In Fig. 8 we show a *schematic* diagram of the projected band structure of the bottom two bands of Si together with the projected band structure of Al. The Fermi levels of the two materials have been set equal. The lowest gap at K is the gap that we are discussing. Silicon surface states existing in this gap cannot be matched with any Al states because there are no Al states with the same energies and $\vec{k}$ vectors. A contour map of the charge density of the interface states at K at −8.5 eV is shown in Fig. 9. The charge density is s-like and highly localized on the outermost Si atoms. The charge is almost completely confined in region IV. Similar states with the same energy and character have been found in Si surface calculations. However the charge for states found in surface calculations is less localized.

IV. SUMMARY AND CONCLUSIONS

We have studied the electronic structure of a metal-semiconductor interface using an Al-Si system as a prototype. A jellium-semiconductor model has been constructed for the Al-Si interface. The electronic structure of the interface was then calculated using a method involving self-consistent pseudopotentials.[18] The model and methods of calculation used in the present paper have wider application than just the Al-Si system; these techniques can be extended straightforwardly to calculate the electronic structure of other metal-semiconductor interfaces.

Four different types of states are found to exist near the interface. The characteristics of these states have been analyzed in detail in terms of their charge densities. Our local-density-of-states results indicate a high density of "gap" states in the Si thermal gap near the Al-Si interface. This implies a pinning of the Fermi level by these gap states which is consistent with experimental results. It is important to note that, in the present calculation, we have used a statistical exchange model for the exchange potential. Hence both the valence bands and conduction bands see the same screening potential. Also, from examining the structure of the local density of states, there does not seem to be a merging of the valence band with the conduction band near the interface. Therefore, the pinning of the Fermi level can be explained without invoking Inkson's argument of merging of the bands due to the difference in the screening of the valence band and the conduction band at the interface. Furthermore, it is not very

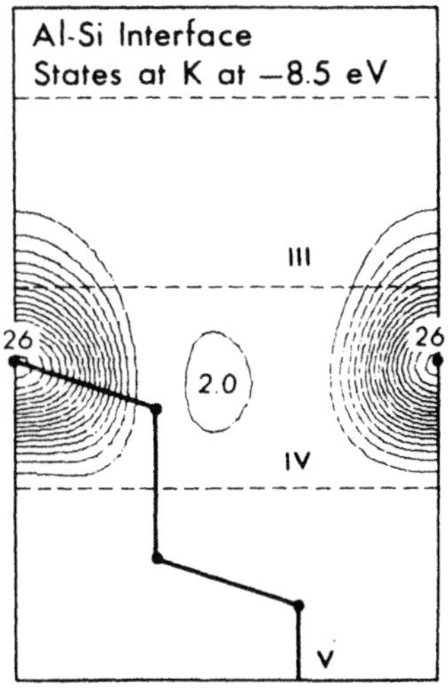

FIG. 9. Charge density contours for the interface states at K in the same plane and normalization as in Fig. 1(a).

meaningful to talk about a band picture as a function of distance away from the interface on such a microscopic scale.

The present calculation is for a high-density metal, Al, in contact with Si. For metals with a low density of *s-p* electrons, interface states can coexist with gap states in the energy range of the Si optical gap, such as in the − 7.2- to − 9.5-eV gap in the present calculation. Under such conditions, one expects that an even higher density of extra states will appear near midgap[18,19,30] and the Fermi level is again pinned in the thermal gap. This may be an explanation of why surface states continue to exist in the GaAs gap when an overlayer of[4,5] Cs or[3] Pd is placed on GaAs. Both Cs and Pd are metals of low *s-p* electron densities.

ACKNOWLEDGMENTS

The authors would like to thank Dr. F. Yndurain, Dr. M. Schlüter, and Dr. J. R. Chelikowsky for stimulating discussions. We would also like to thank Dr. Marc Brodsky for helpful conversations. Part of this work was done under the auspices of the U. S. Energy Research and Development Administration.

---

*Supported in part by the National Science Foundation under Grant No. DMR72-03206-A02.

†Supported by a National Science Foundation fellowship.

[1] A. Thanailakis, J. Phys. C 8, 655 (1975).

[2] A. Thanailakis, in Conference Series 22, *Metal-Semiconductor Contacts*, edited by M. Pepper (The Institute of Physics, London, 1974).

[3] J. L. Freeouf and D. E. Eastman, Phys. Rev. Lett. 34, 1624 (1975).

[4] W. E. Spicer, P. E. Gregory, P. W. Chye, I. A. Babalola, and T. Sukegawa, Appl. Phys. Lett. (to be published).

[5] P. E. Gregory and W. E. Spicer, Phys. Rev. B 12, 2370 (1975).

[6] J. Bardeen, Phys. Rev. 71, 717 (1947).

[7] V. Heine, Phys. Rev. 138, A1689 (1965).

[8] J. C. Inkson, J. Phys. C 5, 2599 (1972); 6, 1350 (1973).

[9] F. Yndurain, J. Phys. C 4, 2849 (1971).

[10] P. W. Anderson, in *Elementary Excitations in Solids, Molecules and Atoms, Part A*, edited by J. Devreese, A. Kunz, and T. Collins (Plenum, New York, 1974), p. 1.

[11] J. C. Phillips, Phys. Rev. B 1, 593 (1970).

[12] J. C. Phillips, J. Vac. Sci. Technol. 11, 947 (1974).

[13] A. J. Bennett and C. B. Duke, Phys. Rev. 160, 541 (1967); 162, 578 (1967).

[14] B. Pellegrini, Phys. Rev. B 7, 5299 (1973).

[15] S. M. Sze, *Physics of Semiconductor Devices* (Wiley, New York, 1969).

[16] S. Kurtin, T. C. McGill, and C. A. Mead, Phys. Rev. Lett. 22, 1433 (1969).

[17] M. L. Cohen, M. Schlüter, J. R. Chelikowsky, and S. G. Louie, Phys. Rev. B 12, 5575 (1975).

[18] M. Schlüter, J. R. Chelikowsky, S. G. Louie, and M. L. Cohen, Phys. Rev. Lett. 34, 1385 (1975).

[19] M. Schlüter, J. R. Chelikowsky, S. G. Louie, and M. L. Cohen, Phys. Rev. B 12, 4200 (1975).

[20] J. R. Chelikwosky, M. Schlüter, S. G. Louie, and M. L. Cohen, Solid State Commun. 17, 1103 (1975).

[21] M. L. Cohen and V. Heine, Solid State Phys. 24, 37 (1970).

[22] J. R. Chelikowsky and M. L. Cohen, Phys. Rev. B 10, 5095 (1974).

[23] J. C. Slater, Phys. Rev. 81, 385 (1951); W. Kohn and L. J. Sham, Phys. Rev. 140, A1133 (1965).

[24] E. P. Wigner, Phys. Rev. 46, 1002 (1934).

[25] I. V. Abarenkov and V. Heine, Philos. Mag. 12, 529 (1965).

[26] N. D. Lang and W. Kohn, Phys. Rev. B 1, 4555 (1970).

[27] G. P. Alldredge and L. Kleinman, Phys. Rev. B 10, 559 (1974).

[28] S. G. Louie and M. L. Cohen, Phys. Rev. Lett. 35, 866 (1975).

[29] J. P. Walter and M. L. Cohen, Phys. Rev. B 4, 1877 (1971).

[30] J. A. Appelbaum and D. R. Hamann, Phys. Rev. Lett. 32, 225 (1974).

PHYSICAL REVIEW B VOLUME 15, NUMBER 4 15 FEBRUARY 1977

Ionicity and the theory of Schottky barriers*

Steven G. Louie,† James R. Chelikowsky,‡§ and Marvin L. Cohen
Department of Physics, University of California, and Materials and Molecular Research Division, Lawrence Berkeley Laboratory, Berkeley, California 94720
(Received 5 August 1976)

We have investigated the role of ionicity in metal-semiconductor Schottky barriers by examining interfaces of increasing semiconductor ionicity. The electronic structure of four separate interfaces consisting of jellium (of Al density) in contact with the (111) surface of Si and the (110) surfaces of GaAs, ZnSe, and ZnS is investigated through the use of a self-consistent pseudopotential method. The barrier height and the surface density of states in the semiconductor band gap are determined. The phenomenological index of interface behavior S (studied by Kurtin, McGill, and Mead for semiconductors of different ionicity) is discussed in terms of a simple model involving metal-induced states in the semiconductor gap.

I. INTRODUCTION

Experimentally the behavior of the Schottky-barrier height ϕ_b for metal-semiconductor (M-s) interfaces as a function of the metal electronegativity is found to be dramatically different depending on whether the semiconductor is covalent or ionic.[1] For covalent semiconductors ϕ_b is approximately constant for all metals, whereas for ionic semiconductors, ϕ_b is strongly dependent on the metal contact. Furthermore, the transition from covalent behavior to ionic behavior appears to be a rather sharp transition which occurs at a critical ionicity.

While there have been a number of theories and speculations[2-7] and various mechanisms have been proposed to explain these properties of the barriers, a definitive explanation has yet to emerge because of the lack of detailed information on the microscopic nature of M-s interfaces. A necessary step toward understanding the properties of Schottky barriers should therefore involve a systematic study of the electronic structure of a series of M-s interfaces as a function of increasing semiconductor ionicity. Our present work is motivated by these considerations.

The electronic structure of a metal-Si interface has been studied recently[7] by two of the authors (S.G.L. and M.L.C.) using a self-consistent pseudopotential method. To circumvent the complex geometric problem of matching two crystal lattices, the metal was replaced by a jellium model in Ref. 7. Energy bands, charge densities, and the local density of states of the interface were calculated and analyzed within the one-electron theory. A barrier height in very good agreement with experiment was obtained. In this paper, we extend the analysis in Ref. 7 to the metal–zinc-blende semiconductor interfaces and present theoretical results on the microscopic origin of the ionicity-dependent behavior of M-s Schottky barriers.

The interfaces studied are interfaces of Al (modeled by a jellium core potential with $r_s = 2.07$) in contact with the ideal (111) surface of Si and the ideal (110) surfaces of GaAs, ZnSe, and ZnS. We find that, within the jellium-semiconductor model, the electronic structure of the four interfaces under investigation is qualitatively similar. Moreover we find that the experimentally observed variation in ϕ_b for different metals in contact with semiconductors of different ionicity can be understood quantitatively in terms of a simple model involving metal-induced states in the semiconductor band gap.

The remainder of the paper is organized as follows: In Sec. II the methods of calculation are discussed briefly. In Sec. III the results for the electronic structure of the metal–zinc-blende semiconductor interfaces are presented. In Sec. IV the ionicity-dependent behavior of the Schottky-barrier height is examined. And in Sec. V some discussion and conclusions are presented.

II. CALCULATIONS

As in Ref. 7 we are considering intimate M-s interfaces; i.e., there is no oxide layer between the metal and the semiconductor. We further approximate the system by replacing the metal with a jellium model and describing the semiconductor in the pseudopotential formalism. This model for the M-s interface and the method we used to calculate the interface electronic structure have been discussed at length in Ref. 7. Hence, in this section, we shall only briefly describe some of the essential features of the method and will be mainly concerned with the parameters needed in the calculations.

The two crucial features of our method are (i) self-consistency in the potential of the valence electrons is imposed to allow for the correct electronic screening near the interface, and (ii) per-

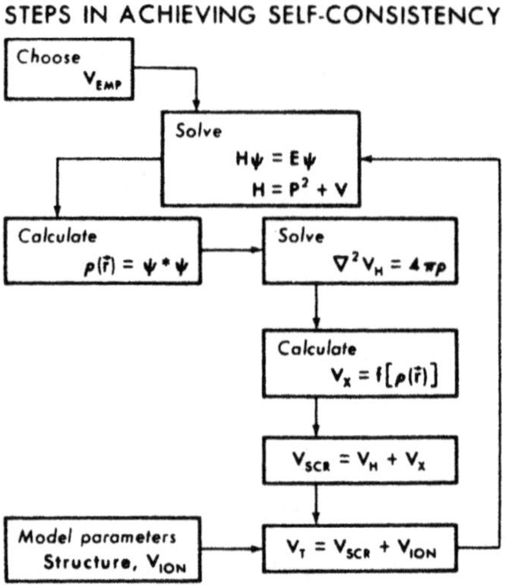

FIG. 1. Self-consistent loop in calculating the electronic structure of a metal-semiconductor interface (after Ref. 7).

iodicity is artificially retained along the direction perpendicular to the interface to allow for the use of the usual pseudopotential techniques.[8]

The calculations were carried out by constructing an elongated unit cell which, in two dimensions, is spanned by the shortest lattice vectors parallel to the appropriate semiconductor surface and, in the third dimension, by a long c axis extending over m atomic layers of the semiconductor and n layers of equivalent thickness of jellium metal. (Here the thickness of one layer is the interatomic distance between planes of semiconductor atoms parallel to the interface; and, the length of the c axis is therefore equal to $m+n$ interplane distances.) The numbers used were $m = 11$ and $n = 7$ for metal-GaAs and metal-ZnSe and $m = 11$ and $n = 9$ for metal-ZnS.

There are no adjustable parameters in the calculations. The only input consists of the structures (i.e., the geometry of the interface) and the ionic pseudopotentials of the semiconductor ion cores which are determined from atomic spectra. Since we use a jellium-semiconductor model, the structure is determined by the crystal structure of the semiconductors except for the placement of the edge of the positive jellium core. This edge has been taken to be at a distance of one-half of an interlayer distance away from the outermost semiconductor atoms. The ion core potentials used are local pseudopotentials whose Fourier transforms are of the form

$$V_{ion}(q) = (a_1/q^2)[\cos(a_2 q) + a_3 q]\exp(a_4 q^4)\,, \qquad (1)$$

where the parameters a_i are fit to a Heine-Abarenkov core potential.[9]

Figure 1 summarizes the self-consistent procedure employed. Screening is achieved using a Hartree potential via Poisson's equation and a Slater statistical exchange. The details of the method are given in Ref. 7. In addition to the ionic core potentials, a starting potential is needed to initiate the self-consistent loop. For this purpose we have used the empirical pseudopotentials obtained from bulk calculations with Fourier transforms expressed in a four-parameter curve of the form

$$V_{emp}(q) = b_1(q^2 - b_2)/\{\exp[b_3(q^2 - b_4)] + 1\}\,. \qquad (2)$$

The parameters a_i and b_i for the various semiconductors used in the calculations are listed in Tables I and II, respectively.

Using the same convergence criteria as in Ref. 7, a basis set of approximately 500 plane waves was employed in expanding the wave functions in the calculations. An additional ~1200 plane waves were also included via Lowdin's perturbation scheme.[8] The total valence charge density $\rho(\vec{r})$ needed for each iteration was determined by a five-point sampling over the irreducible part of the rectangular zone. The points included the symmetry points Γ, X, X', and M and one general point in the center of the irreducible zone.[10] This set of points yields an accurate charge density and, at the same time, allows the use of symmetrized plane waves to reduce the sizes of the Hamiltonian matrices and hence the computation time

TABLE I. Ionic core potential parameters a_i. The potentials are normalized to an atomic volume of 152.3 a.u.[3] The form of the potential is given by Eq. (1). The units for $v(q)$ are Ry if q is given in a.u. (The Ga potential is valid only for $q \le 3$ a.u.)

	Ga	As	Zn	Se	S
a_1	−0.3384	−0.7057	−0.3056	−2.3258	−5.4101
a_2	1.3305	1.0448	1.3412	0.5283	0.3275
a_3	0.4466	0.1662	0.0802	−0.5740	−0.8169
a_4	0.0071	−0.0151	−0.0086	−0.0321	−0.0250

TABLE II. Empirical starting potential parameters b_i. Normalization and units are as in Table I. The form of the potential is given by Eq. (2).

	Ga	As	Zn	Se	S
b_1	1.2214	0.3474	6.7008	0.2334	0.2361
b_2	2.4495	2.6203	1.4983	3.3858	3.3630
b_3	0.5445	0.9335	0.6696	0.7266	0.7243
b_4	−2.7148	1.5677	−4.7128	2.2012	2.1900

for diagonalization.

Upon the completion of the self-consistent loop, the various states near the interface are analyzed in terms of a local density of states (LDOS) and individual charge densities. The LDOS displays the density of states in real space near the interface and, for a given region in real space, it is given by

$$N_i(E) = \sum_{\vec{k}_\parallel, n} \int_{\Omega_i} |\psi_{\vec{k}_\parallel, n}(\vec{r})|^2 d^3r \, \delta(E - E_n(\vec{k}_\parallel)), \quad (3)$$

where $\vec{k}_\parallel$ is the wave vector parallel to the interface, n is the band index, ψ is the electronic wavefunction, and Ω_i is the volume of the chosen region. The physical significance of $N_i(E)$ is that it gives the probability that an electron with energy E will be found in region i.

III. ELECTRONIC STRUCTURE

In this section, the electronic structure of the three metal–zinc-blende semiconductor interfaces is presented. Some of the results have been briefly reported previously.[11]

We have chosen the interfaces M-Si, M-GaAs, M-ZnSe, and M-ZnS to study because the semiconductors composing this series are of the same crystal structure and of increasing ionicity. Within our model, we find that the calculated electronic structure of the four M-s interfaces is qualitatively very similar. In all four cases, as found in Ref. 7, the intrinsic surface states which existed in the fundamental gaps of these semiconductors[9] are removed by the presence of the metal and new types of states occur in this energy range. These metal-induced gap states[7] (MIGS) are bulklike in the metal and decay rapidly into the semiconductor with some of the characteristics of the semiconductor-vacuum surface states (which exist in the absence of the metal) weakly retained at the semiconductor surface. In addition, truly localized interface states which have charge densities decaying in both directions away from the interface are found for energies near the lower part of the semiconductor valence band.

Before we discuss the individual states, let us examine the self-consistent, valence charge densities for the three metal–zinc-blende semiconductor interfaces. They are shown in Figs. 2 to 4. In each figure the total valence charge density is displayed in two different planes containing the two types of semiconductor surface atoms. The units are normalized to one electron per unit cell. Several interesting features are seen from

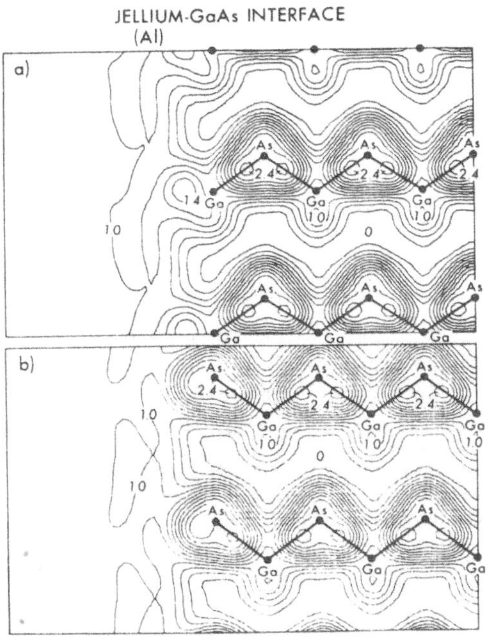

FIG. 2. Total valence charge density for the M-GaAs interface plotted in the (1 $\bar{1}$0) plane containing the (a) Ga surface atom and (b) As surface atom. The charge density has been normalized to one electron per unit cell.

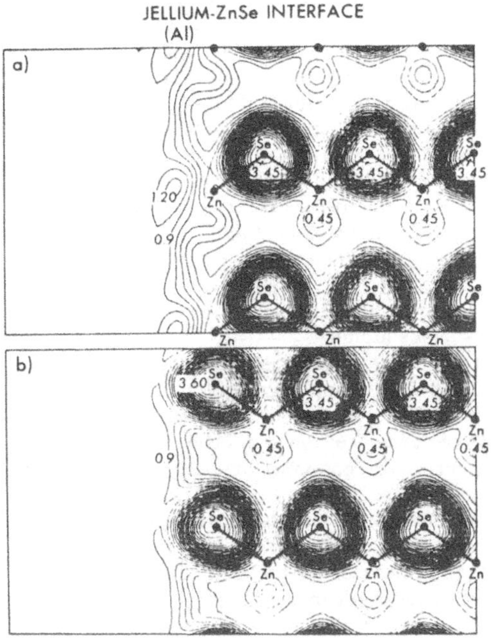

FIG. 3. Total valence charge density for the M-ZnSe interface plotted in the (1$\bar{1}$0) plane containing the (a) Zn surface atom and (b) Se surface atom. Normalization is as in Fig. 2.

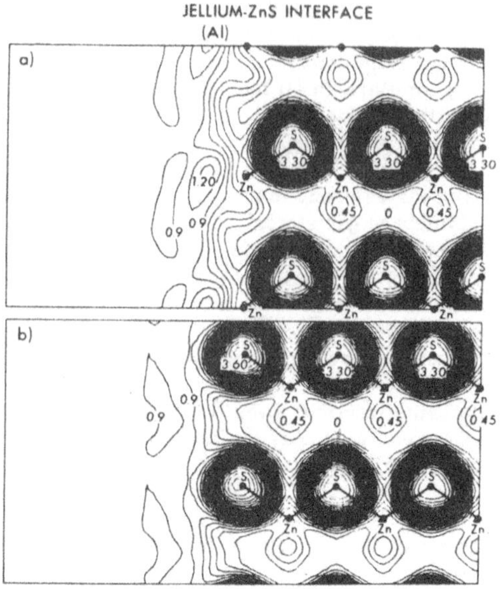

FIG. 4. Total valence charge density for the M-ZnS interface plotted in the $(1\bar{1}0)$ plane containing the (a) Zn surface atom and (b) S surface atom. Normalization is as in Fig. 2.

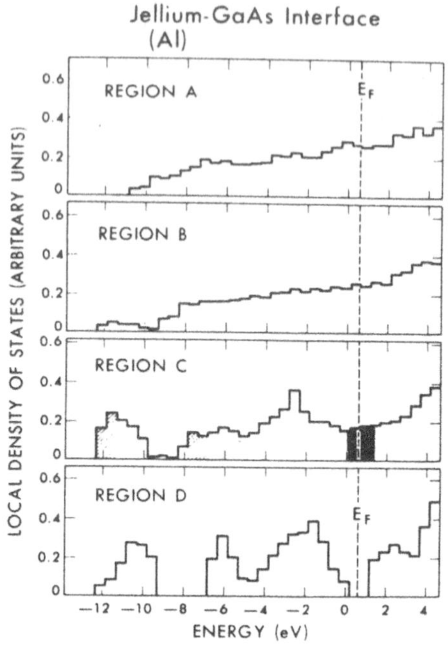

FIG. 5. Local density of states for the M-GaAs interface in arbitrary units as defined by Eq. (3).

the figures: (i) Charge densities in the bulk configuration are essentially the same as those found in the third layer into the semiconductor showing that the significant influence of the interface is quite short range (i.e. in the order of two to three atom layers); (ii) owing to the stronger potential of the anions, charge is increasingly localized on the anions as the ionicity of the semiconductor increases; and (iii) for all three interfaces, the charges on the outermost semiconductor atoms are slightly higher than the charges on the atoms deeper in the slab. This probably results from the presence of the metal which lowers the potential of the surface atoms.

In Fig. 5 the local density of states for the M-GaAs system is displayed for four regions. Each region contains one atomic layer. Region D is at the center of the semiconductor slab. Region C is the layer containing the outermost semiconductor atoms. Region B is adjacent to region C on the metallic side. The boundary between region C and B defines the interface. And finally region A is at the center of the metallic slab. The LDOS was evaluated according to Eq. (3) with five $\vec{k}$ points in the irreducible zone to calculate the histograms.[12] The LDOS for the M-ZnSe and M-ZnS interfaces are calculated in the same manner; these are shown in Figs. 6 and 7, respectively. Although the number of $\vec{k}$ points used is too

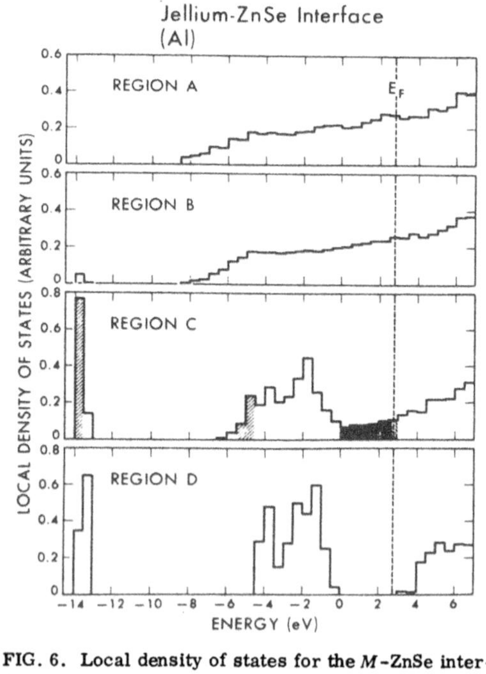

FIG. 6. Local density of states for the M-ZnSe interface.

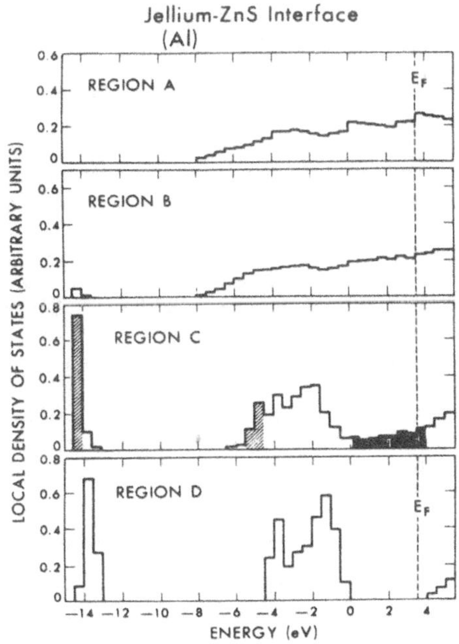

FIG. 7. Local density of states for the M-ZnS interface.

small to reproduce a nice $\sqrt{E}$ curve for the free-electron-gas density of states on the metallic side, it yields most of the prominent features of interest. A LDOS curve of much finer quality for the M-Si interface was obtained in Ref. 7 with the use of 21 $\vec{k}$ points in the irreducible zone.

The region of most physical interest is region C whose LDOS essentially describes the energy spectrum of the electrons on the semiconductor surface. The darkly shaded areas in Figs. 5–7 indicate the MIGS in the semiconductor thermal gaps. Also indicated (by the lighter shaded areas) are the energies of the localized interface states. The MIGS in the thermal gaps have, as we shall show in Sec. IV, a large influence on the Fermi level E_F and thus play a dominant role in determining the behavior of the M-s Schottky barriers.

Figure 8 displays the charge profiles $\bar{\rho}(z)/\bar{\rho}(0)$ of the penetrating tails of the MIGS in the thermal gap for the four M-s interfaces studied as a function of distance z into the semiconductor. Here $\bar{\rho}(z)$ is the charge density for the MIGS averaged over the states in the thermal gap and averaged parallel to the interface with $z=0$ at the interface. We note that the overall behavior of the charge profiles for Si and GaAs is quite similar and that the average penetration distances are considerably shorter than previously believed. The differences in the short range oscillations in the charge pro-

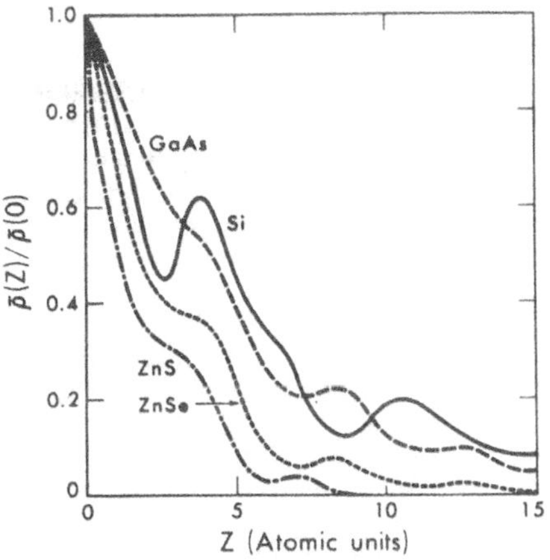

FIG. 8. Charge distributions of the penetrating tails of the MIGS in the semiconductor thermal gap. $\bar{\rho}(z)$ is the total charge density for these states averaged parallel to the interface with $z=0$ at the edge of the jellium core.

files mostly arise from the difference in the atomic arrangement between the two types of semiconductor surfaces [Si(111) and GaAs (110)].

We shall only discuss and illustrate the interface states at the M-ZnS interface to avoid redundancy. The characteristics of the interface states for the other two systems are qualitatively similar. The lowest lying interface states at the M-ZnS interface (see Fig. 7) are localized strongly on the outermost sulfur atoms and have s-like character in their charge distribution. These states split off from the bottom valence band of ZnS and form a narrow interface band extending over the whole Brillouin zone. The corresponding charge density given in Fig. 9 is extremely localized on the sulfur surface atoms with practically zero charge on the zinc atoms. A surface band,

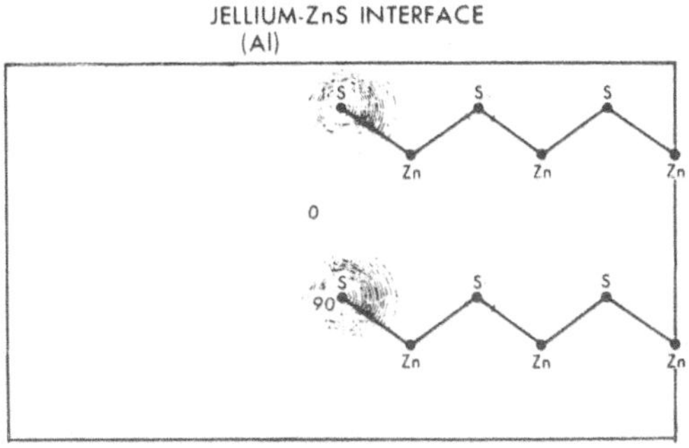

FIG. 9. Charge density of the s-like sulfur interface states in the same plane as Fig. 4(b). The charge density is again normalized to one electron per unit cell.

very similar to this s-like interface band, has been found in calculations on the (110) surface of zinc-blende semiconductors.[13] The surface states are, however, located at higher energies in the antisymmetric gap instead of at the bottom of the spectrum. These s-like interface states therefore appear to be intrinsic to the semiconductor surface with energies shifted because of the presence of the metal.

Two additional interface bands are found at ~−5 eV below the ZnS valence-band maximum. Unlike the previously discussed states, these interface states exist over a rather small region in k space at the zone edge around the point M. The charge distribution differs from the s-like state since it is p-like around the outermost sulfur atoms. The charge density for a state at M in the lower of the two interface bands is given in Fig. 10. Figure 10(a) shows the charge density contours in a $(1\bar{1}0)$ plane containing the surface Zn atoms. Figure 10(b) shows the charge density contours in a (110) plane parallel to the interface containing both types of semiconductor surface atoms. As seen from the figure, the charge is highly localized on the outermost semiconductor layer with the maxima of the p-like lobes lying in the (110) plane. The other interface state at M belonging to the higher of the two bands has a very different charge distribution which is displayed in Fig. 11. The charge is again p-like around the S atom. However, the charge lobes for this state are pointing along the back-bond direction between the first layer S atoms and the second layer Zn atoms. Surface states somewhat similar to these states are also found in surface calculations.[13]

Our results are consistent with recent experiments on metal overlayers which have provided information on the electronic structure of M-s interfaces in the energy range of the semiconductor band gap. Rowe *et al.*[14] have found that the intrinsic surface states on the (111) and (100) semiconductor surfaces are removed by metallic overlayers and extrinsic metal-induced states are found within the band gap. Their findings on the Ge(110) surface is however somewhat ambiguous. Similar extrinsic metal-induced states are found but they are weaker and the intrinsic surface states appear not to be completely removed by the thin metallic overlayers.

IV. IONICITY AND SCHOTTKY BARRIERS

In this section some of the properties of M-s Schottky barriers and their relation to the calculated electronic structure are examined.

The calculated barrier heights for the four M-s interfaces studied are presented in Table III together with the measured ϕ_b.[15,16] The calculated values were obtained by determining the position of the conduction-band minimum of the bulk semiconductor relative to the Fermi level E_F via the

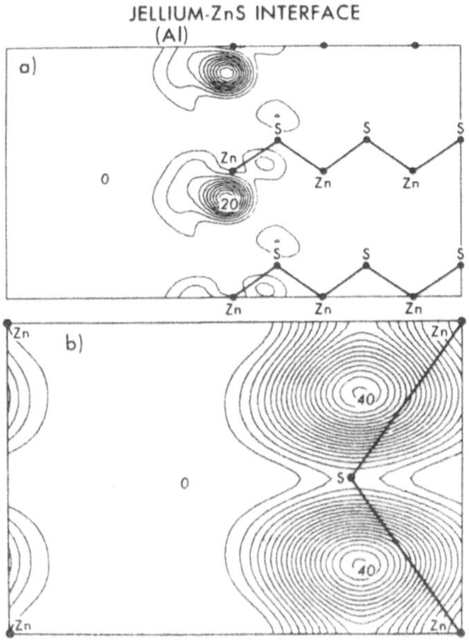

FIG. 10. Charge density of an interface state at M at −5.1 eV (see text).

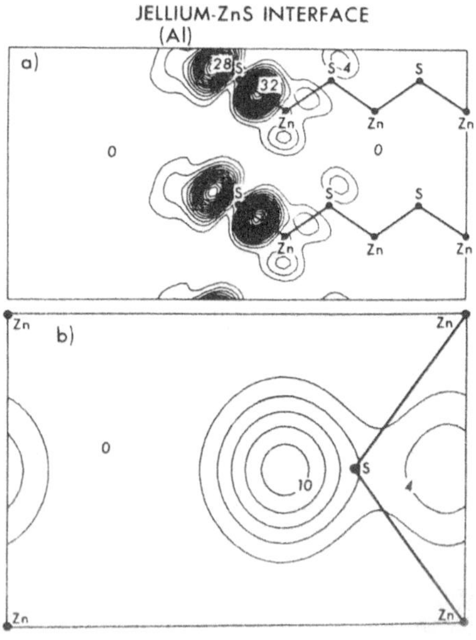

FIG. 11. Charge density of an interface state at M at −4.6 eV (see text).

TABLE III. Theoretical and experimental values for the Schottky-barrier height ϕ_b(eV) and the index of interface behavior S. D_s in units of 10^{14} states/eV cm^2 is the surface density of states used to obtain the calculated S.

	ΔX [a]	ϕ_b (cal)	ϕ_{bn} (expt)	D_s	S (cal)	S (expt) [b]
Al-Si	0	0.6 ± 0.1	0.6 [c]	4.5	0.1	0.1
Al-GaAs	0.4	0.8 ± 0.2	0.8 [d]	5.0	0.1	0.1
Al-ZnSe	0.8	0.2 ± 0.2	...	2.0	0.4	0.5
Al-ZnS	0.9	0.5 ± 0.2	0.8 [d]	1.4	0.7	1.0

[a] Reference 17.
[b] Reference 1.
[c] Reference 16.
[d] Reference 15.

local density of states. Considering the sizes of the thermal gaps of the more ionic crystals, the agreement obtained between theory and experiment is quite good.

Empirically the barrier height $\phi_b(M, s)$ obeys the relation[1]

$$\phi_b(M, s) = S(s)X_M + \phi_0(s) , \qquad (4)$$

where X_M is the Pauling-Gordy electronegativity[17] of the metal and S and ϕ_0 are constants depending on the semiconductor. As an example, the experimental barrier heights[15,16] for our four semiconductors are presented in Fig. 12 as a function of the X_M of various metals. Moreover the slope or "index of interface behavior" S is found to be a smooth function of $\Delta X = X_A - X_B$, the electronegativity difference of the anions and cations in the semiconductor. Since ΔX provides a measure of the ionicity of the semiconductor, S is also a function of the semiconductor ionicity. For ϕ_b expressed in units of electron volts, S is small ~0.1 for semiconductors with $\Delta X < 0.5$ but S is ~1.0 for semiconductor with $\Delta X > 0.9$. In addition, there is a well defined and rather sharp transition in the value of S at $\Delta X \sim 0.7$ to 0.8 (see Fig. 13).

The standard explanation for S relies on the Bardeen model which attributes this behavior of ϕ_b to the density of surface states existing in the semiconductor band gap. However arguments had been presented by Heine[3] which showed that semiconductor surface states do not exist in the fundamental gap for most M-s interfaces and many alternate theories have since been proposed.[3-7]

Physically the barrier height is determined by the requirement that in equilibrium the Fermi levels of two materials in contact are equal. This is achieved by creating an electric dipole potential Δ at the interface. Hence, in the one-electron theory, the density of the MIGS in the semiconductor thermal gap and their penetration into the semiconductor will strongly influence the behavior of ϕ_b. The extent of their penetration can be measured by a penetration depth δ defined by $\bar{\rho}(\delta)/\bar{\rho}(0) = 1/e$. From Fig. 8, δ is equal to ~3.0 and ~2.8 Å for Si and GaAs, respectively. As the ionicity of the semiconductor increases, δ however rapidly reduces to ~1.9 Å for ZnSe and to ~0.9 Å for ZnS.

The other quantity which is relevant to the behavior of ϕ_b and related to the MIGS is the surface density of states $D_s(E)$. For energies in the semiconductor thermal gap, we define

$$D_s(E) = A^{-1} \int_A \int_0^\infty N(E, \vec{r})\, dz\, dA , \quad 0 \leq E \leq E_g , \qquad (5)$$

where A is the interface area, $N(E, \vec{r})$ is the LDOS as defined in Eq. (3), and the integral over z is to be evaluated from the interface to deep into the

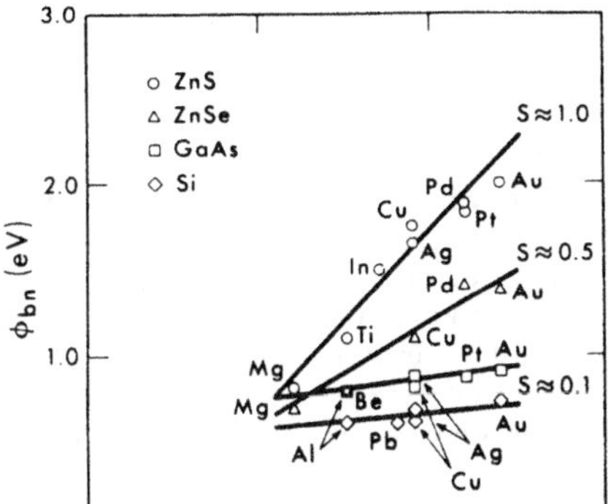

FIG. 12. Experimental values of the barrier heights for four semiconductors in contact with various metals. X_M is the electronegativity of the metal in the Pauling-Gordy scale. Data were taken from Refs. 16 (Si) and 15 (GaAs, ZnSe, ZnS).

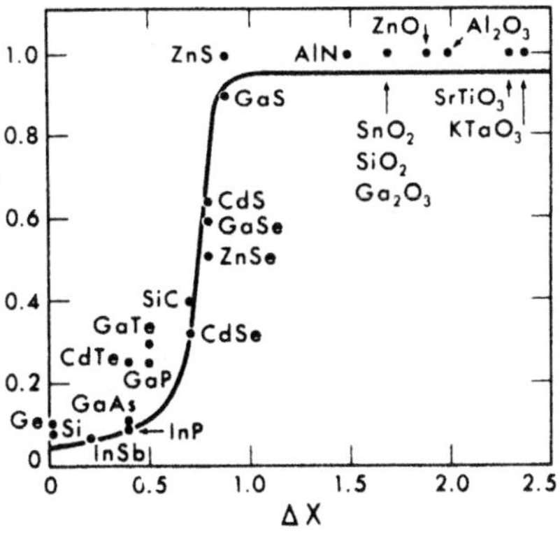

FIG. 13. Index of interface behavior S from Ref. 1.

bulk of the semiconductor. Thus $-eD_s(E)$ gives the density of localized surface charge per unit energy on the semiconductor surface. The calculated $D_s(E)$ are depicted in Fig. 14. The averaged D_s near the center of the gap for Si and GaAs which both have about the same S is approximately the same. Two trends which can be observed from Fig. 14 are that $D_s(E)$ decreases for more ionic semiconductors and $D_s(E)$ has a relatively flat minimum over the center region of the gap. The D_s for M-ZnSe is essentially identical to that of M-ZnS except its magnitude is ~30% higher. Hence it is omitted from Fig. 14.

From an electrostatic point of view, the fact that both δ and D_s decrease for more ionic semiconductor crystals implies that the change in Δ with respect to a change in E_F will be small for ionic semiconductors and larger for covalent semiconductors. Therefore we expect from the calculated δ and D_s that S will be large for ionic crystals and small for covalent crystals.

To estimate the influence of δ and D_s on the barrier height, we use the following simple model to calculate $S(s)$. Cowley and Sze[18] had used a somewhat similar approach to obtain the interface density of states in terms of the experimentally determined S. In this model, δ and D_s are assumed to be quantities intrinsic to the semiconductor (i.e., they are independent of the metal contacts) and also D_s is taken to be approximately constant over the central portion of the thermal gap. Calculations on metal-Si interfaces using surface Green's -function methods have shown that D_s is approximately constant for a wide range of metals.[5,19] In this model, we have also made use of the empirical relation that the metal work function ϕ_M is linear in X_M, i.e., $\phi_M = AX_M + B$ with $A = 2.27$ and $B = 0.34$ for ϕ_M expressed in electron volts.[17,20]

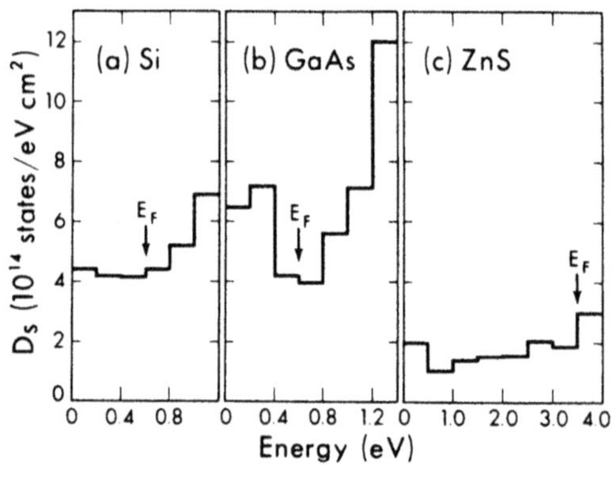

FIG. 14. Surface density of states as defined in Eq. (5).

For a semiconductor of electron affinity χ_s in contact with a metal, the electric-dipole potential established at the interface can be approximated as

$$\Delta = \chi_s + \phi_b - AX_M - B . \tag{6}$$

The change in Δ for a metal of slightly different X_M in contact with the same semiconductor is therefore

$$d\Delta = d\phi_b - AdX_M . \tag{7}$$

Using simple electrostatic arguments, another expression for $d\Delta$ is

$$d\Delta = -4\pi e^2 D_s \delta_{eff} d\phi_b , \tag{8}$$

where δ_{eff} is the effective distance between the center of mass of the negative charge transferred to the semiconductor due to the change in ϕ_b and the center of mass of the positive charge left behind in the metal. This distance is the true distance divided by the appropriate dielectric screening function ϵ, i.e., $\delta_{eff} = (t_M/\epsilon_M + t_s/\epsilon_s)$. We may approximate t_s by our calculated δ and t_M/ϵ_M by the typical screening length in a metal which is ~0.5 Å. Equations (4), (7), and (8) then yield the following S (see Ref. 21):

$$S = 2.3/[1 + 4\pi e^2 D_s(0.5 + \delta/\epsilon_s)] . \tag{9}$$

The dielectric screening for potential fluctuations in the distance of the order of δ has been found to be ≈2 by Walter and Cohen[22] for our four semiconductors. Hence we may evaluate S using the calculated values of δ and D_s and $\epsilon_s = 2$. They are presented in Table III together with the experimentally determined S. The agreement between theory and experiment is surprisingly good for this very simple model. We note here that, within our model, S is a function of *both* D_s and δ. Although our results are for intrinsic semiconductors at zero temperature, the calculated ϕ_b and S will be essentially the same as those for doped semiconductors at finite temperature. The argument is as follows, for typical doping density of $n \leq 10^{17}$ cm^{-3}, a small charge accumulation of ~10^{12} electrons/cm^2 at the semiconductor surface will result in band bending on the order of volts. Hence, with $D_s \sim 10^{14}$/eV cm^2, only a slight change (~0.01 eV) in E_F at the interface is needed to account for the band bendings caused by impurities or thermally excited electrons.

We have also examined the sensitivity of our results to the only uncertain parameter in the calculations, i.e., the placement of the jellium core edge. Our results appear to be quite insensitive to this parameter. In the case of the M-Si interface, a change of 25% in this parameter left ϕ_b and δ essentially unchanged and only changed D_s by a few percent. A similar observation has been made by Louis *et al.*[5,21] They have performed non-self-consistent calculations on M-s interfaces using a Green's-function method.

V. SUMMARY AND CONCLUSIONS

Using a self-consistent pseudopotential method, we have studied the electronic structure of a series of M-s interfaces of increasing semiconductor ionicity. Our results are consistent with recent experiments on metal overlayers which indicated that the intrinsic surface states on the semiconductor surfaces are removed by metallic overlayers and extrinsic metal-induced states are found within the energy range of the band gap. Hence, contrary to the Bardeen model and the recent speculations made by Eastman and Freeouf,[23] intrinsic semiconductor surface states do not appear to play a dominant role in determining ϕ_b. Detailed experimental information on the electronic structure of these M-s is however not available at present for comparison.

We have also examined the question of ionicity in the behavior of Schottky-barrier heights. A simple model involving the MIGS has been constructed to estimate S. We find that both ϕ_b and S can be satisfactorily determined using the self-consistent pseudopotential results for the more covalent semiconductors and somewhat less accurately for the more ionic semiconductors. Our results suggest that the important properties of Schottky barriers are mostly incorporated in the one-electron jellium-semiconductor type of model. Other effects not included in the present calculations such as many-body effects and bonding between metal and semiconductor atoms are most likely necessary before complete agreement between theory and experiments for the more ionic semiconductors can be achieved.

ACKNOWLEDGMENTS

Part of this work was done under the auspices of the U. S. Energy Research and Development Administration.

---

*Supported in part by the NSF Grant No. DMR72-03206-A02.

†Supported by the NSF.

‡Supported by the NSF.

§Present address: Bell Laboratories, Murry Hill, N.J. 07974.

[1]S. Kurtin, T. C. McGill, and C. A. Mead, Phys. Rev. Lett. 22, 1433 (1969); C. A. Mead and T. C. McGill, (unpublished).

[2]J. Bardeen, Phys. Rev. 71, 717 (1947).

[3]V. Heine, Phys. Rev. 138, A1689 (1965).

[4]J. C. Inkson, J. Phys. C 5, 2599 (1972); and 6, 1350 (1973).

[5]F. Yndurain, J. Phys. C 4, 2849 (1971); F. Flores, E. Louis, and F. Yndurain, *ibid.* 6, L465 (1973); E. Louis, F. Yndurain, and F. Flores, Phys. Rev. B 13, 4408 (1976).

[6]J. C. Phillips, Phys. Rev. B 1, 593 (1970); and J. Vac. Sci. Technol. 11, 947 (1974).

[7]S. G. Louie and M. L. Cohen, Phys. Rev. Lett. 35, 866 (1975); and Phys. Rev. B 13, 2461 (1976).

[8]M. L. Cohen and V. Heine, Solid State Phys. 24, 37 (1970).

[9]J. R. Chelikowsky and M. L. Cohen, Phys. Rev. B 13, 826 (1976).

[10]Notation for the high symmetry points is as in C. Kittel, *Quantum Theory of Solids* (Wiley, New York, 1964).

[11]S. G. Louie, J. R. Chelikowsky, and M. L. Cohen, J. Vac. Sci. Technol. 13, 790 (1976).

[12]Each nonzero histogram bin was constructed by averaging its value with 25% of the neighboring bins to remove some of the unphysical sharp structure resulting from the small number of $\vec{k}$ points used.

[13]See, for example, Ref. 9 and references therein.

[14]J. E. Rowe, S. B. Christman, and G. Margaritondo, Phys. Rev. Lett 35, 1471 (1975).

[15]S. M. Sze, *Physics of Semiconductor Devices* (Interscience, New York, 1969).

[16]A. Thanailakis, J. Phys. C 8, 655 (1975).

[17]L. Pauling, *The Nature of the Chemical Bond,* 3rd ed. (Cornell U.P., Ithaca, N. Y., 1960).

[18]A. M. Cowley and S. M. Sze, J. Appl. Phys. 36, 3212 (1966).

[19]For metals with very low density of s-p electrons, localized interface states may occur in the thermal gap and thus change the character of D_s. See Ref. 7 for discussion.

[20]W. Gordy and W. J. O. Thomas, J. Chem. Phys. 24, 439 (1956).

[21]E. Louis *et al.* have used an expression similar to Eq. (9) to analyze their results in Ref. 5. In their paper however the relation between ϕ_M and X_M was neglected which produced a larger discrepancy between their calculated S and experimentally determined values. They have also used a different dielectric screening.

[22]J. P. Walter and M. L. Cohen, Phys. Rev. B 2, 1821 (1970).

[23]D. E. Eastman and J. L. Freeouf, Phys. Rev. Lett. 34, 1624 (1975).

Chemical trends in metal-semiconductor barrier heights

M. Schlüter
Bell Laboratories, Murray Hill, New Jersey 07974
(Received 16 May 1977)

Experimental data on metal-semiconductor interfaces are reexamined. It is found that the previously reported abrupt transition between covalent and ionic semiconductors is not that clearly defined and the outcome is diffused by data scattering. Furthermore, the data indicate *no* saturation of the interface parameter S for $S = 1$. Considering the definition of S, it follows that the true Schottky limit should occur for some number $S \approx 2.0$–3.0 rather than for exactly $S = 1$ as previously claimed.

In 1969, Kurtin *et al.*[1] presented an extensive compilation of experimental metal-semiconductor interface (MSI) data. In classifying the behavior of MSI's the interface slope parameter $S = \partial\Phi_B/\partial X_m$ was used to describe the extent of Fermi-level stabilization for a given semiconductor. Here Φ_B denotes the rectifying barrier height of the MSI and X_m denotes the metal electronegativity. The definition of S as a descriptive interface index is based on a linear interface potential theory developed by Cowley and Sze[2] and by Heine.[3] If correct, this linear theory yields the striking result that the interface behavior of a given semiconductor can be described by a characteristic parameter S independent of the metal with which it is in contact. The theory assumes that a dipole is created at the interface formed by localized states on the semiconductor side, which in turn are more or less filled by electrons spilling over from the metal. Thus, by moving the Fermi level (i.e., changing the metal characterized by its electronegativity X_m) the barrier height is changed at a rate given by S.

Two limiting cases can be considered. The case of complete or nearly complete Fermi-level stabilization ($S \approx 0$) as found in Si or Ge, for example, is attributed to the existence of a high density of interface states in the gap of these semiconductors which "pin" the Fermi level. This limit had been first proposed qualitatively by Bardeen,[4] who related the stabilization to surface states. Later the problem had been formulated more rigorously by Heine,[3] who noted that true surface states should be absent in a metal-semiconductor interface, however, metal states tailing into the semiconductor would play a similar role of Fermi-level stabilization. A number of more-sophisticated calculations of the interface dipole[5-8] as well as theories involving band-narrowing, many-body interactions, and elementary excitation[9,10] confirm the existence of the $S \approx 0$ Bardeen limit.

The opposite limit of little or no Fermi-level stabilization, referred to as the Schottky limit, is reached if virtually no interface states exist in the semiconducting gaps. The important point to note is that the limit (no stabilization) is a $S = A$, where $A \equiv \partial\Phi_m^I/\partial X_m$. The quantity A, which is not in general equal to 1 or may not even be metal independent, relates the metal electronegativity X_m (e.g., on a Pauling scale[11]) to an "internal" metal work function Φ_m^I. The term "internal" emphasizes the difference between Φ_m^I and the usual (measured) metal work function Φ_m which characterizes the metal-vacuum interface and which contains an additional electrostatic contribution D from the metal surface dipole. Even though this latter fact (namely, $\Phi_m = \Phi_m^I + D$) was realized early and led to the use of X_m rather than Φ_m in characterizing the metal,[12] the correct value of $A \neq 1$ was usually omitted.[13] One exception is the theoretical work of Ref. 8, where a value of $A = 2.27$ was used in the calculations. As discussed later, this value though quite different from Ref. 1 may not be completely exact either since it originates from correlating measured work functions Φ_m with electronegativities X_m and thus is affected by the electrostatic surface dipole.

Moreover, in Ref. 1 S values were plotted for a number of semiconductors as a function of their electronegativity difference ΔX in a way which strongly suggested that, for semiconductors with $\Delta X > 1.0$, saturation is reached, i.e., the Schottky limit is reached at $S = 1$, i.e., $A = 1$. This plot, which is reproduced in Fig. 1 also suggested a rather abrupt transition between $S \approx 0$ and $S \approx 1$ around $\Delta X \approx 0.8$, which could not be explained satisfactorily by the calculations of Refs. 5–8. [The data points of SiC and C (diamond) are taken from Ref. 14. The originally published S value for SiC of 0.4 is thus corrected to 0.04.] Instead, many-body effects such as those proposed for electronic states in glasses were invoked to explain the "sudden" transition between covalent ($S \approx 0$) and ionic ($S \approx 1$) regimes.[15]

It is the purpose of this note to show that (a) the data contained a great deal of scattering and if they

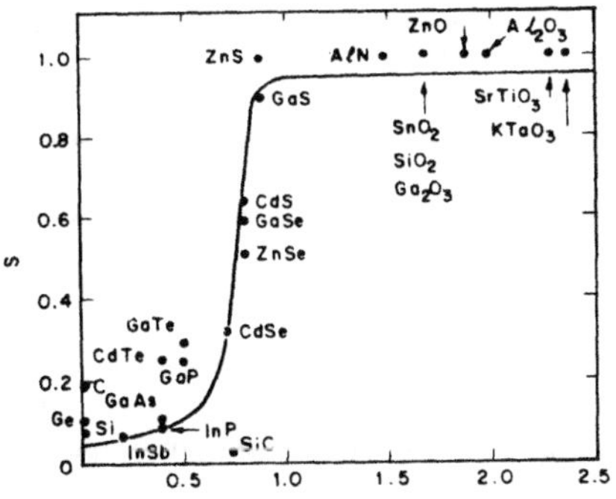

FIG. 1. Original plot of interface slope parameter S versus semiconductor electronegativity difference ΔX as given in Ref. 1.

are least-squares fitted, the suggested "sudden" transition is rather smooth and diffuse. (b) Taking into account the correct definition of S, the Schottky limit is not found for $S=1$ but should rather occur for some value $S \approx 2.0-3.0$. (c) For ionic semiconductors like ZnS, SiO_2, ZnO, SnO_2, Al_2O_3, or $SrTiO_3$ with ΔX ranging between 1.0 and 2.5, this suggested Schottky limit has by far not yet been reached.

As can already be seen from the original data plotted in Fig. 1 of Ref. 1, the scattering of barrier height values Φ_B for a given semiconductor or insulator (like SiO_2) as a function of metal electronegativities is rather high and straight-line fits (whether least squares or not) may become quite questionable in certain cases. The scattering varies from case to case and generally increases with the deduced value of S, i.e., with the ionicity (or band gap) of the semiconducting compound. However, even for a covalent material as Si, significant scattering was reported in the literature.[16] In view of these difficulties, we have reexamined the original data as far as they are available in Refs. 1, 12, 17, and 18 and performed least-squares fits to them. The results (listed in Table I) deviate significantly from those reported in Ref. 1, in some cases by more than 50%. In particular, the newer results for $SrTiO_3$ of Ref. 17 are not at all compatible with the $S=1$ value reported in Ref. 1. Moreover, a χ^2 analysis of the quality of the fits showed that the uncertainty in S is at least ±20% purely on the basis of data statistics. It may thus be still questionable whether the least-squares-fitted values describe the physical situation much better than the values reported in Ref. 1; they have, however, a statistical meaning. The main result is that straight-line fits (i.e., the linear interface theories) only very roughly describe the situation and may contain much arbitrariness like weighting of certain "good quality" metal-semiconductor contacts. The poor quality of the fits also indicates that a number of other effects, such as atomic interdiffusion[10] and formation of interface complexes, are of equal importance.

Bearing in mind these uncertainties, however, one may still attempt to classify MSI's in the spirit of the linear interface theory. Thus, in Figs. 2(a) and 2(b) the least-squares-fitted S values are plotted against the electronegativity difference ΔX and the semiconductor polarizability as proposed in Ref. 10. We used here the total polarizability $\epsilon_0 - 1$ which contains contributions from ion motion, i.e., ion relaxation in the interface region.[19] Though the data show a wide scatter, no saturation in S is obvious. More data of compounds with low polarizabilities would, however, be desirable to establish this point. It should be noted that both features, wide scatter and the absence of any saturation show up equally, whether S is plotted versus ΔX or versus $\epsilon_0 - 1$. However, the $\epsilon_0 - 1$ plot is more consistent with the spirit of the linear interface theory. To distinguish between the validity of ΔX or $\epsilon_0 - 1$ as coordinate, the case of diamond should provide a critical test. The results on diamond reported in Ref. 14 seem to favor the electronegativity coordinate ΔX. However, there is a possibility that graphitelike structures are formed at the C-metal interface, changing the effective polarizability to higher values.

TABLE I. Table of interface slope parameters $S = \partial\Phi_B/\partial X_m$. The quoted experimental values are least-squares fitted to the results of Refs. 1, 12, 14, and 22. Also indicated are theoretical results of Louie, Chelikowsky, and Cohen (LCC) (Ref. 8) using $A=2.86$, and empirical values estimated as described in the text.

	$\Delta S = \pm 20\%$ expt	LCC	Present estimate
C (diamond)	0.20		0.45
Si	0.08	0.13	0.16
Ge	0.09		0.13
GaAs	0.15	0.13	0.36
GaP	0.33		0.50
CdTe	0.16		0.56
CdSe	0.18		0.69
CdS	0.70		0.76
GaTe	0.32		
GaSe	0.57		
GaS	0.95		
SnO_2	0.83		
ZnSe	0.66	0.50	0.64
ZnS	1.18	0.88	0.87
ZnO	0.95		0.86
SiO_2	1.52		
$SrTiO_3$	0.45		

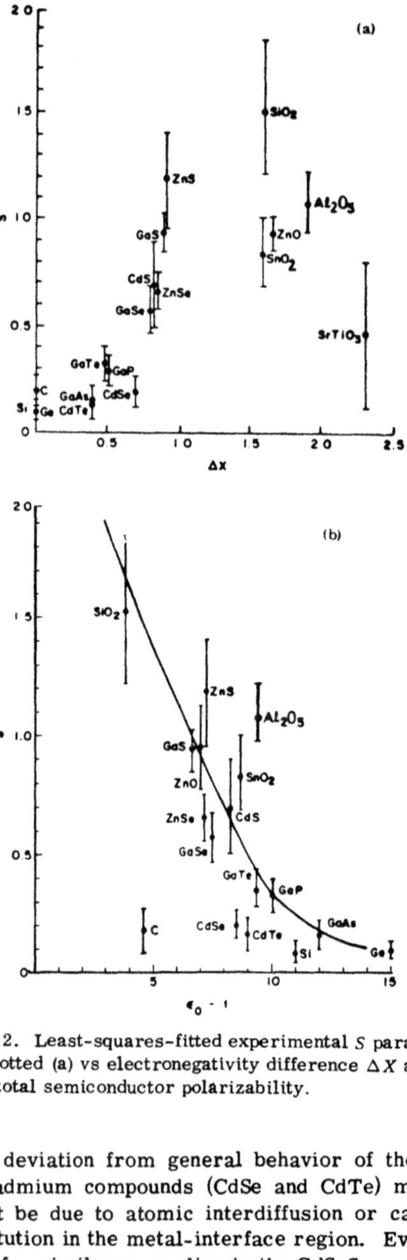

FIG. 2. Least-squares-fitted experimental S parameters plotted (a) vs electronegativity difference ΔX and (b) vs total semiconductor polarizability.

The deviation from general behavior of the two cadmium compounds (CdSe and CdTe) may in fact be due to atomic interdiffusion or cation substitution in the metal-interface region. Evidence for similar anomalies in the CdS_xSe_{1-x} alloy-interface systems have been reported[12] and interpreted in these terms.[10]

We shall now turn to the second problem, the definition of S and its upper saturation limit A. The linear theory predicts saturation at $\partial\Phi_B/\partial\Phi_m^I = 1$ or $S = \partial\Phi_B/\partial X_m = A$. In order to use tabulated metal electronegativities (e.g., Pauling's values[11]), a connection between the internal work function Φ_m^I and the electronegativity X_m has to be established; in other words, an energy scale has to be correlated with X_m. Mulliken[20] first noted that for group-I and group-VII elements the electronegativity X_m should be proportional to the arithmetic mean of the first ionization potential Φ and the electron affinity X:

$$M = \tfrac{1}{2}(\Phi + X) = AX_m, \quad \text{(in eV)}, \tag{1}$$

where M has been called the Mulliken potential[21] and expresses electronegativities in eV. By definition, this potential M may be correlated with the internal work function Φ_m^I. Mulliken's arguments were later extended to all first- and second-row elements and a proportionality factor[22] of $A = 3.15$ was deduced. Using different data, Pauling proposed a factor[11] of $A = 2.71$. A proportionality factor of $A = 2.86$, based on more-recent data, has been used very successfully by Nethercot[21] to predict photoelectric threshold energies for a variety of compound semiconductors.

A correlation of vacuum-metal work functions Φ_m with Pauling electronegativities has been given by Gordy and Thomas[23] and an empirical linear relationship of the form

$$\Phi_m = 2.27X_m + 0.34 \tag{2}$$

has been found. This form is roughly compatible with the values of A given above if one takes into account the wide spread of experimental work-function data. As noted, these are affected by an electrostatic surface dipole which is small for low-density materials such as Cs and which increases with electron density (i.e., increasing X_m) and becomes strongly surface-structure dependent.[18,24] This in fact would suggest that the value of A should be somewhat smaller than 2.27.

All arguments given above suggest that the Schottky limit for MSI behavior is reached at some number $S \approx 2.0-3.0$ which is not clearly defined and not exactly at $S = 1$ as previously proposed. Data on ZnS, ZnO, SnO_2, $SrTiO_3$, Al_2O_3, and SiO_2, previously believed to represent the Schottky limit, are thus far below any limiting value. Thus, we do not expect any saturation at $S = 1$ and in fact do not find any in our least-squares-fitted data. Moreover, calculations based on the linear interface theory[5-8] yield reasonable results of S if proportionality factors of $A \approx 2.0-3.0$ are used (see Table I, in which we used $A = 2.86$ as in Ref. 21).

We have also included in Table I estimated S values for some tetrahedrally coordinated semiconductors, which are empirically derived from dielectric properties using the linear equation[2,3,8]

$$S = A/(1 + 4\pi e^2 D_s \delta). \tag{3}$$

The interface density of states D_s is interpolated

between calculated[8] values for Si, GaAs, ZnSe, and ZnS assuming it to be inversely proportional to the average gap E_g in the semiconductor. The effective penetration depth δ of interface states is estimated to be

$$\delta = t_m/\epsilon_m + t_s/\epsilon_s , \qquad (4)$$

where $t_m/\epsilon_m \approx 0.5$ Å, a typical (rather constant) screening length in metals and where $t_s \approx 2\sqrt{\Delta}/E_g$, given by an effective semiconductor band width Δ and the average gap E_g.[3] The values of Δ vary from ~13.0 eV for Si to ~6.0 eV for ZnO and are interpolated using the electronegativity difference ΔX. The screening in the semiconductor ϵ_s is composed of a strongly wave-vector-dependent electronic part as described in Refs. 8 and 9 in addition to an ionic lattice part assumed to be wave-vector independent that accounts for atom relaxation effects in the interface. With these input data,[25] D_s and δ decrease from 5×10^{14} states/eV cm^2 and ~2 Å for Si to ~2×10^{14} states/eV cm^2 and ~0.6 Å for ZnO, respectively. As can be seen from Table I, this very crude model yields reasonable S values for most of the semiconductors with uncertainties comparable to the experimental errors.

We find from this analysis that the original MSI data show an enormous amount of scattering and that straight-line fits yielding the interface index S are rather uncertain. No saturation at $S = 1$ is found if the data are least-squares fitted. This also is theoretically unexpected since S is defined with respect to the metal electronegativity and not with respect to an *internal* work function. Theoretical considerations rather suggest some limiting value of S between 2.0 and 3.0, which may not clearly be defined.

The linear one-electron MSI theory, while not expected to be quantitative, yeilds reasonable answers for MSI behavior in *qualitative* agreement with experiment, if one keeps in mind the uncertainties in experiment and data analysis. More systematic experimental results are clearly needed before conclusions as to the range of validity of this theory can be drawn, in particular to decide on the appropriateness whether to use the electronegativity difference or the polarizability as a characteristic coordinate.

ACKNOWLEDGMENTS

Helpful discussions with J. C. Phillips, S. G. Louie, J. Appelbaum, and T. C. McGill are gratefully acknowledged. I wish to thank Mrs. R. Fulton for assistance in the data analysis.

---

[1]S. Kurtin, T. C. McGill, and C. A. Mead, Phys. Rev. Lett. 22, 1433 (1969).
[2]A. M. Cowley and S. M. Sze, J. Appl. Phys. 36, 3212 (1965).
[3]V. Heine, Phys. Rev. 138, A1689 (1965).
[4]J. Bardeen, Phys. Rev. 71, 717 (1947).
[5]F. Yndurain, J. Phys. C 4, 2849 (1971).
[6]F. Flores, E. Louis, and F. Yndurain, J. Phys. C 6, L465 (1973).
[7]E. Louis, F. Yndurain, and F. Flores, Phys. Rev. B 13, 4408 (1976).
[8]S. G. Louie, J. R. Chelikowsky, and M. L. Cohen, J. Vac. Sci. Technol. 13, 790 (1976).
[9]J. C. Inkson, J. Phys. C 5, 2599 (1972); 6, 1350 (1973).
[10]J. C. Phillips, Phys. Rev. B 1, 593 (1970); J. Vac. Sci. Technol. 11, 947 (1974).
[11]L. Pauling, *The Nature of the Chemical Bond* (Cornell University, New York, 1960).
[12]C. Mead, Solid State Electron. 9, 1023 (1966).
[13]T. C. McGill, J. Vac. Sci. Technol. 11, 935 (1974).
[14]C. A. Mead and T. C. McGill, Phys. Lett. A 58, 249 (1976).
[15]C. M. Varma and K. C. Pandey (unpublished).
[16]A. Thanailakis, J. Phys. C 8, 655 (1975).
[17]R. C. Neville and C. A. Mead, J. Appl. Phys. 43, 4657 (1972).
[18]V. Heine and C. H. Hodges, J. Phys. C 5, 225 (1972).
[19]K. F. Young and H. P. R. Frederikse, J. Phys. Chem. Ref. Data 2, 313 (1973).
[20]R. S. Mulliken, J. Chem. Phys. 2, 782 (1934); 3, 573 (1935).
[21]A. H. Nethercot, Phys. Rev. Lett. 33, 1088 (1974).
[22]H. A. Skinner and H. O. Pritchard, Trans. Faraday Soc. 49, 1254 (1953).
[23]W. Gordy and W. J. O. Thomas, Phys. Rev. 24, 439 (1956).
[24]N. D. Lang and W. Kohn, Phys. Rev. 1, 4555 (1970).
[25]A. Braunstein, M. Braunstein, G. S. Picus, and C. A. Mead, Phys. Rev. Lett. 14, 219 (1965).

Transition in Schottky Barrier Formation with Chemical Reactivity

L. J. Brillson
Xerox Webster Research Center, Webster, New York 14580
(Received 21 November 1977)

Low-energy electron-loss spectra reveal that a majority of common metals react with CdS, CdSe, and most other covalent semiconductors. Furthermore, barrier heights of metals on individual compound semiconductors exhibit a sharp transition as a function of heat of reaction, increasing dramatically above an experimentally determined critical heat of reaction.

The mechanism of Schottky barrier (SB) formation at metal-semiconductor (M-SC) contacts has been the subject of long-standing controversy. Until recently, the phenomenon of Fermi level pinning associated with interface band bending was thought to be associated with intrinsic surface states of the semiconductor.[1] However, studies of the microscopic II-VI compound semiconductor interfaces by ultrahigh vacuum techniques demonstrate that extrinsic phenomena due to the presence of the deposited metals dominate the SB formation. In particular, reactive metals on these semiconductors form interface compounds,[2] whereas unreactive metals induce new interface states within the band gap.[3] For all metals, dipole layers form which account for the observed electrical behavior of the contact.[4] In this Letter, I present evidence to show that the chemical reactivity of the constituents at a M-SC interface play a central role in determining the SB heights. Kurtin, McGill, and Mead[5] (KMM) have described a fundamental transition in interface behavior related to electronegativity difference of the semiconductor constituents and thereby the degree of lattice ionicity. In turn, this was related to intrinsic surface states via the discontinuity in lattice potential at the surface of covalent versus ionic semiconductors.[6,1] However, this transition can be described equally well in terms of chemical reactivity. Furthermore, the Sb heights of metals on individual semiconductors display a striking correlation with chemical reactivity, exhibiting a sharp transition as a function of heat of reaction ΔH_R at a critical value determined from low-energy electron-loss spectroscopy (ELS). These results emphasize the importance of local charge redistribution associated with interface chemical reactions rather than with intrinsic surface states of the semiconductor.

Figure 1 illustrates the parameter of interface behavior S for various semiconductors, where S is defined by[5] $qV_B = S(\varphi_M - \varphi_{SC}) + C$. Here qV_B, φ_M, and φ_{SC} are the band bending, metal work function, and semiconductor work function, respectively, and C is a constant. The same S values used by KMM are represented here. However, the abscissa of this curve is heat of formation ΔH_f rather than electronegativity difference ΔX. The smooth curve fitted to the data points is *identical* to that in the S vs ΔX curve of KMM. This correlation shows that ΔH_f (and thereby ΔH_R) describe the transition in S at least as well as ΔX. Figure 1 shows that covalent semiconductors have lower ΔH_f than ionic semiconductors and thus are more likely to react with metals at M-SC interfaces. While ΔH_f and ΔX can be related,[7] the chemical heats alone emphasize the microscopic interface's role in determining SB heights.

Figure 1 also shows the representative nature of CdS and CdSe metal interfaces.[6] The wide span of interface behavior between these two systems suggests that their electronic and chemical properties are characteristic of M-SC interfaces in general. ELS measurements of metals on CdS and CdSe reveal the presence of chemically reacted interface layers. For such interfaces, new ELS excitation features appear which are not due to the bulk semiconductor or metal. For Al on CdS and CdSe, these features have been related to coupled interface plasmon modes of the reached interface layer and surrounding media.[2] For unreactive interfaces such as Au on CdSe,[4] only a superposition of ELS features from the semiconductor and metal appear. This contrast between reactive and unreactive interfaces is evident for increasing overlayer thicknesses of Cu vs Ag on CdS(10$\bar{1}$0) as shown in Fig. 2. These spectra were obtained with a cylindrical mirror analyzer (CMA) and a glancing incidence electron gun. Incident beam energy was 100 eV and instrumental resolution was 0.7 eV. The ultrahigh vacuum apparatus has been described elsewhere.[8]

The ELS features observed for the cleaved CdS(10$\bar{1}$0) surfaces correspond to bulk and surface plasmons ω_p and ω_{SP}, a Cd 4d level transition to the conduction band, and interband transitions labeled I_1, I_2, and I_3. For several angstroms of Cu on CdS, new ELS features (shaded) appear which are not due to the bulk Cu or CdS. These features change with increasing metal cov-

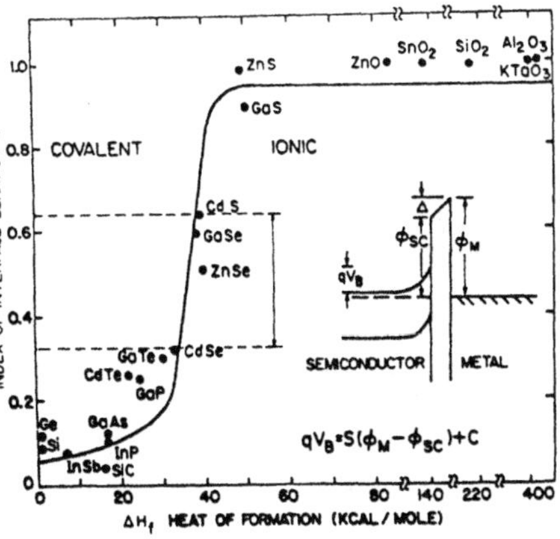

FIG. 1. Index of interface behavior (Ref. 5) S plotted vs heat of compound formation. Dashed lines indicate S range spanned by CdS and CdSe.

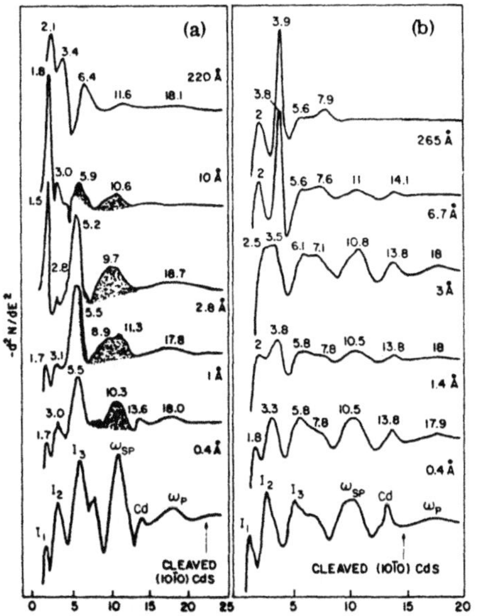

FIG. 2. Electron-loss spectra of reactive vs unreactive interfaces as a function of metal overlayer coverage: (a) Cu on cleaved CdS (10$\bar{1}$0); features unrelated to bulk Cu or CdS are shaded; (b) Ag on cleaved CdS (10$\bar{1}$0).

erage, eventually corresponding to losses of the bulk metal.[9] Analogous features appear for Cu and Ag on CdSe(11$\bar{2}$0). Furthermore, the adsorbed Cu Auger features are chemically shifted from those of bulk Cu. In contrast, no dramatic changes occur for Ag on the same CdS(10$\bar{1}$0) surface under identical conditions. With increasing Ag coverage, the spectra manifest only a slowly changing superposition of Ag and CdS features.

The indication of M-SC reactivity from ELS spectra can be used to estimate a critical heat of reaction ΔH_R^c for metals with CdS and CdSe. Chemical heats of reaction ΔH_R for most common metals with CdS and CdSe are listed in Table I. These ΔH_R were calculated per metal atom for the reaction $M + (1/x)CA \rightarrow (1/x)[M_xA + C]$ from ΔH_f's[10] of compound semiconductor CA and most stable metal-anion product[11] M_xA, analogous to the ΔH_f calculations of transition metals on Si reported by Andrews and Phillips.[12] The stars in Table I indicate the most stable compounds at the interfaces probed with ELS. The extent of reactivity for Al, Ag, Cu, and Au on CdS and CdSe outlines a division between reactive versus unreactive interfaces at $\Delta H_R^c \sim 0.5$ eV per metal atom. Reactions occur for $\Delta H_R > 0$ due to the incomplete dissociation of semiconductor cations from the new interface molecules, which reduces the barriers to chemical reaction. The divisions in Table I demonstrate that most common metals react with CdS and CdSe. Furthermore, since covalent semiconductors are less stable against interface reaction than CdS or CdSe, these metals should react with these compounds as well. Clearly, for such M-SC junctions, the microscopic charge redistribution and atomic rearrangement associated with chemical reactions render any intrinsic surface states of the semiconductor irrelevant to the interfacial electronic properties.

M-SC barrier heights also exhibit a systematic dependence on chemical reactivity. Figure 3 illustrates this relationship between qV_B vs ΔH_R for a variety of metals on four compound semiconductors. Barrier heights were measured primarily by internal photoemission for semiconductors cleaved in vacuum.[6] Each curve yields a fit to the data points at least as good as linear plots of qV_B vs φ_M or X.[6]

In all four graphs, the fitted curves exhibit sharp changes in qV_B at $\Delta H_R \sim \Delta H_R^c$ as determined for CdS and CdSe. The center of this tran-

TABLE I. Heats of reaction ΔH_R for common metals with CdS and CdSe. Stars designate interfaces studied with ELS. The division between reactive and unreactive interfaces indicates a critical heat of reaction.

	METAL SULFIDE	ΔH_R (eV/metal atom)	METAL SELENIDE	ΔH_R (eV/metal atom)
Unreactive	★ AuS	3.94	★ AuSe	1.36
	Bi_2S_3	1.99	HgSe	1.05
	HgS	1.00	WSe_2	1.05
			$FeS_{0.96}$	0.75
			$PtSe_{0.8}$	0.69
	PtS	0.69	SnSe	0.58
	★ Ag_2S	0.61	★ Ag_2Se	0.52
	PbS	0.53	PbSe	0.46
	FeS	0.51	Bi_2Se	0.42
Reactive	★ Cu_2S	0.50	★ Cu_2Se	0.41
	SnS	0.43	In_2Se	0.16
	WS_2	0.41		
	$NiS_{0.67}$	0.29		
	InS	0.16		
	CrS	0.15		
	TaS_2	-0.57	MnSe	-0.10
	ZnS	-0.58	ZnSe	-0.15
	GaS	-0.62	Ga_2Se	-0.20
	MnS	-0.66	★ Al_2Se_3	-0.69
	Cs_2S	-1.00		
	Ti_2S	-1.27	$TiSe_{1.5}$	-1.00
	★ Al_2S_3	-1.43	MgSe	-1.54
	MgS	-2.10	$ThSe_{1.7}$	-2.66
	$ThS_{1.7}$	-3.53		

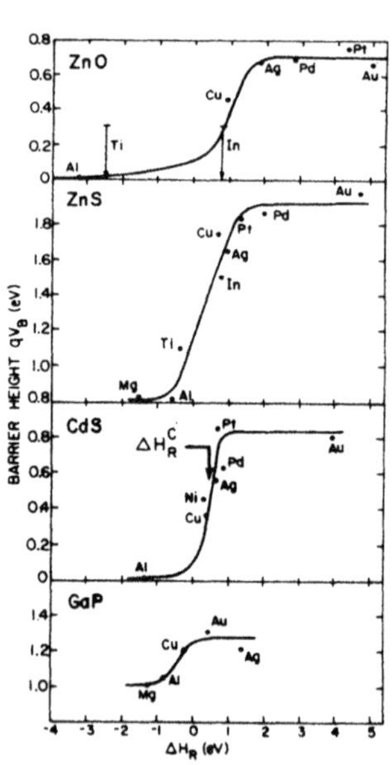

FIG. 3. Barrier heights (Ref. 6) correlated with heats of chemical reactivity for metals on ZnO, ZnS, CdS, and GaP.

sition occurs at 0.3 eV for CdS and ZnS. This transition center is higher for the more stable compound ZnO (0.4 eV) and lower for the less suitable compound GaP (− 0.2 eV). Although these four semiconductors encompass a wide range of interface behavior, their qV_B vs ΔH_R transition slopes vary by only a factor of 2 and exhibit no apparent dependence on chemical stability or ionicity. Similar plots for other metal–compound-semiconductor systems such as ZnSe and GaAs suggest analogous behavior. The linear dependence of qV_B vs ΔH_R for transition metals on elemental Si corresponds to the left side of such a chemical reactivity plot, since $\Delta H_R < 0$ for all associated interfaces.[12] These correlations in terms of metal-semiconductor anion reactivity are also consistent with the qV_B dependence on anion properties of Au-semiconductor junctions.[13] The extension of reactivity plots to other semiconductor systems appears to be limited only by the availability of electronic and thermodynamic information.

The agreement between the transition center of the reactivity plot and ΔH_R^c for metals on CdS suggests that low qV_B on the left-hand side of the curves of Fig. 3 are related to reactive interfaces while high qV_B on the right-hand side are due to unreactive junctions. Charge relocalization and atomic rearrangement can account for this barrier height transition qualitatively. The semiconductor barrier height depends on both microscopic dipole formation at the intimate M-SC contact and charge transfer to or from the semiconductor surface space charge region. Microscopic charge redistribution at the interface is commonplace and substantial, as evidenced by the many low-work-function metals on semiconductors which produce n-type Schottky barriers even though $\varphi_M - \varphi_{SC} < 0$.[6] This charge redistribution is driven by the difference in bulk exchange correlation energies[14] and has been observed[15] as electron tunneling from metal into semiconductor band gap.[16] The position with respect to the anions of metal atoms on or within the outermost semiconductor layer can also affect the direction and magnitude of charge exchange. One may hypothesize that the extent of compound formation controls both the local charge redistribution and atomic positions at the interface, thereby dominating the barrier heights.

In conclusion, ELS results and a striking correlation between M-SC barrier heights and chemical reactivities are reported here which strongly suggest replacing the concept of Fermi level pinning via intrinsic surface states by that of microscopic charge redistribution at the M-SC junction.

The author gratefully acknowledges valuable discussions with Professor J. Bardeen, Dr. R. S. Bauer, and Dr. C. B. Duke.

[1] S. G. Davisson and J. D. Levine, in *Solid State Physics*, edited by H. Ehrenreich, F. Seitz, and D. Turnbull (Academic, New York, 1970), Vol. 25, p. 2, and references therein.

[2] L. J. Brillson, Phys. Rev. Lett. 38, 245 (1977).

[3] L. J. Brillson, in *Proceedings of the Thirteenth International Conference on Physics of Semiconductors, Rome, Italy, 1976*, edited by F. G. Fumi (North-Holland, Amsterdam, 1977), p. 665.

[4] L. J. Brillson, to be published.

[5] S. Kurtin, T. C. McGill, and C. A. Mead, Phys. Rev. Lett. 22, 1433 (1970).

[6] C. A. Mead, Solid State Electron. 9, 1023 (1966), and references therein.

[7] L. Pauling, *The Nature of the Chemical Bond* (Cornell Univ. Press, Ithaca, New York, 1960); J. C. Phillips and J. A. Van Vechten, Phys. Rev. B 2, 2147 (1970). The accuracy of such relationships is also discussed here.

[8] L. J. Brillson, Surf. Sci. 69, 62 (1977).

[9] J. Daniels, C. v. Festenberg, H. Raether, and K. Zeppenfeld, in *Springer Tracts in Modern Physics*, edited by G. Höhler (Springer, New York, 1970), Vol. 54, p. 77.

[10] D. D. Wagman *et al.*, *Selected Values of Chemical Thermodynamic Properties*, National Bureau of Standards Tech. Notes 270-3–270-7 (U. S. GPO, Washington, D. C., 1968–1971).

[11] Alternatively, first nucleation products in transition metal silicide and germanide reactions have been determined from bulk phase diagrams by R. M. Walser and R. W. Bene, Appl. Phys. Lett. 28, 624 (1976).

[12] J. M. Andrews and J. C. Phillips, Phys. Rev. Lett. 35, 56 (1975).

[13] J. O. McCauldin, T. C. McGill, and C. A. Mead, Phys. Rev. Lett. 36, 56 (1976).

[14] C. B. Duke, J. Vac. Sci. Technol. 6, 152 (1969).

[15] F. Steinrisser, L. C. Davis, and C. B. Duke, Phys. Rev. 176, 912 (1968).

[16] V. Heine, Phys. Rev. 138, A1689 (1963).

New and unified model for Schottky barrier and III–V insulator interface states formation

W. E. Spicer, P. W. Chye, P. R. Skeath, C. Y. Su, and I. Lindau

Stanford Electronics Laboratories, Stanford University, Stanford, California 94305

(Received 22 May 1979; accepted 18 October 1979)

For *n*- and *p*-doped III–V compounds, Fermi-level pinning and accompanying phenomena of the (110) cleavage surface have been studied carefully using photoemission at $h\nu \lesssim 300$ eV (so that core as well as valence band levels could be studied). Both the clean surfaces and the changes produced, as metals or oxygen are added to those surfaces in submonolayer quantities, have been examined. It is found that, in general, the Fermi level stabilizes after a small fraction of a monolayer of either metal or oxygen atoms have been placed on the surface. Most strikingly, Fermi-level pinning produced on a given semiconductor by metals and oxygen are similar. However, there is a strong difference in these pinning positions depending on the semiconductor: The pinning position is near (1) the conduction band maximum (CBM) for InP, (2) midgap for GaAs, and (3) the valence band maximum (VBM) for GaSb. The similarity in the pinning position on a given semiconductor produced by both metals and oxygen suggests that the states responsible for the pinning resulted from interaction between the adatoms and the semiconductor. Support for formation of defect levels in the semiconductor at or near the surface is found in the appearance of semiconductor atoms in the metal and in disorder in the valence band with a few percent of oxygen. Based on the available information on Fermi energy pinning, a model is developed for each semiconductor with two different electronic levels which are produced by removal of anions or cations from their normal positions in the surface region of the semiconductors. The pinning levels have the following locations, with respect to the VBM: GaAs, 0.75 and 0.5 eV; InP, 0.9 and 1.2 eV (all levels + 0.1 eV). The first energy given is assocaited with a missing anion and the second with a missing cation. For GaSb, only an acceptor due to a missing Sb has been located at 0.1 eV. Our work is found to correlate well with that on practical Schottky barriers. A detailed comparison is made with interface state positions and densities found by others on practical MIS structures, and it is suggested that the large density of these states on III–V's as compared to Si is due to extrinsic interface states created by stoichiometric deficits of the III–V semiconductor. For GaAs, the dominant state is found at 0.7 eV and is associated with an As deficit. For InP, the major interface level is about 0.1 eV below the CBM. These positions are in good agreement with the existing data obtained from a wide variety of samples.

PACS numbers: 73.30. + y, 71.55.Fr, 79.60.Eq, 73.40.Qv

I. INTRODUCTION

At the 1976 Physics of Compound Semiconductor Interface (PCSI) Meeting, all experimental groups reported[1-3] agreed that, on the (110) faces of GaAs (and other smaller band gap III–V semiconductors), there were no intrinsic[4] surface states in the band gaps. Since the classic work by Mead *et al.*[5] was also done on cleavage (110) faces of these semiconductors, this had strong implications for Schottky-barrier formation. From LEED[6] and bonding considerations,[7] it appeared likely that the surface states were moved out of the band gap region by electronic and lattice rearrangement at the surface. Work since then has firmly established that this is the case.[8,9] Figure 1 schematically shows the rearrangement and the electronic structure.[10] The lack of intrinsic states in the band gap had immense implications for the understanding of Schottky barriers (SB) since it established that the free surface contained no intrinsic surface states which could pin the Fermi level at the metal–semiconductor in the manner suggested by Bardeen[11] and applied by Mead *et al.*[5]

This left quantum mechanical interaction between metal and semiconductor as the most popular mechanism of Schottky barrier pinning. Heine[12] had first pointed out such a possibility and had emphasized the tunneling of electrons from the metal into the band gap of the semiconductor. In the period of approximately 1976–1978, there have been a number of theoretical papers exploring in a more quantitative manner this possibility as well as looking into related many-body effects.[13] These papers had one aspect in common: they assumed an ideal planar junction between the semiconductor and metal. Recent experimental work strongly suggests that the situation is not so simple.[14-16]

In this paper, we will propose a detailed and unified model not only for Schottky barrier pinning but for the interface states at the semiconductor–insulator interface of MIS (metal–insulator semiconductor) structures.[17] This is an ex-

 0022-5355/79/051422-12$01.00

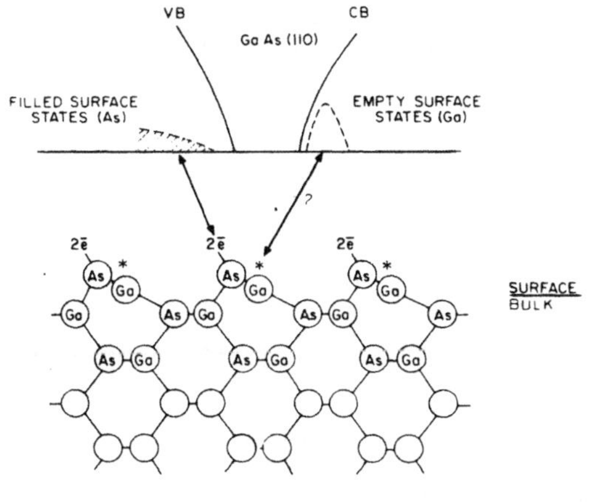

FIG. 1. A schematic drawing of the lattice and electronic structure of the GaAs (110) face after rearrangement (relaxation). The As surface atoms have taken up a p^3 bonding arrangement (with two electrons in a $4s^2$ filled state), while the Ga has gone from the bulk sp^3 to a sp^2 arrangement. Since the p^3 bond angles are more acute than the bulk sp^3 bonds, the As moves outward; conversely, the Ga moves inward. The movements are large (large fractions of an angstrom). This rearrangement moves the filled and empty surface states out of the band gap. However, since the surface lattice is no longer lattice match to the bulk, there is strong strain at the surface.

tension of our work as reported by Lindau *et al.*[15] at last year's PCSI. Our general approach has been to (1) develop experiments in which the interaction between the adatoms (metal or oxygen) and semiconductor can be followed on an atomic scale, particularly in the early stages of Fermi-level pinning, i.e., very low (small fractions of a monolayer) coverages, (2) systematically study the Fermi level pinning at both semiconductor–metal and semiconductor–oxide interfaces on both n- and p-doped semiconductors, and (3) compare results from thick device-type overlayers and those from the atomic scale studies. Thus, in this paper, we will draw extensively on data from our own studies of the first steps in oxygen adsorption[3,10,18–20] on GaAs, InP, and GaSb as well as on our work on Schottky barrier formation.[14,15] In addition, we will draw on work of others[21–27] on thicker "device"-type oxides.

II. DO INTRINSIC STATES MOVE INTO THE BAND GAP?

As emphasized in the introduction, it became clear between 1976 and 1978 that, on the (110) face of GaAs and smaller band gap III–V compounds, the intrinsic surface states were swept out of the band gap due to the rearrangement of the atoms near the surface. At once, we became concerned that the pinning we and others[28,29] had seen consistently on clean-cleaved (110) n-type GaAs might be due to intrinsic surface states brought into the band gap by a new and different reconstruction of the surface.[2]

Starting in 1976,[3] we have continuously attempted to test the hypothesis that intrinsic surface states might move into the band gap due to new surface rearrangements. We will briefly review that work since the negative results obtained played an important role in developing the model of interface states which will be presented here.

First, it was recognized by ourselves and Peter Mark that all of the LEED work had been done on sputter-annealed (110) GaAs surfaces and that such clean surfaces always showed Fermi level pinning near midgap for n-type samples. In contrast, no LEED had been done on cleaved surfaces and these were the only surfaces which showed[1–3] no surface Fermi level pinning (i.e., cleaved surfaces sometimes showed no pinning, whereas sputter-anneal surfaces always showed pinning. Thus, LEED work was done on a number of cleaved GaAs (110) surfaces and compared to the sputter and annealed work.[30,31] No measurable difference was found, establishing that the same surface rearrangement did take place on both types of surfaces. This and related work was reviewed by Skeath *et al.* in last year's PCSI.[30]

The overall conclusion is that the pinning which is often observed on clean cleaved n-GaAs (110) is due to extrinsic states produced by defects formed in the cleaving process. Any comprehensive model for extrinsic states must satisfactorily explain these states.

Monch[32] has pointed out that bulk doping impurities may produce extrinsic states at the surface; however, the density of these is not, in general, sufficient to explain the observed pinning; thus, defect states must be invoked, in addition. However, the bulk doping can reduce the number of defects needed.

Another more general method for studying the empty surface states is the constant final state (CFS) or partial yield spectroscopy developed by Lapeyre[33] and Kunz[34] and their co-workers and ingeniously applied to the surface state

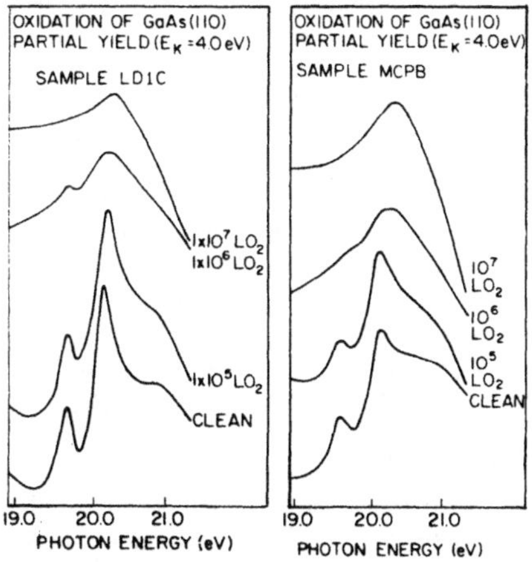

FIG. 2. Measurements of the exciton produced by optical excitation from the Ga 3*d* core level to the lowest lying intrinsic empty states at the surface. For sample LD1C, the Fermi level was not pinned at the surface; whereas, in sample MCPB, it was. As can be seen, the position of surface excitons are the same, independent of pinning. This establishes that the pinning is not due to intrinsic surface states. The upper curves in each panel indicate the change in the exciton as a function of oxygen exposure. By an exposure of 10^6 to 10^7 L (a few percent of a monolayer), the exciton is no longer seen. This is roughly the exposure at which the valence band disorders (as will be shown in Fig. 7).

problem by Eastman and co-workers.[35] It is now clear that the original interpretation of such work[35] erred by not taking excitonic effects into account. However, in the studies we undertook, we were most interested not in the absolute energy of the final state involved in the transition but in decreases in this energy which would be associated with the empty intrinsic surface states moving[36] into the band gap to produce the observed surface Fermi level pinning. This movement, since it should be an electron volt or more, would produce an easily detected decrease in the excitation energy for the exciton. Since these studies have been reported in detail elsewhere, we will restrict ourselves here to giving examples and discussing the results and appropriate references.

A first question (also addressed above) was whether the pinning sometimes observed on n-type cleaved samples was due to intrinsic surface states moving into the band gap region. Our CFS measurements showed no difference in excitation energy for the exciton on pinned surfaces and unpinned surfaces.[3,37] This is taken as definitive evidence that the pinning which occurs on cleaving is due to extrinsic and not intrinsic surface states. As an example of this type of data, Fig. 2 compares spectra taken from cleavage surfaces which did and did not show pinning.[3,37]

The same test was made with regard to pinning due to placing of oxygen or metals on the surface. As the oxygen or metal is added in small quantities to a clean surface which shows no pinning, both the excitonic structure in the CFS spectra and the Fermi level pinning are followed as a function of coverage of the foreign atom.[38] Examples are given in Fig. 2 for oxygen addition. The paper by Skeath *et al.*[39] in these proceedings gives an example for metals as does the work of Chye *et al.*[14] Again, no evidence is found for reduction of the exciton energy[3,18,19] as the overlayer is added, showing that the empty intrinsic surface states are not moving into the band gap. These data will be discussed in detail later in this article. Thus, we conclude that the pinning observed after deposition of metals or oxygen is not due to movement of the intrinsic states characteristic of the free surface into the band gap because of surface rearrangement produced by the addition of oxygen or metals to the surface. Mele and Joannopoulos[40] have proposed new interface states in the band gap due to Al on GaAs. It appears that, to detect these, one must look at the CFS structure using excitation from the Al core levels.

III. A MECHANISM FOR FERMI LEVEL PINNING BY METAL OR OXYGEN DEPOSITION

A. General approach

Using photoemission, the Fermi level position at the surface can be measured directly.[39] It is important to make a clear distinction between such measurements and those (see, for example, Ref. 41) in which Fermi level is not measured directly, i.e., when methods such as contact potential (CPD) measurements are used.[41] In the latter case, the changes in the potential barrier at the surface must be successfully separated from those of band bending. On addition of any adatoms to the surface of a semiconductor, both the Fermi level position and the semiconductor electron affinity will, in general, change.[42] It is the sum of these two effects which is measured

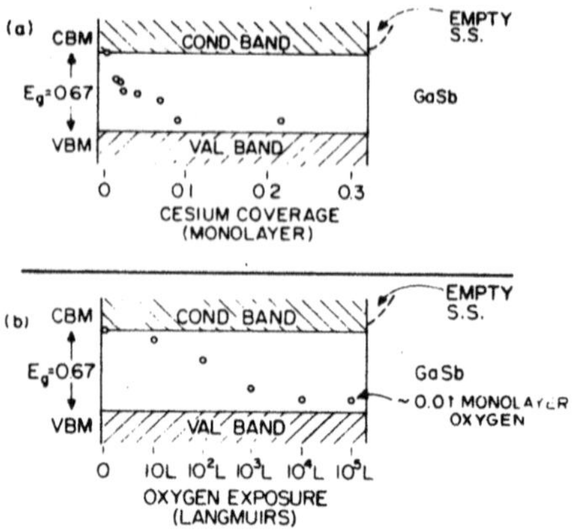

FIG. 3. Starting with a clean cleavage face (110) of an n-type GaSb crystal, Cs (panel a) or oxygen (panel b) is added as indicated on the abscissa. In both cases, the surface Fermi level position moves through the band gap and pins at about 0.1 eV above the VBM for submonolayer coverages. Similar motion of the surface Fermi level is seen in GaAs and InP with oxygen and metals. However, the final pinning positions are different.

by CPD measurements. As the comments after Ref. 41 indicate, these are not always simple to sort out and serious errors can be made in estimating the change in Fermi level at the surface. In contrast, the photoemission measurement, as mentioned above, obtains the Fermi level position directly. This is illustrated in more detail by Skeath *et al.*[39]

B. Fermi level pinning versus metal coverage

Our general approach in this work has been to deposit metal in rather small increments and to determine the change of surface Fermi level position with coverage. Detailed data are present in these proceedings (Ref. 39), in last year's proceedings (Ref. 15), and elsewhere (Refs. 14 and 43). An example is given in Fig. 3(a).[15] As can be seen, the change in Fermi level pinning due to the metal is essentially completed at only about 0.1 of a monolayer coverage. Detailed studies of the 5d bands of Au have indicated that the Au atoms are dispersed across the surface and essentially atomic in nature.[14,43] This is also the case for Cs. Thus, the pinning takes place before a metallic layer is formed. This brings into question those theoretical calculations that start from the model of a metal on a semiconductor.

C. Similarity of Fermi level pinning for oxygen and metals

It is interesting to note that there is surprisingly little difference (~0.3 eV on GaAs, less on GaSb) in the pinning position[14] for Cs and Au despite the very large difference in the nature of the outermost valence wave functions of the atoms of these two metals. Even more striking is the similarity of the Fermi-level pinning position due to oxygen and the metals. In Fig. 3(b), we present Fermi-level position due to oxygen on GaSb. As can be seen, the final Fermi-level pinning position is almost identical to that of the metals and the pinning is completed at a small fraction[44] of a monolayer.[45] Similar results have been obtained on GaAs and InP.[3,18,19,44,51–53,64,65]

In Fig. 4, we present our data on the pinning position for a number of metals and oxygen on GaAs, InP, and GaSb. This figure has two striking characteristics. First, the oxygen pinning position falls within the scatter of surface Fermi-level positions from the metals. Second, the pinning positions for each semiconductor fall in a distinct and separate part of the band gap: InP in the upper part of the gap, GaAs near midgap, and GaSb near the valence band maximum.

D. Are Pinning levels produced indirectly by adatoms?

Why would both metals and oxygen produce levels at approximately the same energy? It is very difficult, if not impossible, to rationalize this in terms of energy levels produced directly by the adatoms. This statement is based on the gross differences in atomic energy levels (as well as chemistry) between the various adatoms. On the other hand, if the pinning states are produced indirectly, for example, by the formation of vacancies or other defects, the energy levels of these defects would not necessarily depend on the details of the adatoms. Thus, it is attractive to think of the pinning produced by the adatoms as being due to defects produced at or near the surface of the semiconductor due to the deposition of the adatoms. These defects must then be created by the interaction of the adatoms with the semiconductor surface.

Such a suggestion raises at least two new questions: (1) Is there any experimental information suggesting the formation of such defects and (2) where does the activation energy come from to form such defects? Last year,[15] we made some suggestions in this regard. In this paper, we will refine and generalize those original suggestions. In addition, we will make use of data in the literature which relate to interface states on thick-device-type oxides in order to test our suggestions. However, first, we will present a model of the interface states produced at the interface of InP, GaAs, and GaSb by metal deposition or oxygen exposure.

IV. THE MODEL FOR INTERFACE STATES FORMED BY ADATOMS

In Fig. 5, we present models for the interface states formed near the surface of GaAs, InP, and GaSb by addition of oxygen or metals to the surface. For each material, two levels are given and three quantities are associated with each of these levels: (1) the energy level [measured with respect to the valence band maximum (VBM)], (2) whether the defect is an acceptor or donor,[46] (3) a suggestion as to the nature of the defect. However, it is important to emphasize that the essence of the model is that pinning is dominated by defect levels formed due to the addition of metals or oxygen to the surface.

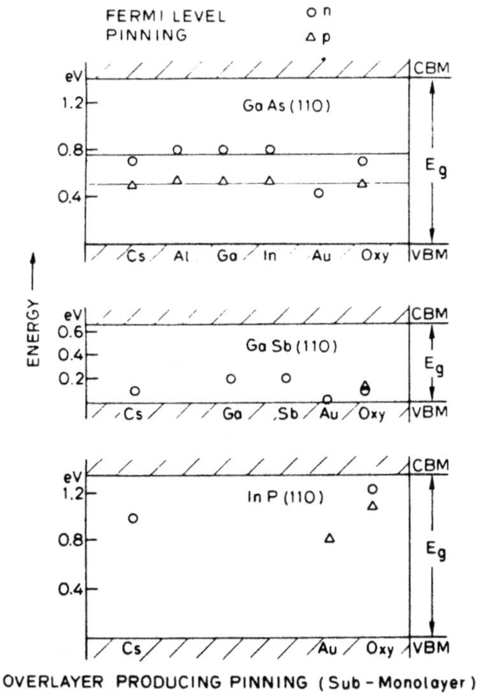

FIG. 4. A summary of the final surface Fermi level pinning positions obtained by experiments similar to those detailed in Fig 3 for a wide variety of metals and oxygen on GaAs, InP, and GaSb. The absolute energy positions are located no better than ±0.1 eV although relative positions may be better. Note the striking differences in the surface Fermi level positions between the three semiconductors.

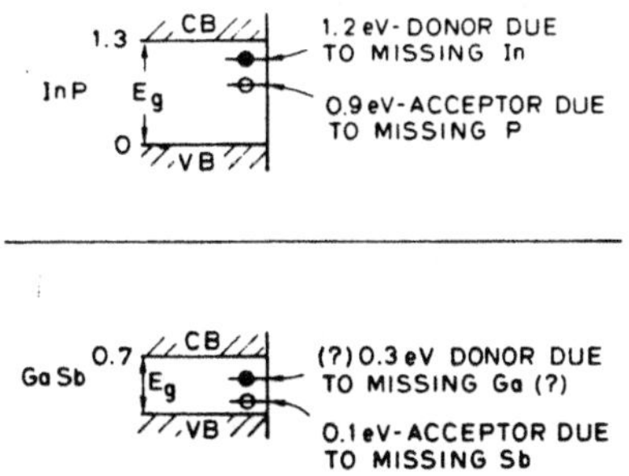

FIG. 5. A model of extrinsic states produced near or at the surface by perturbing the surface through addition of metals or oxygen to the surface. Each level is tentatively associated with a deficit of a anion or cation; however, it is unlikely that the defect is a vacancy. More likely, it is an antisite or more complicated defect.

As Fig. 4 shows and as will be shown when we examine practical oxides, the general systematics of pinning positions is in agreement with the specific energy levels in Fig. 5, i.e., the pinning levels lie in the upper part of the band gap for InP, near mid-gap for GaAs, and near the VBM for GaSb.

Two levels were used in Fig. 4, as this is the smallest number which can begin to give agreement with a large number of different experiments. However, once one starts creating defect levels, it should be recognized that a large number of different defect combinations are possible (e.g., vacancies, anti-site defects, and various combinations of vacancies, etc.).[47] Therefore, the exact energy positions are, in general, not to be taken too seriously within a tenth of an electron volt. However, as emphasized above, the energy position within these limits is to be taken quite seriously.

Also indicated on Fig. 5 is the acceptor or donor nature of the defect. This has been deduced by the dependence of the pinning on doping type of the material. For example, there was a long experience of no pinning on cleaved p-type GaAs but pinning at 0.7–0.8 eV for n-type samples. This indicates an acceptor at 0.75 ± 0.1 eV. Similarly, the pinning of n-type GaSb near the VBM (see Fig. 3) must be due to an acceptor level.

The experimental and theoretical data on bulk defect levels do not seem to be definitive enough to act as a strong guide to the nature of the defect. Thus, we will take a simple model and assume that donors are formed by anion (column 5) atoms in cation (column 3) sites and acceptors are due to the inverse situation with cation atoms in anion sites. These models should be considered as tentative.

As can be seen from the experimental results presented above (and that which will be presented below), one type of defect is often predominant in a given material after a specific treatment, e.g., the 0.75 eV level in GaAs after cleaving and the 0.1 eV level in GaSb after adsorption of oxygen or deposition of a metal. The pinning of the as-cleaved n-type material at 0.75 eV together with the lack of pinning on p-type GaAs give clear evidence of dominance of the 0.75 eV acceptor level for this surface. Figure 3 gives strong evidence for the dominance of the 0.1 acceptor level in GaSb. A similar level has long been associated with a deficit of Sb which occurs in GaSb.[48] The nature of the second defect in GaAs and GaSb is tentatively assigned by the author's liking for symmetry and order. In GaSb, the Sb defect appears to dominate; pinning toward mid-gap only appears after several molecular layers of oxide have been grown. This casts some doubt on the second level for GaSb, as it may be associated with the oxide rather than interface. Thus, it is accompanied by a question mark in Fig. 5.

The second (0.5 eV donor) level in GaAs is deduced by the systematic difference in pinning position due to Al, Ga, and In on n- and p-type GaAs found by Skeath *et al.*[39] and by a similar difference reported earlier for the pinning of n and p GaAs after the addition of a fraction of a monolayer of oxygen.[49] The energy levels on InP were deduced from studies of Chye *et al.*[19,43,44] of the Fermi level position as a function of oxygen or metal coverage. In Fig. 6, we present data for InP exposed to oxygen. Again, we see that, at the larger oxygen exposures, the curves for n- and p-type material are parallel

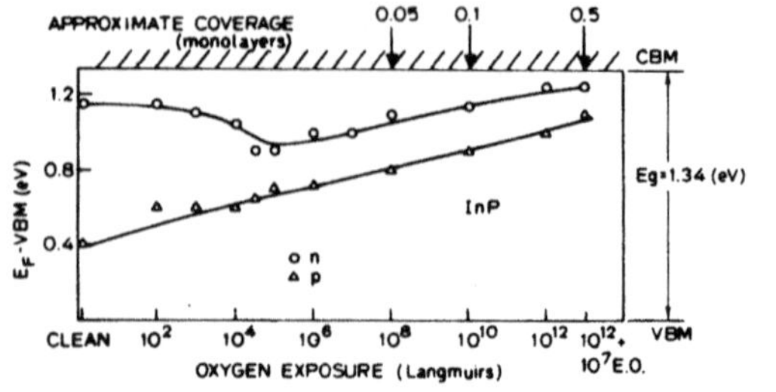

FIG. 6. The surface Fermi level position for n- and p-type InP as a function of oxygen exposure (and approximate coverage). The initial cleaved surfaces were pinned due to extrinsic defects; however, this appears to have no effect on the final pinning position. The last oxygen exposure (10^{12} L, then $+10^7$ L EO) was made with oxygen excited by turning on an ionization gauge (see Refs. 3 and 18).

with the p curve displaced lower in energy by about 0.2 eV. This suggests, again, the presence of two states. Using both the oxygen and Schottky barrier data, the levels are assigned as 1.2 and 0.9 eV. At least two factors suggest that the upper level is a donor. These are the facts that the p-type material rises above 0.9 eV for the highest oxygen exposure and that the Fermi level of the n-type material dips to about 1.0 eV at intermediate coverages before rising to 1.1 eV. Both of these results can only be explained if the donor is placed above the acceptor to make compensation possible. However, these should only be considered tentative conclusions. Using our detailed assignment of donors and acceptors discussed elsewhere, we assign the upper level to P in an In site and the lower to an In in a P site. Williams,[50] on the other hand, would associate the upper level with a missing P based on his results. An open mind must be kept until more definitive data are available.

To summarize, the position of the levels in Fig. 5 is to be taken seriously within 0.1 eV. Under a given set of circumstances, one level may be dominant. The assignments in terms of acceptor or donor and the detailed nature of the defect level should be considered tentative and subject to much future examination.

V. INDIRECT EVIDENCE FOR LATTICE DEFECT FORMATION

A. During oxygen uptake

We were driven to the suggestion of defect formation by evidence we obtained (making full use of the capabilities of synchrotron radiation[3]) of disruption of the surfaces by oxygen or metal adlayers. In Fig. 7, we present energy distribution curves (EDC's) obtained from photoemission studies of the valence electronic states associated within the last few layers of GaAs at the surface[10,18,51,52] as a function of oxygen exposure. Note that there is a rather abrupt transition in the valence band for exposures of 10^6 to 10^7 L of oxygen (a few percent of a monolayer adsorbed oxygen). For exposures below 10^6 L, the sharp valence band structure is relatively unchanged. The transition to the disordered state is relatively abrupt for a given sample, but the point of transition varies by about an order of magnitude from sample to sample. This transition

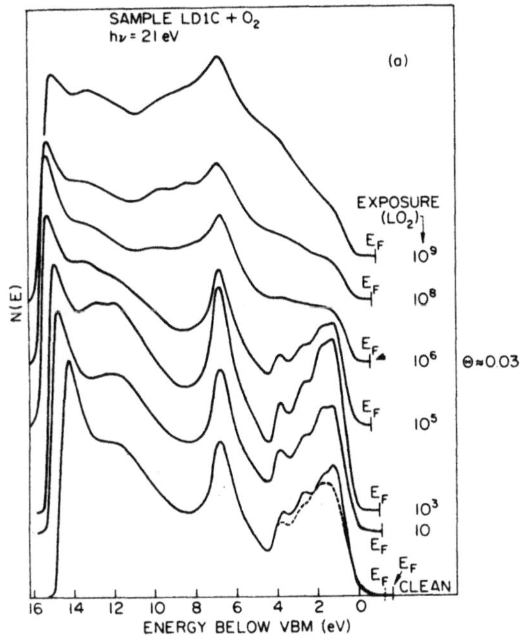

FIG. 7. Energy distribution curves (EDC's) from the valence states of GaAs within a few atomic layers of the surface as a function of oxygen exposure. Up to 10^6 L, the structure is relatively sharp and unchanged by the oxygen. Near 10^6 L, there is an abrupt loss of structure for $E > -5$ eV. This is associated with a partial disordering of the surface in which lattice defects may be formed.

would make the formation of defects possible. It roughly corresponds to the coverage at which the pinning is completed.

The abrupt loss of surface valence band structure suggests strongly that the periodic structure at that surface is highly disturbed by the oxygen adsorption. In the introduction (see Fig. 1), we showed that the free surface is highly rearranged (i.e., distorted in a periodic way) from the bulk and emphasized that there was considerable strain between the surface and bulk of the crystal. Such strain might be relieved by defect formation. In Ref. 10, we discuss the fact that chemisorption of an oxygen atom will cause considerable local lattice rearrangement (driven by the electronic rearrangement) and increased local strain as well as producing local heating due to the heat of adsorption. Apparently, these effects are sufficient to cause a certain amount of lattice disorder to set in at a few percent of a monolayer of oxygen coverage (remember that, on the average, oxygens can be located only 5 or 10 lattice sites apart at this coverage). The oxygen uptake increases markedly at this point, and the Fermi level tends to stabilize its pinning position.[18,52,53] Thus, we have a process in which it is likely that defects will be formed.

In cases where thick oxides are formed, one probably first goes through this stage. In addition, because of the difference in the oxygen chemistry of Ga and As, it is highly likely that these materials will not be removed in exactly stoichiometric quantities from the interface. The complexity of this chemistry is well established.[18,51,53,54] Thus, conditions at the interface between the oxide or chemisorbed oxygen seems very favorable for formation of defects due to local removal of Ga or As atoms.

B. During Schottky barrier formation

Making use of both synchrotron radiation[3,14,43] and sputter Auger techniques, we have consistently found sizable amounts of semiconductor material in the metal of the Schottky barriers. Figure 8 gives EDC's taken at SSRL with $h\nu$ = 120 eV so that both the core levels of the semiconductor constituents as well as the valence levels and core levels of the gold could be followed.

The striking thing about Fig. 8 is that the Sb cores persist strongly even after about 110 monolayers of Au have been added. In contrast, no detectable Ga is found near the surface at this coverage. (It should be emphasized that the sampling depth in these experiments is about two atomic layers.) These experiments show clearly that Sb is removed from the semiconductor by the addition of the metal. (If pin holes in the metal were present, both the Ga and Sb core levels should have shown through with equal strength.) Confirmation of the movement of the semiconductor material into the metal is given by sputter–Auger[3,43] studies.

Closer examination of Fig. 8 shows that the Ga does not disappear as quickly with addition of Au as one would expect

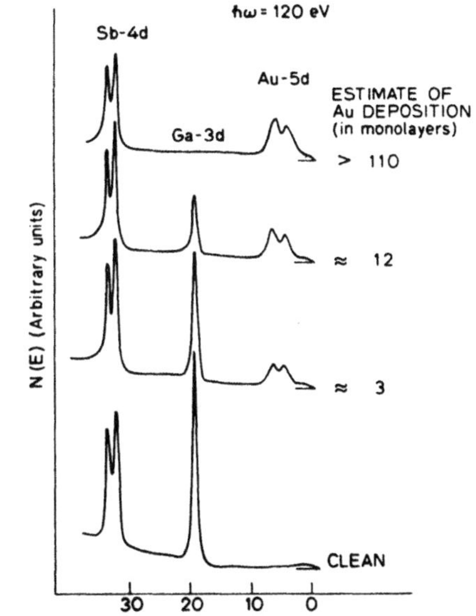

FIG. 8. EDCs taken as Au is very slowly (see Ref. 14) deposited on GaAs. $h\nu$ = 120 eV so that one can clearly see the Sb 4*d* and Ga 3*d* core levels as well as the Au 5*d* valence levels. The sampling depth is about two layers. Note that, even after many layers of Au have been applied, the Sb 4d signal is relatively strong but there is no Ga signal. This establishes the movement of Sb from the semiconductor to the surface. Both anions and cations are seen to move to the metal surface when GaAs and InP are studied in the same manner.

(taking the short sampling depth into account). The sputter-Auger also indicates Ga in the Au well away from the surface for a Au layer several hundred angstroms thick. The fact that Sb is found on the surface of thick layers can be understood in terms of minimizing surface energy (Sb has a much lower heat of evaporation than Au or Ga) and the kinetics which removed the III–V material from the semiconductor originally.

For GaAs and InP, both species are also moved into the metal.[3,14,43] Both species appear on the surface, although there appears to be slightly more As or P.

As was suggested last year,[3] we believe the driving force for dissolution of the III–V and its movement into the metal is the heat of absorption of the metal on the surface of the semiconductor. The heat of absorption or condensation has been habitually overlooked in considering such problems as this. It is the heat of bonding of the metal to the free semiconductor surface. It is not to be directly compared to any bulk compound formation; it is also *much* larger than the kinetic energy associated with the metal atom after evaporation. In general, data are only available for the metallic heat of condensation of metals on themselves. These numbers can be high (~87 kcal/mol for Au). The heat of condensation of Cs on itself is smaller (~20 kcal/mol); however, very beautiful measurements have been made by Derrien and Arnaud D'Avitoya[55] of the heat of condensation of Cs on GaAs; these give about 60 kcal/mol at low coverages. Thus, on being absorbed on the semiconductor surface, the metal gives up energy comparable to or considerably larger than the heat of formation of the semiconductor.[3] This energy must be dissipated and, in so doing, will momentarily excite the semiconductor atoms near it. If a few atoms of the semiconductor are moved into the metal for each hundred metal atoms bound to the surface, this can account for the local departure from stoichiometry and resulting defect states which produce Schottky barrier pinning.

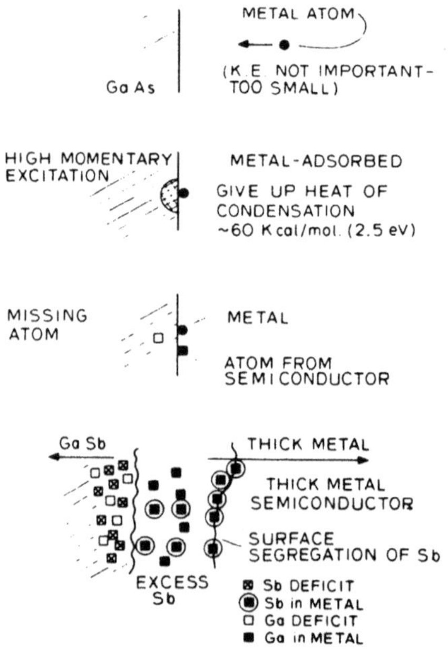

FIG. 9. Schematic of suggested defect mechanism due to deposition of metal atoms on clean III–V surfaces. This process (i.e., a defect must be formed) needs occur only about once for every hundred metal atoms striking the surface in order to explain the Fermi level pinning.

This process is depicted in a pictorial way in Fig. 9. Note that, in that diagram, the defect levels are usually placed at least one lattice site beneath the surface. This is because a metal atom could fill a vacancy at the surface (although not a simple antisite defect). Again, we can not be sure of the nature of the defect, at this time, and thus must allow for various possibilities.

To summarize, experimental evidence for both oxygen uptake and Schottky-barrier formation on the clean semiconductor surface shows that the conditions are present for the formation of defect centers at or near the surface of the semiconductor. The similarity of the Fermi level pinning positions for very different metals and/or oxygen suggest strongly that such centers are formed and produce the Schottky barrier pinning and interface states at the III–V oxide interface.

VI. COMPARISON WITH THICK "DEVICE" OXIDES OR SCHOTTKY BARRIERS

A. Schottky barriers

The pinning positions we obtain are, to a first approximation (within a few tenths of an electron volt), in agreement with both the work obtained by Mead[5] (who used clean surfaces but very high metal deposition rates) and, for practical devices in which the sample, because of etching and transport through air, has at least a thin oxide outer layer. To put the results more emphatically, the pinning position we obtain with only 10% to 20% of a monolayer of metal is, to the first approximation, the same as that obtained for thick (thousands of monolayers of metal) Schottky barriers. Thus, it appears that the same basic pinning mechanism must occur in all cases.

The classical work on Schottky barriers by Mead *et al.*[5] was performed in a slightly different manner than our own work.[14,43,44] The pressure in our preparation chamber was about four orders of magnitude lower than that in Mead's chamber. As a result, we were able to make very "gentle" depositions (e.g., depositions in steps of 0.1 monolayer),[40,43,44] minimizing the chances of heating the crystal or creating a large density of defects (which may interact) in a short period of time. However, the general results of our work and those of Mead *et al.*[5] as to pinning position are, on the average, quite close. The main difference[40] is that, after detailed studies of Al, Ga, and In on GaAs,[56–58] the pinning position on *n*-type material is found to be about 0.25 eV higher than on p-type material. This difference between the previous work and ours may be due to the details of deposition with the result that the defect levels formed are not the same. However, more work must be done to clarify this question.[59]

In the past, it has been hard to understand why the Schottky barrier pinning positions were so similar for metal on atomically clean GaAs and on practical surfaces where the semiconductor was usually covered by at least a few layers of oxide. The present work resolves this question in showing that the oxide pinning position is approximately the same as that of the metal.

There is an interesting sidelight to the work presented here. Garner *et al.*[60,61] have studied the effect of the chemical composition of the oxide present on the forward characteristics of Au Schottky barriers formed on n-type GaAs. His results indicate that the Schottky barrier is greatly reduced if an As oxide is present.[62] This is consistent with the pinning of n-type GaAs being due to an acceptor near 0.7 eV, resulting from a deficit of As. The As oxides are notoriously unstable, and free As may be produced by the thick Au deposition. Some free As might, in turn, be available for incorporation in the GaAs surface region, reducing the pinning states to such a degree that the Schottky barrier height is strongly reduced. The fact that excess As tends to remove the pinning but Ga does not must relate to the detailed hopping probability or other details of defect formation or reduction at room temperature.

To summarize, the model proposed here explains, to first order, a wide range of phenomena extending from Schottky-barrier pinning produced by a small fraction of a monolayer of metal to that obtained on a practical semiconductor surface with a region containing a few layers of oxides after a thick metallic deposition. This lends credence to the underlying physical model; however, much more work must be done to definitively establish this model and fill in all of the details which are now missing.

B. Correlation with practical "device" oxides

One of the most fascinating aspects of this work has been the correlation of our fundamental studies of oxygen chemisorption and oxidation with the data on interface states obtained from "devicelike" thick oxides intended for MOS (metal oxide semiconductor) as well as other device applications. Whereas, Schottky barriers on III–V's have been used practically for many decades, the practical native oxide on III–V's is a new development which has not yet reached the production stage. Thus, a recent object of our work has been to compare our results for oxygen adsorption in relatively

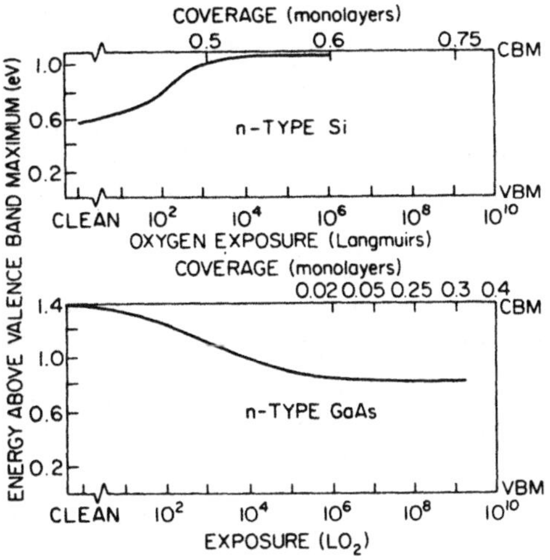

FIG. 10. The movement of the surface Fermi level in n-type cleaved Si and GaAs. The clean Si is pinned by intrinsic surface states which are removed by approximately a monolayer of oxygen and the band bending disappears. In contrast, the clean GaAs (110) is originally unpinned but becomes pinned with the addition of a small amount of oxygen.

small quantities on the clean, cleaved (110) surfaces with the work from MIS-type structures where the device qualities were sufficiently good so that a measure could be made of the interface state density. In Table I, we present a comparison of our data for the "saturated" pinning position obtained after the addition of a few tenths of a monolayer or more of oxygen to the surface[3,5,6,44,53,63,64] to the zero bias or Fermi-level pinning position for the "practical MIS devices.[21]" Typical data of this showing the surface pinning position versus exposure for a clean surface is given in Fig. 10 for n-type GaAs.

In order to gain perspective, Fig. 10 also contains similar data on n-type Si (111). Whereas, the oxygen removes the surface states and pinning from the Si,[63] the situation is the opposite for GaAs (110).[3,18,53] For GaAs, one goes from an unpinned to a pinned situation, indicating creation of states in the gap by the oxygen adsorption. As outlined in Sec. II, this is not due to the movement of the intrinsic surface states back

TABLE I. Fermi energy pinning position with respect to conduction band minimum after indicated surface treatment.

Material	Thick "device" oxide	Submonolayer of chemisorbed oxygen	Submonolayer to several monolayers of: Cs	Al,Ga,In	Au
GaAs (n-type)	0.83 (Ref. 54) 0.8 (Ref. 69) 0.5 (Ref. 24)	0.65–0.8 (Refs. 18, 30, 53) and Fig. 10	0.6–0.8 (Ref. 28) 0.7 (Ref. 73)	0.6–0.7 (Ref. 39)	0.9–1.0 (Ref. 14)
InP (n-type)	0.14 (Ref. 70) 0.087 (Ref. 70) 0.075 (Ref. 70) 0.1 (Ref. 21) 0.1 (Ref. 71)	0.05–0.2 (Ref. 72) and Fig. 6	0.15–0.45 (Ref. 72)		

The similarity between the Fermi-level pinning position for thick oxides formed by various techniques and that obtained by a fraction of a monolayer of oxygen should be noted. It should also be noted that the pinning position for Schottky barriers (the example here is cesium) is close to that for oxygen[15].

into the band gap, but must be due to the formation of extrinsic states in the gap. As Sec. III indicates, the levels do not appear to be associated with the oxygen, but to a level produced indirectly.

In Table I, the "thick practical" data was obtained from Ref. 20; Wieder and his co-workers[65] have recent indications of strong interface states at about 1.1 and 1.3 eV, in reasonable correlation with the pinning positions we obtain with a fraction of a monolayer of oxygen. Note that the spread of pinning position for either GaAs or InP is less than the systematic difference in pinning positions in the band gap between the two materials. Also given in Table I are Schottky barrier pinning positions for a large range of metals on the clean, cleaved (110) faces of InP and GaAs (GaSb is not included in Table I since we could not find data for "device-type" oxides) to, once again, emphasize the similarity between the oxide and Schottky-barrier results.

In Fig. 11, we present plots of interface states density for a thick "device" oxide from the work of Koshiga and Sugano.[22] It is clear that a state near 0.7 eV is playing a dominant role. Measurements of interface density above 0.7 eV have been difficult because of frequency effects. However, it is now clear that, because of the strength of the level near 0.7 eV, it is impossible to make[23–25,66] direct interface state density measurements much above 0.7 eV.

This is in contrast to the situation on InP where interface state density measurements can be made through most of the band gap. As mentioned earlier, there is evidence for interface states about 0.1 and 0.3 eV below the conduction band minimum[65]; however, these seem to be smaller in density than for GaAs since inversion can be achieved in InP but not GaAs.[65] In general, Wieder and his co-workers find the density of interface states in InP to be an order of magnitude less than in GaAs.[65]

Convincing evidence is not found in the work of Hasegama *et al.*[23,66] or Koshiga and Sugano[22] of a peak in the density of interface states near 0.5 eV (there may be some suggestion of such a peak in Koshiga and Sugano's[22] unannealed sample). However, Meiners has obtained evidence of pinning in this

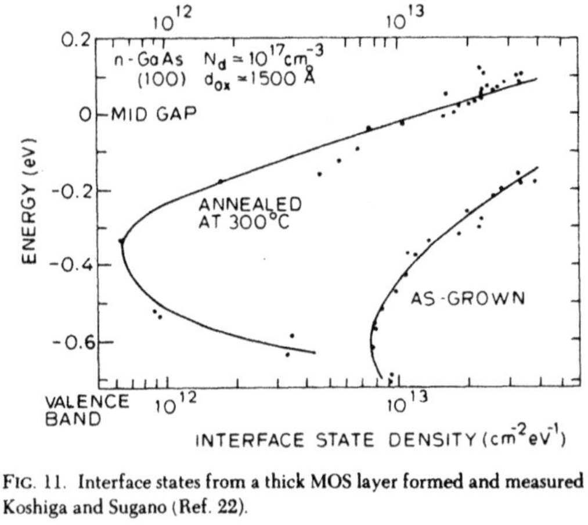

FIG. 11. Interface states from a thick MOS layer formed and measured by Koshiga and Sugano (Ref. 22).

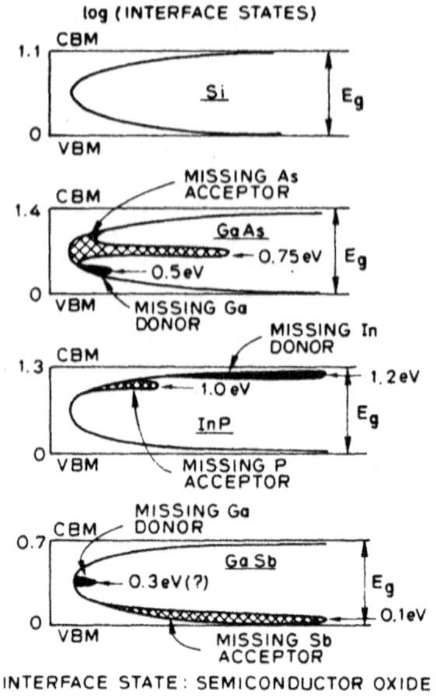

FIG. 12. Model of interface states for GaAs, InP, and GaSb MIS structures based on the present work.

region.[26,27] Thus, it appears that the treatments used to provide insulating layers on GaAs definitely tend to suppress the 0.5 eV level (which we tentatively associate with missing Ga) and enhance the 0.7 eV level which we associate with missing As.

In Fig. 12, we present a proposed density of semiconductor–oxide interface states for GaAs, InP, and GaSb. For the sake of comparison, Si is also included. These density of states are not to be taken seriously in detail; rather, one should concentrate on their general features.

Note that the density of interface states for Si is generally "U" shaped. (These states are usually associated with the strains due to the lattice mismatch between the semiconductor and oxide.) For the III–V's, we superimpose on these "U"-shaped background broad peaks. These are associated with the deficit centers proposed earlier due to missing anions or cations. It should be emphasized that the absolute values of these peaks depend on the method of providing the insulating layer. For example, the state near 0.7 eV in GaAs (associated with missing As) is clearly dominant in the "practical" samples mentioned here[21–27,66]; whereas, clear evidence of a level near 0.5 eV is only found in one study.[26,27] Thus, it appears that it is fairly easy to suppress the 0.5 eV level (associated with a Ga deficit) in preparing an MOS structure on GaAs, but that the level at 0.7 eV associated with a deficit of As is universally strong in the data studied to date.

A similar situation appears to occur for InP in which the level near 1.2 eV (tentatively associated by us with an In deficit) dominates. We found no MOS data for GaSb; however, in our own data, the 0.1 eV level (associated with missing Sb) clearly dominates at low oxygen coverages; however, there is an indication of a level near 0.3 eV at thicker coverages. This is also indicated in Fig. 12 and associated with a missing Ga; however, the lack of certainty is indicated by the question mark.

VII. SUMMARY AND CONCLUSIONS

Despite the fact that Schottky diodes have been of practical importance for over four decades, the mechanism for the Fermi-level pinning which determines the height of the rectifying barrier has not been definitively determined. In this paper and that by Skeath *et al.*[39] as well as in the papers by Lindau *et al.*[15] and Chye *et al.*[14] the previous suggestions have been examined and rejected. Further, we have put forth a new model based on pinning by defect states. The defect formation is caused by the energy released when the metal atom is chemisorbed on the surface (see Fig. 9). One point of support for the model is the appearance of one or more constituents of the semiconductor in the metal (see Fig. 8).

Another critical result is that the pinning position for metals and oxygen is essentially identical. Based on the difference in the electronic configuration for oxygen and metals (as well as the difference in the metals themselves which extend from Cs to Au), it is suggested that this is only consistent with defects being produced indirectly by either oxygen or metal adsorption. Studies of the valence band structure at the surface (see Fig. 7) as a function of oxygen adsorption lend support to this.

Using our experimental data, we propose a general model shown in Fig. 5. In that model, which is the simplest one that is consistent with the available data, two energy levels are present in the band gap; one due to a missing anion is an acceptor, and the other, which is due to a missing cation, is a donor. A simple suggestion for the defects is that they are anti-site defects. In the case of the acceptor for GaSb, however, there is evidence that it arises from a more complex center such as Ga on an Sb site associated with a Ga vacancy.[48] It is clear that a major effort must be made both theoretically and experimentally to identify and understand these defects at or near the surface. Any possible interactions between the metallic wave functions and the defect levels also need investigation. One would think that the understanding of defects in the bulk of III–V's could be drawn upon; however, it appears that the electronic levels associated with bulk defects are not well enough understood to make such correlations. Clearly, the understanding of defects at the surface and in the bulk is an area where much more work is needed. Theory and such experiments such as spin resonance studies seem especially appropriate.

McCaldin *et al.*[67] have noted that the Fermi-level pinning position on p-type samples is determined principally by the anion. This is consistent with our model since a surface donor (needed to pin a *p*-type sample) is produced by removing a cation and/or placing a anion on a cation site, thus forming a center only containing anions. Thus, the pinning will be expected to be determined by the anion. GaSb is a special case, as it seems that Sb is preferentially removed, leaving many more acceptor states which lie below any donor states; thus, there is strong pinning on *n*-type GaSb but not on *p*-type material, in agreement with the findings of McCaldin *et al.*[67] of a zero barrier on *p*-type GaSb.

Since our model by its very nature applies to semiconductor oxide interfaces as well as Schottky barriers, the detailed reasons for the difficulties in passivating the III–V as compared to Si become more apparent. In addition, we find that the peaks in the density of interface states in MIS devices correspond rather well with the states we have identified in our fundamental studies. Our models for the interface states are presented in Figs. 5 and 12. For GaAs, a state near midgap which we associate with an As deficit is found to dominate; whereas, for InP, the dominant state is near the conduction band minimum.

To conclude, there is evidence indicating that Schottky-barrier pinning on III–V's as well as the interface states at III–V oxide interfaces are due to energy levels produced by defects formed by putting down the metal or oxygen on the semiconductor. However, much more work must be done to tie down the detailed nature of these defects as well as to test and quantify the model.

An important aspect of this model is that it suggests ways by which both the Schottky barrier height and the interface state density at the oxide interface may be changed, e.g., by treating the sample to reduce or otherwise modify the interface state density. The work of Garner et al possibly gives some evidence of this.[60,61]

It should be noted that we have compared Fermi-level pinning and interface states on different crystal faces. The fact that agreement is found under such procedures gives further evidence that the states involved are defect states which are not strongly affected by the crystal face. This gives further evidence that the pinning defects probably lie just below the surface and are due to crystal defects. Heime's[68] review of deep "accidental" defects in bulk GaAs shows a broad peak about 0.7 eV. Thus, there may be a close connection to the pinning defects created near the surface by adlayers and defects in the bulk of the semiconductor.

[1]J. van Laar and J. J. Scheer, Surf. Sci. **8**, 342 (1967); J. van Laar and A. Huijser, J. Vac. Sci. Technol. **13**, 769 (1976).

[2]W. Gudat and D. E. Eastman, J. Vac. Sci. Technol. **13**, 831 (1976).

[3]W. E. Spicer, I. Lindau, P. E. Gregory, C. M. Garner, P. Pianetta, and P. Chye, J. Vac. Sci. Technol. **13**, 780 (1976).

[4]In this article, we will use this terminology; by intrinsic surface states, we mean surface states characteristic of the "ideal" rearranged surface. The density of these intrinsic surface states should correspond to the total density of surface atoms. The term "extrinsic surface state" will be used for surface states induced by surface defects or impurities. The density of these states will normally correspond to the density of surface defects, imperfections, and impurities.

[5]C. A. Mead and W. G. Spitzer, Phys. Rev. **134**, A713 (1964); S. Kurten, T. C. McGill and C. A. Mead, Phys. Rev. Lett. **22**, 1433 (1969).

[6]C. B. Duke, A. Lubinsky, B. W. Lee, and P. Mark, J. Vac. Sci. Technol. **13**, 761 (1976); P. Mark, G. Cisneros, M. Bonn, A. Kahn, C. B. Duke, G. Patton, and A. R. Lubinsky, J. Vac. Sci. Technol. **14**, 883 (1977); C. B. Duke, J. Vac. Sci. Technol. **14**, 870 (1977).

[7]A. U. MacRae and G. W. Gobeli, in *Semiconductors and Semimetals*, Vol. 2 of Physics of 3-5 Compounds, edited by R. K. Willardson and A. C. Beer (Academic, New York, 1966), pp. 115–137; W. A. Harrison, Surf. Sci. **55**, 1 (1976).

[8]A. Kahn, E. So, P. Mark, C. B. Duke, and R. J. Meyer, J. Vac. Sci. Technol. **15**, 1223 (1978); B. J. Mrstik, S. Y. Tong, and M. A. Van Hove, J. Vac. Sci. Technol. **16**, 1258 (1979). C. B. Duke, R. J. Meyer, A. Kahn, E. So, and P. Mark, J. Vac. Sci. Technol. **16**, 1252 (1979).

[9]D. J. Chadi, J. Vac. Sci. Technol. **15**, 1244 (1978); A. Huijser, J. van Laar, and T. I. Van Rooy, Phys. Lett. **65A**, 335 (1978); J. R. Chelikowsky, S. G. Louie, and M. L. Cohen, Phys. Rev. B **14**, 4724 (1976).

[10]W. E. Spicer, I. Lindau, J. N. Miller, D. T. Ling, P. Pianetta, P. W. Chye, and C. M. Garner, Physica Scripta **16**, 388 (1977).
[11]J. Bardeen, Phys. Rev. **71**, 717 (1947).
[12]V. Heine, Phys. Rev. **138**, A1689 (1965).
[13]J. C. Inkson, J. Phys. **C5**, 2599 (1972); **C6**, 1350 (1973); J. Vac. Sci. Technol. **11**, 943 (1974); S. G. Louie, J. R. Chelikowsky, and M. L. Cohen, Phys. Rev. **B15**, 2154 (1977); J. M. Andrews and J. C. Phillips, Phys. Rev. Lett. **35**, 56 (1975); C. Tejedor, F. Flores, and E. Louis, J. Phys. **C10**, 2163 (1977); E. J. Mele and J. D. Joannopoulos, J. Vac. Sci. Technol. **15**, 1370 (1978); H. I. Zhang and M. Schlüter, J. Vac. Sci. Technol. **15**, 1384 (1978); W. A. Goddard III and John J. Barton, J. Vac. Sci. Technol. **15**, 1273 (1978).
[14]P. W. Chye, I. Lindau, P. Pianetta, C. M. Garner, and W. E. Spicer, Phys. Rev. **B 17**, 2682 (1978); P. W. Chye, I. Lindau, P. Pianetta, C. M. Garner, C. Y. Su, and W. E. Spicer, Phys. Rev. **B 18**, 5545 (1978).
[15]I. Lindau, P. W. Chye, C. M. Garner, P. Pianetta, C. Y. Su, and W. E. Spicer, J. Vac. Sci. Technol. **15**, 1332 (1978).
[16]L. J. Brillson, Phys. Rev. Lett. **40**, 260 (1978); Phys. Rev. **B 18**, 2431 (1978); J. Vac. Sci. Technol. **15**, 1378 (1977).
[17]Since, in all cases known to the authors, at least several monolayers of native oxide (i.e., oxide formed from the semiconductor material) occur on the semiconductor before a subsequent deposition of a second insulator film, interface states formed in conjunction with native oxide growth will be of universal importance. Thus, we will concentrate our attention on these states in this paper. However, if the native oxide is successfully removed before the new insulator is deposited, it is anticipated that the same general mechanism of interface state formation will occur due to the difficulty of chemically bonding the insulator to the semiconductor.
[18]P. Pianetta, I. Lindau, P. E. Gregory, C. M. Garner, and W. E. Spicer, Surf. Sci. **72**, 298 (1978); P. Pianetta, I. Lindau, C. M. Garner, and W. E. Spicer, Phys. Rev. **B 18**, 2792 (1978).
[19]P. W. Chye, C. Y. Su, P. Skeath, I. Lindau, and W. E. Spicer (submitted for publication).
[20]P. W. Chye, C. Y. Su, I. Lindau, P. Skeath, and W. E. Spicer, J. Vac. Sci. Technol. **16**, 1191 (1979).
[21]H. H. Wieder, Thin Solid Films (in press), and references therein.
[22]Fusrko Koshiga and Takuo Sugano, Surf. Sci. (in press).
[23]H. Hasegawa, T. Sawada, and T. Sakai, Surf. Sci. (in press).
[24]A. Shimano, A. Moritani, and J. Nakai, Jpn. J. Appl. Phys. **15**, 939 (1977).
[25]C. R. Zeisse, L. J. Messick, and D. L. Lile, J. Vac. Sci. Technol. **14**, 957 (1977).
[26]L. G. Meiners, Appl. Phys. Lett, **33**, 747 (1978).
[27]H. H. Weider, L. G. Meiners, and D. L. Lile, (private communication).
[28]P. Gregory and W. E. Spicer, Phys. Rev. **B 12**, 2370 (1975).
[29]G. W. Gobeli and F. G. Allen, Phys. Rev. **137**, A245 (1965); J. H. Dinan, L. K. Galbraith, and T. E. Fisher, Surf. Sci. **26**, 587 (1971); D. E. Eastman and J. L. Freeouf, Phys. Rev. Lett. **34**, 1624 (1975).
[30]P. Skeath, W. A. Saperstein, P. Pianetta, I. Lindau, W. E. Spicer, and P. Mark, J. Vac. Sci. Technol. **15**, 1219 (1978).
[31]P. Mark, P. Pianetta, I. Lindau, and W. E. Spicer, Surf. Sci. **69**, 735 (1977).
[32]W. Monch and H. J. Clemens, these proceedings.
[33]G. J. Lapeyre, R. J. Smith, J. Knapp, and J. Anderson, J. Physique Colloq. **39**, C4–149 (1978).
[34]W. Gudat and C. Kunz, Phys. Rev. Lett. **29**, 169 (1972).
[35]D. E. Eastman and J. L. Freeouf, Phys. Rev. Lett. **33**, 1601 (1974).
[36]The soft x-ray transitions studied by these methods involve formation of an exciton (see the paper by M. Altarelli, G. Bachelet, and R. Del Sole, these proceedings) during excitation from the filled Ga 3d core levels into the lowest available empty states. These final states will certainly be the empty surface states provided that they lie in or near the band gap. The problem is more difficult when the empty states lie above the CBM as appears to be the case for the rearranged GaAs (110). However, if any new surface reconstruction moves the surface states into the band gap, this will certainly produce an easily detected reduction in the photon energy necessary to excite the exciton. For metals and oxygen exposures, the surface Fermi level has moved by large amounts before the excitonic transition is removed by the adsorption.
[37]I. Lindau, P. Pianetta, W. E. Spicer, and C. M. Garner, Proc. Seventh Intl. Vacuum Congress and the Third Intl. Conf. on Solid Surfaces, Vienna, Austria, 12–16 Sep 1977, p. 615 (R. W. Dobrozemsky, F. G. Rüdenauer, F. P. Viehböck, and A. Breth, eds.).
[38]P. W. Chye, P. Pianetta, I. Lindau, and W. E. Spicer, J. Vac. Sci. Technol. **14**, 917 (1977).
[39]P. R. Skeath, C. Y. Su, P. W. Chye, P. Pianetta, I. Lindau, and W. E. Spicer, these proceedings; P. R. Skeath, I. Lindau, P. W. Chye, C. Y. Su, and W. E. Spicer, J. Vac. Sci. Technol. **16**, 1143 (1979).
[40]E. J. Mele and J. D. Joannopoulos, J. Vac. Sci. Technol. **16**, 1154 (1979).
[41]L. J. Brillson, Phys. Rev. Lett. **42**, 397 (1979); J. Vac. Sci. Technol. **16**, 1137 (1979).
[42]For example, the maximum change of electron affinity found on putting Cs on GaAs is about 3.0 eV compared to a maximum change in Fermi level of about 0.7 eV. This is an extreme case, but it illustrates how large the change in electron affinity can be compared to the Fermi level motion at the surface.
[43]P. R. Skeath, I. Lindau, P. Pianetta, P. W. Chye, C. Y. Su, and W. E. Spicer (to be published).
[44]Patrick Chye, Ph.D. dissertation, Stanford University, 1978 (unpublished).
[45]When several monolayers of oxide are grown on GaSb (see Refs. 18 and 44), the oxide position rises to near mid-gap. It is not yet clear whether this represents the oxide surface Fermi level or that at the oxide-semiconductor interface.
[46]The usual definition is used for acceptors and donors, i.e., an acceptor is uncharged when containing no electron(s) and a donor is uncharged when it is filled with electron(s).
[47]In addition, since the defects are near the surface (we will discuss their position with regard to the surface or interface in more detail later), some interaction with the tunneling wave functions of the metal atoms may also be important.
[48]Y. J. Van der Meulen, J. Phys. Chem. Solids **28**, 25 (1967).
[49]See Fig. 8 of Ref. 3.
[50]R. H. Williams, these proceedings.
[51]W. E. Spicer, P. Pianetta, I. Lindau, and P. W. Chye, J. Vac. Sci. Technol. **14**, 885 (1977).
[52]I. Lindau, P. Pianetta, W. E. Spicer, P. E. Gregory, C. M. Garner, and P. W. Chye, J. Electron Spectrosc. Related Phenomena **13**, 155 (1978).
[53]Piero Pianetta, Ph.D. dissertation, Stanford University, 1976 (unpublished). Stanford Synchrotron Radiation Laboratory Rep. 77/17, 1977.
[54]C. C. Chang, P. H. Citrin, and B. Schwartz, J. Vac. Sci. Technol. **14**, 943 (1977); G. P. Swartz, J. E. Griffiths, and B. Schwartz, J. Vac. Sci. Technol. **16**, 1388 (1979).
[55]Jacques Derrien and Francois Arnaud D'Avitoya, Surf. Sci. **65**, 668 (1977).
[56]It is particularly interesting that Al and In act so similar. Al displaces Ga in the first layer of GaAs (forming AlAs and Ga) where In does not. However, the pinning positions are very similar. This and much other information (Ref. 57) establishes that the displacement reaction of Al in GaAs does not play a key role in establishing the Schottky-barrier height as suggested by Brillson (Ref. 16).
[57]See comments by T. McGill and W. E. Spicer after the paper (Ref. 16) presented by Brillson in these proceedings, p. 1137.
[58]As can be seen from the data of Ref. 39, Ga may be different from In and Al in that it takes a large coverage to reach the final pinning point. However, more work must be done before this point is clear.
[59]A second-order but important question is the difference in pinning position between Au and more electro-positive metals and Al and Cs.
[60]C. M. Garner, J. Appl. Phys. (in press).
[61]C. M. Garner, Ph.D. dissertation, Stanford University, 1978 (unpublished).
[62]The Schottky-barrier characteristics become very "soft" and can not be fit by the ideal Schottky-barrier equations. This is probably due to a large amount of "patchiness" in the surface.
[63]L. F. Wagner and W. E. Spicer, Phys. Rev. B **9**, 1512 (1974).
[64]W. E. Spicer, I. Lindau, P. Pianetta, P. W. Chye, and C. M. Garner, Thin Solid Films **56**, 1 (1979); W. E. Spicer, P. W. Chye, C. M. Garner, I. Lindau, and P. Pianetta, Surf. Sci. (in press).
[65]H. H. Wieder (private communication).
[66]H. Hasegawa and T. Savada, J. Vac. Sci. Technol. **16**, 1483 (1979).
[67]J. O. McCaldin, T. G. McGill, and C. A. Mead, Phys. Rev. Lett. **36**, 56 (1976); J. Vac. Sci. Technol. **13**, 802 (1976).
[68]K. Heime, *Proc. NATO Summer School*, "Non-destructive Testing of

Semiconductor Materials," Frescati, Italy, 1978 (J. Zemel, ed.) (in press).
[69]L. Meiners, J. Vac. Sci. Technol. **15**, (4) (1978).
[70]Y. W. Lam, J. Phys. D. **4**, 1370 (1971).
[71]L. Messick, J. Appl. Phys. **47**, 4949, (1976).
[72]P. W. Chye, I. A. Babalola, T. Sukegawa, and W. E. Spicer, Phys. Rev. B **13**, 4439 (1976). P. W. Chye *et al.* (to be published).
[73]J. J. Uebbing and R. L. Bell, Appl. Phys. Lett. **11**, 357 (1967).

Schottky barriers: An effective work function model

J. L. Freeouf and J. M. Woodall
IBM Thomas J. Watson Research Center, Yorktown Heights, New York 10598

(Received 2 July 1981; accepted for publication 18 August 1981)

The experimental observations of metallurgical interactions between compound semiconductor substrates and metallic or oxide overlayers have stimulated a new model of Fermi level "pinning" at these interfaces. This model assumes the standard Schottky picture of interface band alignment, but that the interface phases involved are not the pure metal or oxide normally assumed by other models. For both III-V and II-VI compounds, the barrier height to gold is found to correlate well with the anion work function, suggesting the interface phases are often anion rich. This correlation holds even for cases in which the "common anion rule" fails, and explains both successes and failures of this earlier model.

PACS numbers: 73.30. + y, 85.30.Hi, 68.48. + f

Metal-semiconductor contacts, while crucial to semiconductor devices and studies, are still not well understood. Models relating Schottky barrier heights to metal workfunction, electronegativity, and heats of condensation and reaction with substrate constituents, as well as semiconductor properties such as surface and interface states, heats of formation, polarizability, ionicity, band gap, and defect energy levels can all be found in the recent literature. Some of these models assume the interface to occur abruptly between the two desired phases, while other models require the occurrence of the metallurgical interactions recently observed. The wealth of models available, and the diversity of assumptions they invoke, imply that the fundamentally important aspects of Schottky barrier formation have not yet been established.

In spite of the rich array of various models there are some notable experimental results which remain unexplained. One is the fact that liquid gallium will make a temporary ohmic contact to lightly doped n-type GaAs under the conditions in which the native oxide to GaAs is disrupted exposing clean gallium to an oxide free GaAs surface.[1] With time and exposure to air the contact will become rectifying as predicted by previous models. The second and more convincing result is the Okamoto *et al.* study[2] of Schottky barrier heights for the Al-(GaAs-AlAs) interface prepared by molecular beam epitaxy. They find barrier heights, particularly to AlAs, which are significantly different from those predicted by previous models and which are significantly different from those for Au-AlAs.[3] We have reexamined earlier models in light of the recent observations of interface intermixing and propose that the simple Schottky picture of work-function matching—if coupled with mixed phases at the interface—appears to account for a large amount of experimental data and suggests directions for research in controlling Fermi level pinning.

Our model begins with that of Schottky,[4] which assumes an ideal metal-semiconductor interface, i.e., one in which the interface is inert and there are no appreciable surface or induced interface states in the semiconductor. The Schottky barrier height is given by[4]

$$\phi_{bn} = \Phi_M - \chi,$$

$$\phi_{bp} = (E_G/q) + \chi - \Phi_M,$$

where ϕ_{bn} (ϕ_{bp}) is the Schottky barrier height to an n-type (p-type) semiconductor, Φ_M is the metal work function, q is the electron's charge, and χ is the electron affinity of the semiconductor. Thus, for the ideal case and for a given semiconductor, ϕ_b should be determined by the metal work function. Unfortunately, this is not the case for GaAs and many other semiconductors.[3]

Our model, called the effective work function model (EWF), suggests that the Fermi level at the surface (or interface) is not fixed by surface states but rather is related to the work functions of microclusters of the one or more interface phases resulting from either oxygen contamination or metal-semiconductor reactions which occur during metalization. The theory requires that when a metal is deposited, or an oxide is formed, there is a region at the interface which contains a mixture of microclusters of different phases, each having its own work function. We should therefore modify the "ideal" surface discussion as follows:

$$\phi_{bn} = \Phi_{\text{eff}} - \chi,$$

where Φ_{eff} is an appropriately weighted average of the work functions of the different interface phases. Thus the measured ϕ_{bn} can depend somewhat on the measurement technique, i.e., C-V or I-V.

For most of the compounds under discussion, metalization and/or oxidation results in a condition in which Φ_{eff} is due mainly to Φ_{Anion}, the work function of the anion; we suggest that this occurs as a result of one or both of the following reactions:

Anion oxide + Compound→Anion + Cation oxide,

M + compound→(Anion or Metal-Anion complex) + (M-Cation).

The condition for driving this reaction to the right and hence generating excess Anion at the interface is that the Gibbs free energy ΔF is negative. Such oxide reactions have been examined,[5] and excess group V anions have been experimentally observed when ΔF is negative, i.e., for GaAs, InAs, and InSb.[6,7] This has not been observed when ΔF is positive, i.e., for GaP.[7] It is interesting to note that for InP, $\Delta F \approx 0$; it has been possible to form metal-oxide semiconductor field-effect transistor (MOSFET) structures using SiO_2, which exhibit a low interface state density[8] on this

TABLE I.[a] Au Schottky barriers.

Compound	$E_G/q+\chi$	ϕ_{bp}	$\Phi_{Au}=5.1\text{–}5.5$ eV[b] $E_G/q+\chi-\phi_{bp}$	Φ_{Anion}
GaP	5.86[c]	0.96[j]	4.9	5.0[r]
InP	5.75[c]	0.85[k]	4.9	5.0[r]
AlAs*	5.6–6.0[d,e]	0.9[l]	4.7–5.1	5.0[r](4.8)[s]
		(1.4)**[m]	(4.2–4.6)**	($\Phi_{Al}=4.0\text{–}4.3$[b,r])
GaAs	5.5[f]	0.5[l]	5.0	5.0[r](4.8)[s]
InAs	5.3[f]	0.3–0.5[n,o]	4.8–5.0	5.0[r](4.8)[s]
AlSb*[g]	5.2[g]	0.54[l]	4.7	4.8[r](4.7)[t]
GaSb	4.76[f]	0.1[l]	4.7	4.8[r](4.7)[t]
InSb	4.77[f]	≈0.1[l]	4.8(77 K)	4.8[r](4.7)[t]
ZnO	7.92[h]	2.7[l]	5.2	7.3[r]
ZnS	7.5[h]	1.6[l]	5.9	5.74[r]
CdS	7.21[h]	1.63[l]	5.58	5.74[r]
GaS*	6.5[i]	0.75[p]	5.75	5.74[r]
ZnSe	6.76[h]	1.31[l]	5.45	5.7[r]
CdSe	6.65[h]	1.21[l]	5.44	5.7[r]
GaSe*	5.4[i]	0.5[p]	4.9	5.7[r]
ZnTe	5.79[h]	0.65[q]	5.14	4.88[r]
CdTe	5.72[h]	0.78[l]	4.94	4.88[r]
GaTe*	4.95[i]	0.45[p]	4.5	4.88[r]

*Does not obey common anion rule.
**Al-AlAs barriers.
[a]Band gaps were taken from A. G. Milnes and D. L. Feucht, *Heterojunctions and Metal-Semiconductor Junctions* (Academic, New York, 1972), p. 8.
[b]Reference 14.
[c]J. Van Laar, A. Huijser, and T. L. Van Rooy, J. Vac. Sci. Technol. **14**, 894 (1977).
[d]R. Dingle, A. C. Gossard, and W. Wiegmann, Phys. Rev. Lett. **4**, 1327 (1975).
[e]A. H. Nethercot, Jr., Phys. Rev. Lett. **33**, 1088 (1974).
[f]G. W. Gobeli and F. G. Allen, Phys. Rev. **137**, A245 (1965).
[g]T. E. Fischer, Phys. Rev. **139**, A1228 (1965).
[h]R. K. Swank, Phys. Rev. **153**, 844 (1967).
[i]R. H. Williams and A. J. McEvoy, Phys. Status Solidi A **12**, 277 (1972).
[j]B. L. Smith and M. Abbott, Solid-State Electron. **15**, 361 (1972).
[k]B. L. Smith, J. Phys. D **6**, 1358 (1973).
[l]C. A. Mead, Solid State Electron. **9**, 1023 (1966).
[m]Reference 2.
[n]K. Kajiyama, Y. Mizushima, and S. Sakata, Appl. Phys. Lett. **23**, 458 (1973).
[o]J. N. Walpole and K. W. Nill, J. Appl. Phys. **42**, 5609 (1971).
[p]S. Kurtin and C. A. Mead, J. Phys. Chem. Solids **30**, 2007 (1969).
[q]W. D. Baker and A. G. Milnes, J. Appl. Phys. **43**, 5152 (1972).
[r]K. W. Frese, Jr., J. Vac. Sci. Technol. **16**, 1042 (1979).
[s]As on native oxide of GaAs, J. L. Freeouf and J. M. Woodall (unpublished).
[t]J. L. Freeouf, M. Aono, F. J. Himpsel, and D. E. Eastman, J. Vac. Sci. Technol. (to be published).

material. This is consistent with our model that would predict either no or very little excess free phosphorus at the interface. A GaP MOSFET structure with low interface-state densities would be predicted, since no free P is expected at this interface. It should also be noted that for GaAs it is well known that MOSFET structures have notoriously high "interface-state densities" (10^{13}–10^{14} cm^{-2}) and that excess arsenic is usually observed at the interface.[9] Again this is consistent with the model, since the ϕ_{bn} expected for the As-GaAs interface is about 0.8 eV (the usually observed barrier height for most metal depositions as well). Since workers have reported a large density of mid-gap states for MOSFET GaAs structures, the model would ascribe these "states" to arsenic clusters at the interface which act as Schottky barrier contacts with $\phi_{bn}\approx 0.8$ eV embedded in an oxide matrix.

Excess anions can also be generated by reaction of metals with the substrate. For example, it is known that Au deposited on GaAs and GaP results in excess Ga in the Au film.[10] Also preliminary phase diagram data[11] show that an arsenic phase is expected at equilibrium for Au-GaAs and Au-InSb. Thus a knowledge of both oxide and reactive metal chemistry should enable accurate predictions of the transport properties of metal-semiconductor devices (including Schottky barrier heights).

The current status of this model[12] is shown in Table I, which lists the experimentally derived values of ϕ_{bp} and $E_G/q+\chi-\phi_{bp}$ for Au/III-V and Au/II-V contacts. There are three points to note in this table. First, the Schottky model ($E_G/q+\chi-\phi_{bp}=\Phi_{Au}\equiv 5.1\text{–}5.5$ eV) is not obeyed. Second, the EWF model agrees well, as expected, for these data by assuming Φ_{eff} to be dominated by Φ_{Anion}, i.e., $\Phi_{Anion}=E_G/q+\chi-\phi_{bp}$. Third, the common anion rule[13] is not obeyed for AlAs and AlSb. We believe that the common anion rule followed more directly from the anion than initially suggested; in fact, we believe that this rule followed from the formation of microclusters of anions at the interface which dominated the Fermi level position determinations cited . The common anion rule asserts that ϕ_{bp} depends only upon the semiconductor anion. Since, in our model, $\phi_{bp} = ((E_G/q) + \chi) - \Phi_{\rm Anion}$, a common anion would lead to a constant ϕ_{bp} only for a constant $E_G/q + \chi$; Table I shows that those cases following the common anion rule also obey that constraint.

The EWF model also explains such departures from "normal" behavior as the Al–AlAs result,[2] also shown in Table I. For the Al-AlAs case, the metalization was performed in an ultrahigh vacuum molecular beam epitaxy system, where the AlAs surface was very clean, and subsequently annealed. Under these conditions, excess As should react with Al rather than forming microclusters of As. Thus, it is expected that $\Phi_{\rm eff}$ should be dominated by $\Phi_{\rm Al} = 4.0$–4.3 eV. We believe that this explanation is correct, since $\chi + \phi_{bn} \approx 4.2$–$4.6$ for this case, which is much closer to $\Phi_{\rm Al}$ than to $\Phi_{\rm As}$. Similarly, the Ga-GaAs ohmic contact mentioned earlier can be explained since $\Phi_{\rm Ga} = 4.36$ eV (Ref. 14) and $\phi_{bn} = 0$–0.3 (for ohmic behavior); $\chi_{\rm GaAs} + \phi_{bn} = 4.1$–$4.4 \approx \Phi_{\rm Ga}$.

The electrical behavior of most covalent semiconductor interfaces is dominated by the apparent pinning of the Fermi energy level at the interface. We are proposing a model of this behavior which assumes work function matching and (typically) mixed phase behavior at the interface; "pinning" normally observed is shown to follow naturally from microclusters of anions at the interface, which are expected from chemical arguments and observed in some recent experiments.

[1]J. M. Woodall, C. Lanza, and J. L. Freeouf, J. Vac. Sci. Technol. **15**, 1436 (1978).
[2]K. Okamoto, C. E. C. Wood, and L. F. Eastman, Appl. Phys. Lett. **15**, 636 (1981).
[3]C. A. Mead and W. G. Spitzer, Phys. REv. **134**, A713 (1964).
[4]W. Schottky, Zeitschrift fur Physik **118**, 539 (1942).
[5]J. L. Freeouf, J. M. Woodall, and L. M. Foster, Bull. Am. Phys. Soc. **26**, 285 (1981).
[6]R. L. Farrow, R. K. Chang, S. Mroczkowski, and F. H. Pollak, Appl. Phys. Lett. **31**, 768 (1977).
[7]G. P. Schwartz, G. L. Gualtieri, J. E. Griffiths, C. D. Thurmond, and B. Schwartz, J. Electrochem. Soc. **127**, 2488 (1980).
[8]L. G. Meiners, D. L. Lile, and D. A. Collins, J. Vac. Sci. Technol. **16**, 1458 (1979).
[9]H. H. Wieder, J. Vac. Sci. Technol. **15**, 1498 (1978).
[10]A. Hiraki, K. Shuto, S. Kim, W. Kammura, and M. Iwami, Appl. Phys. Lett. **31**, 611 (1977).
[11]R. S. Williams (private communication).
[12]A similar model has been applied to the silicide-silicon interface; see J. L. Freeouf, Sol. State Commun. **33**, 1059 (1980).
[13]J. O. McCaldin, T. C. McGill, and C. A. Mead, Phys. Rev. Lett. **36**, (1976).
[14]H. B. Michaelson, J. Appl. Phys. **48**, 4729 (1977).

The formation of the Schottky barrier at the V/Si interface [a)]

J. G. Clabes,[b)] G. W. Rubloff, B. Reihl, R. J. Purtell, P. S. Ho, A. Zartner, F. J. Himpsel, and D. E. Eastman

IBM Thomas J. Watson Research Center, Yorktown Heights, New York 10598

(Received 14 September 1981; accepted 29 October 1981)

Synchrotron radiation photoemission measurements have been used to study the behavior of the Schottky barrier height ϕ_{bn} and electronic structure of the V/Si interface for both cleaved Si(111)-(2×1) and sputter-cleaned Si(111)-(7×7) surfaces. Although the Schottky barrier height ϕ_{bn} of the clean surface is influenced by surface reconstruction (and by steps), the barrier becomes pinned at low (~2 Å) V coverage at a position essentially independent of the initial clean surface structure. Formation of the bulk V metal band structure is not complete until ~30–40 Å V coverage, indicating coverage-dependent structural effects in the growth of the metal film. These effects also seem to produce secondary influences on ϕ_{bn}, which are manifested as a gradual increase of ϕ_{bn} with higher coverage or mild (~200 °C) annealing. However, upon higher temperature annealing (≥350 °C) the trend reverses, with ϕ_{bn} decreasing to a value (~0.64 eV) characteristic of the bulk VSi_2 contact which is formed at 500–550 °C; this change in the behavior of ϕ_{bn} is directly correlated with the onset of atomic mixing across the interface.

PACS numbers: 73.30. + y, 81.40.Ef

To achieve a basic understanding of what mechanisms determine the Schottky barrier height of the metal/Si interface, it is necessary to establish correlations between the electrical characteristics of the contact and other properties such as interface electronic structure, composition, and microstructure. Over the past few years, a number of investigations have revealed important aspects of the overall interface behavior, particularly for near-noble metals which form silicide compounds.[1–5] These studies have also shown specific chemical and electronic mechanisms which might determine the Fermi level pinning position, including specific interface chemical bonds[6,7] and metal impurities in the Si lattice.[8] However, these studies have not clarified the relation of such behavior to the interface electrical properties.

Recent surface spectroscopy studies have identified the chemistry which occurs at the refractory-metal/Si interface.[9–11] With this information available, we have undertaken synchrotron radiation studies of core-level/band-bending shifts during Schottky barrier formation at low metal coverage[12] to assess which interface characteristics influence the Schottky barrier height ϕ_{bn}. These initial investigations already suggest how ϕ_{bn} is affected by interface microstructure and electronic structure and by the progress of the silicide formation reaction.

Synchrotron radiation from the 240 MeV Tantalus–I electron storage ring at the University of Wisconsin–Madison was used with a UHV toroidal grating monochromator and a display-type electron spectrometer[13] to measure angle-integrated $h\nu$-dependent photoemission spectra for the V/Si(111) interface. The total instrumental energy resolution was ≤0.2 eV. The Si(2*p*) core levels were measured with a resolution ~150 meV, so that core level shifts (and thereby changes in ϕ_{bn}) as small as ~10 meV could be detected. Interface chemical bonding and electronic structure were observed in valence band spectra at various $h\nu$, while the metal-induced changes in Si band-bending or Schottky barrier height were determined from shifts in the substrate Si(2*p*) core level binding energy.[12] The Si samples were prepared by either (i) cleavage at room temperature (using *p*-Si samples), which gives the metastable (2×1) surface reconstruction, or (ii) Ar^+ ion bombardment and thermal annealing (using *n*-Si wafer samples 0.010 in. thick), which produces the stable (7×7) reconstruction. Thin V overlayers (1–15 Å) were deposited at room temperature by direct sublimation from a resistively-heated V wire, with coverage determined by a quartz crystal thickness monitor. For the Si(111)-(7×7) wafer samples, annealing was carried out by direct resistive heating of the sample, and the annealing temperature was determined with an infrared pyrometer.

The valence band photoemission spectra, shown for $h\nu = 21$ eV in Fig. 1, demonstrate the changes in interface electronic structure which occur upon deposition of thin V layers. The spectrum for the clean-cleaved Si(111)-(2×1) surface shows a strong intrinsic surface state at ~ −0.9 eV.[14] Although the surface states are fully quenched upon deposition of 1 Å V, the metallic character of the overlayer film is not developed until ~4 Å V coverage, when the Fermi–Dirac cutoff lineshape appears at the Fermi energy E_F. The nonmetallic character of the surface at low coverage suggests that the chemical bonding between V and Si in this range is essentially like atomic chemisorption rather than the metallic band formation observed at higher coverage. The coverage-dependence of the Auger line intensities[15] indicates that formation of three-dimensional islands on the surface is unlikely.

More detailed valence band studies[15] show clearly that, in contrast to the behavior for near-noble metals, silicide formation and atomic mixing across the interface do not occur at room temperature even at very low coverage. Furthermore, the final V metal bulk band structure, seen in Fig. 1 for 75 Å V, is not established until ~30–40 Å coverage. This slow evolution of the metal band structure may represent the

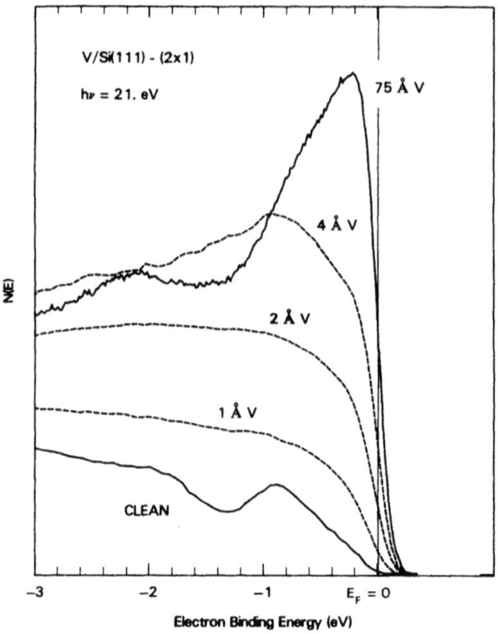

FIG. 1. UPS ($h\nu = 21$ eV) valence band spectra for the clean-cleaved Si(111)-(2×1) surface and for V deposited at room temperature.

growth of a metal film which is homogeneous overall but has coverage-dependent changes in its atomic structure. Although the specific nature of these structural changes is not known at present, they might include, e.g., conversion of an initially amorphorous layer to a polycrystalline state or longer range microstructural effects such as changes in grain size or strain.

The Si(2p) core level spectrum has been measured using two different photon energies. At $h\nu = 120$ eV the escape depth of the photoexcited core electrons is ~5 Å (surface-sensitive mode), while for $h\nu = 108$ eV the escape depth is ~13 Å (bulk-sensitive mode) and provides a better monitor of metal-induced band-bending changes in the underlying Si substrate. As seen in the surface-sensitive mode, the Si(2p) core level peaks broaden with metal coverage, indicating electronic interaction between the Si surface atoms and the deposited V atoms, while the shape of the bulk-sensitive spectrum is basically unchanged. In principle the measured binding energy shifts could include a contribution from metal-induced changes in the screening of the core hole by surface image charges; these effects should depend strongly on the depth of the core hole below the surface. Since we observed that the bulk Si(2p) core levels appear at the same binding energy (to ~ ±20 meV) for both bulk- and surface-sensitive modes, we neglect these surface image charge screening effects in the analysis which follows.

We concentrate here on the changes in the (n-type) Schottky barrier height $\phi_{bn} = E_C - E_F$, where E_C and E_F refer respectively to the energies of the Si conduction band minimum and the Fermi level at the interface. Since band-bending in the Si shifts the entire Si band structure at the interface (including core levels and valence and conduction band edges) rigidly up or down together, changes in ϕ_{bn} can be directly monitored from changes in the Si(2p) core level binding energy. In the bulk-sensitive mode, changes as small as ~10 meV can be detected. Conventional techniques (I–V, etc.) to determine ϕ_{bn} cannot be applied to such thin metal overlayers or to the clean surface, and the core-level binding energy ($E_F - E_{core}$) does not give an accurate absolute measure of ϕ_{bn}. Therefore, the changes in ϕ_{bn} observed here from core level shifts are referenced[12,16] to absolute ϕ_{bn}'s using an independent measurement[17] of ϕ_{bn} for the clean-cleaved Si(111)-(2×1) surface. We emphasize that the main conclusions of the present work depend on the relative changes in ϕ_{bn} (directly observed from the core level shifts), not on their absolute values.

The Schottky barrier heights for the clean Si surfaces and for thin V layers deposited on Si at room temperature are shown in Fig. 2. For these measurements, which involved either fresh cleaves or sample transfer (still under UHV) to the evaporation position and subsequent realignment, we estimate the reproducibility as ~ ±20 meV. For four measurements on the clean-cleaved Si(111)-(2×1) surfaces, three different ϕ_{bn} values were obtained. We take the lowest binding energy (largest ϕ_{bn} value) as the reference point $\phi_{bn} = 0.79$ eV for the "ideal" cleaved surface because: (i) it gives the separation $\phi_{bn}(2\times1) - \phi_{bn}(7\times7) = 180$ meV in agreement with previous work[16]; (ii) it was obtained twice; and (iii) for these cleaves a sharp (2×1) LEED pattern was observed, while for the other two cleaves the LEED pattern showed streaked half-order spots and splitting of the integral order spots in one direction, indicating a relatively high density of atomic steps on the surface. Thus these differences in surface microstructure seem to influence the clean surface barrier

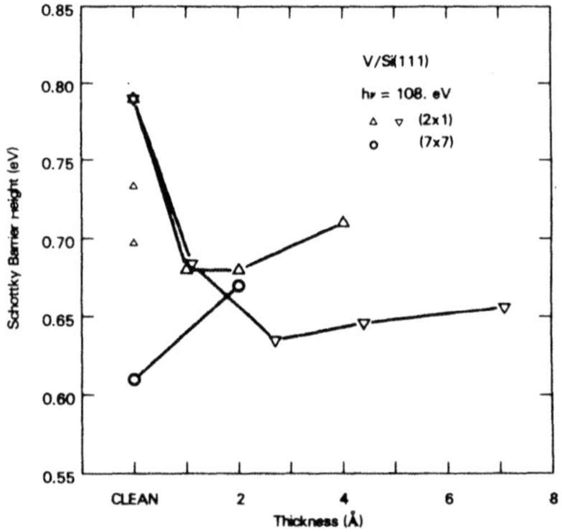

FIG. 2. Schottky barrier height ϕ_{bn} for clean Si(111)-(2×1) and Si(111)-(7×7) surfaces and for thin V overlayers deposited at room temperature, as obtained from the Si(2p) core level binding energy in the bulk-sensitive ($h\nu = 108$ eV) mode. The absolute ϕ_{bn} values here and in Fig. 3 are determined from absolute measurements of ϕ_{bn} (Ref. 17) for clean-cleaved Si(111)-(2×1) surfaces of high quality.

height considerably. In contrast, these structural differences appear not to significantly alter the overall surface electronic structure: for all these cleaves, sharp intrinsic surface states were observed at the same energy position.

Starting from the "ideal" cleaved (2×1) surface, ϕ_{bn} decreases sharply with the first 1–2 Å V coverage to ~0.67 eV. Although ϕ_{bn} for the sputter-cleaned (7×7) surface begins at a considerably lower value (~0.61 eV) than that of the clean (2×1) surface, upon deposition of ~2 Å V, it rises to about the same ϕ_{bn} value (~0.67 eV) as observed for a similar V coverage on the (2×1) surface. This demonstrates that the different initial Fermi level pinning conditions of the clean surface have been overcome and that metal-induced electronic states associated with the V–Si bonding now determine the Fermi level position at the interface.

After its initial sharp decrease on the (2×1) surface, ϕ_{bn} then rises more slowly with higher V coverage. This increase may be related to the coverage-dependent atomic structure changes of the V overlayer inferred above from the valence band photoemission spectra. If this is so, the different ϕ_{bn}'s attained on the two (2×1) cleaved surfaces (e.g., at 4 Å V) may indicate that the metal film growth proceeded somewhat differently on these two surfaces. Although it is interesting and perhaps significant that the barrier height is changed at higher coverage, these appear to be more complicated secondary effects compared to the initial Fermi level pinning associated with the basic V–Si bonding produced at lower coverage.

The dependence of the Schottky barrier height on the state of interfacial reaction is shown in Fig. 3 for the sputter-cleaned Si(111)-(7×7) surface. Starting from ϕ_{bn} for the 25 °C deposition, the reproducibility of ϕ_{bn} was better (±10 meV) than before since the annealing could be done in the measurement position without moving the sample.

Upon annealing to 200 °C, ϕ_{bn} increases further to ~0.70 eV, a trend like that seen for higher coverage at 25 °C (Fig. 2).

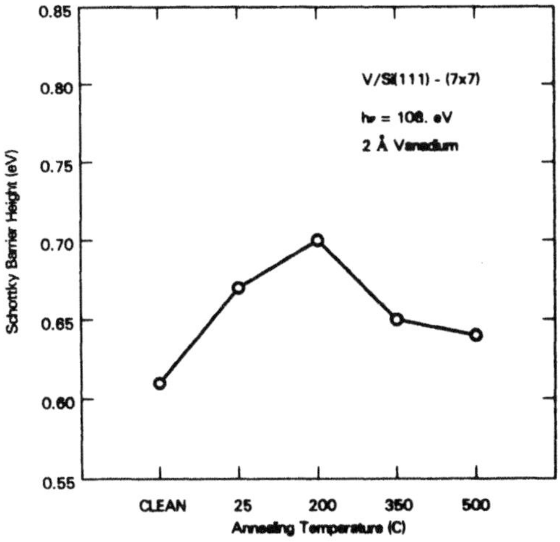

FIG. 3. Schottky barrier height ϕ_{bn} for annealing of 2 Å V film deposited on the clean Si(111)-(7×7) surface, as obtained from the Si(2p) core level binding energy in the bulk-sensitive ($h\nu$ = 108 eV) mode.

At higher annealing temperatures, the trend clearly reverses and ϕ_{bn} decreases markedly, essentially reaching a saturation value ~0.64 eV already at 350 °C. Other investigations[11,15] have established that atomic mixing across the V/Si interface does not occur at 200 °C but begins at 350 °C, while formation of the bulk silicide (VSi_2) contact does not take place until 500 °C. Therefore, we conclude that the behavior (qualitative as well as quantitative) of ϕ_{bn} is dramatically influenced by strong interfacial reaction (atomic mixing). This interface chemical process at 350 °C already pins ϕ_{bn} at essentially the final value observed for VSi_2 formation at 500 °C. Finally, we note that this Schottky barrier height agrees well with that determined from I–V characteristics of much thicker VSi_2 Schottky diodes produced under standard thin film conditions.[18]

These synchrotron radiation studies provide a valuable tool for determining ϕ_{bn} with high resolution (compared to the range of barrier heights observed from different metal/Si contacts) during the important initial stages of interface formation which determine the Fermi level pinning position of the final metal contact. Furthermore, such measurements of ϕ_{bn} can be correlated with the interface electronic structure and chemical reactivity which are simultaneously observed in these investigations.

Our findings can be summarized in the following way: The clean Si Fermi level position is clearly influenced by reconstruction of and steps on the surface. However, a thin metal overlayer (~1–2 Å) pins the Schottky barrier height at a position which seems to be associated with the basic V–Si bonding at the interface, independent of the initial clean surface structural conditions. This thin metal layer has not yet developed full metallic character. Our results also indicate that changes in the atomic structure of the unreacted V film with coverage influence the electronic structure, and they provide a secondary influence on ϕ_{bn}. Finally, the qualitative as well as quantitative behavior of the Schottky barrier height is strongly altered by interface chemical reactions which involve atomic mixing across the initially unreacted metal/Si interface.

Acknowledgment: We appreciate the able assistance of the Synchrotron Radiation Center in Stoughton, Wisconsin.

[a)]Supported in part by the Office of Naval Research.
[b)]Present address: University of Hannover, Institut B für Experimentalphysik, F. R. Germany.

[1]G. W. Rubloff, P. S. Ho, J. L. Freeouf, and J. E. Lewis, Phys. Rev. B 23, 4183 (1981); J. L. Freeouf, G. W. Rubloff, P. S. Ho, and T. S. Kuan, Phys. Rev. Lett. 43, 1836 (1979).
[2]P. E. Schmid, P. S. Ho, H. Föll, and G. W. Rubloff, J. Vac. Sci. Technol. 18, 937 (1981).
[3]N. W. Cheung, R. J. Culbertson, L. C. Feldman, P. J. Silverman, K. W. West, and J. W. Mayer, Phys. Rev. Lett. 45, 120 (1980).
[4]P. J. Grunthaner, F. J. Grunthaner, and J. W. Mayer, J. Vac. Sci. Technol. 17, 924 (1980).
[5]L. Braicovich, I. Abbati, J. N. Miller, I. Lindau, S. Schwarz, P. R. Skeath, C. Y. Su, and W. E. Spicer, J. Vac. Sci. Technol. 17, 1005 (1980); J. N.

Miller, S. A. Schwarz, I. Lindau, W. E. Spicer, B. deMichelis, I. Abbati, and L. Braicovich, J. Vac. Sci. Technol. **17**, 920 (1980).
[6]P. S. Ho, P. E. Schmid, and H. Föll, Phys. Rev. Lett. **46**, 782 (1981).
[7]G. W. Rubloff (to be published).
[8]N. W. Cheung and J. W. Mayer, Phys. Rev. Lett. **46**, 671 (1981).
[9]G. W. Rubloff, in *Proc. of the 8th Int. Vacuum Congress*, Cannes, France, edited by F. Abeles and M. Croset (Suppl. la Revue "Le Vide, les Couches Minces" No. 201), Vol. I, p. 562.
[10]A. Franciosi, D. J. Peterman, and J. H. Weaver, J. Vac. Sci. Technol. **19**, 657 (1981); J. H. Weaver, V. L. Moruzzi, and F. A. Schmidt, Phys. Rev. B **23**, 2916 (1981).
[11]J. G. Clabes and G. W. Rubloff, J. Vac. Sci. Technol. **19**, 262 (1981).
[12]J. L. Freeouf, M. Aono, F. J. Himpsel, and D. E. Eastman, J. Vac. Sci. Technol. **19**, 681 (1981).
[13]D. E. Eastman, J. J. Donelon, N. C. Hien, and F. J. Himpsel, Nucl. Instrum. Methods **172**, 327 (1980).
[14]F. J. Himpsel, P. Heimann, and D. E. Eastman, Phys. Rev. B **24**, 2003 (1981).
[15]J. G. Clabes, G. W. Rubloff, and T. Y. Tan (to be published).
[16]F. J. Himpsel, P. Heimann, T.-C. Chiang, and D. E. Eastman, Phys. Rev. Lett. **45**, 1112 (1980).
[17]Absolute values for $E_F - E_V (E_V =$ valence band maximum) have been reported as 0.33 eV($\phi_{bn} = 0.79$ eV) by F. G. Allen and G. W. Gobeli, Phys. Rev. **127**, 150 (1962) and as 0.48 eV ($\phi_{bn} = 0.64$ eV) by C. Sebenne, D. Bolmont, G. Guichar, and M. Balkanski, Phys. Rev. B **12**, 3280 (1975). We use the former value here to facilitate comparison with previous studies (Refs. 12, 16) which chose this reference point and to be consistent with the conclusions from surface photovoltage studies by J. Clabes and M. Henzler, Phys. Rev. B **21**, 625 (1980).
[18]K. N. Tu, R. D. Thompson, and B. Y. Tsaur, Appl. Phys. Lett. **38**, 626 (1981).

Formation of Ultrathin Single-Crystal Silicide Films on Si: Surface and Interfacial Stabilization of Si-$NiSi_2$ Epitaxial Structures

R. T. Tung, J. M. Gibson, and J. M. Poate
Bell Laboratories, Murray Hill, New Jersey 07974
(Received 22 November 1982)

Continuous and planar single-crystal $NiSi_2$ films (<60 Å thick) have been grown on Si(100) and (111) by ultrahigh-vacuum techniques. Nickel deposition (<20 Å) on atomically clean Si at room temperature, followed by low-temperature heating at ~450 °C, results in the growth of epitaxial $NiSi_2$ films whose structure depends critically on the starting Ni thickness. The films are coherent, without misfit dislocations. The orientation of the (111) silicide can be controlled by the initial Ni thickness.

PACS numbers: 68.45.+f, 68.55.+b, 73.40.Ns

The reaction of deposited metals with silicon is a subject of considerable interest. A large number of publications on this matter exist but most deal with just two aspects of the problem. Firstly, surface properties, such as structure and electronic states, are usually emphasized in the studies of submonolayer or monolayer(s) of metal under ultrahigh-vacuum (UHV) conditions. In particular, it has been shown that Ni reacts readily with clean Si surfaces even at room temperature for the first few monolayers of Ni deposition.[1,2] A second area of extensive research lies in the electronic and structural properties of thick metal silicides on Si and their interfaces.[3-4] Silicides are usually formed by depositing and heating thick (>200-Å) layers of metal films under poor vacuum conditions (~10^{-7} Torr). In this Letter, we show that the room-temperature Ni-Si reaction, although involving only a few monolayers of material, strongly affects the subsequent growth and epitaxy of very thin (<60-Å) $NiSi_2$ layers on Si. Novel structures and interfaces can be stabilized as a result.

Disilicides of Ni and Co are ideal choices for epitaxy as they have the cubic CaF_2 structure with bulk lattice parameters within 1.2% of Si.[4] Recently we demonstrated that thick single-crystalline $CoSi_2$ films can be grown with an unusual orientation on Si(111), by cold deposition and heating under UHV conditions.[5] The films are rotated 180° about the normal to the Si surface (we label this type-*B* orientation, as opposed to the unrotated, type-*A* orientation). The $NiSi_2$ films grown on Si(111) by the same techniques are mixtures of type-*A* and type-*B* orientations. The (111) films with thicknesses less than 500 Å are not continuous. It has not been possible with these techniques to grow uniform epitaxial films on Si(100). The interface breaks up into {111} facets and the films tend to be discontinuous for thicknesses <1000 Å.[6] In this Letter we show that by use of low-temperature surface reaction and novel epitaxy techniques, very thin and continuous single-crystal silicide films can be grown with quite novel structures. Defect-free metallic layers with thicknesses <50 Å have been grown on a semiconductor for the first time.

We will firstly describe epitaxy on Si(111). Polished Si(111) substrates of both *n* and *p* type were degreased and dipped in buffered HF before mounting on a sample holder. Samples are cleaned by resistive heating and, when necessary, sputtering in a UHV chamber with base pressure of 1×10^{-10} Torr. At the onset of each experiment, samples were heated to 100 °C for ~2 min and allowed to cool slowly. Surfaces so prepared always displayed sharp (7×7) low-energy electron diffraction (LEED) patterns characteristic of a clean Si(111) surface, and no impurities could be detected by Auger-electron spectroscopy (AES). Nickel was deposited by *e*-gun evaporation at rates of ~0.5–3 Å/s. During evaporation, the background pressure never exceeded 2×10^{-9} Torr.

Thin layers (~1–30 Å) of nickel were deposited at room temperature on clean Si surfaces, causing the disappearance of the (7×7) LEED pattern when the Ni coverage exceeded ~3 Å. The AES line from the Ni *LMM* transition (848 eV) is lowered in energy by about ~1 eV for the initial ~5 Å, which indicates room-temperature reaction of Ni on Si. The temperature of the samples was then raised quickly to ~450–550 °C for 3–5 min. Immediately, a LEED pattern appeared with a sharp (1×1) array having the same two-dimensional unit-cell length as that of the unreconstructed Si(111). The AES peak-to-peak height ratio of the Ni *LMM* and Si *LVV* lines reached that of thick $NiSi_2$ for Ni coverage of more than 12 Å. The LEED pattern was always threefold symmetric, having different intensities for the (1, 0)- and ($\bar{1}$, 0)-type beams. The relative intensities of the two types of spots, however, depend critically and reproducibly on the amount of deposited Ni. At all incident electron energies, the LEED pattern of the annealed surface with ~1–7 Å of deposited Ni conforms exactly to that obtained with 16–20 Å of Ni, but with a 180° rotation.

The LEED and Auger results indicate that annealing has resulted in the formation of continuous, epitaxial $NiSi_2$ layers. Thin layers formed with 1-7-Å Ni are type-*B* oriented and those with 16–20 Å of Ni are type-*A* oriented. This conclusion is also verified by Rutherford backscattering (RBS) and channeling analysis and by transmission electron microscopy (TEM) lattice imaging. The volume percentage of type-*A* orientation in these $NiSi_2$ layers, determined by LEED, RBS and channeling, and TEM, is plotted in Fig. 1. The remaining volume of these films is type *B*. Cross-sectioned and ion-thinned samples were analyzed with an ultrahigh-resolution transmission electron microscope. Shown in Fig. 2 are axial bright-field electron micrographs of thin $NiSi_2$ layers viewed in cross section, in which the white dots correlate with tunnels in the $\langle 1\bar{1}0 \rangle$ projected atomic arrangements. The layer formed with 18 Å Ni is uniform in thickness and clearly epitaxial with type-*A* orientation. The $NiSi_2$/Si(111) interface is flat and well localized to one interplanar spacing. Figure 2(a) shows the continuous (~3–6-Å) thin type-*B* layer resulting from deposition of 5 Å of Ni. The excess Ni is found in the form of $NiSi_2$ islands which constitute <5% of the total film area. The orientation of the very thin (~3–6-Å) layer is hard to identify. Laser diffraction from various portions of the negatives of these micrographs is consistent with this very thin layer being type *B*, agreeing with the LEED and RBS results. No misfit dislocations are observed in any of the thin layers (<60 Å), either in cross section or in plan view. The epitaxial growth is therefore coherent and the misfit has not been relieved.

The existing $NiSi_2$ orientation can be maintained in further silicide growth by depositing thin Ni layers at temperatures <100 °C and then annealing at 450 °C. For example, the orientation of a type-*B* layer (initially formed with 1-7-Å Ni) will always stay type *B* even if successive depositions (~4 Å of Ni at a time) and annealings incrementally increase the silicide thickness to greater than a Ni coverage of ~20 Å. The initial (1×1) LEED pattern changes to a $(\sqrt{3}\times\sqrt{3})$-*R*30° after deposition of ~4 Å of Ni, and subsequent annealing restores the sharp type-*B* (1×1) array. The type-*A* orientation is also preserved after such incremental growth from an initial type-*A* layer. The thickness of these thin $NiSi_2$ layers can be extended much more expeditiously by depositing Ni at elevated temperature ($\lesssim$ 650 °C) or by molecular-beam epitaxy (MBE). In this manner, thick (~1000-Å) and uniform layers of single-crystalline type-*A* or type-*B* epitaxial

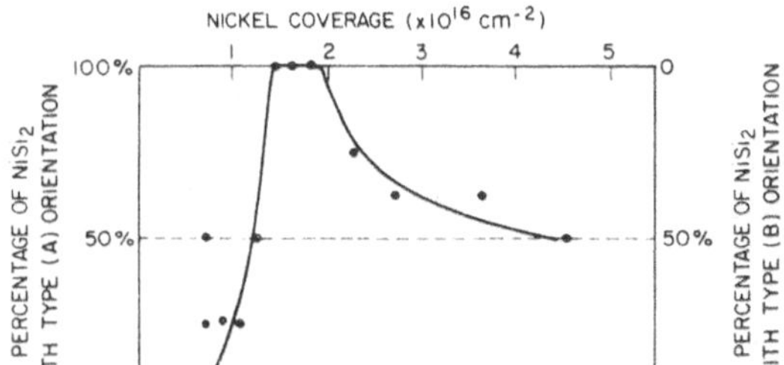

FIG. 1. The percentage of $NiSi_2$ volume with type-*A* orientation in films grown by annealing room-temperature–deposited Ni on Si(111), as determined by LEED, RBS and channeling, and TEM. [A 1-Å-thick nickel film has areal density of 9.1×10^{14} atoms/cm^2, about 1.2 monolayers on Si(111), and reacts with Si to form a $NiSi^2$ layer of 3.65-Å average thickness.]

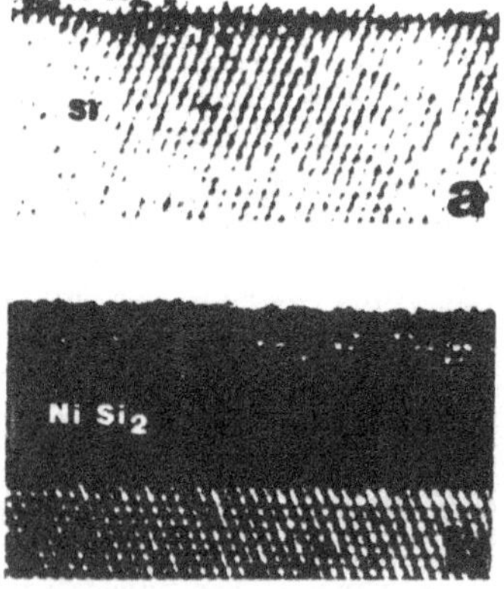

FIG. 2. Ultrahigh-resolution TEM images of $NiSi_2$ on Si(111) interface viewed in a $\langle 1\bar{1}0 \rangle$ cross section. These layers are formed by UHV annealing (a) 5 Å of Ni, and (b) 18 Å on Ni. Interface structure is not directly interpretable in (b) which was chosen to most dramatically illustrate the layer uniformity.

$NiSi_2$ can be grown[7] on top of these thin "templates."

The formation of these thin layers should be contrasted with the reaction of thick metal films to form silicides. The annealing of thick (>200-Å) layers of Ni on Si (Ref. 3) always results in the growth of Ni_2Si, followed by NiSi, and finally the formation of $NiSi_2$ [of mixed type-A and type-B orientations on a (111) substrate]. However, the reaction of a few monolayers of Ni with clean Si substrate need not follow the same pattern as that of thick layers, because of the greatly reduced diffusion length. As an example, the temperature used in this study (~450 °C) for $NiSi_2$ reaction is far below the minimum formation temperature of $NiSi_2$ (~775 °C) for thick Ni layers.[8] The room-temperature precursor Ni-Si reaction is limited to a few monolayers of nickel and results in a disordered surface layer. The composition of this layer can be indirectly deduced from photoelectron studies to be more silicon rich than Ni_2Si.[2,9] Recent surface extended x-ray-absorption fine-structure studies[10] show that the short-range environment surrounding Ni atoms in this disordered layer is similar to that encountered in a $NiSi_2$ lattice. The formation of epitaxial $NiSi_2$ from such a disordered layer does not need significant transport of matter. The type-B material forms probably because of the lower free energy of its interface. The remarkable continuity and uniformity of the very thin $NiSi_2$ layer in our work is most likely due to the precursor reaction covering the whole surface. Complete coverage is possible because the high Ni surface diffusivity[11] allows unreacted Ni to migrate at room temperature to regions of exposed Si.

The silicide reaction sequence will change radically when unreacted Ni is present on top of the disordered precursor layer for deposited Ni layers greater than 10 Å. The precursor layer is erased at the beginning of the anneal from diffusion of Ni to form Ni_2Si and NiSi. The subsequent $NiSi_2$ nucleation presumably occurs at the interface between a thin NiSi layer and the Si lattice. Since Ni is the dominant diffusing species at lower temperatures,[3] $NiSi_2$ will grow into the Si lattice. The reason why growth of type A is favored for the thicker deposition is intriguing and we propose the following explanation. Movement of a planar type-A interface one interplanar spacing (3.1 Å) towards Si requires the rearrangement of only one layer of Si atoms with the concomitant diffusion of one atomic layer of Ni. The positions of all other atoms, either in the Si or in the $NiSi_2$, can stay unchanged. However, for a sevenfold or fivefold coordinated type-B interface,[12] the advancement toward Si can only be made by three interplanar spacings at a time (~9.4 Å) to avoid significant displacement of other atoms. Planar growth breaks down for deposited thicknesses greater than 25 Å, and both A- and B-type materials grow. A possible reason for this behavior is that significant Ni and Si diffusion is required for the thicker-film growth. There is some evidence for this interpretation from studies of thick- (>200-Å) film deposition and reaction where silicide growth is a nucleation-limited process.[8,13] At the formation temperature (~775 °C) of these thick films both Ni and Si diffusions are significant. Columns of $NiSi_2$ are observed to form which then spread laterally with heating. These columns are of both A- and B-type material.

We have demonstrated that epitaxial growth can be controlled on a (111) surface by template formation. The question can then be asked whether these low-temperature templates can be used to stabilize the (100) silicide-Si interface which is unstable at high temperatures. Templates were formed on the Si(100) surface in essentially the same fashion as described previously on Si(111). Thin Ni layers (3–50 Å) were deposited at room temperature on clean Si(100) surfaces which displayed the (2×1) LEED pattern. They were then heated to ~500 °C for 3–5 min. Surfaces formed from annealing of ~10–30 Å Ni display LEED patterns similar to that reported for the $NiSi_2$(100) surface.[6] LEED patterns from surfaces annealed with either >40 or <10 Å of Ni consisted of this pattern, characteristic of $NiSi_2$(100), and, in addition, half-order spots. This suggests a surface of $NiSi_2$(100) plus some areas of bare Si(100). High-resolution TEM micrographs of $NiSi_2$ layers, viewed in the ⟨011⟩ cross section, are shown in Fig. 3. The $NiSi_2$/Si interface is heavily faceted on inclined {111} planes for $t_{Ni} < 10$ Å; the films are epitaxial but discontinuous. However, epitaxial $NiSi_2$ films formed from Ni deposition in the range $10 < t_{Ni} < 30$ Å are continuous and of uniform thickness. Moreover, the interface is relatively flat, running along (100) planes. The reason why a range of thickness exists in which the epitaxial $NiSi_2$ formed on Si(100) is uniform and continuous is not clear. We suspect that the room-temperature precursor reaction and the growth process are key factors. Moreover, these uniform and

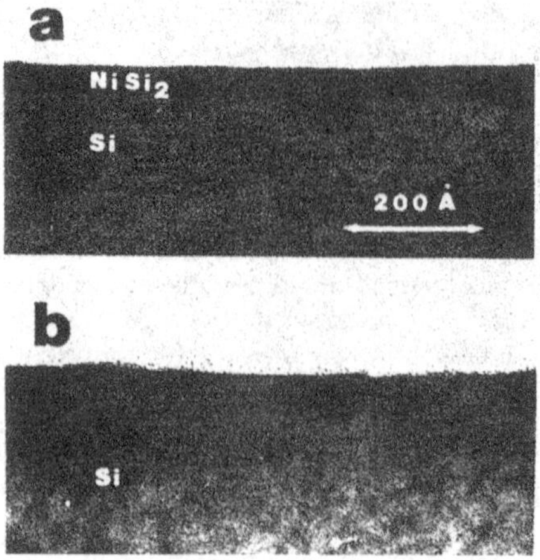

FIG. 3. Ultrahigh-resolution TEM images of $NiSi_2$ on Si(100). These layers are formed with UHV annealing of (a) 11 Å of Ni, and (b) 6 Å of Ni. Dark "spikes" near the surface in (b) are $NiSi_2$ crystals.

continuous (100) layers can also be used as templates for the growth of thicker, yet still uniform and perfectly epitaxial, $NiSi_2$ (Ref. 7) by deposition of Ni at ~600–700 °C or by MBE.

In this Letter we have demonstrated that novel epitaxial $NiSi_2$ films can be grown by making use of the unusual surface reactivity of Ni and Si at low temperatures. In this way a continuous, surface-disordered phase can be used to pin and control subsequent epitaxial growth. Remarkably thin and continuous single-crystal metal films can be fabricated with novel structural properties.

We thank J. E. Rowe, J. W. Mayer, P. H. Citrin, F. Comin, and S. Mahajan for unpublished results and discussions. We also thank M. L. McDonald for preparing TEM samples.

[1]N. W. Cheung, R. J. Culbertson, L. C. Feldman, P. J. Silverman, K. W. West, and J. W. Mayer, Phys. Rev. Lett. 45, 120 (1980).

[2]P. J. Grunthaner, F. J. Grunthaner, and J. W. Mayer, J. Vac. Sci. Technol. 17, 924 (1980), and 19, 649 (1981).

[3]K. N. Tu and J. W. Mayer, in *Thin Films—Interdiffusion and Reactions*, edited by J.M. Poate, K. N. Tu, and J. W. Mayer (Wiley, New York, 1978).

[4]R. T. Tung, J. M. Poate, J. C Bean, J. M. Gibson, and D. C. Jacobson, Thin Solid Films 93, 77 (1982).

[5]R. T. Tung, J. C. Bean, J. M. Gibson, J. M. Poate, and D. C. Jacobson, Appl. Phys. Lett. 40, 684 (1982).

[6]K. C. R. Chiu, J. M. Poate, J. E. Rowe, T. T. Sheng, and A. G. Cullis, Appl. Phys. Lett. 38, 988 (1981).

[7]R. T. Tung, J. M. Gibson, and J. M. Poate, to be published.

[8]J. Baglin, F. d'Heurle, and S. Petersson, in *Thin Film Interfaces and Interactions*, edited by J. E. E. Baglin and J. M. Poate (Electrochemical Society, Princeton, N. J., 1980).

[9]A. Franciosi, J. H. Weaver, D. G. O'Neill, Y. J. Chabal, J. E. Rowe, J. M. Poate, O. Bisi, and C. Calandra, J. Vac. Sci. Technol. 21, 627 (1982).

[10]F. Comin, J. E. Rowe, and P. H. Citrin, to be published.

[11]R. T. Tung and W. R. Graham, Surf. Sci. 97, 73 (1980).

[12]J. M. Gibson, J. C. Bean, J. M. Poate, and R. T. Tung, Appl. Phys. Lett. 41, 818 (1982).

[13]S. S. Lau and N. W. Cheung, Thin Solid Films 71, 117 (1980).

Schottky Barrier Heights and the Continuum of Gap States

J. Tersoff
AT&T Bell Laboratories, Murray Hill, New Jersey 07974
(Received 7 November 1983)

Simple physical considerations of local charge neutrality suggest that near a metal-semiconductor interface, the Fermi level in the semiconductor is pinned near an effective gap center, which is simply related to the bulk semiconductor band structure. In this way "canonical" Schottky barrier heights are calculated for several semiconductors. These are in excellent agreement with experiment for interfaces with a variety of metals.

PACS numbers: 73.30.+y, 73.40.Ns

Despite decades of intense study, there exists no quantitative, predictive theory of Schottky barrier heights. Simple models[1-3] and phenomenological theories[4-7] have had some success in explaining barrier formation and chemical trends in barrier heights. Also, a few calculations for model systems[8,9] have yielded reasonable agreement with experiment. In general though, the complexity of real interfaces and the subtlety of the effects involved have frustrated attempts at a truly predictive theory. In view of the intrinsic interest and technological importance of metal-semiconductor (M-S) interfaces, this inability to understand their most crucial electronic characteristic is quite disappointing.

For the more covalent semiconductors, the barrier height is independent of the metal used to within ± 0.1 eV for metals of practical interest.[4] Thus, one hopes for a rather simple explanation of the roughly "universal" barrier heights for M-S interfaces with these semiconductors. Explanations so far have focused on the possible pinning of the Fermi level (E_F) by states associated with defects in the semiconductor.[10-12] I argue below that these explanations, while successful in describing surfaces with submonolayer metal coverages, are inappropriate for bulk interfaces.

Here I show that a simple parameter-free model for Fermi-level pinning by metal-induced gap states (MIGS) can predict quantitatively the observed values of the "universal" barrier heights, as well as explaining why more-ionic semiconductors do not exhibit such universality. Such MIGS pinning has been found in numerical calculations by Louie, Chelikowsky, and Cohen[8,9] but the simplicity and generality of the mechanism has not been recognized. In particular, the behavior seen is by no means peculiar to the ideal planar interface as has been suggested.[10] While following Heine[3] in spirit, I stress here the continuum nature of gap states, and the resulting locally metallic character of the semiconductor near the interface.

The various models of Schottky barrier formation are discussed in several excellent reviews.[13,14] The crucial point is simply that the barrier height is determined by the position of E_F within the semiconductor gap. The barrier is the energy needed to excite an electron from E_F to the conduction minimum (for n-type semiconductors). Band bending due to doping can be neglected in the region of interest, which extends only ~ 10 Å from the interface.

At a M-S interface there is a continuum of states around E_F because of the metal. As first discussed by Heine,[3] those states within the gap decay exponentially inside the semiconductor, but still have significant amplitude a few layers from the interface. Any deviation from local charge neutrality in this region results in "metallic" screening by the MIGS.

A very small density of MIGS is sufficient to pin E_F. With use of Thomas-Fermi screening, a local density of only 0.02 state/atom eV in the gap gives a screening length of about 3 Å. Numerical results of Louie and Cohen[8] show a density much greater than this for the first 6 Å or so at a metal-Si interface.

It is therefore convenient to consider the barrier height as having two contributions—a short-range part, which may be related to surface dipoles,[1] the M-S electronegativity difference,[4] or more subtle details of bonding; and an additional dipole from metallic screening by MIGS, which tends to pin E_F so as to maintain local charge neutrality. Here I argue that whether short-range or screening effects dominate (E_F unpinned or pinned) depends simply on bulk semiconductor properties, as does the barrier height in the pinned limit. Moreover in the strongly pinned limit appropriate to Si, Ge, and GaAs, the pinning occurs relatively deep in the semiconductor.

It is important to remember that the MIGS are actually Bloch states of the bulk semiconductor with complex wave vector.[15-17] The formal properties of gap states have been studied extensively.[15-17] There is a sum rule[18,19] on the density of states (DOS) whereby any weight in gap states must come from the valence and conduction bands. Gap states take their weight primarily from those bands that are nearest in energy (allowing for wave-vector and symmetry selection rules). Charge neutrality thus requires occupation of those MIGS which come primarily from the valence band, while leaving those of mainly conduction-band character empty.

I therefore propose that E_F must fall at or near the energy where the gap states cross over from valence- to conduction-band character. In one dimension this energy corresponds[16] to the branch point E_B of the complex band structure, as discussed by Kohn and Rehr.[15,16] The generalization to three dimensions is discussed below. (Of course, there is no discontinuous change in the character of the wave functions at E_B.[19] Rather, states at E_B derive their weight equally from valence and conduction bands. The net effect is still to pin E_F at or near E_B.) For covalent semiconductors, E_B is closely related to the surface-state and vacancy levels, explaining why different theoretical approaches[3,12] yield similar results.

By finding the branch point in the complex energy bands, we immediately have a "canonical" barrier height for the given semiconductor. The barrier heights which are determined in this way from the bulk band structure of several semiconductors are in excellent agreement with experimental values for interfaces with a variety of metals.

The expected behavior can be seen in the self-consistent calculation of Louie and Cohen[8] for a "jellium"-Si(111) interface. At Si atoms a few layers from the metal, states "spill over" into the gap from the conduction and valence bands above and below. In between there is a minimum in the calculated local DOS, presumably at E_B (where the MIGS decay length is shortest). The Fermi level is pinned precisely at this minimum; this is viewed as a natural consequence of the principle of local charge neutrality.

One begins by defining the cell-averaged real-space Green's function,

$$G(\vec{R},E) = \int d^3r \sum_{nk} \frac{\psi_{nk}^*(\vec{r})\psi_{nk}(\vec{r}+\vec{R})}{E-E_{nk}} = \sum_{nk} \frac{e^{i\vec{k}\cdot\vec{R}}}{E-E_{nk}}, \quad (1)$$

where $\vec{k}$ is the Bloch wave vector, n is the band index, and ψ and E are the corresponding wave function and energy.

In one dimension, for sufficiently large R, $G(R,E)$ changes sign at the energy of the branch point.[17] In higher dimensions we pick a direction by specifying $\vec{R}$; then for each $k_\parallel$ there is a branch point as the longitudinal wave vector is varied. By integrating over the entire Brillouin zone, for large $\vec{R}$ we automatically pick out the contribution to $G(\vec{R},E)$ with longest range.

For an ideal interface, $\mathrm{Im}(\vec{k})$ must be normal to the interface. It would therefore appear that we should pick $\vec{R}$ in that direction. However, for a disordered interface it seems preferable to assume that all directions are permitted, as in an impurity problem. Then the direction which gives the most slowly decaying MIGS is the important one. Experimentally there appears to be no dependence of barrier height on orientation for interfaces prepared in the usual manner. On the other hand, there is evidence of strong orientation dependence for ideal epitaxial interfaces,[20] consistent with the model here but not with defect models.

Fortunately, the appropriate choice of direction is obvious. The fcc (cell) nearest-neighbor lattice vector is $(a/2)(110)$. This is also the direction along the chains of bonded atoms in the diamond and zinc-blende structures. Numerical calculations have shown that charge disturbances propagate farthest along these (110) chains.[21] We therefore consider $\vec{R}_m = m(a/2)(110)$. For various $\vec{R}_m$ ($m = 1,\ldots,10$) we calculate $G(\vec{R},E)$ and locate the energy in the gap where this changes sign. This energy approaches a constant value for large $\vec{R}_m$ ($m > 3$). The direction dependence is discussed below.

In general, E_F must depend on the details of the interface, since the density of MIGS is determined by the boundary condition. For example, if the metal continuum were replaced by a few discrete levels in the gap (e.g., defect levels), then E_F would be pinned at one of these levels, possibly quite far from E_B. Such an effect has been seen at surfaces.[10] If, however, the metal DOS is relatively featureless throughout the gap, and the MIGS penetrate deep enough to screen the interface, then the position of E_F in the semiconductor gap will be determined primarily by the complex band structure. We ignore conservation of $k_\parallel$ across the interface, since microscopically M-S interfaces are disordered.[20]

The calculation was carried out using energy

bands obtained with the linearized augmented plane-wave method, summing 152 points in the irreducible wedge of the bulk Brillouin zone. Because the local density approximation used for correlation and exchange gives poor values for the absolute gap but good dispersions, I rigidly shift the conduction bands to give the correct room-temperature energy gap, following Baraff and Schluter.[22] I stress that it is unimportant how the band structure used in (1) is obtained, as long as it is reasonably accurate.

The results are tabulated in Table I, which gives the calculated barrier height for n-type semiconductors, i.e., the difference between the inferred Fermi level and the conduction-band minimum; and the asymptotic charge decay length λ for states at this energy. Results are numerically accurate to better than 0.05 eV (0.4 for ZnS) and 0.5 Å, respectively.

Only results for (110) lattice vectors are given in Table I. The (100) and (111) directions require much larger wave-vector samples for good convergence. Preliminary results with a 1186-point sample indicate a smaller barrier for Si(100), suggesting a possible explanation for new experimental results on orientation dependence for epitaxial films.[20] Results also confirm that the (110) direction gives the longest decay length, justifying the choice of (110) to determine the barrier height for disordered interfaces.

Also shown are experimental Schottky barrier heights.[23-25] I give both the range for a number of "ordinary" metals, and specific values for Au and Al, which have been extensively studied and represent the normal range in metal electronegativity. [C/V measurements for intimate contacts were chosen where available (for Si and GaAs).] In all cases the theoretical value falls within the scatter of barrier heights for typical metals. The decay lengths found here are consistent with those calculated by Louie, Chelikowsky, and Cohen for semiconductor-jellium interfaces.[9] For the elemental semiconductors the measured barriers are smaller (larger) than given by the theory for metals with lower (higher) electronegativity than the semiconductor. Thus, the deviations from the canonical barrier heights calculated here may be attributed, at least in part, to the M-S electronegativity difference (cf. Ref. 4).

TABLE I. Schottky barrier heights.

	Gap	Barrier heights (eV)				λ
	(eV)	Au	Al	Other	Theory	(Å)
Si [a]	1.12	0.83	0.70	0.70–0.82	0.76	3.0
Ge [b]	0.66	0.59	0.48	0.38–0.64	0.48	4.0
GaAs [c]	1.42	0.94	0.78	0.71–0.94	0.74	3.0
ZnS [b]	3.60	2.00	0.80	0.80–2.00	1.40	1.5

[a]Ref. 23. [b]Ref. 24. [c]Ref. 25.

In fact, the predicted barrier heights are in as good agreement with experiment as any calculations reported to date, despite the fact that no allowance has been made for the properties of the metal or the geometry of the interface. I believe that this strongly supports the correctness of the underlying physical idea, that is, the necessity of occupation of the MIGS according to their degree of valence or conduction character in order to maintain layer-by-layer charge neutrality.

Large-gap ionic semiconductors have barrier heights which vary considerably depending upon the metal used.[4] Results in Table I suggest an explanation consistent with that of Louie, Chelikowsky, and Cohen.[9] The short decay length of MIGS in midgap for ZnS results in a negligible DOS in the gap except very near the interface. The MIGS are therefore unable to screen the effect of the metal electronegativity.

If, however, the MIGS decay length is large, the pinning is metallic (Thomas-Fermi-like), and any deviation of E_F from its canonical position is screened exponentially with distance from the interface. The screening is cut off effectively at the MIGS decay length. Reexamining results of Louie and Cohen[8] for jellium-Si(111) in this light clarifies the mechanism at work there. The first Si layer sees a self-consistent potential significantly lowered by proximity to the metal, and the corresponding local DOS is shifted downward in energy (i.e., E_F is near the local conduction minimum). However, band bending (screening by MIGS) between the first and second double layer restores E_F to its canonical midgap position in the Si by moving the conduction minimum up 0.3–0.4 eV. Thus even for the ideal planar interface, pinning takes place far inside the Si, explaining the relative insensitivity to interface details.

Other models of barrier formation have been proposed based on pinning by defect levels.[10-12] At the free surface a very small number of states in the gap (defect or intrinsic surface states) can pin the Fermi level.[2] Since the screening length is hundreds or thousands of angstroms (depending on doping), any charging gives rise to an enormous dipole, which shifts E_F so as to maintain neutrality. At a M-S interface, however, the metal will screen any defects nearby. Since the screening charge is only a few angstroms away, each defect contributes at most a modest local dipole. Defects are therefore orders of magnitude less effective at pinning E_F at a M-S interface than at a surface.

Experimentally there is evidence for defect pinning of E_F on surfaces with submonolayer metal coverage.[10] However, results of these experiments are inconsistent with bulk barrier-height measurements[24,25] and must reflect a different mechanism than the true bulk interface. In particular, submonolayer metal coverages neither screen the defect charge effectively nor provide a continuum of states in the gap. The metal atoms themselves are in effect merely local defects.

Defect models of Schottky barrier formation, while appropriate for bare surfaces, have not established any direct relevance to bulk M-S interfaces. In contrast, the continuum model described here is specifically appropriate to bulk M-S interfaces; it has immediate predictive value, and is in excellent agreement with experiment.

It is a pleasure to acknowledge many stimulating discussions and helpful suggestions by D. R. Hamann. Conversations with E. O. Kane, J. C. Phillips, M. Schlüter, and R. T. Tung are also gratefully acknowledged.

[1]W. Schottky, Phys. Z. 41, 570 (1940).
[2]J. Bardeen, Phys. Rev. 71, 717 (1947).
[3]V. Heine, Phys. Rev. A 138, 1689 (1965).
[4]S. Kurtin, T. C. McGill, and C. A. Mead, Phys. Rev. Lett. 22, 1433 (1969); also M. Schlüter, Phys. Rev. B 17, 5044 (1978).
[5]J. C. Phillips, J. Vac. Sci. Technol. 11, 947 (1974).
[6]J. M. Andrews and J. C. Phillips, Phys. Rev. Lett. 35, 56 (1975).
[7]L. J. Brillson, Phys. Rev. Lett. 40, 260 (1978).
[8]S. G. Louie and M. L. Cohen, Phys. Rev. B 13, 2461 (1976).
[9]S. G. Louie, J. R. Chelikowsky, and M. L. Cohen, Phys. Rev. B 15, 2154 (1977).
[10]W. E. Spicer, I. Lindau, P. R. Skeath, C. Y. Su, and P. W. Chye, Phys. Rev. Lett. 44, 420 (1980); W. E. Spicer, P. W. Chye, P. R. Skeath, C. Y. Su, and I. Lindau, J. Vac. Sci. Technol. 16, 1422 (1979).
[11]R. E. Allen and J. D. Dow, Phys. Rev. B 25, 1423 (1982).
[12]O. F. Sankey, R. E. Allen, and J. D. Dow, to be published.
[13]M. Schulter, Thin Solid Films 93, 3 (1982).
[14]E. H. Rhoderick, *Metal-Semiconductor Contacts* (Clarendon, Oxford, 1978).
[15]W. Kohn, Phys. Rev. 115, 809 (1959).
[16]J. J. Rehr and W. Kohn, Phys. Rev. B 9, 1981 (1974), and 10, 448 (1974).
[17]R. E. Allen, Phys. Rev. B 20, 1454 (1979).
[18]J. A. Appelbaum and D. R. Hamann, Phys. Rev. B 10, 4973 (1974).
[19]F. Claro, Phys. Rev. B 17, 699 (1978).
[20]R. T. Tung, preceding Letter [Phys. Rev. Lett. 52, 461 (1984)]. Perfectly epitaxial interfaces, prepared by novel techniques, may exhibit barrier heights which are quite different from normal interfaces, and which depend strongly on the crystal orientation. For an ideal interface those approximations made here, which were justified by interfacial disorder, break down; it remains to be seen what light the present results may shed on epitaxial M-S interfaces.
[21]E. O. Kane, private communication, and unpublished.
[22]G. A. Baraff and M. Schluter, in *Defects and Radiation Effects in Semiconductors—1980*, edited by R. R. Hasiguti, The Institute of Physics Conference Series No. 59 (The Institute of Physics, London, 1981), p. 287.
[23]A. Thanailakis and A. Rasul, J. Phys. C 9, 337 (1976), and references therein.
[24]S. M. Sze, *Physics of Semiconductor Devices* (Wiley, New York, 1969).
[25]W. G. Spitzer and C. A. Mead, J. Appl. Phys. 34, 3061 (1963); B. L. Smith, Ph.D. thesis, Manchester University, 1969 (unpublished) (from Ref. 14 above).

Schottky-Barrier Formation at Single-Crystal Metal-Semiconductor Interfaces

R. T. Tung
AT & T Bell Laboratories, Murray Hill, New Jersey 07974
(Received 30 September 1983)

Electrical behaviors at two single-crystal metal-semiconductor interfaces are studied. Schottky-barrier heights of $NiSi_2$ layers grown under ultrahigh-vacuum conditions on n-type Si(111) are found to be 0.65 and 0.79 eV for type-*A* and type-*B* epitaxial systems, respectively. These results are compared with the proposed theoretical models of Schottky barriers.

PACS numbers: 73.30.+y, 68.48.+f, 73.40.Ns

The formation of Schottky barriers (SB) at metal-semiconductor interfaces has attracted much attention for over thirty years.[1-3] Various models were proposed to explain how these rectifying barriers are determined. But thus far no theory emerges which gives explicit account of the experimental results. The details of the structure, impurity, stoichiometry, defects, etc., at the metal-semiconductor interface are not known in most studies, making progress in this field difficult. In the very few cases where careful characterization of the interface was made, the structure was too complicated because of polycrystallinity to provide a reasonable basis for a first-principle calculation. With a few exceptions, almost all metal films grown on semiconductor are multicrystalline. A homogeneous interface, of known atomic structure, between a single-crystal metal and a single-crystal semiconductor is an ideal system for detailed theoretical investigation. In this Letter, measurements of Schottky-barrier height (SBH) are presented for just such a system.

Thin films of single-crystal $NiSi_2$ and $CoSi_2$ can be grown with very high degree of perfection on silicon under ultrahigh-vacuum (UHV) conditions.[4] Single-crystal Al films[5,6] can also be grown on differently reconstructed GaAs(100) surfaces, but the structures at such interfaces are largely unknown. The $NiSi_2$Si epitaxy is quite unique in that uniform and continuous single-crystal metal layers can be grown on Si(111) and (100). Moreover, the orientation of the $NiSi_2$ films grown on Si(111) can be controlled to be either identical to that of the substrate (type-A orientation), or to be rotated 180° about the surface-normal direction with respect to the substrate (type-B orientation).[7,8] Lattice imaging of such silicide-silicon interfaces by high-resolution transmission-electron microscopy (TEM) has demonstrated the sharpness of the interface and revealed the most likely interfacial atomic structure.[9,10] In this Letter, electrical properties of type-A and type-B $NiSi_2$Si(111) junctions are determined. It is shown that SBH is higher for the type-B interface by 0.14 eV over that of type A. Such a large difference is not expected on the basis of previous SBH results obtained largely from polycrystalline metal films. This shows the importance of having well-characterized, homogeneous interfaces as an experimental basis for understanding the SB formation mechanism(s).

Arrays of circular windows with diameters of 127–508 μm are photolithographically defined in the 3000–5000-Å-thick SiO_2 layer grown on n-type Si(111) wafers. Also interspersed on the patterned wafers are 3-mm-wide stripe openings for better characterization by low-energy electron diffraction (LEED), Auger-electron spectroscopy, Rutherford backscattering spectroscopy (RBS) and channeling, TEM, etc. Samples are degreased and exposed to HF fume for ~5 sec to remove oxide in the windows, and rinsed with methanol. They are cleaned by a heat treatment (1000 °C for 2 min) in a UHV chamber with base pressure of 1×10^{-10} Torr. The exposed Si region displays a good 7×7 LEED pattern even though a small amount of carbon is detected by Auger analysis. Thin (60–70 Å thick) single-crystal $NiSi_2$ template layers are then grown in exposed Si regions by deposition of Ni and annealing to 500 °C.[7] The orientation (A or B) of such layers is controlled by varying the initial coverage of deposited Ni. Silicon deposition is used in conjunction with Ni deposition in the growth of some type-B templates.[10] Thick (700–1000 Å) $NiSi_2$ type-A and type-B layers are grown by deposition of nickel onto the template layer of the corresponding orientation while the sample temperature is kept at ~700 – 750 °C.[8] The orientation of these silicide layers is determined by LEED and RBS and channeling. The perfection of these single-crystal layers has already been described.[10] The unreacted Ni on top of the SiO_2 is removed by a chemical etch. Ohmic contact is made to the back side of the sample without heat-

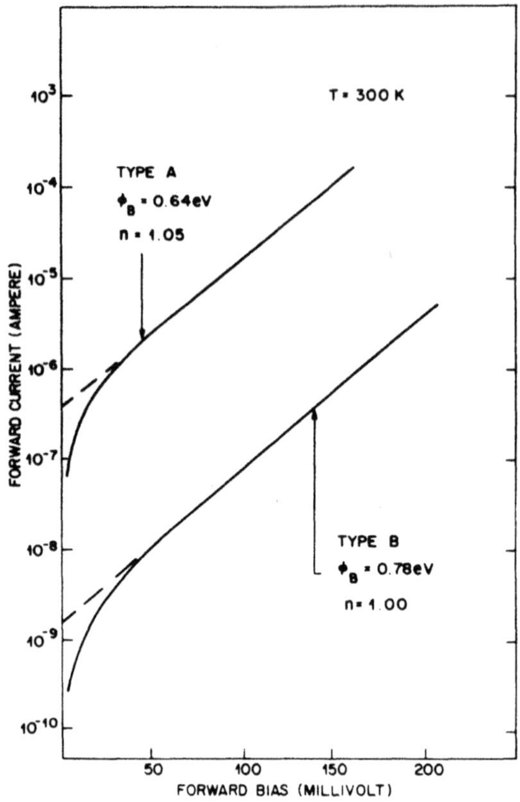

FIG. 1. Forward current characteristics of thin (~ 70 Å) type-A and type-B $NiSi_2$ layers on Si(111). Spot size is 394 μm.

ing the sample above room temperature, by antimony deposition, pulsed-laser melting of Sb and surface Si, and aluminum deposition. SBH's are determined[11] by current-voltage (I-V) and capacitance-voltage (C-V) scans on an x-y recorder. Metallized openings are probed with thin gold wire to avoid puncturing the thin layers.

Forward current characteristics of a type-A and a type-B template layers are shown in Fig. 1. These data are for junctions which show close to ideal electrical behaviors, approximately 50% of the samples studied. All dots which show a linear dependence of log I on V always display good ideality factor (< 1.06) and extrapolate to yield the same barrier height (± 0.01 eV) of the corresponding orientation. Capacitances are measured mainly at 1 MHz; a few select samples are studied with different frequencies ranging from 20 kHz to 2 MHz. Linear C^{-2} dependence on V is found for almost all diodes studied, with the slope agreeing well with the nominal doping concentration. For a given orientation, extrapolations consistently yield the same SBH as obtained by the I-V method. Results of these electrical measurements are summarized in Table I. Thick (700-1000 Å) and thin (~70 Å) junctions of the same orientation yield the same SBH (± 0.01 eV); this demonstrates that misfit dislocations which exist at the interfaces of the thick layers (~10^5 per cm^2), but not at the thin pseudomorphic template layer interfaces,[10] do not contribute actively to the formation of SBH. Recently published results[12] suggest that misfit dislocations may pin

TABLE I. Schottky-barrier heights of UHV grown $NiSi_2$ single crystals to n-type Si(111).

Orientation	Film thickness (Å)	Formation temperature (°C)	Misfit dislocations	SBH (eV) [a] I-V [b]	C-V [c]
B	~ 70	500	no	0.79	0.79
A	~ 70	500	no	0.65	0.65
B	700–1000	750	yes	0.78	0.79
A	700–1000	750	yes	0.65	0.65

[a]Tabulated values are the intrinsic barrier heights (flat bands) with no Schottky lowering. The experimental error of every entry is estimated to be 0.01 eV.

[b]Only junctions with ideality factor of less than 1.06 and linear semilogarithmic behavior over two decades are included. The effective Richardson's constant is 100 A cm^{-2} K^{-2}. Schottky lowerings of 18 to 30 mV are added.

[c]Obtained from over 150 different spots on 6×10^{15} cm^{-3} P and 3×10^{16} cm^{-3} As substrates.

the Fermi level at semiconductor interfaces.

SBH's of metal silicides to Si were previously reported to be nearly constant for a given metal, independent of the structure and stoichiometry of the silicides.[13,14] For instance, Ni_2Si, NiSi, and $NiSi_2$ all have the same SBH, 0.66 ± 0.02 eV, on n-type (111) and (100) substrates. This has led to speculations of an interfacial layer, interstitial metal atoms, or Si vacancies being responsible for the SBH to Si. With the present results, a change in viewpoint is suggested.

The difference between SBH's of types A and B is rather significant considering the subtle difference between the atomic structures at the two interfaces. Atomic structures of the two sevenfold coordinated interfaces are depicted in Fig. 2. Computer simulations based on these unrelaxed structures agree well with high-resolution TEM images.[10] The structures of these two interfaces are very similar, differing only in the positions of the third- and higher-nearest neighbors to the last nickel layer. The key feature is the unpaired electrons which because of Ni-Si hybridization are directional,[15] similar to dangling bonds.

Beside the original Schottky description, there are two models proposed to account for how the Fermi level arrives at its value at the interface. One involves the filling of the surface/interface states,[16,17] and the other depends on local chemical bonding[18] and/or the formation of defects.[19,20] No experimental results are known regarding the electronic structure of the A and the B interfaces. Self-consistent calculations are being performed because the most straightforward explanation for the difference in SBH is the existence of electronic states positioned higher in the band gap for

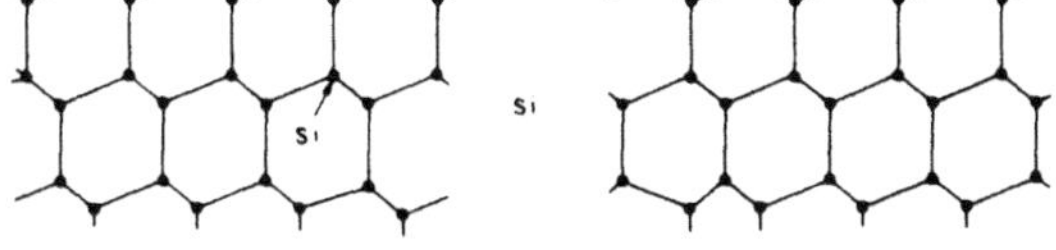

FIG. 2. Balls and sticks models of the two $NiSi_2$-Si(111) interfaces viewed in the $\langle 1\bar{1}0 \rangle$ direction. Orbitals of unpaired electrons are also indicated.

type A than for type B. Such calculations may have to include atomic relaxations which may exist at the interfaces but are outside the TEM resolution. An alternative explanation due to Tersoff[21] suggests that the difference in SBH may be due to electronic properties characteristic of hexagonal Si (a lower conduction-band minimum) at the type-B interface. Defects 3 Å away from the interface should have identical energy levels for the A and the B and hence cannot account for the present results. I am not aware of any defect sites right at the interface which would give rise to energy levels differing by 0.14 eV. High-resolution TEM[10] does not show any evidence of point defects of the high density[22] required for Fermi-level pinning.

The published $NiSi_2$ SBH to n-type Si(111) is 0.64–0.7 eV for non-UHV-grown layers.[14,23,24] Electron microscopy has shown such layers to have both type-A and type-B grains.[25] An interface which consists of regions of high barrier and regions of low barrier would display electrical behavior intermediate between the two, but more closely resembling the lower barrier. So it can be argued that 0.65 and 0.79 eV are the barrier heights associated with type-A and type-B interfaces, respectively, and that the reported SBH of non-UHV films is the result of polycrystallinity. However, the SBH of type-A $NiSi_2$ (0.65 eV) is close to the 0.66 eV usually associated with polycrystalline Ni silicides (Ni_2Si, NiSi, and $NiSi_2$).[14] This suggests that if the observed SBH insensitivity to stoichiometry, structure, etc., is due to some particular mechanism, other than simple averaging from polycrystallinity, then this mechanism may also be in effect at type-A $NiSi_2$ interface. The existence of a thin type-A $NiSi_2$ layer at every (Ni_2Si and NiSi) interface can almost be ruled out on the basis of TEM[25] and the fact that there is no such equivalent layer at a (100) interface.[26] If this mechanism is defects, then the type-A $NiSi_2$ interface may also have these defects. The SBH of the type-B interface would then be determined by a different mechanism, most likely by the intrinsic electronic states. There is evidence that type B has lower interfacial energy than type A, because $NiSi_2$ will crystallize overwhelmingly in the B phase upon pulsed laser melting and rapid recrystallization.[27] So one can even argue that the type-B interface is more stable and hence has a negligible density of defects; it is a known fact that they have very low density of steps.[10]

Many research efforts have been directed at

the understanding of phenomena occurring at the metal-semiconductor interfaces. Understanding the formation of the SBH on a fundamental basis still remains the most challenging problem. With the discovery of the stabilization of ultrathin $NiSi_2$-Si epitaxy under UHV conditions, sharp and homogeneous metal-semiconductor interfaces are now available. At the $NiSi_2$Si(111) interface, either of the two "double position" orientations, perfect in its own right, can be singled out and studied separately, providing a new variable and a unique comparison. Moreover, the interface structure has already been determined. Thus, this is an ideal system to better our understanding of the SB. We have performed the first measurement of barrier height formed by a single-crystal metal on silicon. Experimental results indicate the importance of crystallographic orientation and, hence, the interface structure.[26] This dependence suggests an intrinsic SB mechanism; however, defects associated with only one interface cannot be ruled out.

The author wishes to thank J. M. Poate, K. Ng, and G. N. Taylor for making available to him experimental equipments and F. C. Unterwald and D. C. Jacobson for their contributions to the experiment. He is also indebted to J. M. Gibson, M. T. Yin, L. C. Feldman, J. Tersoff, and J. C. Hensel for helpful discussions.

---

[1]M. Schlüter, Thin Solid Films 93, 3 (1982).

[2]L. J. Brillson, Surf. Sci. Rep. 2, 123 (1982).

[3]E. H. Rhoderick, *Metal-Semiconductor Contacts* (Clarendon, Oxford, 1978).

[4]R. T. Tung, J. C. Bean, J. M. Gibson, J. M. Poate, and D. C. Jacobson, Appl. Phys. Lett. 40, 684 (1982), and Thin Solid Films 93, 77 (1982).

[5]A. Y. Cho and P. D. Dernier, J. Appl. Phys. 49, 3328 (1978).

[6]S. P. Svensson, G. Landgren, and T. G. Andersson, J. Appl. Phys. 54, 4474 (1983).

[7]R. T. Tung, J. M. Gibson, and J. M. Poate, Phys. Rev. Lett. 50, 429 (1983).

[8]R. T. Tung, J. M. Gibson, and J. M. Poate, Appl. Phys. Lett. 42, 888 (1983).

[9]D. Cherns, G. R. Anstis, J. L. Hutchinson, and J. C. H. Spence, Philos. Mag. 46A, 849 (1982).

[10]J. M. Gibson, R. T. Tung, and J. M. Poate, in *Defects in Semiconductors II*, edited by S. Mahajan and James W. Corbett, Proceedings of the Materials Research Society Symposium, Boston, 1982 (North-Holland, New York, 1983), Vol. 14, p. 395.

[11]S. M. Sze, *Physics of Semiconductor Devices* (Wiley, New York, 1981).

[12]J. M. Woodall, G. D. Pettit, T. N. Jackson, C. Lanza, K. L. Kavanagh, and J. W. Mayer, Phys. Rev. Lett. 51, 1783 (1983).

[13]G. Ottaviani, K. N. Tu, and J. W. Mayer, Phys. Rev. B 15, 3354 (1981).

[14]P. E. Schmid, P. S. Ho, and T. Y. Tan, J. Vac. Sci. Technol. 20, 688 (1982), and Phys. Rev. B 28, 4593 (1983).

[15]Y. J. Chabal, D. R. Hamann, J. E. Rowe, and M. Schlüter, Phys. Rev. B 25, 7598 (1982).

[16]J. Bardeen, Phys. Rev. 71, 717 (1947).

[17]V. Heine, Phys. Rev. A 138, 1689 (1965).

[18]J. M. Andrews and J. C. Phillips, Phys. Rev. Lett. 35, 56 (1975).

[19]W. E. Spicer, I. Lindau, P. Skeath, and C. Y. Su, J. Vac. Sci. Technol. 17, 1019 (1980).

[20]O. F. Sankey, R. E. Allen, and J. D. Dow, to be published.

[21]J. Tersoff, following Letter [Phys. Rev. Lett. 52, 465 (1984)].

[22]A. Zur, T. C. McGill, and D. L. Smith, Phys. Rev. B 28, 2060 (1983).

[23]M. Iwami, K. Okuno, S. Kamei, T. Ito, and A. Hiraki, J. Electrochem. Soc. 80, 102 (1980).

[24]T. R. Harrison, A. M. Johnson, P. K. Tien, and A. H. Dayem, Appl. Phys. Lett. 41, 734 (1982).

[25]H. Föll, P. S. Ho, and K. N. Tu, J. Appl. Phys. 52, 250 (1980).

[26]Preliminary results (R. T. Tung, to be published) show that the SBH of single-crystal $NiSi_2$ on Si(100) differs significantly from that of either of the two $NiSi_2$-Si(111) interfaces. This again points to the importance of interfacial structure in the SB formation.

[27]R. T. Tung, J. M. Gibson, D. C. Jacobson, and J. M. Poate, Appl. Phys. Lett. 43, 476 (1983).

Reflection high-energy electron diffraction study of the growth of In on GaAs(110) at different temperatures

D. E. Savage and M. G. Lagally
Department of Metallurgical Engineering and Materials Science Center, University of Wisconsin, Madison, Wisconsin 53706

(Received 13 March 1986; accepted 23 April 1986)

Measurements have been made of the growth of In on GaAs(110) at two temperatures using reflection high-energy electron diffraction (RHEED). Growth occurs by 3D island formation. Island orientations, sizes, volumes, separations, and densities are extracted from the angular profiles of transmission diffraction beams from the In. The dose and temperature dependencies of these quantities are investigated. The growth microstructure is related to concepts of barrier formation. It is suggested that Fermi level pinning measurements in this system depend on band bending in the substrate laterally away from the edges of the island that are separated by the bare substrate.

I. INTRODUCTION

Metal–semiconductor interfaces are an essential feature of semiconductor devices. The properties of such interfaces probably play the decisive role in the formation and behavior of Schottky or Ohmic contacts. It is recognized that contact formation involves both electronic and structural properties at the interface, but the relationship between them is so far not understood. The growth mode of the overlayer film is the usual starting point in discussions of the atomic structure at the interface. Three film growth modes, Frank–van der Merwe, Volmer–Weber, and Stranski–Krastanov, have been postulated.[1] They correspond, respectively, to atomic layer-by-layer growth, growth of three-dimensional (3D) islands, and monolayer growth followed by 3D island growth. These growth modes derive from relatively simple considerations of adsorbate–adsorbate and adsorbate–substrate interactions, and of relative lattice constants.[2] Despite the existence of these models, it is generally not possible to predict how an overlayer will grow even if the interactions are known, because these models do not take into account the role of structural defects, trace impurities, and the possibility of kinetic limitations in the growth rates.

Electronic properties at the interface are usually discussed assuming atomic layer-by-layer growth, i.e., neglecting any lateral inhomogeneities in structure. However, even in the simplest cases this assumption may not be justified. The importance of the relationship between microstructure and electronic properties is recognized in at least some of the current models of Schottky barrier formation. The classical model advanced by Schottky[3] is based simply on contact potential differences between metal and semiconductor and takes no account of structural inhomogeneities. It does not explain Fermi level pinning in most systems. In the unified defect model[4] it is proposed that structural defects formed in the semiconductor either during deposition or in the process of condensation of 3D islands of the metal adsorbate[5] cause localized charges that pin the Fermi level. In the metal-induced-gap-states (MIGS) model[6] it is assumed that the metal induces localized states in the band gap. The model that most directly acknowledges the importance of structural inhomogeneities is the effective-work-function model,[7] in which it is proposed that laterally inhomogeneous interfacial phases with differing work functions cause variations in the barrier height, and that therefore the local chemistry or microstructure at the interface will determine the measured barrier height. The applicability of these models to various metal–semiconductor systems continues to generate considerable interest. In particular, for metal deposition on GaAs(110) a variety of measurements have been made with frequently conflicting results[4,8–10] and with no clear evidence that any one of the above models is correct. We contribute to this confused picture with a study of growth of In on GaAs(110) at different temperatures using reflection high-energy electron diffraction (RHEED). The purpose of this paper is twofold. First, we provide quantitative crystallographic information for this system, including island sizes, shapes, orientations, and separations. Second, we consider the implications of our results for the different models of Fermi level pinning. We suggest that at least one additional feature not inherent in the above models, namely the separation of 3D metal islands in an incomplete film on the substrate, can affect Fermi level pinning.[11,12]

II. EXPERIMENTAL

The majority of the work reported here was done in an ion and turbopumped UHV chamber that operates at a base pressure at $<8\times10^{-11}$ Torr. The chamber contains two MBE sources, a quartz crystal oscillator deposition rate monitor, a sample manipulator with sample cooling capability, a sample cleavage device, a cylindrical mirror analyzer ordinarily used for scanning Auger microscopy (SAM), and a fluorescent screen for viewing the RHEED pattern. The internal gun in the SAM analyzer is used as the source for RHEED. When focused at its normal SAM position, it produces a beam of $\sim 3\ \mu$m. In the RHEED configuration, in which the sample is farther from the analyzer, the beam is larger. The gun is operated at energies between 5 and 10 keV. This relatively low energy allows the use of grazing angles of incidence between 1° and 4° for the RHEED measurements, while still maintaining high surface sensitivity. Detection is done via the fluorescent screen and a vidicon camera interfaced to a computer. In this manner individual beam profiles can be measured in any direction and stored for subsequent analysis. Complete diffraction patterns are photographed with high-speed film.

GaAs(110) surfaces are produced by *in situ* cleavage of an oriented notched bar of *n*-type GaAs (Te doped to a level of $\sim 3\times 10^{18}/\mathrm{cm}^3$). Multiple cleaves are possible. The crystallographic and morphological quality of the resultant surface is investigated visually and with RHEED. It is by now well established[13,14] that a high-quality surface with a low density of structural defects gives a series of sharp diffraction spots that fall on arcs centered about the shadow edge. An example of a RHEED pattern from a surface produced from a good cleave is shown in Fig. 1(a). Only surfaces that give sharp diffraction patterns such as that shown in Fig. 1(a) are used for In depositions and subsequent measurements.

In is deposited from a MBE source of standard design, consisting of a C crucible wound with W wire, several heat shields, a water-cooled shroud, and a shutter. A typical operating temperature is 700 °C. The shutter is kept close for 20 min while the source material outgasses and the source equilibrates. The pressure in the vacuum chamber while the source is in operation and the shutter is open is $< 2\times 10^{-10}$ Torr. We have found that the ambient pressure can significantly affect the growth mode, especially for growth at low temperatures. It is therefore essential to maintain the background pressure as low as possible, and to cleave the sample only immediately before deposition. Deposition rates range from 0.05 to 3.5 ML/min. A higher deposition rate reduces the contamination problem, but at higher rates one needs to be more concerned about the influence on the growth mode of competition between arrival rate and surface diffusion rate of In atoms.

Depositions are made with the sample held at room temperature or at lower temperatures. To determine the number of atoms deposited per unit area on the sample (i.e., the dose or nominal coverage) a quartz crystal oscillator is placed at the sample position to monitor the flux for a given source temperature. Source temperatures can be controlled to within ± 1 K using a Chromel–Alumel thermocouple and a temperature controller. A given deposition rate can be reproduced to $\pm 10\%$. We quote the dose in equivalent monolayers, where one equivalent monolayer is defined to be equal to the sum of Ga and As atoms per unit area of GaAs(110).

The actual surface coverage is not the same as the dose or nominal coverage if 3D structures form in the deposit. We measure the actual surface coverage using RHEED line shape analysis, Auger electron spectroscopy (AES), and scanning electron microscopy (SEM). In AES, we monitor the *MNN* In and *LMM* Ga and As lines as a function of dose. From the RHEED results it is evident that the In islands very rapidly become thick, so that Auger electrons cannot penetrate them, except possibly right at the edges. Therefore we assume as a limiting case that the Ga or As Auger signal all comes from bare patches of GaAs. This assumption provides a lower limit for the actual surface coverage. The comparison of coverage measurements by RHEED and AES is presented later in this paper.

Quantitative RHEED analysis on the deposited In and the

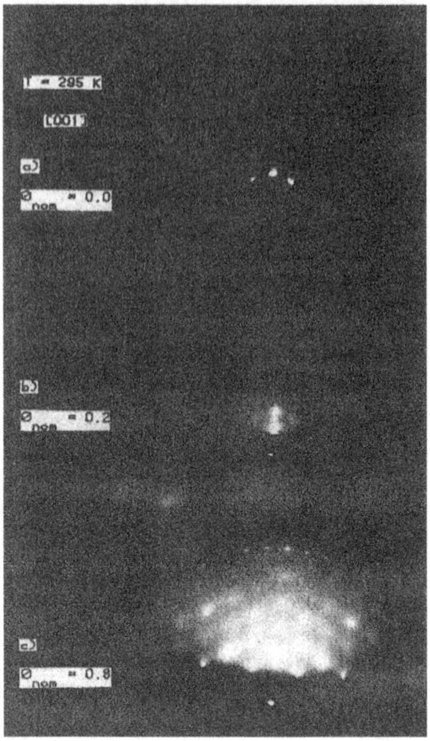

FIG. 1. RHEED patterns for In deposited onto GaAs(110) at room temperature. (a) Clean cleaved surface; (b) In dose of $\theta_{\mathrm{nom}} = 0.2$; (c) $\theta_{\mathrm{nom}} = 0.8$. All patterns are taken with the incident beam along GaAs[001]. The large spots are the transmission diffraction pattern from 3D In islands. Their width reflects the In island size. The substrate diffraction is visible as the arc of sharp spots. The sample is slightly tipped to the right. The undeflected part of the incident beam appears as the bright spot below the shadow edge.

GaAs substrate is performed by measuring the shapes of the diffracted beams and fitting these shapes with models for diffraction from an array of domains that have a given distribution of sizes. Such calculations have been made for LEED from a number of systems,[15] and recently also for RHEED.[12,16] An important input into such modeling is the response function of the instrument, which causes a broadening of all diffraction features. A limited instrument response is caused in general by incident-beam divergence and energy spread, beam size, and detector aperture size.[17] For the diffraction measurements here the resolution is limited only by the beam size at the detector.[13] We determine the instrument response function by measuring the narrowest angular profile we can obtain from a well-cleaved GaAs(110) surface and assuming that the width of this profile is all due to limited instrumental response. Because of the grazing incidence in RHEED, there are two quite different values of the resolving power,[18] one corresponding to measurement perpendicular to the plane of the surface and one parallel to it.[13,14] The two measurements are related by the inverse of sin θ, where θ is the exit angle with respect to the surface plane. For transmission diffraction the resolving power measured parallel to the surface plane [a profile horizontally across a beam in Fig. 1(a)] is the important one. It applies for all measurements in transmission. A typical resolving power determined from a profile measurement is of the order of 150 Å. This implies that island or domain sizes smaller than 150 Å can be measured, but larger ones look perfect to the instrument. In the other plane, the resolving power for reflection diffraction is much larger, of the order of 3000 Å; however, that value is not relevant in the present measurements.

III. RESULTS

In this section we present the RHEED measurements for In deposits on cleaved GaAs(110) at various coverages and at two temperatures. We also show scanning electron micrographs of the In films to corroborate the RHEED results.[19]

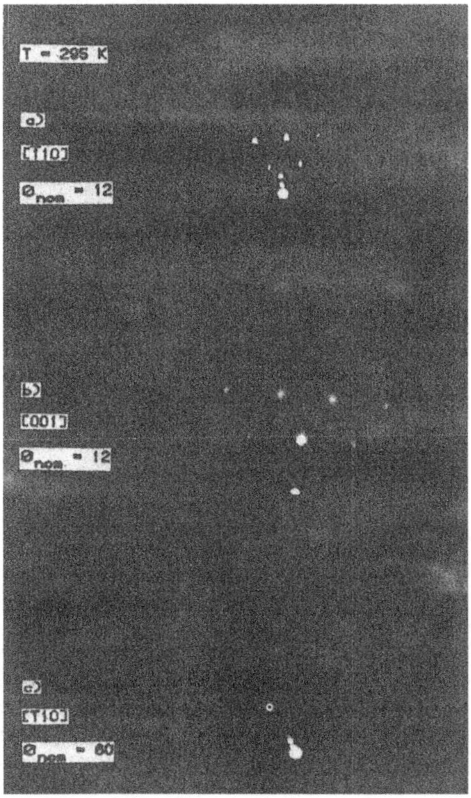

FIG. 2. RHEED patterns for In deposited onto GaAs(110) at room temperature. (a) $\theta_{nom} = 12$, incident beam along GaAs[$\bar{1}$10]; (b) $\theta_{nom} = 12$, incident beam along GaAs[001]; (c) $\theta_{nom} = 60$, incident beam along GaAs[$\bar{1}$10]. The streaks in (a) and (c) are reflections off facets on the In islands. From their orientation the facet planes can be determined. The substrate reflection pattern is still visible at $\theta_{nom} = 12$ along GaAs[001].

A. In depositions at room temperature

Figures 1 and 2 show the evolution of the RHEED patterns for various doses of In. Figure 1(a) is the pattern for clean cleaved GaAs(110). It shows arcs of sharp spots. Contrary to popular belief, a high-quality surface does not produce streaks, but rather spots. Streaking of RHEED features in a clean surface is a consequence of surface defects such as steps or long-range waviness of the surface, which creates a range of surface orientations that the incident beam illuminates. Streak can be easily produced by sputter etching and annealing the surface. An example is shown in Fig. 3. Streaks cannot subsequently be completely removed. In growth on sputter-etched and annealed surfaces is considerably different. Domains are smaller and a third orientation is present.

The evolution of the RHEED features with the addition of In is clear evidence that In forms 3D islands.[12] One observes a transmission pattern through the In particles. The indication of a transmission pattern is a series of spots on straight lines (not arcs as for reflection diffraction features). A simple Ewald construction, Fig. 4, shows how reflection and transmission patterns arise.[20] For room temperature deposition, 3D island growth is observed in the RHEED pattern at nominal coverages as low as $\theta_{nom} = 0.2$ [Fig. 1(b)]. The 3D In island features become stronger with increasing dose, until the reflection diffraction features from the substrate disappear. (This does not mean the surface is covered, only that the shadowing from 3D islands at grazing angles prevents the beam from seeing the substrate.) Continued deposition of In, Fig. 2, produces streaks in the RHEED pattern for specific orientations. These streaks can be identified as reflections off facets in the growing In islands.[21]

By analyzing the distances between diffracted beams from the In islands and their positions relative to the GaAs reflection diffraction spots, one can determine the structure of the In islands and their orientation relative to the substrate. We find[12] that the islands always have the bulk In structure (In is face-centered tetragonal with $a = b = 4.588$ Å and $c = 4.938$ Å). At room temperature, we find two orientations of In relative to the substrate. For one domain orientation the (3$\bar{1}$1) plane of In contacts the GaAs(110) plane, with In[$\bar{2}\bar{3}$3] parallel to GaAs[001]. For the other orientation, the (1$\bar{1}$3) plane of In contacts the GaAs(110) surface,

FIG. 3. RHEED pattern from cleaved GaAs(110) that was sputter-etched and subsequently annealed. The streaks are due to steps and other finite-size effects that cannot be eliminated by annealing.

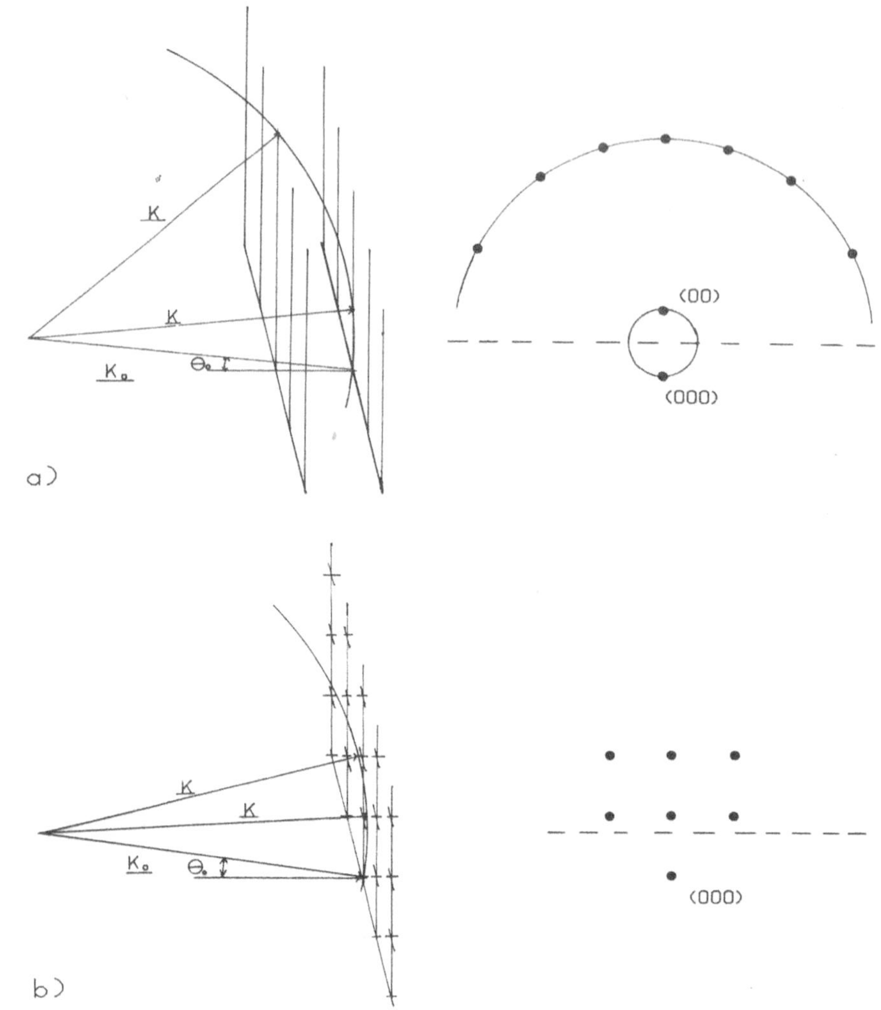

FIG. 4. Ewald constructions for (a) reflection diffraction from a perfect surface and (b) transmission diffraction from small particles. The shapes of the reciprocal-lattice points in (b) reflect the Fourier transform of the shape of the particle. Note that θ represents a grazing angle for reflection, and a nearly normal angle for transmission. In reflection, the first arc of spots is on a circle including the unscattered beam; the shadow edge bisects this circle. Transmitted beams may appear within this circle; the (000) beam coincides with the unscattered beam.

with In[$3\bar{3}\bar{2}$] parallel to GaAs[001]. The latter domain is prevalent, but in all cases we have observed both domains for room temperature depositions. The alignment along GaAs[001] produces a 1.8% misfit for domain type I, and a 3.5% misfit for domain type II. Alignment along GaAs[$1\bar{1}0$] is forced by the In crystal structure, producing much greater misfits, of the order of 16% and 19% for the two domain types, respectively, and a rotational misfit as well for domain type I.

The streaks in well defined directions shown in Fig. 2 for specific diffraction geometries are due to reflection diffraction features off facets in the In islands. From the streak directions, the facet orientations and hence the shape of the In islands can be determined. The In islands look like prisms having one side in contact with the GaAs(110) substrate and the gable of the prism aligned roughly along GaAs[$1\bar{1}0$]. The faces of the prisms are ($1\bar{1}1$) and (100) or ($1\bar{1}1$) and (001) planes for the two In domain types.

The shapes and orientations of the In islands can be verified at higher coverages with SEM. Figure 5 shows a typical result, for $\theta_{\rm nom} = 150$. In was deposited on GaAs(110) and the sample subsequently removed from the MBE analysis chamber and inserted into an SEM. The exposure to atmosphere of course oxidizes the In, but In forms a passive thin oxide that does not affect the shape or arrangement of the islands. This was verified for a sample with 20 ML dose. A RHEED pattern taken after the sample was exposed to air showed the same general features as the pattern for the as-deposited layer, including facet streaks for the ($1\bar{1}1$) facets. Figure 5 also shows a cleavage step on the GaAs surface, along a [001] direction. The prismatic shape of the islands and their preferential alignment along the [$1\bar{1}0$] direction

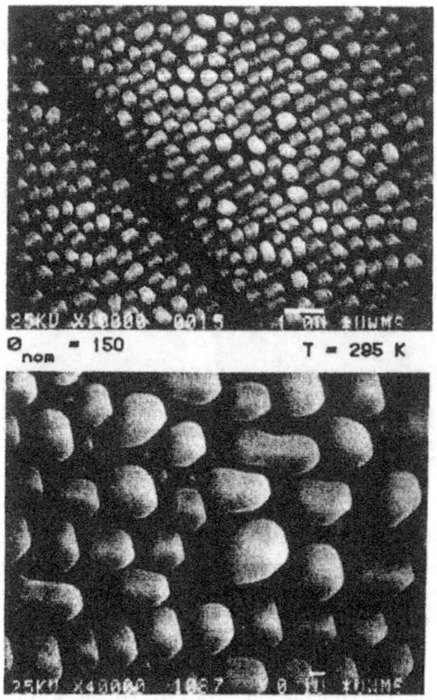

FIG. 5. Scanning electron micrographs of In islands on cleaved GaAs(110) at a dose of 150 ML at two magnifications showing the shape, orientation, and relatively uniform size of the large islands. Coalescence events are observed as larger irregular or dumbbell shapes. The direction of the cleavage step is along GaAs[100]. Secondary nucleation of smaller islands can be seen.

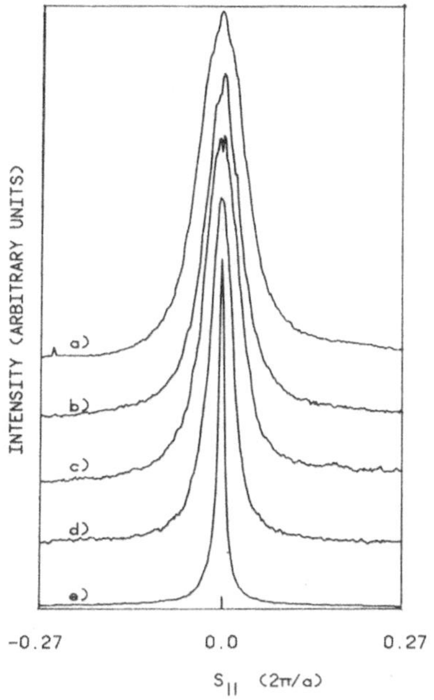

FIG. 6. Angular profiles of the $(1\bar{1}1)$ beam along GaAs[110] in the In transmission pattern for several doses deposited at room temperature. (a) $\theta_{nom} = 0.6$, (b) 2, (c) 4, (d) 8, (e) instrument response function profile. The curves below $\theta_{nom} = 4$ are essentially identical. The profiles are measured at the Bragg condition for the 3D structure.

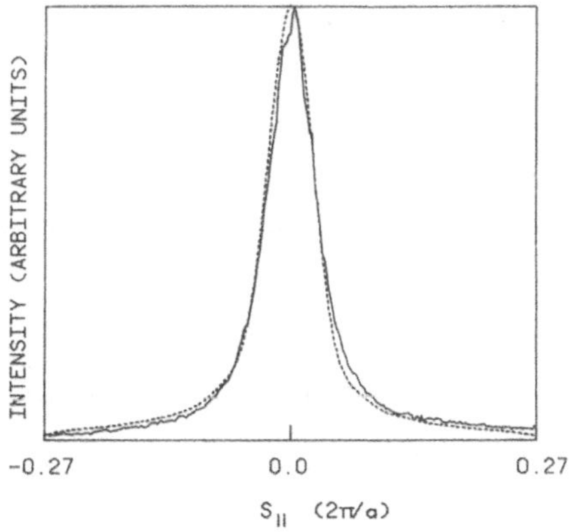

FIG. 7. Fit to an angular profile of a transmission diffraction beam from In. The model calculation is for a single island size of height 30 Å. Experimental curve: curve (b) from Fig. 6.

are clearly evident. Figure 5 also shows coalescence events for some of the large islands, which were the ones that nucleated initially. The result of coalescence is a larger island with poorly defined shape, but with greater empty space surrounding the island. At this high dose, secondary nucleation of islands has begun in the interstices between the large islands. A particle size analysis shows that the size distribution of islands at this dose is bimodal with large, nearly equal-size islands that have an average size of ~3000 Å, and small islands (< 500 Å) that are due to secondary nucleation. At a dose of 60 ML the bimodal distribution is also evident, the large islands now having an average size of ~1800 Å. At 20 ML, the bimodal distribution is no longer observable. The islands have an average size of ~700 Å. Because of the quality of our SEM, we have been able to

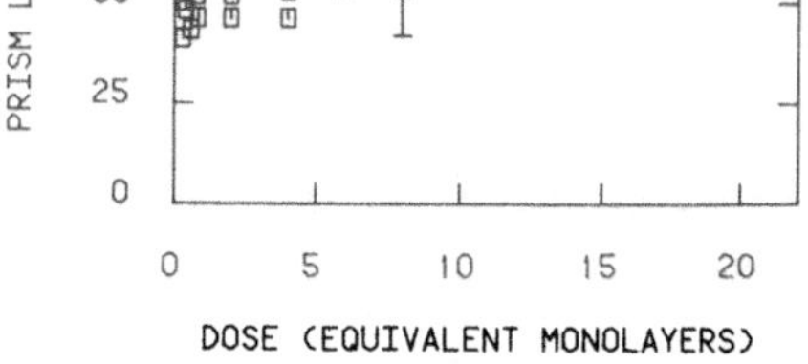

FIG. 8. Dependence of average size of In islands on dose for deposition at room temperature. The size along GaAs$[1\bar{1}0]$ (the gable length) is plotted; see Fig. 5. The average size does not change significantly up to a dose of ~8 ML.

obtain usable micrographs only above doses of ~ 12 ML.

The evolution of the average size of these islands with dose can be determined by measuring angular profiles of the diffracted beams that are part of the transmission diffraction pattern from the In islands. As long as these beams are broader than the instrumental limit, one can determine the size and size distribution of islands. Figure 6 shows angular profiles of the In$(1\bar{1}1)$ beam in the GaAs[110] direction (the height of the islands, see discussion later) for various doses deposited at room temperature. Figure 7 shows a fit to a profile using a single island size for the distribution function. We are able to fit all profiles with such a single-island size distribution. Figure 8 shows the evolution of the average In island size with coverage in the other direction, i.e., parallel to the gables of the prisms (see figure caption). It can be seen that this "prism length" stays effectively constant up to $\theta_{nom} \simeq 8$.

B. In depositions at low temperatures

Measurements similar to those described above were performed for In deposited on GaAs(110) surfaces held at 190 K. The growth is quite clearly different at low temperatures. 3D island formation still takes place,[19] but 3D nucleation is delayed until a dose of $\theta_{nom} \simeq 0.8$. The average island volume at any dose is smaller at 190 K than it is at room temperature. The growth mode is different. Two orientations of domains are present. The main one again has the In$(1\bar{1}3)$ plane in contact with GaAs(110) and has $(1\bar{1}1)$ and (100) facets. The other is less prevalent and has a $(1\bar{6}6)$ contact plane. It is very possibly a twin of the $(1\bar{1}3)$ plane. Some comparisons for high and low temperatures are shown in Figs. 9–11. Figure 9 shows RHEED patterns at 190 K for the same orientations as, and coverages similar to, those of Figs. 2(a) and 2(b). The low-temperature data show the extra diffraction beams of the $(1\bar{6}6)$ plane but lack the beams of the $(3\bar{1}1)$ domains. Figure 10 shows SEM micrographs at $\theta_{nom} = 60$. The island density at 190 K is greater. At these coverages the density differences are more obvious than the size differences, because a small size difference can imply a large volume difference. At 190 K and at $\theta_{nom} = 60$, the large islands have a size of 1200 Å while at $\theta_{nom} = 20$, this size is ~ 600 Å. Figure 11 shows the average size of In islands at 190 K in the direction of the gables (the prism length) as a function of dose. This figure is directly analogous to Fig. 8.

Depositions made at low temperatures and subsequently raised to and annealed at room temperature do not change their RHEED patterns or profiles to those for depositions made at room temperature. It can be concluded that the size distribution incorporated at the low temperature is stable at

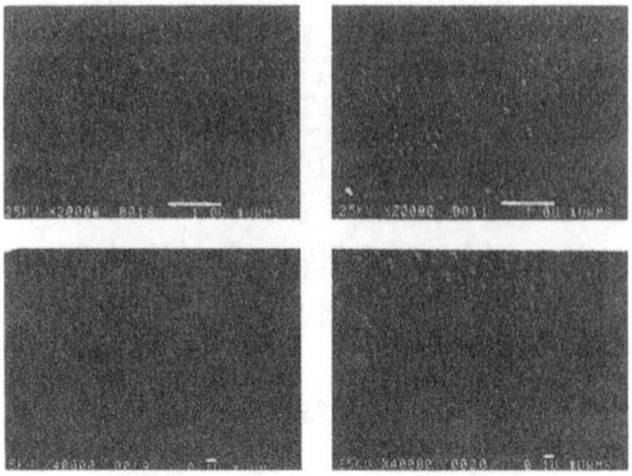

FIG. 10. Scanning electron micrographs of In islands for $\theta_{nom} = 60$ deposited at low and high temperatures. The density is lower and average size greater for the room-temperature depositions. The lower panel in each case is a magnified region from the upper panel.

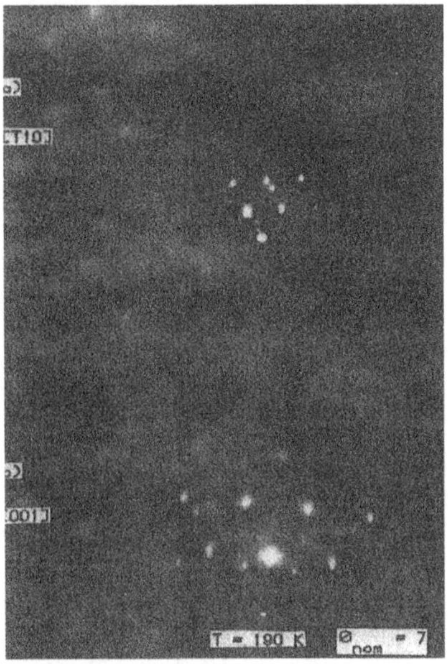

FIG. 9. RHEED patterns for In deposited at low temperature, $\theta_{nom} = 7$; incident beam along (a) GaAs $[\bar{1}10]$, (b) GaAs [001]. Compare Figs. 2(a) and 2(b).

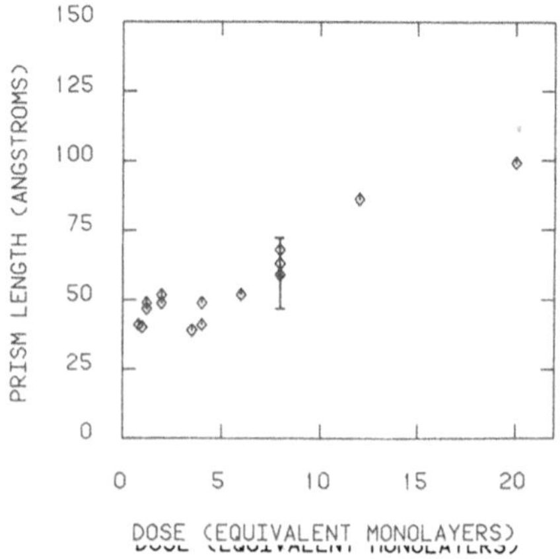

FIG. 11. Dependence of the average size of In islands on dose for deposition at 190 K. Compare to Fig. 8.

295 K. All SEM measurements were made at room temperature.

Finally, it should be pointed out that the contamination from the ambient vacuum environment can significantly affect the In growth at low temperatures. A background pressure of 1×10^{-9} Torr causes additional phases to form [In(111) and mirror planes in contact with the substrate] and causes the initial island size to be smaller. We do not know if the background pressure of $<2\times10^{-10}$ Torr during depositions in the present experiments affects the growth mode, and if possibly the same growth as at 295 K would be obtained with even lower background pressures. We do know that GaAs(110) remains clean for hours at room temperature and pressures of 2×10^{-10} Torr. For the In-covered surface, C and O contamination appears, presumably on the In. At low temperatures, both the substrate and the In-covered surface contaminate more rapidly than at room temperature. In any case, for the purposes of comparison with other data on this system, these speculations are irrelevant, because background pressures in other systems are no better. However, there can be no question that other structures are possible and that great care must be taken with contamination at low temperatures.

IV. ANALYSIS

In this section we present a quantitative analysis of the measurements of the last section and discuss the results in terms of their possible impact on the development of the electronic properties at the metal–GaAs interface. We speculate on the consequences of our measurements for the different models for Fermi level pinning for In on GaAs(110).

It is evident from the discussion in the last section that 3D islands of In form on GaAs(110) at both high and low temperatures, but that they do so in different ways. We have demonstrated that it is possible to determine the size, size distribution, and shape of the growing islands from the RHEED pattern and have presented in Figs. 8 and 11 the size in one direction of islands deposited at high and low temperatures. It is possible also to determine other island dimensions, from these the volume, and finally to calculate an average separation of islands.

We begin by detailing how the island volume is determined. From the orientation of the RHEED streaks at high coverages, it is possible to determine the facet planes and hence the shape of the triangle that represents a cross section through the prism shape. The length of the prism is determined directly from a profile measurement in the appropriate direction (GaAs[$1\bar{1}0$], see Figs. 8 and 11). To calculate the island volume the height of the prism must be known. It is determined by measuring the diffracted-beam profile in a direction normal to the surface plane (GaAs[110]). This profile reflects the average length of all the vertical strings of atoms in the triangle. Examples of such profiles were shown in Fig. 6 for room-temperature depositions. Figure 12 shows profiles for low-temperature depositions. They are always broader at the same dose (and hence the island height is less) than those for room-temperature depositions. Because the shape of this triangle is known, model diffracted-beam profiles for any prism height can be uniquely calculated. We do this and compare the resulting calculated profiles to the measurements. An example calculated assuming a single height for all islands at a given dose was shown in Fig. 7. Once the height is known the base of the triangle and the average island volume are easily obtained. The dependence of the average volume on dose is compared for high and low-

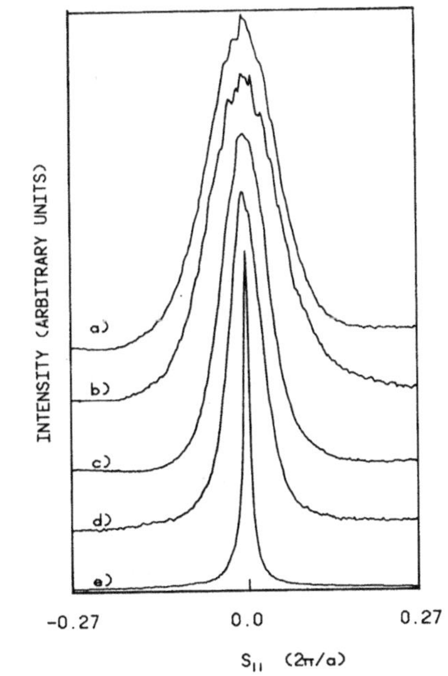

FIG. 12. Angular profiles of the ($1\bar{1}1$) reflection from In as a function of dose for deposition at 190 K. (a) $\theta_{nom}=0.8$, (b) 2, (c) 4, (d) 8, (e) instrument response. The profiles have the same shape below $\theta_{nom}=2$ and begin to narrow above that value. Beam profiles are along GaAs[110].

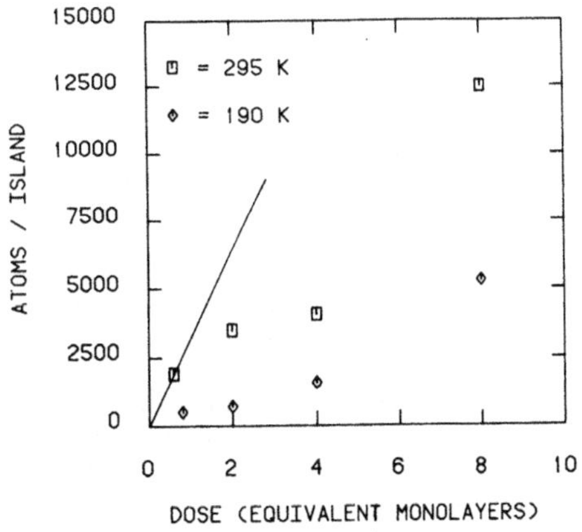

FIG. 13. Comparison of island volumes as a function of dose at 295 K (□) and 190 K (◇). If the density of islands were fixed at the initial density, measured at the lowest observable coverages, the number of atoms per island would follow a straight line through the origin and the first data point for each temperature. Such a line is shown for the 295 K data.

TABLE I. Structural parameters of In islands deposited at two temperatures.[a]

	Dose (ML)	Island height (Å)	No. of atoms per island	Island density $10^{11}/cm^2$
295 K	0.6	22	2 000	2.8
	2	30	3 500	5.0
	4	32	4 000	9.0
	8	50	12 500	6.0
190 K	0.8	13	500	14
	2	15	700	25
	4	22	1 600	23
	8	35	5 300	14

[a] Values assume that all atoms are in 3D islands.

temperature depositions in Fig. 13. Table I compares island heights, densities, and average volumes for the two temperatures.

A perusal of Figs. 8, 11, and 13 provides some interesting conclusions. Figures 8 and 11 show that the gable length initially is essentially the same for depositions at high and low temperatures. It stays constant up to doses of at least 5 ML, and higher for the room-temperature depositions. Figure 13 shows that the volume increases only slowly for doses below 5 ML, a consequence of the fact that the height is not changing significantly (see Fig. 6) and the gable length is not changing (see Fig. 8). The growth that does take place occurs by accretion of mass on the facets, effectively making larger the triangle that represents a cross section of the prism. One concludes that after rapidly reaching an initial size, the islands stop growing in one direction, which happens to be the direction in which the lattice mismatch is larger. It appears that a larger island dimension in this direction is energetically unfavorable.

The average volume of islands formed at low temperature is considerably smaller (Fig. 13) than that at room temperature. This is a consequence of the lesser height, the prism lengths being nearly the same at high and low temperatures. Because the facets in the In islands at high and low temperatures are the same, the base of the prism-cross-section triangle must also be smaller. Again, the average island volume initially changes only slowly with dose.

The slow change of average island volume with dose below ~5 ML is an important aspect of the growth of In on GaAs(110). In this dose range the increased dose goes mostly into increasing the density of In islands that all look very similar. This phenomenon has been observed earlier in other systems[22] and is described by the "diffusion zone model."[23] The process can be pictured as follows. Initially In forms two-dimensional arrays that presumably nucleate at defects. As the dose increases, a 2D to 3D transition takes place. All atoms within a zone of a specific radius are captured by the growing 3D island. When a 3D island reaches a particular size the driving force for acquiring additional atoms diminishes. The details of this process are not understood but presumably reflect free-energy considerations, i.e., the balancing of surface and strain energy against bulk cohesive energy. It then becomes favorable to nucleate new islands, which undergo the same process. Although all islands continue to grow slowly due to direct impingement on them, the preference is toward creation of new islands that very rapidly attain the minimum size mentioned above. Thus over a certain coverage range, the observed average size of islands does not vary significantly, but their density increases. When the density becomes too great for the nucleation of new islands, the existing ones begin to grow. Island coalescence also begins to take place, reducing the density. The phenomenon is observable in the SEM micrographs as dumbbell-shaped particles with extra space around them in which new particles can nucleate, and in Fig. 13 as a rapid increase in island volume above $\theta_{nom} \simeq 5$.

One can further conclude that at low temperatures the nucleation of 3D islands occurs with more difficulty, the capture zone of an island is smaller, and the energetics or kinetics favor smaller islands. It may be simply a matter of kinetics. Atoms arrive and may form 2D patches. There is some evidence from the RHEED pattern that 2D islands form at low coverages.[24] It is not until the 2D island density gets quite high that 3D island nucleation occurs. Possibly because of limited diffusion kinetics, many small 3D islands form. The fact that there is a dose range in which the average island volume changes very slowly indicates that a diffusion zone model still applies; however, with a smaller capture radius.

If the island shape and average volume per island at a given dose are known, it is possible to calculate an actual surface coverage as a function of dose. This value can be compared with AES measurements. We have done this for the room-temperature measurements with the assumption that all atoms are in islands. For example, for $\theta_{nom} = 0.6$, $\theta_{actual} = 0.13$, and for $\theta_{nom} = 2$, $\theta_{actual} = 0.31$ from the RHEED results. If the islands are assumed to be impenetrable to substrate Auger electrons the AES value for $\theta_{nom} = 2$ is $\theta_{actual} = 0.17$. Allowing penetration through the edges of the islands using a realistic mean free path of 16 Å increases θ_{actual} to 0.24. Thus both methods provide similar values, within a factor of 2.[25]

We now know the base dimensions of the growing islands and their average volume at any given dose. We also know that the islands all have a similar size at any dose. From these parameters it is possible to determine the density of islands. Values of density are shown in Table I. One can then also calculate an average separation of islands as a function of dose. This is done in the following way. The projected area of the islands is approximated by a circle whose effective radius r is related to the average of the length and width of the prism base. An average surface area per island, πR^2, where $2R$ is the average island center separation, is assigned to each island by dividing the island volume by the dose (i.e., determining an inverse island density):

$$\pi R^2 = \text{island volume}/\theta_{nom}, \quad (1)$$

$$\text{island volume} = 0.428\ (l \times w \times h/2), \quad (2)$$

where l, w, and h refer to the island dimensions, and the coefficient is a conversion factor between thickness in Å and equivalent monolayers. Equations (1) and (2) determine

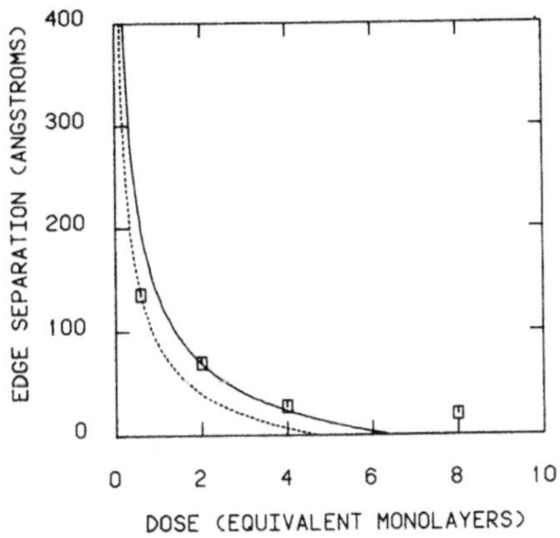

FIG. 14. Calculated edge separations of In islands as a function of dose for room-temperature depositions. The data points are the values of island separation based on measured values of the island size from Fig. 8 and volumes from Fig. 13. The curves are calculated for two fixed volumes that correspond to the first two data points in Fig. 13. Because the volumes vary slightly with dose, the two curves do not fall on top of each other. The solid curve almost passes through the third data point because the volume is almost the same for the second and third points in Fig. 13. Because of coalescence, the edge separation is still of the order of 25 Å at doses of 8 ML. Coalescence is not included in the calculated curves. The dashed curve gives a lower limit to edge separations as a function of dose at room temperature.

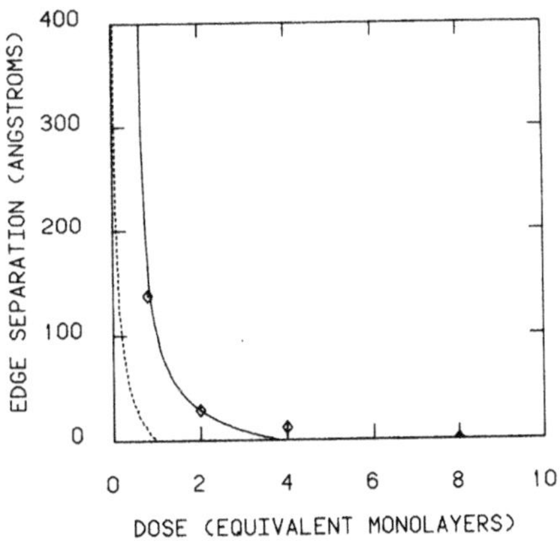

FIG. 15. Calculated edge separations of In islands as a function of dose for low-temperature deposition. The dashed curve is a calculation for the separation of islands that grow as 2D patches with 85 Å diam. A rapid decay of separations with dose occurs. The data points are the values of island separation based on measured values of island size from Fig. 11 and volumes from Fig. 13, for 3D island formation that begins at $\theta_{nom} = 0.8$. It is assumed that only atoms above $\theta_{nom} = 0.8$ participate in 3D island formation. The solid curve is calculated assuming a constant island volume above $\theta_{nom} = 0.8$. The value used in the calculation corresponds to the measured island volume at $\theta_{nom} = 2$ (see Fig. 13). A discontinuity in the separation occurs at the 2D to 3D transition.

the center separation. The average edge separation is then

$$d = 2(R - r). \tag{3}$$

The result of carrying out this procedure for the room-temperature data in Fig. 13 is shown in Fig. 14 as the data points. Figure 14 also shows two curves that were calculated with the assumption that the volume does not change at all, and only the density does. The dashed and solid curves use, respectively, the volumes for the first and second data points in Fig. 13.

This type of calculation can be performed for any array of patches on a surface as long as the volume, patch area, and distribution of patch areas is known. As an example, we have calculated the average separation versus dose for 2D patches that are 85 Å in diameter and 1 ML thick. This is shown as the dashed curve in Fig. 15. Clearly, a much lower dose is required for a given separation if less volume can be accommodated, as it is in a 2D patch. It is also evident that the separation must go to zero at a dose of 1 ML.

For a system that forms first as a 2D array and later transforms to a 3D array, the picture is somewhat more complicated. If the system initially grows as 2D islands, a 2D curve will be followed. When the system transforms to 3D islands at a given dose, the separation of islands should suddenly increase. How far it increases depends on the details of the transformation, i.e., does every 2D patch transform to a 3D island; do all patches in a 3D capture zone combine, does the 2D patch act like a new substrate, etc.? In any case, there should be an observable transition in the separations at the dose at which the overlayer transforms. In the room-temperature data, we observe 3D islands for $\theta_{nom} \geqslant 0.2$. Hence we have calculated the separation of 3D islands only above this dose, but including the 0.2 ML dose already on the surface in the island formation. For the low-temperature data, this transition occurs at $\theta_{nom} = 0.8$. Therefore, a separation function for 3D islands can be plotted only above this dose. We have done that in two limits: (1) that all the atoms present on the surface at $\theta_{nom} = 0.8$ combine into 3D islands, and (2) that 2D patches remain on the surface and that atoms form 3D islands on top of the 2D layer. To be consistent with the room-temperature formation of 3D islands we allow a 0.2 ML dose to participate in 3D island formation, i.e., the initiation is shifted to $\theta_{nom} = 0.6$. Because the available number of atoms is less in limit (2), larger separations are obtained than in limit (1) if some minimum island size must be achieved for stable 3D island formation. Limit (2) is plotted as the solid curve in Fig. 15. Limit (1) is plotted as the dashed curve in Fig. 16. We have no direct evidence for either limit. One might expect substrate Auger lines to reflect the transition in limit (1) but not in limit (2) [in limit (1) surface actually gets uncovered] and thus be able to distinguish between them. However, a realistic assessment of the Auger intensities indicates that the effect would be small.

Figure 16 compares the dose dependencies of separations at high and low temperatures with the same set of assumptions, namely that all atoms are in 3D islands, all islands at a given dose and temperature have the same size, the projected island size can be approximated by a circle of radius r, and island centers at any given dose are separated by a distance $2R$ [Eqs. (1)–(3)]. Measured volumes at $\theta_{nom} = 2$ are used in the model to calculate the solid and dashed curves. Be-

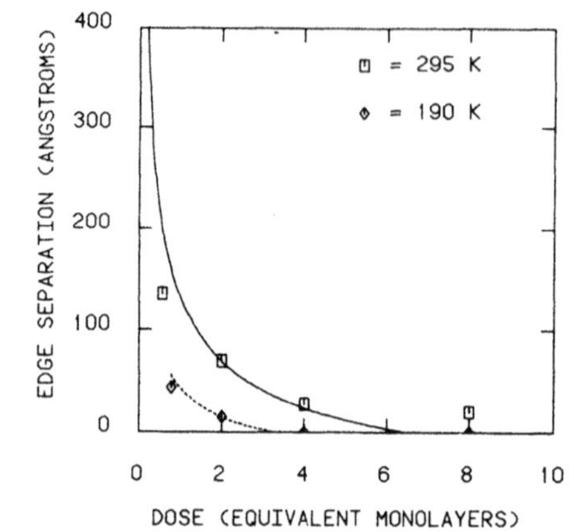

FIG. 16. Comparison of island separation dependence on dose for high and low temperatures. □: 295 K (same as in Fig. 14). ◇: 195 K, calculated for limit (1) discussed in text. The solid and dashed curves are calculated fits based on the data point at $\theta_{nom} = 2$.

cause the islands are smaller at low temperatures, their average separation at any given dose is less.

There are obviously a large number of assumptions involved in the model that can change the details of these curves.[26] We are certain about the shape of the islands. Although we can fit the angular profiles with the assumption of a single island length and a single height (and the SEM micrographs seem to corroborate this), there is no assurance that we could not also fit the measured profiles with a different size distribution that would give a different average length or height. However, experience with other systems[15] has shown that any other distribution that provides a satisfactory fit also gives a similar average size. We assume that the islands have circular bases when actually they are rectangular. This causes negligible uncertainties. We estimate that in the worst case, the values for the separation of islands at a particular dose are uncertain by $\pm 50\%$.

We conclude by speculating on the implications of these measurements for contact formation in In on GaAs(110) and for the several models for Fermi level pinning. Clearly In does not grow smoothly layer by layer. It grows in 3D islands that already at low doses have many hundreds of atoms in them. The density of these islands increases with coverage. Table I listed island densities and average number of atoms per island as a function of dose at 295 and 190 K. It is clear that a simple Schottky model is not applicable because of the laterally inhomogeneous structure. In concord with the effective work function model,[7] one can imagine that underneath each 3D island a Schottky barrier forms that is determined by the contact potential difference between In and GaAs (or between whatever phases are in contact at the interface). Away from the interfaces, into the GaAs, there will then be a band bending that depends on the Debye length and hence the doping level in the GaAs. Such band bending should also occur laterally away from each island, so that the work function in the bare GaAs should vary laterally. The amount of band bending laterally depends on the doping level and the separation of islands. Thus as the dose increases and hence the average separation of islands decreases, the average work function of the remaining bare GaAs changes from that of clean GaAs to that of In. The latter value cannot be achieved until the layer covers the surface completely, a situation that does not occur at room temperature at any dose below at least 150 ML.

Average work function measurements for metals on semiconductors are typically made with photoelectron spectroscopy (PES) by determining the energy of a substrate core-level photoelectron. The energy of this photoelectron for the clean surface fixes the clean-surface value. The addition of a metal overlayer then can cause the appearance of a line shifted to a somewhat different energy, a broadening of the original line, or both. Fermi level pinning measurements are made by observing the ratio of shifted and unshifted components. The Fermi level is said to pin when no further changes occur. In on GaAs(110) has an "anomalous" pinning behavior in that it takes doses of the order of 5–10 ML to "pin" the Fermi level, in contrast to several other metals deposited on GaAs(110).[4,8]

The observation that In exists as large, well-separated islands can explain the extended dose range required for Fermi level pinning, independent of the choice of model for barrier formation. The only assumption is that Fermi level pinning is associated with the existence of In islands. In PES measurements, a substrate core level line is monitored as a function of dose. The intensity of this line comes from bare GaAs or GaAs that is covered with a thin enough layer of In so that core lines can penetrate. If one assumes for the moment that no band bending in the lateral directions occurs and that the core level lines cannot penetrate any area covered with In, then there should be no line shift and just a continuous attenuation of the clean-surface line. If it is assumed that some penetration of core level photoelectrons through the edges of In islands occurs, then as the density of islands increases, there will be more island edges, and the line should have two components, an unshifted one from the bare GaAs and a shifted one from the GaAs covered with In thin enough so that the core level electrons can penetrate. Because the islands have sloped sides, there should be a greater intensity in the shifted component at any given dose for a longer mean free path than for a shorter one, because more of the area underneath an island can participate in observable photoelectron production. This analysis suggests that there should be a dependence of the relative intensities of shifted and unshifted components and hence of the interpretation of Fermi level pinning on mean free path (hence photon energy). Some of the discrepancies in different photoemission measurements[4,8,9] may be attributable to a mean-free-path difference. Pinning, in the sense that there is no unshifted peak, should not occur below 150 ML (the highest we have measured, and still showing bare substrate). However, the ratio of the shifted to unshifted peaks may become nearly constant much before that, when coalescence begins and the actual surface coverage changes very slowly. The situation is more complicated in reality, because, as a consequence of the small Debye lengths in GaAs, the band bending is also depth-dependent over the range of photoelectron mean free paths that have been used.[11]

Lateral band bending away from the islands leads to a component of the core level lines that is shifted and broadened. Only those photoelectrons coming from bare GaAs that lies outside a screening length from the edge of the island produce an unshifted component. Those photoelectrons coming from inside a screening length produce a broadened peak that starts at the unshifted line and has its most shifted component coming from photoelectrons created at the edge of the island or at the interface underneath the island. When the separations of islands get small enough so that the screening lengths overlap, there should no longer be an unshifted component. As the separations decrease, the intensity should move preferentially to the most shifted part of the line. This is the same as the situation considered above, except that it allows the extinction of the unshifted line at less than complete coverages and increases the intensity of all the shifted components at any coverage. Where this occurs depends on the screening length. A limiting value for the screening length is the Debye length. A simple calculation for the Debye length in these samples shows that it is on the order of 35 Å, in the range of the minimum separations measured at room temperature. Thus one may expect that the unshifted line is weak or absent at the corresponding doses. If the separations were to continue to decrease (which they do not appear to do because of coalescence) the shifted component should continue to gain intensity in its most shifted part at the expense of its least shifted part. For the low-temperature depositions, the separations are smaller, and the above phenomena should occur at lower doses, producing an apparent pinning at lower doses. In neither case does the system really reach its final contact potential difference until a complete layer has formed. How well these effects can be observed depends on resolution in the PES experiment as well as areal considerations. An experiment that should allow an assessment of the importance of lateral band bending suggests itself at this point. Substrates with different doping levels would show a different intensity ratio development with dose for the shifted and unshifted peaks, a lower doping level causing the unshifted peak to disappear more rapidly as dose increases. Because the Debye length is greater, less of the bare surface would have the clean-surface work function.

It seems plausible that a lateral band bending will affect the local work functions and hence the interpretation of the dose dependence of Fermi level pinning measurements in systems that grow by 3D island formation, but this has not been proved. A separate issue in the "final" pinning position for In/GaAs(110). All that we have said so far would suggest that it is possible to interpret measurements in terms of a band bending that is less than the final, true value, because a complete barrier has not yet formed. To do so within the constraints of a Schottky model that includes lateral inhomogeneities would require that the observed value be less than the contact potential difference between In($1\bar{1}3$) and GaAs(110). Whereas the average work functions of bulk and GaAs indicate that the contact should be almost Ohmic, this is not the case. Work functions for different crystal orientations and for small particles may differ from the average work function, and hence a small mismatch may exist for In/GaAs(110) even if the average work functions are similar. A lattice mismatch and possible strain are also present. There is also the question whether the size of the In islands is sufficient to supply enough charge to establish a full Schottky barrier underneath the island. A simple calculation would suggest that there is sufficient charge even for the smallest islands we observe. Again, this is doping level dependent. If an incomplete barrier is formed, then the band bending will be size dependent and hence different at high and at low temperatures. Also the final pinning position after some dose such as 10 ML (where the islands are still relatively small) may not reflect the position of the Fermi level for much higher coverages. To address some of these questions may require extensive calculations.[11] If it can be shown that the contact potential difference is greater than any observed band bending then the need for any more complicated model of Fermi level pinning disappears.

The existence of 3D islands has consequences for other models of Fermi level pinning. The unified defect model was proposed to account for the fact that the pinning energy for most metals on GaAs is nearly the same. A simple Schottky model cannot explain this phenomenon. In is an exception to the general trend, however. It neither pins as fast nor at the same level as other metals. As we have mentioned, both effects can be caused by islands that are large and well separated. If defect formation is involved in the band bending, and if it is assumed that one defect is associated with each island, then the Fermi level should become pinned more rapidly at low temperature, where there are more islands. There are insufficient data for In; low-temperature measurements extend only to doses of 1 ML.[9] If one assumes that the behavior for In is like that for Al, then the trend is in the opposite direction. For Al, the measurements are interpreted in terms of slower pinning at low temperatures.[27] This result would suggest that the size of the islands is more important than their number, i.e., the number of defects per island depends on the island size. Measurements of Fermi level pinning for higher doses at low temperatures should resolve this question for In. Furthermore, if the trend to smaller islands continues to lower temperatures, eventually no islands may form. This would help distinguish if the heat of condensation of a 3D island is important in defect formation.

The effect is in the same direction in the MIGS model. At low temperatures the coverage is more complete for any given dose, and one would expect pinning to occur more rapidly with increasing dose.

One may also be able to distinguish whether a defect model, MIGS model, or modified Schottky model is applicable on the basis of reversibility. If the adsorbed metal were removable by evaporation and one could verify that the substrate surface structure and composition have not changed, one should recover the original position of the bands in clean GaAs(110). If the band positions do not return to their original value, point defects probably have been created in the surface. Another possibility for distinguishing between models is to look in this system for changes in the photoelectron core level spectrum at the 2D to 3D growth transition.

V. SUMMARY

We have shown that there are strong spatially inhomogeneous effects in the growth of In on GaAs. These effects depend on temperature. We have extracted quantitative island sizes, volumes, separations, densities, and orientations for the In as a function of dose and temperature. We have discussed these results in terms of their implications for Fermi-level pinning models. Our results allow us to draw conclusions about important parameters in models for Fermi level pinning. We suggest that lateral band bending away from the edges of islands is important in Fermi level pinning in the In/GaAs(110) system because the "coarseness" of the spatial inhomogeneity in the growing film is of the order or greater than the screening length.

ACKNOWLEDGMENTS

This work was supported by ONR and by a gift from the IBM Corporation. We acknowledge useful discussions with G. Margaritondo and J. D. Wiley.

[1]R. Kern, G. Le Lay, and J. J. Metois, in *Current Topics in Materials Science*, edited by E. Kaldis (North-Holland, Amsterdam, 1979), Vol. 3, p. 131.

[2]R. J. H. Voorhoeve, in *Treatise on Solid State Chemistry*, edited by N. B. Hannay (Plenum, New York, 1976), Vol. 6A, p. 241.

[3]W. Schottky, Z. Phys. **118**, 539 (1942).

[4]W. E. Spicer, I. Lindau, P. Skeath, C. Y. Su, and P. Chye, Phys. Rev. Lett. **44**, 420 (1980); J. Vac. Sci. Technol. **16**, 1422 (1979).

[5]A. Zunger, Phys. Rev. B **24**, (1981).

[6]J. Tersoff, Phys. Rev. Lett. **52**, 465 (1984).

[7]J. L. Freeouf and J. M. Woodall, Appl. Phys. Lett. **39**, 727 (1981); J. Vac. Sci. Technol. **21**, 574 (1982).

[8]R. R. Daniels, T. X. Zhao, and G. Margaritondo, J. Vac. Sci. Technol. A **2**, 831 (1984).

[9]K. K. Chin and I. Lindau, Phys. Rev. B **32**, 6902 (1985); K. K. Chin, T. Kendelewicz, C. McCants, R. Cao, K. Miyano, I. Lindau, and W. E. Spicer, J. Vac. Sci. Technol. A **4**, 969 (1986).

[10]W. E. Spicer, I. Landau, P. Skeath, C. Y. Su, and P. Chye, J. Vac. Sci. Technol. **16**, 1143 (1979).

[11]J. Y. F. Tang and J. L. Freeouf, J. Vac. Sci. Technol. B **2**, 459 (1984).

[12]D. E. Savage and M. G. Lagally, Phys. Rev. Lett. **55**, 959 (1985).

[13]For a review, see M. G. Lagally, in *Methods of Experimental Physics*, edited by R. L. Park and M. G. Lagally (Academic, Orlando, FL, 1985), Vol. 22, p. 237.

[14]J. M. Van Hove, P. Pukite, and P. I. Cohen, J. Vac. Sci. Technol. A **1**, 609 (1983).

[15]P. Hahn, J. Clabes, and M. Henzler, J. Appl. Phys. **51**, 2079 (1980); D. G. Welkie and M. G. Lagally, Thin Solid Films **93**, 219 (1982); P. O. Hahn and M. Henzler, J. Appl. Phys. **52**, 4122 (1981); H. M. Clearfield and M. G. Lagally, J. Vac. Sci. Technol. A **2**, 844 (1984); D. Saloner and M. G. Lagally, Mater. Res. Soc. Symp. Proc. **41**, 179 (1985).

[16]J. M. Van Hove, C. S. Lent, P. R. Pukite, and P. I. Cohen, J. Vac. Sci. Technol. B **1**, 741 (1983).

[17]R. L. Park, J. E. Houston, and D. G. Schreiner, Rev. Sci. Instrum. **42**, 60 (1971).

[18]T. M. Lu and M. G. Lagally, Surf. Sci. **9**, 695 (1980).

[19]Electron energy loss measurements have been made for this system at different temperatures. C. R. Bonapace, D. W. Tu, K. Li, and A. Kahn, J. Vac. Sci. Technol. B **3**, 1099 (1985).

[20]A variety of Ewald constructions for reflection are shown in B. A. Joyce, J. H. Neave, P. J. Dobson, and P. K. Larsen, Phys. Rev. B **29**, 814 (1984).

[21]G. W. Simmons, D. F. Mitchell, and K. R. Lawless, Surf. Sci. **8**, 130 (1967). This paper deals with O on Cu. Interestingly, the facets that form on Cu after reaction with O have nearly the same orientation as the In facets in our work, so that the diffraction patterns look very similar.

[22]J. J. Metois, J. C. Zanchi, and R. Kern, Philos. Mag. **33**, 133 (1976).

[23]B. Lewis and D. S. Campbell, J. Vac. Sci. Technol. **4**, 209 (1967); J. A. Venables, Philos. Mag. **27**, 697 (1973).

[24]D. Savage and M. G. Lagally (to be published).

[25]Coverages at higher doses listed in Ref. 19 agree with ours; see Ref. 12.

[26]D. Saloner, P. K. Wu, and M. G. Lagally, J. Vac. Sci. Technol. A **3**, 1531 (1985).

[27]M. D. Williams, K. K. Chin, C. E. McCants, P. H. Mahowald, and W. F. Spicer, J. Vac. Sci. Technol. B **4**, 939 (1986).

Direct variation of metal-GaAs Schottky barrier height by the influence of interface S, Se, and Te

J. R. Waldrop
Rockwell International Corporation, Microelectronics Research & Development Center, Thousand Oaks, California 91360

(Received 30 August 1985; accepted for publication 26 September 1985)

The electrical properties and interface chemistry of Schottky barrier contacts (Ag, Al, Au, Mn, Pd, and Ti) formed on *n*-type GaAs (100) surfaces that had prior exposure to elemental S, Se, and Te are compared to those of ideal (metal deposited onto a clean surface) contacts. The interface Fermi energy E_F^i and interface chemistry during contact formation were investigated by x-ray photoemission spectroscopy; the accompanying Schottky barrier height ϕ_B was measured by current-voltage and capacitance-voltage techniques. A substantial decrease in ϕ_B (and correlated change in E_F^i) for Al and Mn contacts is associated with a contact metal-chalcogen chemical reaction at the Schottky barrier interface, while some nonreactive noble metal-chalcogen interfaces have a ϕ_B increase. By choice of contact metal and interface chalcogen a ϕ_B range of >0.6 eV is obtained (~ 0.35 to 1.0 eV, which is $>40\%$ of the GaAs band gap) via corresponding changes in E_F^i.

The electrical properties of Schottky barrier contacts, especially as determined by the barrier height ϕ_B, are the basis of many GaAs semiconductor devices. The barrier heights of ideal (metal deposited onto a clean surface) contacts exhibit limited variation compared to the 1.43 eV GaAs band gap.[1] Indeed, data for a large number of metals indicate that the intrinsic ϕ_B range for *n*-type material is approximately 0.6–0.95 eV, independent of the contact work function, electronegativity, and metal-GaAs chemical reactivity.[2,3] Hence, if a greater ϕ_B range is desired for device purposes, a departure from the ideal contact interface is required. In an experiment of this kind, which emphasized an ohmic contact application, exposure of a clean *n*-type GaAs surface to H_2S prior to Al deposition significantly lowered the effective ϕ_B for current transport to ~ 0.4 eV.[4] More recently, it has also been shown that for Au contacts ϕ_B is increased ~ 0.1 to 1.0 eV by prior exposure of a GaAs surface to either elemental S or Se.[5] How chalcogens at the metal-GaAs interface can in general change ϕ_B thus needs clarification.

This letter reports the electrical properties and interface chemistry for Ag, Al, Au, Mn, Pd, and Ti Schottky barrier contacts to GaAs surfaces that had prior exposure to either elemental S, Se, or Te. The interface Fermi energy E_F^i and the interface chemistry during contact formation were observed by x-ray photoemission spectroscopy (XPS). The Schottky barrier height was measured by current-voltage (*I-V*) and capacitance-voltage (*C-V*) techniques on the same interfaces characterized by XPS.

The interfaces were prepared and characterized under ultrahigh vacuum (UHV) conditions (10^{-10} Torr range base pressure) in an XPS system that consists of a HP5950A electron spectrometer ($h\nu = 1486.6$ eV monochromatic x-ray source) coupled to a custom sample preparation chamber. The details of the XPS procedure used to correlate E_F^i and interface chemistry can be found elsewhere.[5,6] In essence, E_F^i, which is the Fermi level to valence-band energy at the interface, is obtained by measuring the As 3*d* and Ga 3*d* core-level peak binding energy E_B from the GaAs substrate as a function of chalcogen and metal deposition (note that $\Delta E_F^i = \Delta E_B$ and $\phi_B = 1.43\ \text{eV} - E_F^i$). Interface chemistry is determined both by the appearance of new features in these peaks and by an examination of core-level peaks associated with the deposited chalcogens and metals. The electrical measurements were made with a computerized system that includes a HP4140B pA meter/voltage source and a HP4275A capacitance meter. *I-V* data were taken in forward bias in 0.01-V increments to a current density limit of 2 A cm^{-2}; *C-V* data were taken at 1 MHz in 0.1-V reverse bias increments.

All the GaAs material was liquid encapsulated Czochralski grown *n* type ($\sim 6\times10^{16}\ \text{cm}^{-3}$ Se) that was cut and polished into (100) oriented wafers. To prepare a sample, a $\sim 8\times8$ mm piece is etched for ~ 30 s in freshly prepared 4:1:1 H_2SO_4:H_2O_2:H_2O solution to remove polishing damage, mounted with molten In onto a Mo plate, and immediately inserted into the XPS system. The inevitable ~ 10-Å superficial native oxide layer is removed by momentary heating to the minimum required temperature (~ 550 °C, which also creates a bottom ohmic contact to the GaAs). The thermally cleaned surface is ordered (exhibits a characteristic low-energy diffraction pattern) and is confirmed by XPS analysis to be free of oxygen, carbon, and other contaminants. Clean *room-temperature* GaAs surfaces were exposed in a UHV background to either a S, Se, or Te vapor stream by heating a small quartz oven, located ~ 2 cm away, filled with high-purity elemental material. Metal evaporations were from a W basket. After XPS interface analysis of thin overlayers, a final metal layer of >2000 Å was deposited. Circular 2.54×10^{-2} cm diameter contacts were defined by using photolithography and chemical etching.

Figure 1 shows a comparison of Ga 3*d* and As 3*d* XPS core-level spectra that highlights several general chemical and E_F^i features of Schottky barrier formation with and without an interface chalcogen, in this example Se. The top pair of Ga 3*d* and As 3*d* spectra is for a GaAs clean surface; the vertical reference lines run through the half-width at half-maximum of these peaks. When 5 Å Se is deposited onto

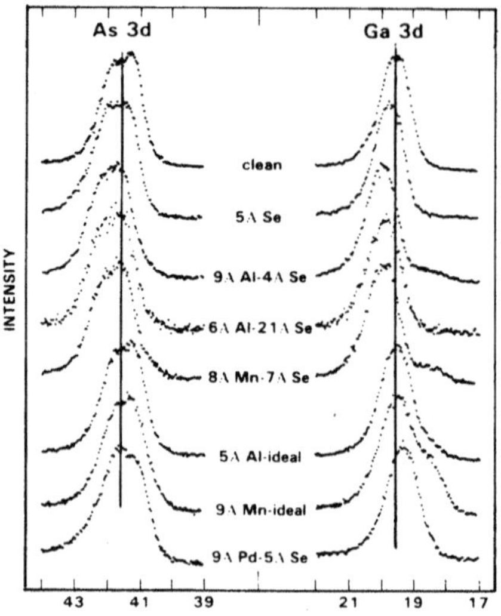

FIG. 1. Comparison of Ga $3d$ and As $3d$ XPS core-level spectra during interface formation for several metal-Se-GaAs contacts and ideal contacts.

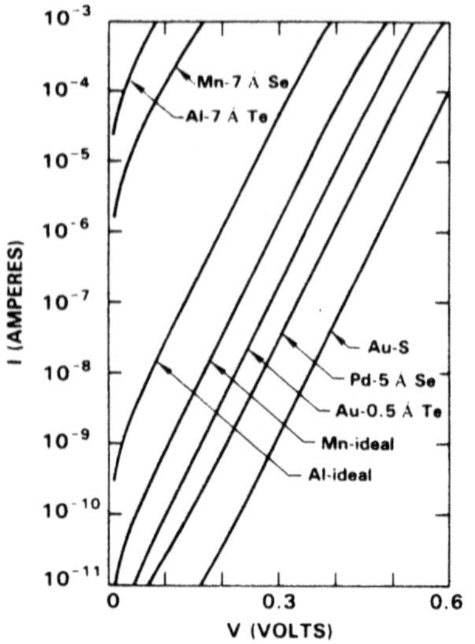

FIG. 2. Representative I-V data for a selection of GaAs Schottky barrier contacts, with and without an interface chalcogen (contact area $= 5.07 \times 10^{-4}$ cm^2).

clean GaAs, the Ga $3d$ and As $3d$ peaks both shift $\Delta E_B = \sim 0.15$ eV to higher binding energy[5] (all peak heights in Fig. 1 are normalized for ease of comparison). The corresponding E_F^i shift is from ~ 0.7 eV for the clean surface to ~ 0.85 eV for the Se covered surface (less band bending). A similar ΔE_F^i shift also occurs with S and with Te exposures.[5] The other six Ga $3d$ and As $3d$ spectral pairs in Fig. 1 are for different samples that divide into two groups of three according to the E_F^i (E_B) shift during metal deposition. The spectra were taken after the designated metal and Se deposition; that is, 9 Å Al-4 Å Se means 4 Å Se was first deposited followed by 9 Å Al. The first group (9 Å Al-4 Å Se, 6 Å Al-21 Å Se, and 8 Å Mn-7 Å Se) is shifted to higher E_B (larger E_F^i), and the second group (5 Å Al-ideal, 9 Å Mn-ideal, and 9 Å Pd-5 Å Se) is shifted to lower E_B (smaller E_F^i). The difference in E_F^i between the two groups is $\Delta E_F^i = 0.4$–0.5 eV. The crucial point is that the final E_F^i (ϕ_B) values are obtained *only after the metal depositions* and that the Al-Se and Mn-Se interfaces have similar high E_F^i values (low ϕ_B) compared to the low E_F^i (high ϕ_B) values of the Pd-Se and ideal interfaces. Thus, the variation in E_F^i, and hence ϕ_B, for ideal versus chalcogen interfaces is due to a metal-chalcogen interaction *not* a chalcogen-GaAs interaction.

The interface chemistry associated with these metal-chalcogen-GaAs interfaces is highly complex. Both S and Se react[5,7] with the clean surface to form a thin $\sim$5-Å-wide region of As and Ga chalcogenides; no Te-GaAs chemical reaction was detected. Although the GaAs substrate is the dominant intensity in the Fig. 1 core-level peaks, an obvious broadening or shoulder at low binding energy appears to a varying degree in the Ga $3d$ peaks (and, in some cases, As $3d$ broadening) when a metal is present which represents a metal-GaAs dissociation chemical reaction of corresponding varying degree.[8,9] No correlation has been found between E_F^i and the metal-GaAs chemical reactions.[2,3] It must be emphasized that there is a correlation between E_F^i variations and contact metal-chalcogen chemical reactions. This effect is exemplified at the 6 Å Al-21 Å Se interface (Fig. 1), where the thin Al deposit was found to have completely reacted with the Se overlayer to form Al_2Se_3; no metallic Al was present, yet the E_F^i increase has occurred. A chemical reaction between Mn and Se was also observed at the 8 Å Mn-7 Å Se interface.

Do the XPS observations of E_F^i shifts correlate with electrical measurements on the same interfaces? Figure 2 shows representative I-V data for a selection of such Schottky barrier interfaces. A considerable range in ϕ_B is evident. Figure 3 shows representative C-V data (plotted as $1/C^2$ vs V) that contrast Au-S, Al-ideal, and Al-1.5 Å Se interfaces; the effect of the chalcogen is also clear. The I-V

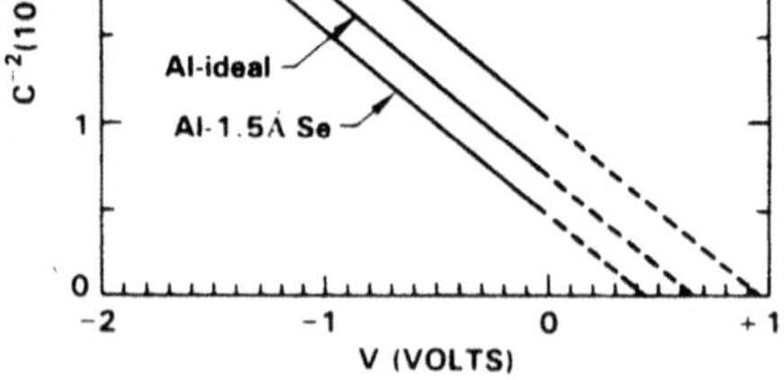

FIG. 3. Representative C-V data that contrast Au-S, Al-1.5 Å Se, and Al-ideal interfaces. Dashed line is a least squares fit extrapolation to V_i.

TABLE I. Influence of interface S, Se, and Te on the Schottky barrier height of various metal contacts to GaAs.

Metal	Interface	n	ϕ_B^{IV} (eV)	ϕ_B^{CV} (eV)
Al	S	1.05	0.52	0.45
	1.5 Å Se	1.03	0.53	0.52
	4 Å Se	...	~0.35	...
	21 Å Se	1.19	0.48	0.39
	3 Å Te	1.04	0.51	0.53
	7 Å Te	1.06	0.44	0.40
	22 Å Te	1.06	0.52	0.50
	ideal	1.04	0.74	0.75
	ideal[a]	1.07	0.85	0.84
Mn	7 Å Se	1.08	0.51	0.61
	ideal	1.03	0.82	0.82
Ti	11 Å Te	1.02	0.72	0.72
	ideal[a]	1.03	0.83	0.83
Au	S[b]	1.03	1.00	1.02
	20 Å Se[b]	1.08	0.97	0.96
	0.5 Å Te	1.03	0.89	0.87
	10 Å Te	1.03	0.83	0.81
	~100 Å Te	1.02	0.79	0.78
	ideal[a,b]	1.03	0.89	0.89
Ag	S[b]	1.05	0.96	0.99
	3 Å Te	1.04	0.84	0.84
	ideal[a,b]	1.03	0.90	0.89
Pd	5 Å Se	1.04	0.93	0.94
	ideal[a]	1.03	0.91	0.93

[a] Ref. 2.
[b] Ref. 5.

data were analyzed by using a form of the thermionic emission model[10] $I = I_s \exp(qV/nkT)[1 - \exp(-qV/kT)]$ A cm^{-2} which has the property that $\log\{I/[1-\exp(-qV/kT)]\}$ vs V is linear for *all* V so that I_s and ideality factor n ($n \sim 1.02$ is ideal) can be determined accurately by a linear least-squares fit, especially for small V. The Schottky barrier height ϕ_B^{IV} is extracted from I_s by $I_s = A^*T^2 \exp[-q(\phi_B^{IV} - \Delta\phi)/kT]$ A cm^{-2}, where $A^* = 8.16$ is the effective Richardson constant and $\Delta\phi$ is the calculated[10] image force correction ($\Delta\phi = +0.04$ eV for $\phi_B \geqslant 0.7$ eV and $+0.03$ eV for $\phi_B < 0.7$ eV). The C-V data were analyzed according to the conventional $1/C^2$ vs V relation[11] that gives ϕ_B as $\phi_B^{CV} = V_i + \delta + kT/q = V_i + 0.08$ eV, where V_i is the intercept on the voltage axis (determined by least-squares fit) of a $1/C^2$ vs V plot and δ is the Fermi energy in the bulk with respect to the conduction-band minimum.

Table I lists the average ϕ_B^{IV}, n, and ϕ_B^{CV} values for a number of metal-chalcogen-GaAs interfaces (~ 10 contacts per sample, $< \pm 0.01$ eV measurement accuracy). The Al and Au contacts were considered as a function of chalcogen type and thickness (no thickness is shown for S because elemental S does not remain on the GaAs surface after exposure, see Ref. 5). The other metals indicate the generality of the effects.

The ϕ_B for Al-chalcogen interfaces is 0.35–0.5 eV compared to 0.75–0.85 eV for the Al-ideal interface (a previous Al-ideal measurement is quoted to indicate that the Al-ideal interface does not appear to have a unique ϕ_B value[12]). Thus, either S, Se, or Te causes a ~ 0.4 eV decrease in ϕ_B for Al Schottky barrier contacts, independent of the amount of chalcogen deposited. The ϕ_B for the Mn-Se interface is also markedly lower than the Mn-ideal interface. A smaller ~ 0.1 eV difference in ϕ_B is observed between the Ti-ideal and Ti-Te interfaces.

The metals Au, Ag, and Pd do not exhibit a large decrease in ϕ_B in the presence of interface chalcogens. In fact, Au and Ag contacts have a surprising ~ 0.1 eV increase to a ϕ_B of ~ 1.0 eV for interface S and Se, while the Pd value is essentially unaffected. A partial monolayer (0.5 Å) of Te does not affect the Au ϕ_B, but for 10-Å and $\sim$100-Å layers there is a ~ 0.1 eV decrease (with no change of n). A slight decrease (< 0.1 eV) is seen for the Ag-3 Å Te interface.

The chalcogenide-forming reactions associated with Al and Mn have relatively large negative heats of formation[13] ΔH_f that lie in the range $\Delta H_f(Al_2S_3) = -173$ kcal mol^{-1} to $\Delta H_f(MnSe_2) = -40$ kcal mol^{-1}. The possible chalcogenides that involve Ag, Au, and Pd have ΔH_f more positive than -20 kcal mol^{-1}. Thus, a plausible guide for predicting contact metals that have substantial ϕ_B lowering is that ΔH_f for chalcogenide formation be fairly large and negative. Based on this criterion, a number of metals are candidates for the effect, but, as with the Ti-Te interface [$\Delta H_f(TiTe_2) = -50.9$ kcal mol^{-1}], the ϕ_B lowering may be modest.

The present results are compatible with a GaAs Schottky barrier model that involves defect related energy levels that move in energy in response to changes in interface chemistry.[14,15] In any case, ϕ_B (and E_F^i) cannot be considered to be restricted (pinned) to a narrow range for all GaAs Schottky barrier contacts; a variation over $> 40\%$ of the band gap has been demonstrated. Finally, the minimum quantity of chalcogen necessary to induce a ϕ_B change has not been determined; it may be that minute amounts unintentionally introduced at the Schottky barrier interface could result in undesirable ϕ_B variations.

[1] W. G. Spitzer and C. A. Mead, J. Appl. Phys. **34**, 3061 (1963).
[2] J. R. Waldrop, J. Vac. Sci. Technol. B **2**, 445 (1984); Appl. Phys. Lett. **44**, 1002 (1984).
[3] J. R. Waldrop, Appl. Phys. Lett. **46**, 864 (1985).
[4] J. Massies, J. Chaplart, M. Laviron, and N. T. Linh, Appl. Phys. Lett. **38**, 693 (1981).
[5] J. R. Waldrop, J. Vac. Sci. Technol. B **3**, 1197 (1985).
[6] R. W. Grant, E. A. Kraut, S. P. Kowalczyk, and J. R. Waldrop, J. Vac. Sci. Technol. B **1**, 320 (1983); E. A. Kraut, R. W. Grant, J. R. Waldrop, and S. P. Kowalczyk, Phys. Rev. B **28**, 1965 (1983).
[7] D.-W. Tu and A. Kahn, J. Vac. Sci. Technol. A **3**, 922 (1985).
[8] J. R. Waldrop and R. W. Grant, Appl. Phys. Lett. **34**, 630 (1979).
[9] A recent general review of metal-semiconductor interfaces is L. J. Brillson, Surf. Sci. Rept. **2**, 123 (1982).
[10] For a discussion of this form of the I-V equation, see E. H. Rhoderick, *Metal-Semiconductor Contacts* (Clarendon, Oxford, 1977).
[11] A. M. Goodman, J. Appl. Phys. **34**, 329 (1963).
[12] Nearly all published values of Al-ideal ϕ_B fall in this range.
[13] K. C. Mills, *Thermodynamic Data for Inorganic Sulphides, Selenides anc Tellurides* (Butterworths, London, 1974).
[14] R. W. Grant, J. R. Waldrop, S. P. Kowalczyk, and E. A. Kraut, J. Vac Sci. Technol. **19**, 477 (1981).
[15] For a recent review of defect type Schottky barrier models, see W. Mönch Surf. Sci. **132**, 92 (1983) and references therein.

Interface potential changes and Schottky barriers

S. B. Zhang, Marvin L. Cohen, and Steven G. Louie
Department of Physics, University of California and Center for Advanced Materials, Lawrence Berkeley Laboratory, Berkeley, California 94720
(Received 24 April 1985)

By investigating local potential changes at the Si(111)/Al interface, we find restrictions on the use of "canonical" Schottky-barrier heights. It is demonstrated that the "metallic" behavior of the metal-induced gap states is insufficient to completely screen out *strong* local interface potential effects on Schottky-barrier heights.

The detailed nature of the electronic structure of semiconductor-metal interfaces and the origin of the Schottky barrier are not completely understood.[1-3] Experimentally, Schottky-barrier heights are pinned inside the gaps of most covalent semiconductors and are nearly independent of the metal contacts.[4,5] Theoretical models have been based largely on the pinning mechanisms of the intrinsic metal-induced gap states[6-9] or the native defects.[10-13] Because of the complexity of most interfaces, a generally accepted model does not exist. Recently, Tersoff[6] has suggested a new model to explain Schottky-barrier-height behavior and in this study we present calculations attempting to test this model.

Tersoff's model is based on local charge neutrality and *metallic* screening near the interfacial region in the semiconductor side (a region about ≈ 10 Å from the interface boundary). The screening is attributed to the metal-induced gap (MIG) states.[8] By reexamining the study by Louie and Cohen on the Si/Al interface[8] and other interfaces,[9] he concludes (also previously observed by others) that the large decay length and high density of MIG states at the Si interfacial region will pin the Fermi level. Moreover, he further postulates that the Fermi-level pinning occurs deep (≥ 3–4 atomic layers) inside the semiconductor and is caused by the tails of the MIG states. Thus, the Fermi-level position is relatively independent of the details of the interfacial region and is mainly a property of the semiconductor alone. Using this idea, Tersoff calculated some "canonical" Fermi levels which determine the Schottky-barrier heights.

Since the Schottky barrier arises from a semiconductor-metal contact, and the canonical Fermi level is only a property of the bulk semiconductor, this model would be valid if the screening of the MIG states is highly effective. Otherwise, the Fermi level is determined by local interface effects on the semiconductor side, which may or may not put the Fermi level at the bulk canonical value as inferred in Tersoff's paper. We define "local" to refer to the first one or two semiconductor layers at the interface; any deeper region is called "bulk". The local effects can be either changes in surface geometry or defects with appreciable density or both. To simulate local effects, we use potential changes near the interface in our calculations. For the case where the local perturbation is comparable to the atomic potential, though well localized in the first two layers, we find that the MIG state tails are unable to pin the Fermi level.

In the present study, we model the interface between Si and Al with a supercell[8] and use *ab initio* pseudopotentials.[14] The calculated band gap of Si in this model is underestimated because of the local-density-functional approximation[15] which adds some uncertainty to the numerical results. Twelve layers of Si atoms and roughly the same size for the jellium region with $r_s = 2.07$ (for Al) are incorporated in one unit cell. The formulation of the jellium ionic potential of Louie and Cohen[8] is employed, and the local density of states is calculated using the tetrahedron method[16] modified for two-dimensional problems. Three k points ($\bar{\Gamma}$, $\bar{M}$, and $\bar{K}$) with high symmetry in the irreducible Brillouin zone are used in the self-consistent calculations to reduce the computing time. For the calculation of the local density of states and the determination of the Fermi level position we use 36 k points. The energy cutoff is 4 Ry which corresponds to ≈ 420 plane waves in the basis set.

To illustrate the local dependence of the Fermi level, we introduce a localized perturbation by changing the first atomic layer of Si closest to the jellium boundary. Two cases will be discussed in this paper. First, the Si layer is replaced by a layer of "donor" atoms [a fictitious group-V atom, $U_{\mathrm{V}}^{\mathrm{ion}}(r) = \frac{5}{4} U_{\mathrm{Si}}^{\mathrm{ion}}(r)$], which "dopes" the interface strongly. The Fermi level shifts up to 0.28 eV (Ref. 17) corresponding to an equivalent decrease in the Schottky-barrier height for n-type samples. The self-consistent charge densities averaged in the plane parallel to the interface in this case and in the case of an ideal planar Si interface are calculated; their difference is plotted in Fig. 1(a), integration over the charge difference gives one electron. It is clear that the charge disturbance (within ≈ 2.5 Å from the jellium boundary) is localized and well within the decay length of the MIG states (≈ 3.0 Å for Si). The charge contour plot in Fig. 1(b) gives another view of the charge distribution, and shows that the jellium edge is well defined even though a local change has been introduced. The backbond charge is centered closer to the first layer atoms and a considerable charge density increase in the Si dangling bond region is found. The change in the Schottky-barrier height arises from the charge transfer from the "donor" atoms to the metal. The asymmetric charge distribution results because all the backbonds are

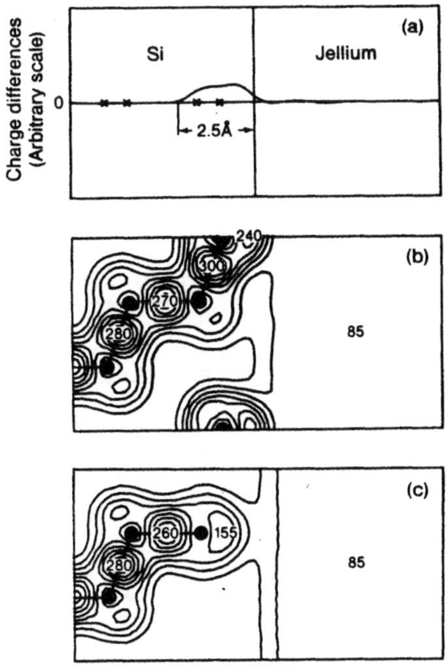

FIG. 1. (a) Total valence charge density changes plotted along the direction perpendicular to the interface after the "doping" of the interface. The charge densities are averaged parallel to the interface. Si atom positions are marked by crosses. For details, see the text. (b) Total valence charge density contours in the (1$\bar{1}$0) plane for the "doped" interface in units of electrons/supercell. The supercell volume is 3191 a.u.[3] and the contour spacing is 30. (c) Total valence charge-density contours in the (1$\bar{1}$0) plane when the first layer of Si atoms are absent. The contour spacing is the same as in (b).

already saturated. Hence, the extra electron charge of the donor is not centered on the donor atom. This creates an additional dipole potential across the Si-jellium boundary which causes an increase in the Fermi level in the metal.

We define a parameter $\kappa = d\phi_b/dV_s$, which measures the degree of "pinning" of the Fermi level, E_F, with respect to the *local* potential change. The self-consistent potential change in the bond was found to be -4 eV, which leads to a lower bound of $\kappa_L = 0.07$. The averaged value of potential change is only -1.4 eV (averaged in the plane parallel to the interface), therefore, the upper bound is $\kappa_U = 0.2$. These bounds are consistent with early estimates of the change in ϕ_b with potential.

The second case is opposite to the one discussed above. Now we remove the first layer of Si, creating a potential barrier at the interface. This is not equivalent to moving the jellium boundary further away since along the (111) direction Si has double layer periodicity. The Fermi level shifts down by 0.27 eV closer to the top of valence band (in our calculation the band gap is smaller than the experimental value). This change is attributed to the large number of unpaired dangling bonds which attract more electrons from the metal. Figure 1(c) gives the contour plot which shows the charge distribution for this case.

The insufficient screening by the MIG states which explains the above Schottky-barrier-height changes can be understood in the following way. The penetration depth of the MIG states at the Fermi level is significantly smaller than the gap averaged value ($\lambda = 3$ Å). Only when the energy is close to the top of valence band or the bottom of conduction band is the decay length in general longer and the density higher.[8,18] When a local perturbation is introduced in a metal, the electron gas deforms to create screening charges which reduce the overall influence. However, the finite size of the MIG states prevents them from creating long-range screening dipoles on the semiconductor side because these electrons are restricted within a range of λ. The effective screening due to the MIG states is limited to $\epsilon(q_\lambda) = 1 + k_s^2/q_\lambda^2$, where q_λ is an effective wave-vector cutoff determined by the MIG states decay length λ. Thus the usual exponential damping behavior from metallic screening does not persist even inside the MIG state tails. Our picture is different from a model where metallic screening is perfect except for a cutoff at λ. The present model is consistent with the model in Ref. 9, where it was pointed out that since the electrons from the MIG states cannot flow freely in the direction perpendicular to the interface, their polarization in this direction resembles semiconductor behavior. Therefore, there is no significant difference between metallic and nonmetallic screening as far as MIG states are concerned for potential disturbances perpendicular to the interface.

Our results reveal that the local charge transfer across the interface (or the local interface dipole change) is at least as important as the "bulk" local charge neutrality consideration in the determination of the Fermi level position, even though there are no bonding details used in our model. The experimental results[19] which establish different Fermi levels arising from the different orientations of epitaxial metal silicide layers are evidently a result of local differences. The local structures at the interface thus appear to play an important role.

In summary, we have demonstrated that the Schottky-barrier height can be shifted substantially by changing the interfacial atomic layer, or introducing a high-density local defects ($\sim 10^{14}/\text{cm}^2$ or higher). In the case of more covalent semiconductors, the density of MIG states is considerably high. This provides a large charge "reservoir" which is important in pinning the Fermi level. Surface defects with *lower* density in that region are screened. Their influence on the Fermi level therefore can be neglected. However, there are still possibilities to be discussed when the density of defects is high or extends into a larger interfacial region.[20] For large-gap ionic semiconductors, the influence of the termination of the crystal periodicity at the surface is small[21] leading to a small density of MIG states in the semiconductor gap when the metal is deposited.[8,9] As a consequence, the Fermi level is easier to move and varies with different metal electronegativities. It appears that the Fermi level is not well pinned deep inside the bulk of more covalent semiconductors but rather by charge dipole rearrangement in the first 1 or 2 layers.

One of us (S.B.Z.) would like to thank Dr. S. L. Richardson for discussions; helpful discussions with M. Y. Chou and her programing assistance are appreciated. We also acknowledge useful conversations with Dr. M. Schlüter and Dr. J. Tersoff. Support for this work was provided by the National Science Foundation under Grant No. DMR-83-19024 and by the Director, Office of Energy Research, Office of Basic Energy Sciences, Materials Sciences Division of the U.S. Department of Energy under Contract No. DE-AC03-76SF00098.

[1]*Metal-Semiconductor Schottky Barrier Junctions and Their Applications*, edited by B. L. Sharma (Plenum, New York, 1984).

[2]M. L. Cohen, *Advances in Electronics and Electron Physics*, edited by L. Marton and C. Marton (Academic, New York, 1980), Vol. 51, pp. 1–62.

[3]L. J. Brillson, Surf. Sci. Rep. **2**, 123 (1982).

[4]S. M. Sze, *Physics of Semiconductor Devices* (Interscience, New York, 1969).

[5]A. Thanailakis, J. Phys. C **8**, 655 (1975).

[6]J. Tersoff, Phys. Rev. Lett. **52**, 465 (1984).

[7]V. Heine, Phys. Rev. A **138**, 1689 (1965).

[8]S. G. Louie and M. L. Cohen, Phys. Rev. B **13**, 2461 (1976).

[9]S. G. Louie, J. R. Chelikowsky, and M. L. Cohen, Phys. Rev. B **15**, 2154 (1977).

[10]W. E. Spicer, I. Lindau, P. R. Skeath, C. Y. Su, and P. W. Chye, Phys. Rev. Lett. **44**, 420 (1980).

[11]W. E. Spicer, P. W. Chye, P. R. Skeath, C. Y. Su, and I. Lindau, J. Vac. Sci. Technol. **16**, 1422 (1979).

[12]O. F. Sankey, R. E. Allen, and J. D. Dow, Solid State Commun. **49**, 1 (1984).

[13]O. F. Sankey, R. E. Allen, and J. D. Dow (unpublished).

[14]D. R. Hamann, M. Schluter, and C. Chiang, Phys. Rev. Lett. **43**, 1494 (1979).

[15]J. P. Perdew and A. Zunger, Phys. Rev. B **23**, 5048 (1981).

[16]G. Lehmann and M. Taut, Phys. Status Solidi B **54**, 469 (1972).

[17]Our accuracy is ± 0.2 eV and is limited by the number of k points used in the self-consistence cycles. A calculation with energy cutoff 3 Ry and six k points in the two-dimensional irreducible Brillouin zone shows this result is stable to within 0.1 eV, reducing the limitation to ± 0.1 eV.

[18]E. Louis and F. Yndurain, J. Phys. C **6**, L465 (1973).

[19]R. T. Tung, Phys. Rev. Lett. **52**, 461 (1984).

[20]C. B. Duke and C. Mailhiot (unpublished).

[21]C. A. Mead, Appl. Phys. Lett. **6**, 103 (1965).

Ruthenium-induced surface states on n-GaAs surfaces

Matthias Ludwig and Günter Heymann[a)]
Department of Electronics, Humboldt-University of Berlin, Invalidenstr. 110, 1040 Berlin, German Democratic Republic
Peter Janietz
Department of Chemistry, Humboldt-University of Berlin, Bunsenstr. 1, 104 Berlin, German Democratic Republic

(Received 3 June 1985; accepted 10 November 1985)

Studies were conducted to investigate the properties of the Ru–GaAs interface. An analysis was made of two sets of experimental findings, namely the distribution of surface states of GaAs covered by submonolayers of ruthenium, as well as of electrical measurements on Ru–GaAs contacts. It was found for the present case that a deep state near $E_t \approx 0.3$ eV above valence band maximum is induced at the surface. Thus the influence of the intrinsic defects is at least compensated. Using a simple model, it was investigated to which degree surface states and their energy would govern the barrier height. It was further studied how Fermi level pinning depends on the doping concentration of the semiconductor.

I. INTRODUCTION

In recent years a substantial amount of investigations involving the barrier formation on metal–semiconductor junctions have been published. The technological importance of Schottky barrier formation on III–V semiconductors has long been recognized. However, a fundamental understanding of the interface structure and device properties of the Schottky barrier height is still incomplete.

According to the Schottky model[1] as the simplest starting point to describe the system, the barrier height is equal to the difference between the metal work function and the electron affinity of the semiconductor. Therefore the value of the Schottky barrier height should depend on the contact metal used. In contrast to that we can see that the barrier height is largely independent of the type of metal electrodes regarding covalent compound semiconductors as, e.g., GaAs.[2–5] The Fermi level at the p- or n-type GaAs–metal interfaces becomes pinned with a position about 0.6 or 0.8 eV, respectively, above the valence band maximum (VBM).

To explain these deviating characteristics, interface states are under consideration which lie energetically within the forbidden band gap and heavily modify the Schottky relation.[6] Among the processes which can cause the existence of interface states the influence of defects is considered predominant.[3,5,7,8] The unified defect model (UDM) of Spicer and co-workers[5] proceeds from the assumption that native defects at or near the semiconductor surface are created by the interaction of either metal or oxygen beams with the surface. The defects associated with vacancies, e.g., gallium and arsenic vacancies in GaAs, cause surface states with energy levels within the band gap. Because surface defect concentrations as small as $\approx 0.1\%$ can produce pinning, only submonolayer coverages are sufficient for a fixed band bending.

However, some (noble and transition) metals on n-GaAs[9–12] cause an E_F pinning position which falls at that of the donor defects (VBM + 0.5 eV) according to the UDM.[5] This is in contrast to an electrostatic point of view predicting an acceptor state as responsible for pinning. Therefore, the unusual properties of the Ag–GaAs interface were interpreted in terms of Ag-derived interface states[9] or attributed to states induced by screened Ag clusters.[10]

A similar deviating reaction was observed at the Ru–GaAs system. The barrier height of n-GaAs electrolytical junctions changes significantly after a surface treatment of GaAs with Ru species[13,14] and a modified surface state distribution was determined.[15] Other results concerned with the reduction of the surface recombination velocity[16] or barrier height variations of Ru–GaAs contacts upon exposure to hydrogen[11] indicate further irregularities of this system.

Our objective is to analyze the extent to which the interface states and their different energy levels determine the barrier height and which dependencies on the doping concentration of the semiconductor exist. For this purpose a model development by Landsberg and Abrahams[17] was transferred to the situation of the n-GaAs surface. The barrier height model is based on an analytical solution of the Poisson equation which jettisons the depletion approximation by the introduction of parallel quasi-Fermi levels.

For comparison, investigations were made with a view to getting results about the specific electronic conditions of the Ru–GaAs interface. It is suggested from the analysis of energetic surface state distributions at real GaAs surfaces after treatment with Ru species as well as from I–V and C–V measurements at Ru–GaAs contacts that through system-conditioned interactions a deep state with $E_t \approx 0.3$ eV above VBM is induced at the interface. The origin of the state is attributed to disturbed As bonds of the host.

II. BARRIER HEIGHT MODEL

We have studied the general dependencies between the barrier height, the doping concentration, the density of acceptor surface states, and their position in the forbidden gap. By adopting the model of Landsberg *et al.*[17] we are confined, in contrast, to n-type GaAs and emphasize only the main phases.

Proceeding from the Poisson equation we regard the bulk

concentration on acceptors as negligible and so we presuppose a nondegenerated system:

$$\frac{d^2\phi(z)}{dz^2} = \frac{e}{\epsilon\epsilon_0}\left[n(z) - p(z) - N_d^+(z)\right]. \tag{1}$$

Here N_d^+ is the concentration of charged donors, n and p are electron and hole concentrations, and ϵ is the dielectric constant $[\epsilon\epsilon_0 = (12.85\times 8.85)\times 10^{-14}$ As V^{-1} cm^{-1} for GaAs]. N_d^+, n, and p were defined by introducing quasi-Fermi levels $F_{n,p}$ in a similar manner as done in Ref. 17.

The necessary boundary conditions to integrate Eq. (1) are given by

$$\left.\frac{d\phi}{dz}\right|_{z=w} = 0; \text{ with } \phi(w) = 0 \tag{2}$$

at the end of the space-charge region and

$$\left.\frac{d\phi}{dz}\right|_{z=0} = -Q_{ss}/2\epsilon\epsilon_0; \text{ with } \phi\ (0) = \phi_s \tag{3}$$

at the surface of the crystal.

Let us assume that the charge per area in the surface states Q_{ss} is due to one discrete surface acceptor level E_{AS} with state density N_{AS} and that these values are independent of bulk doping.

It follows that

$$Q_{ss} = -eN_{AS}(E_{AS})f_s, \tag{4}$$

where the occupation probability f_s of the surface states for charge carriers can be expressed when limiting on stationary conditions by using the Shockley–Read relation.[18] Through the integration of Eq. (1) within the limits of the space-charge region we obtain a transcendental equation for the surface potential or built-in potential $V_{bi} = (kT/e)\ln\mu(0)$ given by

$$n(w)[1-\mu(0)] + p(w)[1-\mu(0)^{-1}] + N_d\left[\ln\mu(0) - \ln\left(\frac{1+\mu(0)\exp\{[E_d(w)-F_n]/kT\}}{1+\exp\{[E_d(w)-F_n]/kT\}}\right)\right] = -\frac{e^2}{8\epsilon\epsilon_0 kT}[N_{AS}(E_{AS})f_s]^2, \tag{5}$$

where k is the Boltzmann constant and T is the temperature. Numerical calculations using an iterative routine were performed to extract V_{bi} and according to

$$V_B = V_{bi} - kT/e - V_{Fc} - \Delta F, \tag{6}$$

the barrier height V_B was derived. In this expression V_{Fc} is the difference between the Fermi level and the bulk conduction band edge and ΔF is an image force lowering correction.[19] All calculations were performed for 300 K.

In Fig. 1 we show the position of the surface Fermi level versus the density of surface acceptors with doping as a parameter. This is consistent with a submonolayer coverage. The chosen energy level of the surface state ($E_{AS} = 0.75$ eV) corresponds to that of acceptor defects.[3,5] The relevance of the second energy level will be shown below. In accordance with the UDM suggested by Spicer *et al.*[5] for a doping of 1×10^{17} cm^{-3} the Fermi level position is almost stabilized when the state density exceeds a value of about 2×10^{12} cm^{-2}. Approximately 3 times the state density is required to pin the Fermi level for a doping of 1×10^{18} cm^{-3}. The onset of pinning is nearly independent of the selected positions of the surface states in the band gap.

Owing to different mechanisms of interaction a 1 : 1 ratio between surface state and coverage density should not be expected. Therefore, a considerably large amount of weakly bonded adatoms may cause only a small band bending. Especially for high doping concentrations, the shifting of the Fermi level will be only a weak function of the coverage density (cf. Fig. 5 of Ref. 10).

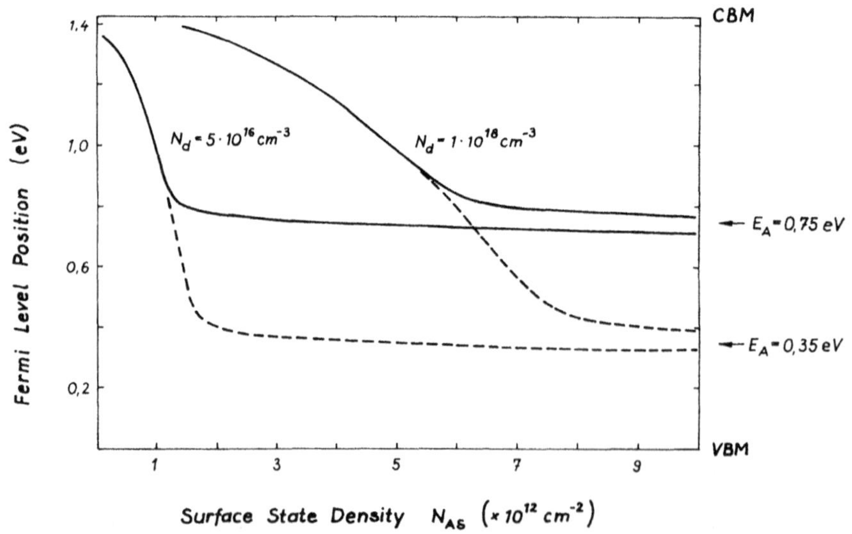

FIG. 1. Position of the Fermi level E_F at the surface vs density of surface acceptors N_{AS} with a single energy level $E_{AS} = 0.75$ and 0.35 eV above VBM for various doping concentrations N_d of the n-type GaAs substrate.

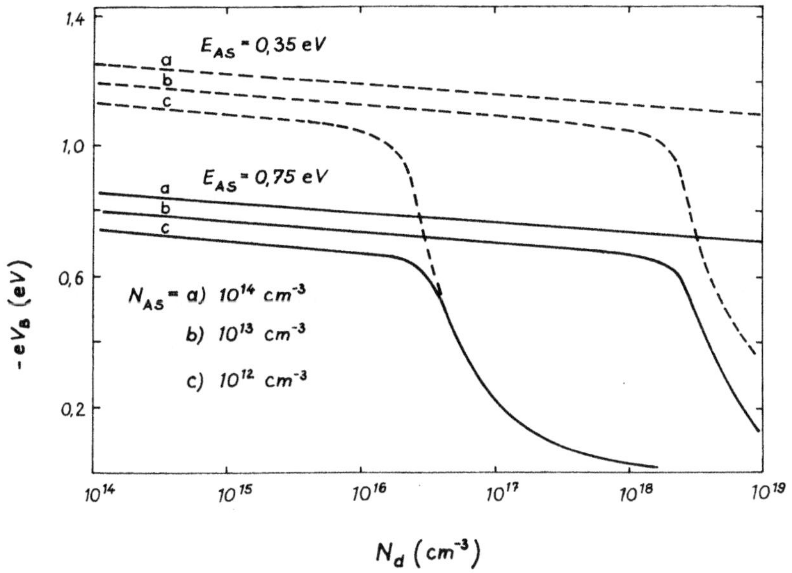

FIG. 2. Electrostatic barrier height eV_B as a function of the donor concentration N_d of the GaAs substrate with the density of surface states N_{AS} at $E_{AS} = 0.35$ and 0.75 eV above VBM as parameter.

The electrostatic barrier heights shown in Fig. 2 as a function of the doping concentration with the state density of surface acceptors as a parameter exhibit a comparatively low variation, as far as the state density lies in the range of a monolayer. This means that the Fermi level is completely pinned. On the other hand lower state densities will be compensated by increasing doping concentrations, which leads to a sharp decrease of the barrier height.

In conclusion, we find that even for a discrete acceptor state on the n-GaAs surface a definite value of the barrier height cannot be given. The barrier height depends on the state density as well as on the doping concentration. This may be one cause for the scatter of Schottky barrier data of individual metals.

III. EXPERIMENTAL CONDITIONS

The GaAs used in the present study was (100) oriented, Si- or Te-doped n-type material with carrier concentrations of 6×10^{16} and 1×10^{18} cm^{-3}. After alcohol cleaning the chemo-mechanically polished front surfaces were etched in an ultrasonic HCl bath to eliminate inhomogeneous oxide layers. Stripping in 3 : 1 : 1 NH_3 : H_2O_2 : H_2O was carried out to remove polishing residues. After rinsing in deionized water the surface was finally treated with 1 : 10 HCl : ethanol for removing arsenic and gallium oxides.[20]

For studying the effect of ruthenium on the GaAs surface ruthenium atoms were deposited by electrochemical methods. As described by Parkinson *et al.*[13–15] dipping of the prepared GaAs surface into solutions of $RuCl_3$ results in an irreversible deposition of ruthenium. The observed coverages are at a submonolayer level. For this treatment ~0.01 M ruthenium solutions were prepared in 0.1 M HNO_3 from $RuCl_3$.

The examination of the Ru concentration of treated GaAs surfaces was carried through with the help of AES and RBS analyses.

Based on the method of the field-modulated large-signal photovoltage the band bending of treated and untreated surfaces was evaluated by measuring photovoltage pulses during strong flashes of light. The photovoltage pulses appearing between a potential-sensing electrode before the surface and a part of the sample in thermodynamic equilibrium were used for the determination of surface potentials. The surface potential Y was thereby modulated by means of and measured as a function of an applied field voltage V_f which was varied at each cycle between 0 and 1000 V. Following Lam[21] the energetic distribution $N_{ss}(E)$ of surface states can be deduced by

$$eN_{ss}(E) = C_i\left(\frac{dV_f}{dY} - 1\right) - \frac{dQ_{sc}}{dY}, \qquad (7)$$

where C_i is the insulator capacitance of the detachable MIS structure and Q_{sc} the charge inside the space-charge layer, both per unit area. A detailed description of the experimental configuration used here and the method of evaluation is given by Heilig *et al.*[22,23] Cleaned samples with and without dipping in Ru solutions were carefully dried and analyzed with respect to how adsorbed ruthenium influences the results.

Real contacts for current–voltage (I–V) and capacitance–voltage (C–V) measurements were made by electrodeposition of about 300-nm-thick Ru layers. The plating solutions (an amidosulfonic acid system) also contained $RuCl_3$. The samples were prepared as follows. After Ohmic contacts were generated by annealing of evaporated AuGe-metallizations on the back surface of the wafers the samples were covered by a photoresist mask. The mask on the front surface was structured to obtain free circular areas with di-

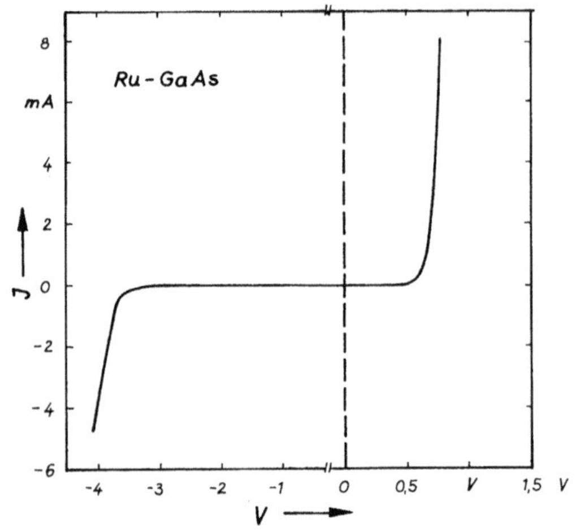

FIG. 3. Rectifying I–V behavior of galvanically deposited Ru contacts on n-GaAs.

ameters of 100–900 μm. The deposition of ruthenium was performed in an electrolytic cell under galvanic separation from the amidosulfonic acid system. The adhesive contacts showed reproducible properties and were characterized at room temperature. Heat treatments were avoided to preserve the initial interface for investigation without thermally stimulated reactions known for other transition metals on GaAs.

The I–V data were obtained in forward bias and analyzed in terms of the thermionic field emission model of current transport.[24] For $V > 3kT/e$ the current density can be expressed as

$$j = A^{*}T^{2}\{\pi R\,[e(V_B - V) - (E_F - E_c)]/kT\cosh^2 R\}^{1/2} \times \exp - \left\{ \frac{\tanh R}{kTR}[e(V_B - V) - (E_F - E_c)] + (E_F + E_C)/kT \right\} \tag{8a}$$

with the dimensionless material parameter

$$R = \frac{h}{4\pi kT}(N_d/m_n\epsilon\epsilon_0)^{1/2}. \tag{8b}$$

In this equation j stands for current density, A^{x} is the effective Richardson constant ($A^{x} = 8.16\ \text{A cm}^{-2}\,\text{K}^{-2}$ for GaAs), m_n is the conduction band effective mass ($m_n = 0.068m_e$ for GaAs), h is Planck's constant, E_F is the Fermi level, and E_c stands for the conduction band minimum. The I–V data were fit to Eq. (8) for forward biases using an iterative calculating routine to extract V_B. The empirical ideality factor n determined according to[19]

$$n = \frac{2}{kT}\frac{\partial V}{\partial(\ln I)} \tag{9}$$

was used as a further indicator for the dominating transport mechanism.

The C–V data (1 MHz) were obtained over a reverse bias range of 0 to -2 V. The characteristics were analyzed as plots of $(A/V)^2$ vs V. According to the depletion model of an abrupt, ideal contact, these quantities are related for an n-type semiconductor by[19]

$$(A/C)^2 = 2(V_{\rm bi} - V - kT/e)/e\epsilon\epsilon_0 N_d, \tag{10}$$

where A is the contact area. The slope and voltage intercept of the plot yield N_d and $V_{\rm bi} - kT/e$. Use was made of Eq. (6) to evaluate the barrier height V_B.

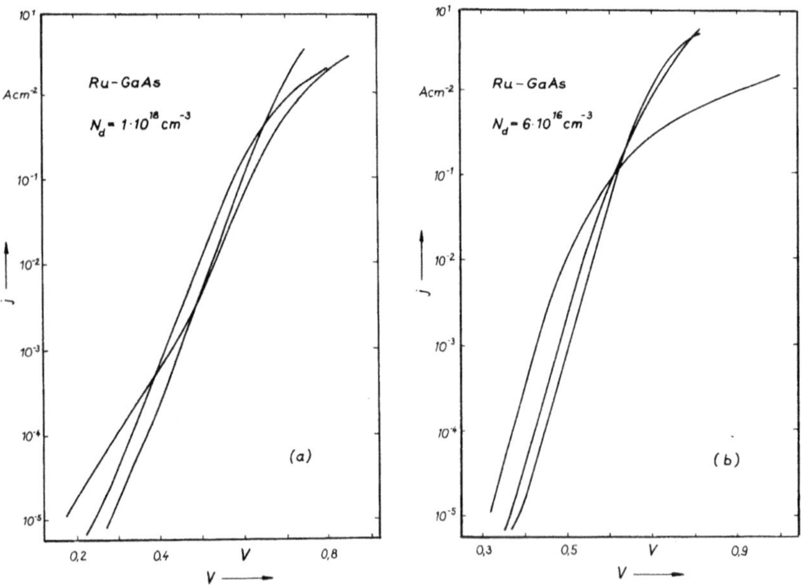

FIG. 4. Detailed forward I–V characteristics of Ru–GaAs contacts: (a) $N_d = 1\times10^{18}\ \text{cm}^{-3}$, (b) $N_d = 6\times10^{16}\ \text{cm}^{-3}$.

TABLE I. Electrical properties of Ru contacts on (100) n-GaAs.

Samples	eV_B^{I-V} (eV)	n	eV_B^{C-V} (eV)	N_d^{C-V} (cm^{-3})
Series A	1.17	1.44	1.24	1.3×10^{18}
Series B	1.09	1.03	1.15	6.4×10^{16}

IV. RESULTS

Figure 3 shows the $I-V$ characteristics of an electrodeposited Ru–GaAs contact on a linear scale. The rectifying behavior is accompanied by the onset of forward conduction at about 0.5 V independent of the contact diameters. A linear $I-V$ dependence in case of large forward bias as an indicator for thin intermediate insulating layers is not pronounced. Some detailed $I-V$ characteristics are given on a logarithmic scale in Fig. 4. The linear portions of the plots were assessed according to Eq. (8). On all n-type GaAs samples of the doping range used, the Ru–GaAs contacts have been found by us to be strongly rectifying. Table I summarizes the measured parameters V_B and n.

For the lower doped samples the values of the ideality factor are very close to the ideal value of 1.02. Therefore the thermionic emission model of current transport is appropriate and the presence of intermediate insulating layers may be ruled out.

Otherwise, we found higher n values for the doping range of 10^{18} cm^{-3}. Even if we cannot give a detailed statement at present concerning how much recombination processes contribute to an additional current density with low forward bias, we conclude from the good conformity of the $I-V$ and $C-V$ barrier heights that thermionic field emission dominates $n > 1.02$.

Figure 5 shows the $C-V$ characteristics of electrodeposited Ru contacts. The determined values of V_B and N_d are listed in Table I. The submonolayer coverage with ruthenium observed after dipping prevents direct $I-V$ measurements. We have found, however, that adsorbed Ru atoms on

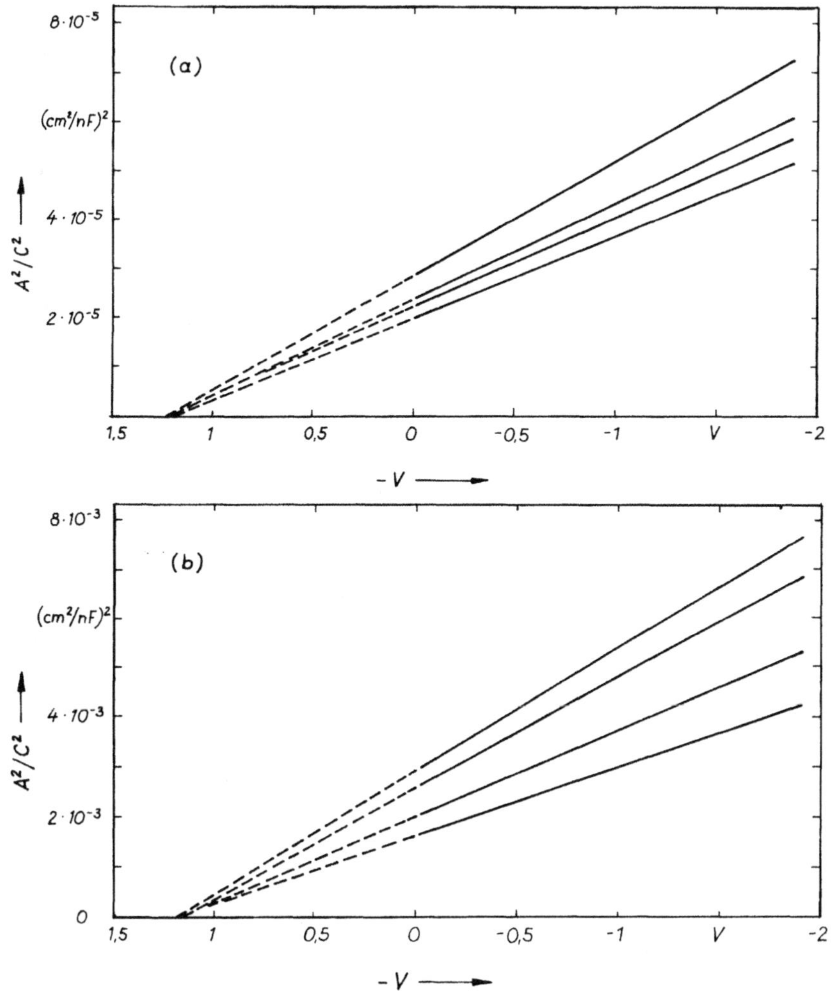

FIG. 5. $C-V$ characteristics of Ru–GaAs contacts: (a) $N_d = 1\times10^{18}$ cm^{-3}, (b) $N_d = 6\times10^{16}$ cm^{-3}.

the GaAs surface act as activators for an electroless (autocatalytic) gold plating process. The electroless gold plating on GaAs is known to require an activation step with normally $PdCl_2$-containing solutions.[25,26] Following the principal procedure as described for electrodeposited Ru contacts electroless gold plating was carried through on Ru-activated, and for comparison, on Pd-activated areas of the surface. In both cases all conditions except those of the activator solutions have been kept constant.

The $I–V$ and $C–V$ data of such Au contacts on GaAs with thin intermediate Ru or Pd submonolayers result in very different V_B values which are summarized in Table II. The barrier height of $eV_B = 0.83$ eV for the Pd-activated surface corresponds to known values measured at thick (bulk) Pd contacts on GaAs.[11,12] On the contrary the Au (Ru)–GaAs system shows a V_B value which is similar to that of galvanically made Ru contacts.

The influence of submonolayer Ru coverages on the distribution of fast states in the forbidden gap at GaAs surfaces can be seen in Fig. 6. For the chemically cleaned GaAs surface we obtained nearly U-shaped distributions which have a minimum at $E - E_i \approx -0.3$ eV with a concentration of about 1×10^{11} cm^{-2} eV^{-1}.

The attachment of ruthenium at the surface by dipping causes a general increase of the surface state density and a lower possibility to modulate the surface potential by the applied field voltage.

Longer treatments result in an increasing disorder of the surface region which was seen from RBS yields. Furthermore, we have found from AES depth profiling that gallium ions go into solution, leaving an arsenic-rich surface. This partial consumption of the semiconductor surface is consistent with an anodic corrosion reaction of GaAs, while the deposition of ruthenium occurs on local cathodic sites.[14] As dipping is therefore accompanied by a structural disorder of the surface which gives rise to an increasing surface state density, the dipping time or coverage density has to be minimized for getting reliable potential modulations.

From the spectra with differing coverage densities (shown in Fig. 6 for two Ru concentrations) it can be seen that the minimum shifts towards E_i and the density of surface states increases at $E < E_i$ simultaneously. To fit the spectrum the tailing distributions of intrinsic band states into the gap have been approached over a Boltzmann function while superimposed Gaussian distributions are used to describe an additional extrinsic defect center.[27] At best conformity by the adoption to the experimental data we assume from the peak positions the presence of a deep state at $E_t \approx 0.35$ eV above VBM. Additionally, the use of the Gaussian function allows the determination of the densities of the

TABLE II. Electrical properties of Au contacts on (100) n-GaAs ($N_d = 1\times10^{18}$ cm^{-3}) with intermediate Ru or Pd layers.

Metallization Structure	eV_B^{I-V} (eV)	n	eV_B^{C-N} (eV)
Au(Pd)–GaAs	0.83	1.30	0.88
Au(Ru)–GaAs	1.14	1.28	1.12

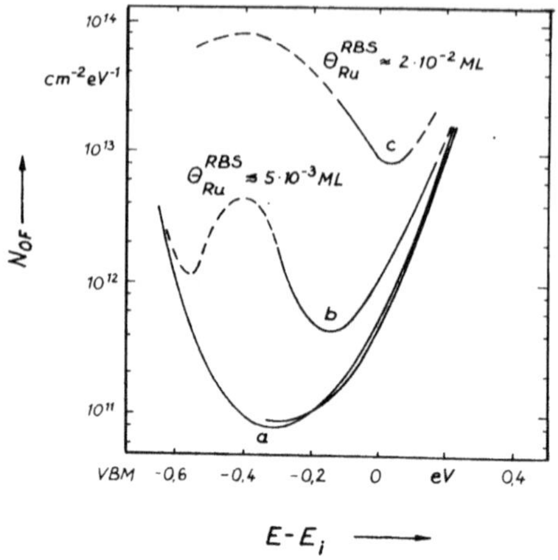

FIG. 6. Distribution of fast states at chemically cleaned surfaces of n-GaAs (curves a and b) and after dipping in Ru-containing solutions (c and d). RBS was used for determination of Ru coverage densities in terms of monolayers (ML). Dashed curves display a hypothetical distribution of states fitted to experimental values.

states which corresponds with Ru concentrations found by RBS. Therefore, it seems reasonable to assume that the variations in the distributions of the surface states are directly related to adsorbed ruthenium on the GaAs surface causing an extrinsic state in the gap. Because the structural disorder for longer dipping treatments broadens the Gaussian distribution an extrapolation to higher coverage densities is limited.

V. DISCUSSION

For GaAs surfaces it is known that the general distribution $N_{ss}(E)$ of fast surface states is continuous throughout the forbidden band gap. The states are attributed to be of intrinsic character by showing a minimum about the middle of the gap and an increase toward CBM and VBM.[28] As described by model calculations[27] it is believed that states of this type result from bond distortions at the GaAs boundary causing a scatter of a number of volume states into the band gap.

A result of the effective mass two-band approximation[27] is that the minimum of N_{ss} should appear at an energy E_0 according to

$$E_0 - E_i = (1 - c/1 + c)E_g/2\,, \tag{11}$$

with E_g as the band gap and $c^2 = m_p/m_n$ as the ratio of the density of effective masses. For GaAs with $c = 2.71$ follows a value $E_0 - E_i = -0.33$ eV, which is in satisfactory accordance with the experimental results of the chemically cleaned surface given above.

As the photovoltage measurements were carried through with samples which were exposed to dried air the presence of an oxide layer is evident. The nonoccurrence of deep states around E_i may be interpreted by only weak distortions of the chemical bonding in the surface during oxidation. This results in a minimized density of defect states. The effect of ruthenium atoms deposited on the surface may then be explained by the additional appearance of extrinsic states at $E_t = 0.3$–0.4 eV above VBM.

If we summarize the V_B values of the Ru–GaAs interface including thick (bulk) film and intermediate layer measurements, the barrier height model gives a pinning position of 0.3 eV above VBM. This is in good agreement with the photovoltage results.

No essential differences for V_B were observed between thick metal layers and submonolayer coverages as pointed out by Zur, McGill, and Smith.[29] The calculated defect density of $\approx 10^{14}$ cm^{-2} required for pinning in the case of coverages by a thick metal[29] is in contrast to not only our barrier height model, but also the Schottky barrier data of other noble metals[12] showing the same pinning position for thin and thick (bulk) coverages.

Although the consistency of the data suggest an additional interface state introduced by the presence of ruthenium on the GaAs surface, the origin of such a state remains unexplained. The pinning position cannot be interpreted within the context of the unified defect model.[5]

For a first approximation we have tried to explain the nature of the Ru-induced state by assuming a direct interaction between ruthenium and arsenic on an atomic scale. This approach is supported by the observed interfacial chemical reactions of refractory metals with arsenic.[12]

Proceeding from the sp^3 model of the bulk electronic structure we used the solid state table and the matrix elements given by Harrison[30] as well as the orbital parameters of ruthenium[31,32] to find out the energy level of a Ru–As bond.[33] In the model we consider a Ru atom substituting for Ga and the resulting molecule consists of the defect Ru and its As neighbors. The calculated energy eigenvalue[30] was related to the GaAs structure and the values from the two-band calculations were simply subtracted.

If with the inclusion of ruthenium into the surface region of GaAs such a bond is built up, we get from the computations an energy level with $E_t = 0.26$ eV above VBM considering the strong tendency to preserve bond lengths in the crystal. The appearing level in the band gap is a bonding state which is mainly built up by the fraction of the disturbed electronic charge associated with the anion. Therefore we find that the energy of the level is determined primarily by the host As atoms surrounding the Ru impurity due to a shifted valence band state and not by the impurity itself.

By applying the same arguments to other metals we calculated the energies of induced levels for Au, Ag, and Cu on GaAs.[33] Table III summarizes the resulting values as well as the barriers according to our barrier height model. The predicted pinning positions are coincident with a wide range of experimental data[5,9,12,19,34] and emphasize the influence of the chemical composition of the interface. It is noteworthy that in all cases the modified bonds of arsenic are dominant for the defect level creation. Only the slight but systematic shift is due to the individual metals. We suggest that because there is only a weak interaction between the (near) noble metals and GaAs this leads to a modification of the semiconductor bonds without releasing the large amount of energy

TABLE III. Calculated defect levels and barrier heights for various metals on n-GaAs ($N_d = 5 \times 10^{17}$ cm^{-3}).

Metal on GaAs	E_t (eV) above VBM	eV_B (eV)
Cu	0.57	0.86
Ag	0.55	0.88
Au	0.50	0.93
Ru	0.26	1.17

necessary for the formation of pinning defects according to the unified defect model.[5] Therefore, the defect model is not incompatible with our results, if one takes into account the experimental conditions of the interface creation (e.g., deposition rates) and/or the formation kinetics. While the deposition over the gas phase is carried through far from equilibrium conditions for the condensate, we come to the conclusion that with the electrolytic deposition the released energy is lower accompanied by a specific chemical interaction between the substrate and the metal atom.

As a consequence, the energy released during the deposition process is decisive for the relative weight of defects due to chemically induced bond distortions or defects according to the unified defect model. So, for instance, large evaporation rates favor the creation of the latter one leading to metal independent pinning positions. In this context the lower barrier heights of 0.90 and 0.98 eV found by Aspnes and Heller[11] for the Ru–GaAs contact system from I–V and C–V measurements, respectively, may be caused by the electron beam evaporation of the metal. Otherwise, very low deposition rates preserve the specific interactions as, e.g., reported by Ludeke *et al.*[9,10] for the Ag–GaAs system with $eV_B = 0.89$ eV, which coincides with our prediction. Thus we assume a transition of the Fermi level pinning to occur between the different defect levels depending on the densities of the different types of defects. This may be a further cause for the scatter of Schottky barrier data.

In summary, we have presented results on the Schottky barrier formation of Ru on the GaAs (100) surface which can be explained in terms of chemically induced interface states. These states depend only weakly on the individual metal and are attributed to disturbed arsenic bonding states. Values for the Schottky barrier of about 1.1 eV on n-type GaAs were obtained for different coverage densities. Schottky barrier heights calculated for other noble metals suggest a similar mechanism of barrier formation.

ACKNOWLEDGMENTS

The authors wish to thank Dr. H. Flietner, Dr. K. Heilig, and Dr. H. Kerkow for helpful support.

[a)] Temporary address: University of Florida, Department of Materials Science and Engineering, Gainesville, Florida 32611.

[1] W. Schottky, Z. Phys. **113**, 367 (1939).
[2] J. M. Woodall and J. L. Freeouf, J. Vac. Sci. Technol. **19**, 794 (1981).
[3] R. H. Williams, J. Vac. Sci. Technol. **18**, 929 (1981).
[4] J. L. Freeouf, Appl. Phys. Lett. **41**, 285 (1982).

[5]W. E. Spicer, I. Lindau, P. Skeath, and C. Y. Su, J. Vac. Sci. Technol. **17**, 1019 (1980); W. E. Spicer, I. Lindau, P. Skeath, C. Y. Su, and P. Chye, Phys. Rev. **44**, 420 (1980).
[6]J. Bardeen, Phys. Rev. **71**, 717 (1947).
[7]R. W. Grant, J. R. Waldrop, S. P. Kowalczyk, and E. A. Kraut, J. Vac. Sci. Technol. **19**, 477 (1981).
[8]A. Nedoluha, J. Vac. Sci. Technol. **21**, 429 (1982).
[9]R. Ludecke, T. C. Chiang, and D. E. Eastman, J. Vac. Sci. Technol. **21**, 599 (1982).
[10]R. Ludeke, T.-C. Chiang, T. Miller, J. Vac. Sci. Technol. B **1**, 581 (1983).
[11]D. E. Aspnes and H. Heller, J. Vac. Sci. Technol. B **1**, 602 (1983).
[12]J. R. Waldrop, S. P. Kowalczyk, and R. W. Grant, J. Vac. Sci. Technol. **21**, 607 (1982); J. R. Waldrop, J. Vac. Sci. Technol. B **2**, 445 (1984).
[13]A. Parkinson, A. Heller, and B. Miller, J. Electrochem. Soc. **126**, 954 (1979).
[14]K. W. Frese, M. J. Madou, and S. R. Morrison, J. Electrochem. Soc. **128**, 1939 (1981).
[15]P. Janietz, J. Electroanal. Chem. Interfacial Electrochem. **106**, 23 (1980).
[16]A. Heller, *IUPAC Frontiers of Chemistry* (Pergamon, Oxford, 1982), p. 27.
[17]P. T. Landsberg and M. S. Abrahams, J. Appl. Phys. **55**, 4284 (1984).
[18]V. L. Bonč-Bruevič, S. G. Kalasnikov, *Halbleiterphysik* (Wissenschaften, Berlin, 1982).
[19]S. N. Sze, *Physics of Semiconductor Devices* (Wiley, New York, 1981).
[20]R. P. Vasquez, B. F. Lewis, and F. J. Grunthaner, J. Vac. Sci. Technol. B **1**, 791 (1983).
[21]Y. Lam, J. Phys. D **4**, 1370 (1971).
[22]K. Heilig, H. Flietner, and J. Reinecke, J. Phys. D **12**, 927 (1979).
[23]K. Heilig, Solid-State Electron. **27**, 395 (1984).
[24]F. A. Padovani and R. Stratton, Solid-State Electron. **9**, 695 (1966).
[25]L. A. D'Asaro, S. Nakahara, and Y. Okinako, J. Electrochem. Soc. **127**, 1935 (1980).
[26]M. Hannemann, Feingerätetechnik **31**, 309 (1982).
[27]H. Flietner, Surf. Sci. **46**, 251 (1974).
[28]E. W. Kreutz, Phys. Status Solidi A **56**, 687 (1979).
[29]A. Zur, T. C. McGill, and D. L. Smith, Phys. Rev. B **28**, 2060 (1983).
[30]W. A. Harrison, *Electronic Structure and the Properties of Solids* (Freeman, San Francisco, 1980).
[31]S. Fraga, J. Karwonski, and K. M. S. Saxana, *Handbook of Atomic Data* (Elsevier, Amsterdam, 1976).
[32]R. Latter, Phys. Rev. **99**, 510 (1955).
[33]M. Ludwig and G. Heymann (in preparation).
[34]W. E. Spicer, S. Pan, D. Mo, N. Newman, P. Mahowald, T. Kendelewicz, and S. Eglash, J. Vac. Sci. Technol. B **2**, 476 (1984).

PHYSICAL REVIEW B VOLUME 33, NUMBER 6 15 MARCH 1986

Metallization and Schottky-barrier formation

Inder P. Batra*
IBM Zurich Research Laboratory, CH-8803 Rüschlikon, Switzerland

S. Ciraci†
IBM Research Laboratory, San Jose, California 95193

(Received 22 October 1985; revised manuscript received 2 December 1985)

Results of self-consistent electronic-structure calculations of aluminum deposited on Ge, as a function of metal coverage, provide strong evidence for overlayer metallization. At monolayer metal coverages, the overlayer changes into a (quasi-) two-dimensional metal characterized by a (modulated) ladder-type density of states. The origin of this transition is traced to the metal-semiconductor interlayer-distance relaxation upon metallization. Our results are in agreement with recent experimental observations, and have important implications for Schottky-barrier formation.

In spite of the recent progress made towards developing a microscopic understanding of the metal-semiconductor interfaces[1-8] and Schottky-barrier (SB) formation, the nature and origin of the electronic states responsible for the Fermi-level pinning are still not fully known. This is due to a large number of complicating factors which influence the Fermi-level position. The pinning, for example, may depend on geometrical structure, metal coverage and reactivity, chemical composition, doping and defect levels, interdiffusion, and surface orientation. It is perhaps a useless exercise to attempt to find a universal pinning mechanism.

Surface states invoked[9-11] at metal-semiconductor interfaces for pinning the Fermi level, in some cases, are eliminated[8] from the gap region owing to the surface reconstruction, and/or owing to the formation of chemisorption bonds.[12,13] The metal-induced states as put forward by Heine[14] require at least several metal layers on the semiconductor surface, and as such are not necessarily the pinning states because a SB gets established[15] at much lower coverages. It is now evident that in many metal-semiconductor systems, the sublayer and multilayer metal coverages lead to different pictures of the interface. The multilayer stage was investigated by Louie and Cohen[16] for an Al-Si junction. They were able to distinguish several types of interface states including metal-induced gap states identified as the pinning states. On the other hand, the binding configuration of metal atoms and chemisorption states have been emphasized in studies[17,18] involving monolayer coverages. In contrast to the chemisorption picture relevant at low coverages, Zunger[19] has proposed an alternate weakly interacting metal-cluster model for Al deposited on GaAs(110).

We believe that, depending on a particular metal-semiconductor system, more than one mechanism may be responsible for determining the position of the Fermi level (E_F) relative to the semiconductor band gap. More importantly, for a given system one may switch from one mechanism to another depending upon the coverage, especially if E_F is not strongly pinned. A crucial parameter[20] which governs the growth sequence may also play a significant role here. This is the so-called γ parameter which is the ratio of the interfacial binding energy to the binding energy of the overgrowth particles. If γ is of the order of unity, then at low coverages one may see the formation of chemisorption bonds. With increased coverage, the overlayer might metallize and relax outwards, resulting in weakening of chemisorption bonds and eventually to the preponderance of metallic states at the Fermi level. Such a picture has emerged from our Al-Ge prototype calculations and is qualitatively consistent with observations[21] on the Al-Si system; both systems have $\gamma \simeq 1$.

In this paper, we have studied the transition from the chemisorbed to metallic overlayer by calculating self-consistent charge density and local density of states for a commensurate Al-Ge interface. Our results present strong support[21] for the metallization of the overlayer, and its origin is the interlayer distance relaxation upon metallization. Furthermore, from this analysis we can bring together a number of concepts about SB formation, which are known in isolation, but have not been previously presented in a consolidated form. Our calculations are performed within the framework of local-density functional theory, applied in the momentum space formalism[22,23] using *ab initio* pseudopotentials.[24] The metal-semiconductor system is represented by a repeating slab geometry; other calculational details can be found elsewhere.[25]

The total energy calculations[20,25] for the prototype Al(001)-Ge(001) have favored the bridge site; it is about 1 eV lower in energy than the on-top site. For the bridge site, it was further noted that the metal-semiconductor interlayer distance, $R_\perp$, relaxed outwards significantly in going from half a monolayer (one Al atom per unit cell—$\theta = 0.5$) to a full monolayer (two Al atoms per unit cell—$\theta = 1$) coverage of metal. It should be noted that our calculations have used an ideal arrangement of the substrate atoms. There is some evidence[26] from low-energy electron diffraction experiments for Ag-Ge(001) that the fractional order spots do indeed disappear at $\theta = 0.5$. Moreover, the general conclusions drawn here are not sensitively dependent on such details of the calculations. Important physical consequences of the change in the value of $R_\perp$ in connection with metallization emerge upon performing charge-density and local-density-of-state (LDOS) analysis. Therefore, in Fig. 1 we show our calculated pseudocharge densities in the $(\bar{1}10)$ plane passing through Al and Ge atoms for the two coverages, $\theta = 0.5$ and 1. The charge densities were calculated self-consistently at equilibrium distances,[20,25] $R_\perp = 1.2$ Å for $\theta = 0.5$ and $R_\perp = 1.8$ Å for $\theta = 1$. The atomic positions are indicated below the charge-density contours.

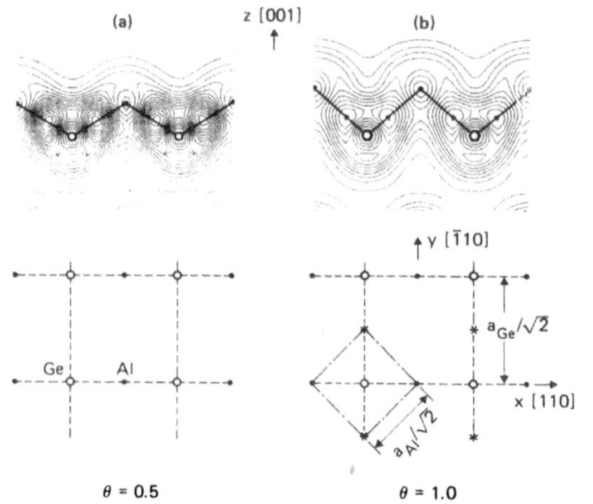

FIG. 1. Charge-density contours in the ($\bar{1}$10) plane of the Al(001)-Ge(001) interface corresponding to aluminum in bridge-bonded sites for $\theta=0.5$ and 1. The contour spacings are 0.005 a.u. Atomic arrangements are also shown for the two coverages; the stars and filled circles represent Al atoms, and open circles Ge atoms.

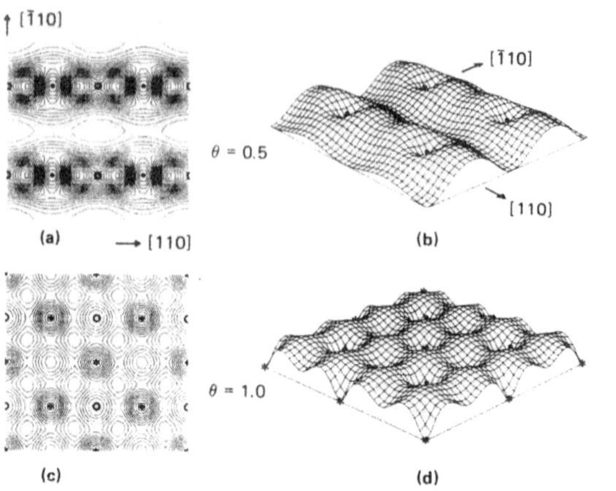

FIG. 2. Charge-density contours (spacings 0.0025 a.u.) and the surface maps in the (001) plane passing through aluminum atoms of the Al-Ge interface corresponding to aluminum in bridge-bonded sites for $\theta=0.5$ and 1. Notice the transition from anisotropic chemisorption state to metallization of the overlayer.

contours. The maximum charge density for $\theta=0.5$ reaches a value of 0.079 a.u. in the middle of the Al–Ge bond direction; the corresponding value for $\theta=1$ is only 0.056 a.u. suggesting a weaker bond. It is thus clear that at the higher coverage, the chemisorption bond has weakened and more charge is oozing out into the vacuum [see Fig. 1(b)]. This indicates the onset, where the directional covalent bonds begin to be replaced by more delocalized metallic-type[21] interaction. The chemisorption bonds are still visible at this coverage, but the metallic bonding can be clearly seen from the charge distribution. We may mention that in going to $\theta=1$, we have placed the additional Al atoms in an epitaxial relationship. The epitaxial geometry is found to be adequate from our force calculations leading to small forces acting on metal and semiconductor atoms at the interface.

The metallization of the aluminum overlayer is displayed more dramatically in Fig. 2. Here, we have plotted charge-density contours in the (001) plane passing through aluminum atoms. Also shown are the corresponding charge-density surfaces. For $\theta=0.5$, there is a valley parallel to the [110] direction showing lack of interaction between aluminum atoms along the [$\bar{1}$10] direction. However, the aluminum atoms located along the [110] direction show highly directed interaction mediated through the substrate. It is important to recall that the interatomic distances in both directions are identical. Hence, at the low coverage we see the formation of directional bonds responsible for such anisotropic behavior. The maximum charge-density contour has a value of 0.048 a.u. and occurs at a distance of $a_{Al}/4$ away from aluminum atoms along the [110] direction. The complete metal overlayer ($\theta=1$) shows a more isotropic charge density as is evident from Figs. 2(c) and 2(d). Maxima in charge density are now located along [010] directions, at halfway between nearest-neighbor Al atoms. The aluminum layer shows a tendency towards metallization, and chemisorption bonds with the substrate have weakened.

In Fig. 3, the LDOS at the Al overlayer for the two coverages under consideration are shown. New major peaks appear, in addition to the usual structure of the bulk Ge, and are connected with the interactions at the interface. At $\theta=0.5$, the Ge–Al–Ge bridge bonds are produced by the bonding combination of Ge-dangling bonds and Al sp^2 orbitals. The chemisorption states appear near the Fermi level and also at about 3 eV below it. The states connected with one sp^2 orbital overlap with the chemisorption bond states near the Fermi level, and contribute to the pinning at this coverage. States having the Al-$3p_y$ character would couple the metal atoms along the [$\bar{1}$10] direction, but since these states are unoccupied, the Al overlayer displays a nearly one-dimensional (1D) character. For $\theta=1$, the LDOS may be viewed as a modulation on top of the ladder-type density

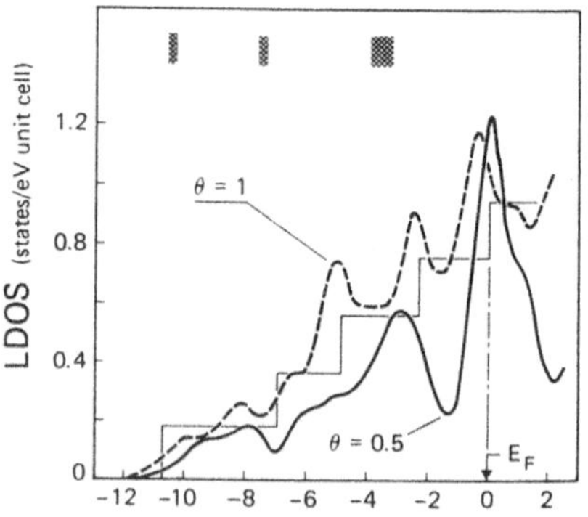

FIG. 3. LDOS at the Al layer for $\theta=0.5$ and 1, calculated by appropriately partitioning the slab. The ladder-type density of states is for the quasi-2D system with the same areal charge density as for $\theta=1$. The shaded zones indicate major structure of bulk Ge.

The maximum charge density for $\theta = 0.5$ reaches a value of 0.079 a.u. in the middle of the Al–Ge bond direction; the corresponding value for $\theta = 1$ is only 0.056 a.u. suggesting a weaker bond. It is thus clear that at the higher coverage, the chemisorption bond has weakened and more charge is oozing out into the vacuum [see Fig. 1(b)]. This indicates the onset, where the directional covalent bonds begin to be replaced by more delocalized metallic-type[21] interaction. The chemisorption bonds are still visible at this coverage, but the metallic bonding can be clearly seen from the charge distribution. We may mention that in going to $\theta = 1$, we have placed the additional Al atoms in an epitaxial relationship. The epitaxial geometry is found to be adequate from our force calculations leading to small forces acting on metal and semiconductor atoms at the interface.

Our proposal to reconcile diverse viewpoints in SB formation is simply that the SB, at least in metal-covalent semiconductors, is fully developed only after a deposition of one to two monolayers[21] of metals and not at submonolayer coverages. With this ansatz, we can make the following observations. At submonolayers of unsupported metal films,[25] macroscopic metallic properties, e.g., work function, are drastically different from the bulk values. At one to two monolayers, on the other hand, we see[25] an evolution towards bulk properties. Inasmuch as it matters which metal is deposited[29] on a substrate for SB formation, one should at least have one to two monolayers of metal. Also, in those cases where nondirectional interactions are sought,[19] a metallized phase is a more likely candidate, rather than a chemisorbed layer.

In a recent study,[30] it was found that a seven-layer thick Al film exhibited a quasi-2D behavior. The states showed a free-electronlike solution in the (x,y) direction, but decayed into vacuum from both surfaces of the film. A similar behavior is expected from the states of a metal overlayer: The metal states may decay into the semiconductor, providing a preponderance of empty states across the gap. Consequently, without invoking either defect states or metal-induced states as described by Heine, a thin metal layer (one to two layers) is already capable of providing pinning states. We have thus also presented a theoretical confirmation of the novel finding[21] that in the Al-Si case the metallic character appears at coverages above one monolayer. The origin is traced to the interfacial distance relaxation upon metallization for the metal-covalent semiconductor system. The pinning states at one to two monolayers can also be thought of as a hybrid[16,29] of Bardeen[9] and Heine's[14] pictures, in which both the intrinsic surface states as well as the metal-induced states play a crucial role. Bardeen's picture by itself is inapplicable when intrinsic states are absent, e.g., in GaAs, and Heine's picture requires thick metal overlayers unnecessary for the SB formation.

We are grateful to Dr. A. Baratoff, Dr. E. Courtens, and Dr. K. H. Rieder for many constructive comments on the manuscript.

---

*Permanent address: IBM Almaden Research Center, San Jose, CA 95120-6099.

†Permanent address: Department of Physics, Middle East Technical University, Ankara, Turkey.

[1] *Thin Films and Interfaces*, edited by P. S. Ho and K. Tu (North-Holland, New York, 1982); *Thin Films and Interfaces*, edited by J. E. E. Baglin, D. Campbell, and W. K. Chu (North-Holland, New York, 1984).

[2] C. Calandra, O. Bisi, and G. Ottaviani, Surf. Sci. Rep. **4**, 271 (1984).

[3] R. Z. Bachrach, in *Metal-Semiconductor Schottky Barrier Junctions and their Applications*, edited by B. L. Sharma (Plenum, New York, 1984).

[4] G. W. Rubloff, in *Festkörperprobleme*, edited by J. Treusch, Advances in Physics, Vol. XXIII (Vieweg, Braunschweig, 1983), p. 179.

[5] A. Zunger, Thin Solid Films **104**, 301 (1983).

[6] R. H. Williams, Contemp. Phys. **23**, 329 (1982).

[7] L. J. Brillson, Surf. Sci. Rep. **2**, 123 (1982); M. Schlüter, Thin Solid Films **93**, 3 (1982).

[8] W. E. Spicer, I. Lindau, P. R. Skeath, and C. Y. Su, J. Vac. Sci. Technol. **17**, 1019 (1980); O. F. Sankey, R. E. Allen, S.-F. Ren, and J. D. Dow, *ibid.* B **3**, 1162 (1985).

[9] J. Bardeen, Phys. Rev. **71**, 717 (1947).

[10] A. Deneuville and B. K. Chakraverty, Phys. Rev. Lett. **28**, 1258 (1972); D. E. Eastman and J. L. Freeouf, *ibid.* **34**, 1624 (1975).

[11] E. J. Mele and J. D. Joannopoulos, Phys. Rev. B **17**, 1528 (1978).

[12] J. C. Phillips, J. Vac. Sci. Technol. **11**, 947 (1974).

[13] J. E. Rowe, S. B. Christman, and G. Margaritondo, Phys. Rev. Lett. **35**, 1471 (1975).

[14] V. Heine, Phys. Rev. **138**, A1689 (1965).

[15] G. Margaritondo, J. E. Rowe, and S. B. Christman, Phys. Rev. B **14**, 5396 (1976).

[16] S. G. Louie and M. L. Cohen, Phys. Rev. Lett. **35**, 866 (1975); Phys. Rev. B **13**, 2461 (1976).

[17] J. R. Chelikowsky, Phys. Rev. B **16**, 3618 (1977); H. I. Zhang and M. Schlüter, *ibid.* **18**, 1923 (1978).

[18] J. E. Northrup, Phys. Rev. Lett. **53**, 683 (1984).

[19] A. Zunger, Phys. Rev. B **24**, 4372 (1981).

[20] I. P. Batra, Phys. Rev. B **29**, 7108 (1984).

[21] M. K. Kelly, E. Colavita, G. Margaritondo, J. Anderson, L. Papagno, D. J. Frankel, and G. J. Lapeyre, Phys. Rev. B **32**, 2693 (1985).

[22] J. Ihm, A. Zunger, and M. L. Cohen, J. Phys. C **12**, 4409 (1979).

[23] M. T. Yin and M. L. Cohen, Phys. Rev. Lett. **45**, 1004 (1980); K. C. Pandey, *ibid.* **49**, 223 (1982).

[24] G. B. Bachelet, D. R. Hamann, and M. Schlüter, Phys. Rev. B **26**, 223 (1982); see also, D. R. Hamann, M. Schlüter, and C. Chiang, Phys. Rev. Lett. **43**, 1494 (1979).

[25] I. P. Batra and S. Ciraci, Phys. Rev. B **29**, 6419 (1984); J. Vac. Sci. Technol. B **3**, 427 (1984).

[26] J. R. Lince, J. G. Nelson, and R. S. Williams, J. Vac. Sci. Technol. B **1**, 553 (1983).

[27] F. K. Schulte, Surf. Sci. **55**, 427 (1976).

[28] P. Chaudhari, H.-U. Habermeier, and S. Maekawa, Phys. Rev. Lett. **55**, 430 (1985).

[29] J. Tersoff, Phys. Rev. B **32**, 6968 (1985); J. Vac. Sci. Technol. B **3**, 1157 (1985).

[30] S. Ciraci and I. P. Batra, this issue, Phys. Rev. B **33**, 4294 (1986).

On the Present Understanding of Schottky Contacts

Winfried Mönch

Laboratorium für Festkörperphysik, Universität Duisburg, D-4100 Duisburg, Federal Republic of Germany

Summary: In 1874 F. Braun observed that "bei einer großen Anzahl natürlicher und künstlicher Schwefelmetalle ... der Widerstand derselben verschieden war mit Richtung, Intensität und Dauer des Stromes". W. Schottky then, in 1938, explained the rectifying behaviour of such metal-semiconductor contacts by a depletion layer which is characterized by the "Austrittsarbeit der Überschuß- oder der Defektelektronen an der Grenze Metall-Halbleiter". Mott and most probably also Schottky postulated that the barrier height, as it is called now, equals the difference of the work function of the metal and the electron affinity of the semiconductor. Experimentally, the barrier heights are not found to obey this simple rule. The deviations were first assigned to electronic interface states by Bardeen. Following a later proposal of Heine's, the metal wave-functions tail into the virtual gap states of the semiconductor and by this determine the barrier height at the metal-semiconductor interface. This model very well describes all chemical trends observed with the barrier heights of metal-semiconductor contacts.

1 Historical Introduction

"Bei einer großen Anzahl natürlicher und künstlicher Schwefelmetalle und sehr verschiedenen Stücken, sowohl Krystallen von so vollkommener Ausbildung, wie ich überhaupt bekommen konnte, als derben Stücken habe ich gefunden, daß der Widerstand derselben verschieden war mit Richtung, Intensität und Dauer des Stromes. Die Unterschiede betragen bis zu 30 pCt. des ganzen Werthes. ... Ich habe benutzt Quecksilbercontact, stark gegen gepreßte Kupfer-, Platin- und Silberdrähte und endlich bei einem Stück eine bereits vorhandene Fassung mit dicken Neusilberbügeln ..."*

With these words Ferdinand Braun summarized his observations "On the electrical transport in sulfides" which he published in 1873 [1] and which described the rectifying behaviour of metal-semiconductor junctions for the first time. He investigated such compound semiconductors as chalcopyrite, iron pyrite, galena, and fahlore while Adams [2] and Siemens (see [3]) then confirmed these observations of unipolar current flow, which is a strong deviation from Ohm's law, with selenium.

* "In a large number of natural and synthetic sulfides and with very different samples, crystals so perfectly shaped as I could get them as well as rough pieces, I have found that their resistance varies by up to 30 % depending on direction, intensity, and duration of the current ... I have used mercury contacts, wires of copper, platinum, and silver, which were heavily pressed against the samples, and a mounting made of German silver with another piece ..."

It took another 40 to 50 years until metal-semiconductor contacts were applied as solid-state rectifiers in the then rapidly growing field of broadcasting. The "cat's whisker" rectifiers, which consisted of a metal point pressed against a crystal usually of lead sulfide and still were typical devices of the type first investigated by Braun, were finally replaced by plate rectifiers based on cuprous oxide and later on selenium. In 1925, the first patent for cuprous-oxide plate rectifiers was issued to Grondahl [4] in the US. However, despite the successful applications of solid-state rectifiers a physical understanding of the basic phenomena in metal-semiconductor junctions was still lacking until the end of 1938.

Already in his very early papers, Braun [1, 3] carried out a number of experiments to find a physical explanation for the rectifying properties of metal-semiconductor contacts which he had discovered. He excluded thermoelectric as well as polarization effects and concluded that the "anomalous phenomena" are intimately related to the contact itself and must originate from a very thin surface layer of extremely high resistance. He hesitated to attribute his observations to "films of air which, even when they are not electrically conductive, might play a secondary, quasi catalytic role" since he had obtained too many results which clearly contradicted this opinion [3].

In 1929 a huge step forward was made by Schottky and his coworkers [5]. By using electrostatic probes they scanned the potential gradient along a cuprous-oxide rectifier biased in reverse direction and demonstrated that the voltage drop occurs across a very narrow blocking layer at the Cu-Cu_2O interface. The width of this layer was estimated from the differential capacitance measured as a function of applied voltage. Schottky and Deutschmann [5] found the differential capacitance to decrease with inverse but to increase with forward bias applied. They modelled the blocking layer by a parallel-plate condenser, i.e.

$$C = \epsilon_s \epsilon_0 \frac{F}{d} . \qquad (1)$$

Here, F and d are the area and the distance of the two "condenser plates" and ϵ_s is the dielectric constant of the semiconductor. The capacitance C and thus the distance d depend on the applied voltage U_a. From the capacitance measured at zero bias they obtained $d(0) = 3 \cdot 10^{-5}$ cm. Already at that time, they definitely excluded a transport by tunneling, since the barrier was that wide, and they concluded that the conduction mechanism in the blocking layer has to be similar to the one in bulk Cu_2O. However, the transport mechanism has to be changed such that the resistivity is drastically increased when the rectifier is biased in reverse direction, as had already been inferred by Braun. Schottky and Deutschmann already discussed the possible presence of a space charge in the blocking layer but they could not arrive at the correct model for several reasons.

First of all, the quantum theory of semiconductors was given by Wilson only in 1931 [6], and it was then that the concept of defect electrons was introduced. Second, the two semiconductors most important at that time, cuprous oxide and selenium, were only poorly characterized. For Cu_2O the correct, positive sign of the Hall coefficient was determined only in 1934 [7] while even in 1940 no such measurements had been carried out for selenium (see [13]). The p-type character of Se was only concluded from the observation that both Cu_2O- and Se-based rectifiers needed the same polarity of applied voltage. And third, the difference between chemical and physical depletion had to be understood. Cu_2O rectifiers were prepared by oxidizing a copper base plate and, as a consequence, the composition of the oxide film grown varied with distance from the interface. Close to the Cu-Cu_2O boundary the oxide is stoichiometric and thus poorly conducting while the "bulk" of the oxide film contains an excess of oxygen which causes p-type doping. Selenium rectifiers, on the other hand, showed an almost constant doping level of acceptors, the nature of which was unknown.

In their correspondence, Schottky and Peierls had already arrived at the concept, that in a rectifier biased in reverse direction the barrier layer is depleted from mobile carriers, in 1932. It took, however, until the end of 1938 that Schottky published his "Halbleitertheorie der Sperrschicht" [8]. At the same time, Mott [9] treated the case of the chemical depletion layer in Cu_2O rectifiers. The long and "thorny" development of experimental results and theoretical concepts that finally led to the correct description of the barrier layer in metal-semiconductor junctions was traced back in detail by Schottky himself in his first extended paper "Zur Halbleitertheorie der Sperrschicht- und Spitzengleichrichter" [10].

2 On the Schottky Barrier

The band scheme close to the interface in an intimate, abrupt, and rectifying contact between a metal and a semiconductor, which is homogeneously doped n-type, is schematically shown in Fig. 1. Since the work Φ_{Bn} needed to excite an electron

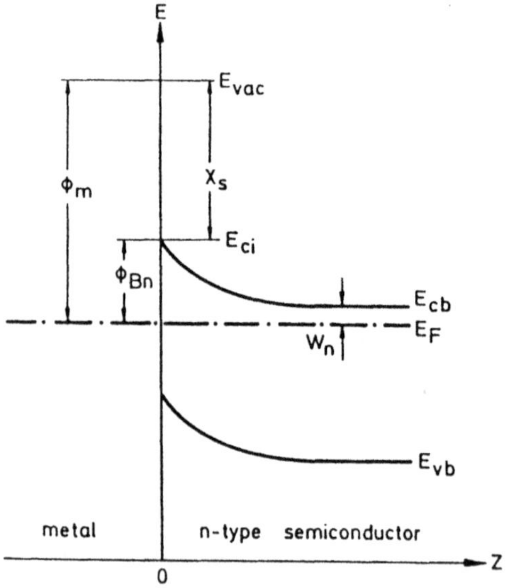

Fig. 1
Schematical band scheme of a Schottky-Mott barrier in a rectifying metal-semiconductor contact.

from the metal into the semiconductor is larger than the distance W_n between the Fermi level and the conduction-band bottom in the bulk of the semiconductor, the bands are bent "upwards". Consequently, close to the interface the semiconductor is depleted of mobile electrons in the conduction band. This depletion layer has a large resistance, and it carries a space charge which is made up by charged donors which are no longer balanced by electrons. With a negative (positive) bias applied to the semiconductor relative to the metal the barrier is decreased (increased), and by this the resistivity is lowered (increased). In honor of Walter Schottky who first described the blocking behaviour of metal-semiconductor contacts by a depletion layer [8, 10] it is called the Schottky barrier.

With the assumptions that the semiconductor only contains shallow donors of density N_D and that even the tail of mobile electrons at the bulk side of the depletion layer is neglected the space charge in the depletion layer is easily obtained as a function of applied voltage U_a by solving Poisson's equation as

$$Q_{sc} = \{2\epsilon_s \epsilon_0 N_D e_0 (V_D + U_a)\}^{1/2}. \tag{2}$$

The differential capacitance of the space charge layer then results as

$$C = \frac{dQ_{sc}}{dU_a} = \left\{\frac{\epsilon_s \epsilon_0 e_0 N_D}{2(V_D + U_a)}\right\}^{1/2}. \tag{3}$$

This means that $1/C^2$ is expected to vary proportional to $(V_D + U_a)$.

The success of Schottky's approach is obviously demonstrated by the experimental data given in Fig. 2. From these plots not only the doping of the selenium films used but also the band bending V_D at the interface could be determined.

As Fig. 2 reveals the interface band-bending depends on the metal used in the rectifier. This behaviour was already known from an extensive study by Schweikert [14] which was never published but was cited by Schottky [15]. Schweikert had prepared metal-selenium rectifiers with 21 different metals and had measured their resistance R_0 at zero bias. As Fig. 3 shows, he found R_0 to decrease exponentially over four decades as a function of the work function of the metals used. In his theory of transport in metal-semiconductor junctions, Schottky assumed thermio-

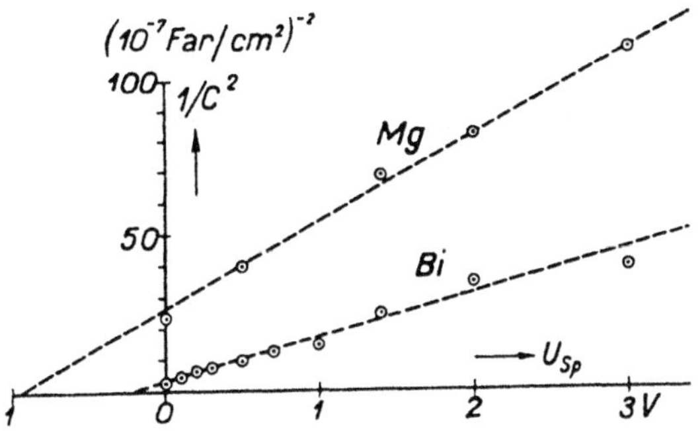

Fig. 2
Differential capacitance of metal-selenium rectifiers as a function of applied voltage. From [13].

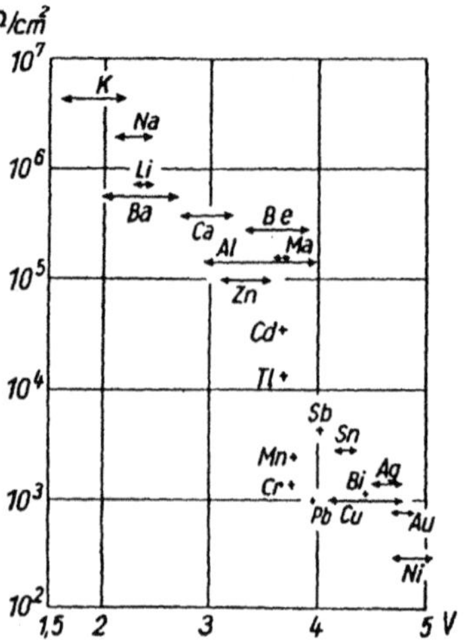

Fig. 3

Resistance at zero bias as measured with metal-selenium rectifiers plotted versus the work functions of the metals used. From [15].

nic excitation of carriers over the interface barrier V_D, and this then gives the zero-bias resistance as

$$R_0 \sim V_D^{1/2} \exp(e_0 V_D/k_B T). \tag{4}$$

Schweikert's results then mean that the barrier height in metal-selenium contacts linearly depends on the work function of the metal.

In his first "Note on the contact between a metal and an insulator or semi-conductor" Mott [9] had proposed that the "work necessary to take an electron from the metal into the conduction band of the insulator is $\Phi_m - \chi_s$ where Φ_m is the work function of the metal. We denote by χ_s the work necessary to remove an electron from the lowest level of the conduction band to a point outside the crystal". He had assumed that the vacuum levels of the metal and the semiconductor adjust when forming a metal-semiconductor contact. A possible dipole layer at the interface is neglected in this approach. With the present notations this means

$$\Phi_{Bn} = \Phi_m - \chi_s, \tag{5a}$$

$$\Phi_{Bp} = I - \Phi_m, \tag{5b}$$

for semiconductors doped n- and p-type, respectively. Here, $I = E_g + \chi_s$ is the ionization energy of the semiconductor. The chemical trend demonstrated by Schweikert's results as plotted in Fig. 3 is clearly confirmed by Eq. (5b), but the barrier heights obviously vary less as a function of metal work-function than suggested by Mott's model. Schweikert's experimental data may, however, be explained when the relation between the barrier height and the metal work function is modified as

$$\Phi_{Bp} = S_\Phi (I - \Phi_m). \tag{6}$$

From Fig. 3 the slope S_Φ then results as approximately 0.1 to 0.5 for selenium.

Relation (5) is not to be found in any of Schottky's early publications although his interpretations of Schweikert's experimental results imply a relationship similar to Eq. (5) between the barrier height of a metal-semiconductor contact and the work function of the metal used. The reason for this might be that Schottky had realized that a relation like (6) is more appropriate. This, however, is mere speculation. Nevertheless, Eq. (5) is referred to as the Schottky-Mott relation.

Already with metal-selenium rectifiers, chemical interactions between metals and the semiconductor were detected, and their influence on the electronic properties was studied in detail. Here, the most prominent example of cadmium on selenium shall be mentioned only. At elevated temperatures, an intermediate layer of the compound CdSe forms which is an n-type semiconductor. The layered structure Cd-CdSe-Se exhibits improved rectifying properties: The threshold voltage is increased by 0.1 V from 0.45 V to 0.56 V, and the breakdown voltage now amounts to 14.5 V compared with 5.7 V measured with no intermediate CdSe layer [16]. But it is physically more important that the rectifying barrier is now found at the interface of the semiconductor heterostructure n-CdSe : p-Se rather than at the metal-semiconductor contact Cd : CdSe. This was first shown by Poganski [17].

3 On the Barrier Height: Some Experimental Data

Approximately 30 years ago, selenium-based rectifiers have lost technical importance, and it is since then that both the scientific and the technical interests have focussed on silicon and the III–V compound semiconductors. With the rapid development of surface science many new experimental tools became available and have been applied to studies of metal-semiconductor contacts. However, the problems dealt with have not changed since the discovery of rectifying metal-semiconductor junctions. Especially, the surface-sensitive measurements have provided a huge body of detailed experimental data but many of the investigations are case studies. Thus, for a long time none of the more general theoretical concepts seemed to apply. In the following chapters, I will first present some of the experimental results which I consider to be typical examples. Readers, who are interested in more details, are referred to some of the recent reviews [18, 19, 20]. Finally, I will then discuss some of the models which have been proposed to explain the experimentally observed barrier heights.

As an example, Fig. 4 shows experimental barrier heights measured with metal-silicon junctions plotted versus the metal work-functions [21]. Such plots are motivated by relation (5). The contacts were all prepared by evaporating a specific metal onto silicon (111) surfaces which were cleaved in situ and held at room temperature. Obviously, the experimental data do not follow the Schottky-Mott relation (5). However, the experimental data may be fitted to a straight line which has a slope $S_\Phi = 0.15$ (see Eq. (6)).

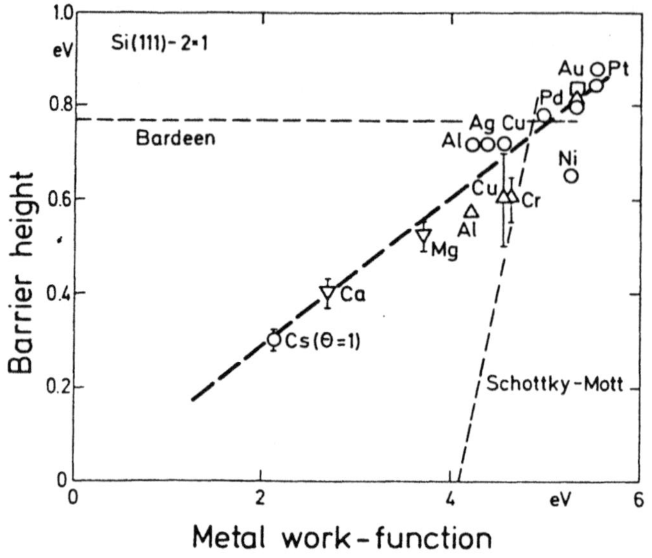

Fig. 4 Barrier heights measured with metal contacts prepared by evaporation on silicon(111) surfaces cleaved in situ. After [21].

Kurtin et al. [22] were the first to compile and analyse the barrier heights measured with Schottky barriers on some 20 semiconductors and insulators. They evaluated the slopes $S_X = d\Phi_{Bn}/dX_m$, where X_m is the electronegativity of the metal. Gordy and Thomas [23] empirically found a linear relationship between the work functions and the electronegativities of metals which has been revised [24] based on a compilation of newer work-function data [25]:

$$\Phi_m = 1.79\,X_m + 1.11\,\text{eV}. \tag{7}$$

The slopes S_X as recently evaluated by Schlüter [26] are plotted in Fig. 5 versus the electronegativity difference of the atoms constituting the compound semiconductors and insulators. The sometimes large error bars indicate that the original data exhibit much scatter in these cases and straight-line fits then become quite questionable. However, Fig. 5 shows a chemical trend. For small differences in electronegativity of the atoms in the compounds, the slope S_X is quite low but at $\Delta X \approx 0.8$eV a steep increase of S_X is observed. Whether or not a saturation at $S_X = 1$, as originally proposed by Kurtin et al. [22], or at $S_X = 1.5$ to 2.0, as evaluated by Schlüter [26] occurs, cannot be clearly decided from the experimental data presently available, although a value of $A = 1.79$ is more plausible (see Eqs. (7) and (13)). This point will be discussed at the end of the present paper.

The barrier heights considerably vary when the same metal is deposited on different semiconductors. As an example, barrier heights measured with gold/semiconductor contacts are plotted in Fig. 6 versus the energies of the fundamental,

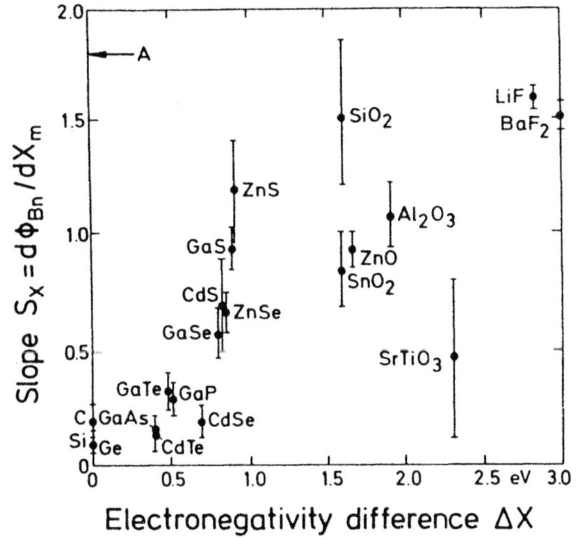

Fig. 5
Slopes $S_X = d\Phi_{Bn}/dX_m$, as obtained from straight-line fits to Φ_{Bn}-versus-X_m plots, as a function of the electronegativity difference of the atoms constituting the semiconductors. After [26].

i. e. optical, and the indirect band gaps of the semiconductors. The barrier heights have been recently compiled [27]. It is clearly seen that neither the $\Phi_{Bp} = 1/3\,E_g$ rule [29] nor the common-anion rule [30], which assumes Φ_{Bp} to be same for compounds sharing the same anion, are obeyed by the experimental data. While the barrier heights do not systematically vary with the energy of the fundamental band gaps, which are direct ones for some of the semiconductors, they show a good but non-linear correlation with the indirect band-gap energies. This finding will be again discussed in a later chapter.

All of the earlier barrier heights have been obtained from I/V or C/V characteristics measured with device-like diodes. With the development of surface science, however, the formation of metal-semiconductor junctions could be followed under clean and reproducible conditions by sequential depositions onto a semiconductor surface cleaned in situ prior to the first deposition. Already in 1970, the change of surface conductivity of cleaved silicon surfaces was measured as a function of cesium coverage [31]. The change of the position of the Fermi level as a function of cesium coverage is shown in Fig. 7. Already some 10^{12} cesium atoms deposited per cm^2 were found to be sufficient to pin the Fermi level at 0.3 eV below the conduction-band bottom. The position of the Fermi level $E_{cs} - E_F$, which corresponds to the barrier height $\Phi_{Bn} = E_{ci} - E_F$ in a Schottky barrier, as measured at monolayer coverage was also entered in Fig. 4. This data point almost perfectly fits the straight line which was obtained by a least-square fit to the barrier heights determined from device-like diodes with thick metal films. From this finding it was concluded that "the properties of the metal-semiconductor contact are already determined after the deposition of the first monolayer of the metal on the clean

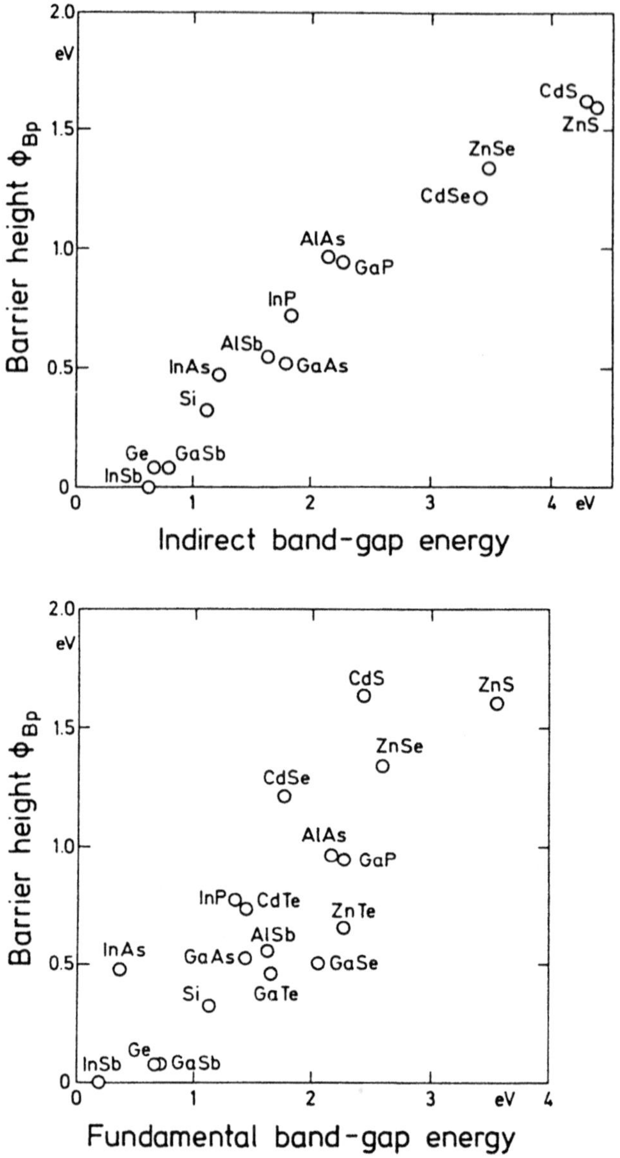

Fig. 6
Barrier heights measured with gold-semiconductor contacts plotted over the widths of the fundamental (below) and of the indirect band gaps of the semiconductors (above).

semiconductor surface" [21]. This finding is well established by now. Furthermore, it was concluded that the deposition of cesium annihilates the intrinsic surface states of the clean silicon surface and the Fermi level is then pinned by adsorption-induced surface states [21].

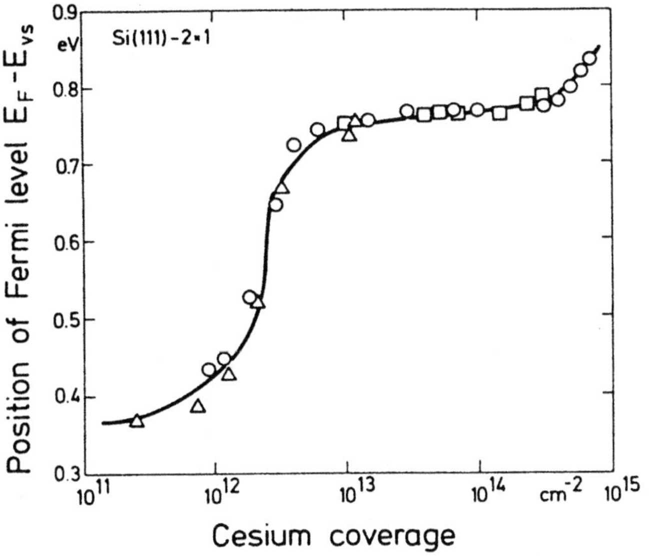

Fig. 7
Position of the Fermi level relative to the top of the valence band as a function of cesium coverage on a silicon(111) surface cleaved in situ. After [31].

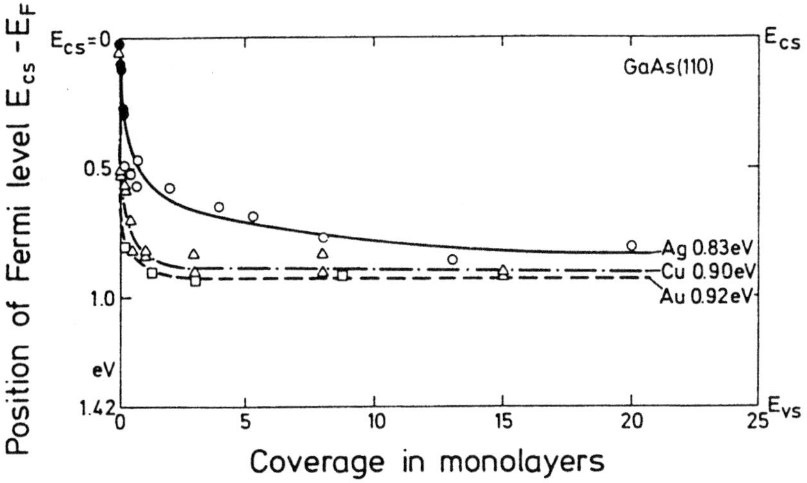

Fig. 8 Position of the Fermi level relative to the bottom of the conduction band on n-type GaAs as a function of noble-metal coverage. After [32].

As another example, Fig. 8 shows results obtained with the noble metals Cu, Ag, and Au deposited on cleaved GaAs(110) surfaces [32]. Here, the position of the Fermi level was determined by photoemission spectroscopy. At clean GaAs(110) surfaces the bands are flat up to the surface since the reconstruction has shifted the occupied surface states to below the top of the valence band and the empty ones to above the conduction-band bottom (for a review see [33]). The depositions of the metal atoms induce surface band-bending such that the Fermi level is now positioned within the band gap, i. e. depletion layers have formed. Since with GaAs samples doped p-type depletion layers are also found, the metal adsorption induces surface states in the gap of both acceptor and donor type. With Au and Cu coverages well below a monolayer the Fermi level reaches a constant position in the band gap which does not change even under thick overlayers. When Ag is deposited the final position of the Fermi level, which is almost identical with those observed with Cu and Au, is approached much slower as a function of nominal coverage. This is attributed to the pronounced formation of islands which initially leave part of the surface uncovered and coalesce only at larger coverages.

The interaction of these three metals with GaAs is quite different. From photoemission spectroscopy of substrate and deposite core-levels it was found that Cu and Au strongly interact with the GaAs substrate, which results in alloy formation ([32] and references therein), while the Ag:GaAs(110) interface is mostly abrupt and no intermixing is observed [34, 35]. Nevertheless, the final band bending at the interface under thick overlayers is almost the same for the three so differently reacting metals. This means that details of the interface chemistry must not necessarily play an important role in determining the barrier height in a metal-semiconductor junction.

As an example for interface reactions, aluminium on cleaved GaAs(110) surfaces shall be briefly considered. Fig. 9 shows Al(2p) and Ga(3d) core-level spectra excited with 130 eV photons from the clean GaAs surface and after successive depositions of Al [36]. It is clearly seen that at low coverages the Al(2p) signal is observed at binding energies 0.7 eV larger compared with thick and then definitely

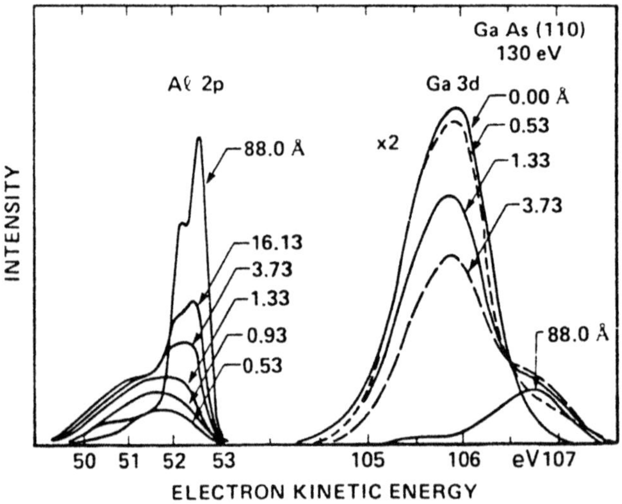

Fig. 9
Core level photoemission spectra excited by 130eV photons of chemically shifted Al(2p) and Ga(3d) core levels for various Al-overlayer thickness on GaAs(110). From [36].

77

metallic films while the Ga(3d) line shows a shoulder shifted by 0.95 eV to smaller binding energies. This is easily understood by the formation of AlAs and metallic Ga by a cation exchange at the interface. Such reaction is to be expected since AlAs has a larger formation energy than GaAs, 29.3 versus 19.5 kcal/mole. The chemical shifts observed with the core levels are also easily to be explained. Due to the larger electronegativity of arsenic compared to the one of Ga and Al, charge is transferred from the metal to the non-metal atoms in the partly ionic III–V compounds. This charge transfer reduces the screening of the core potential of the metal atoms, and their core electrons are thus more tightly bound in the III–V compounds compared with the neutral atoms in their metallic form (see [37] and references given therein).

4 On the Barrier Height: Models

Several models have been proposed to explain the deviations of the barrier heights observed in metal-semiconductor contacts from the most simple Schottky-Mott model which resulted in relation (5).

First, Bardeen [38] has proposed that the surface states of the clean surface might persist under a metal overlayer and are then responsible for the pinning of the Fermi level. This was indicated by the horizontal line in Fig. 4. However, while clean silicon surfaces exhibit a large density of surface states in the fundamental band gap the intrinsic surface states at clean and well-cleaved surfaces of III–V compound semiconductors are shifted to below the top of the valence and to above the bottom of the conduction band due to a reconstruction of these surfaces. Despite the fact that the atomic arrangement changes during the formation of interfaces, Heine [39] has argued that under metal overlayer the surface states of the clean surface have vanished. However, it is owing to Bardeen that surface or interface states are considered to be responsible for the position of the Fermi level at metal-semiconductor interfaces.

Second, Heine [39] has discussed that at metal-semiconductor interfaces the metal wave-functions will tail into the semiconductor in that energy range where the conduction band of the metal overlaps the band gap of the semiconductor. The tails of the metal wave-function then "occupy" virtual gap states (ViGS) of the complex band structure of the semiconductor which are an intrinsic property of their bulk band-structure. These virtual gap states are split off partly from the valence and partly from the conduction band and thus exhibit donor and acceptor character, respectively. Their charge neutrality level is expected close to mid-gap position, but the relevant gap is not necessarily the fundamental one. This concept of Heine's has been further pursued by Yndurain [40], Flietner [41], Flores and coworkers [42], and quite recently by Tersoff [43] who all concluded that at metal-semiconductor interfaces the Fermi level position should be close to the charge neutrality level of the ViGS.

As discussed by Bennett and Duke [44] not only the tailing of the metal wave-functions into the band gap of the semiconductor but also the interactions of the electrons in the valence bands of the metal and of the semiconductor as well as the tailing of the semiconductor valence-electrons, which have energies below the bottom of the metal conduction band, have to be considered. However, for real cases the computational difficulties are very large so that only jellium-silicon junctions have been treated in the past [45]. This study again revealed the importance of the virtual gap states which were then denoted as metal-induced gap states (MIGS).

Third, Freeouf and Woodall [46] have proposed to replace the metal work-function in the Schottky-Mott relation (5) by an effective work function. For the III–V, II–VI, and some III–VI compound semiconductors they assumed the presence of anion microclusters at the interface which are supposedly created during the formation of the interface either by oxide contaminations or by metal-semiconductor interactions. These anion microclusters are then thought to modify the work function of the semiconductor substrate.

Fourth, Spicer et al. [47], Wieder [48], and Williams et al. [49] have suggested that the electronic states of native defects generated during the early stages of interface formation cause the pinning of the Fermi level in metal-semiconductor interfaces. Vacancies, antisite defects, and also defect complexes (for a review see [50] and references therein) have been considered in this respect. The density of defects necessary to reach Fermi-level pinning has been calculated by Zur et al. [51] and by Duke and Mailhiot [52]. In both studies, the metal was modelled as jellium and the defects were placed by 5 and 10 Å, respectively, away from the interface into the semiconductor so that they could not be screened by the electrons of the metal tailing into the semiconductor. The result of both calculations was that 10^{14} defects per cm^2 are needed for an effective pinning of the Fermi level as observed experimentally.

Up to now, no signal which could be directly related to such adsorption-induced defects has been observed. However, 10^{14} antisite defects per cm^2, for example, should be detectable by using photoemission spectroscopy. Arsenic-on-gallium-site antisite defects, which have been proposed to pin the Fermi level at metal-GaAs interfaces, consist of an As atom which is surrounded by four As atoms. In GaAs the As(3d) core levels are shifted by approximately 0.9 eV to larger binding energies with respect to metallic arsenic [53, 54]. Thus, it is to be expected that As_{Ga} antisite defects should give a core-level signal which is comparably shifted to lower binding energies with regard to As atoms in the bulk of GaAs which have four Ga atoms as nearest neighbours. However, no such signals have been reported till now and, at least, the formation of As_{Ga} antisite defects but also of As microclusters at metal-GaAs interfaces seems to be very improbable.

Therefore, electronic interface states most likely play a dominant role at metal-semiconductor interfaces, and in the following the virtual gap states will thus be considered more closely.

5 Virtual Gap States and the Barrier Height

In the band gaps of solids electronic surface states, which exponentially decay into the vacuum as well as into the bulk, may appear. These surface states then have a complex wave vector where the imaginary part is the inverse of the decay length of the wave function. This shall be briefly discussed in a simple one-dimensional model.

In a linear chain of lattice constant a, the potential is assumed to vary as

$$V(z) = E_{pot} + V_1 \cos\frac{2\pi}{a}z. \tag{8}$$

At the boundary of the first Brillouin zone, $k_1 = \pm\,\pi/a$, then a gap of width $2|V_1|$, which is centered around $E_{pot} + E_1 = E_{pot} + \frac{\hbar^2}{2m_0}\cdot\left(\frac{\pi}{a}\right)^2$, opens. Within this gap, solutions of the Schrödinger equation are obtained for wave functions with complex wave vector

$$k = \frac{\pi}{a} - iq, \tag{9}$$

where q varies across the gap as

$$\frac{\hbar^2}{2m_0}q^2 = \pm\sqrt{V_1^2 - 4E_1(E - E_{pot})} - (E - E_{pot} + E_1). \tag{10}$$

These states with complex wave vector within the energy gap are denoted as virtual gap states and are an intrinsic property of the solid. Equation (10) is illustrated by Fig. 10 which also schematically shows the density of the virtual gap states.

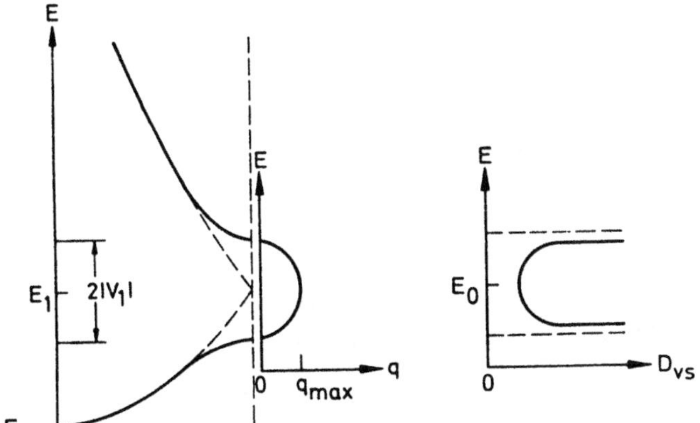

Fig. 10 Band scheme showing virtual gap states and their density of states for a one-dimensional model solid (linear chain).

When a solid is terminated the virtual gap states may lead to surface states. The Bloch function of a surface state decays with exp(− qz) into the solid, i. e. the terminated chain in the example just discussed, and must be joined to an exponential tail into the vacuum. This condition then gives the energy of the surface state. Here, it shall only be mentioned that the existence of surface states needs an attractive potential at the surface, i. e. V_1 has to be positive. For further details the reader is referred to textbooks on solid state physics (see, for example, [55]).

The virtual gap states are derived from valence- as well as conduction-band states and consequently have partly donor and partly acceptor character. In a Gedanken-experiment the virtual gap states may be charged. The total charge stored in them is then positive, zero, or negative when the Fermi level is below, at, or above their charge neutrality level E_0. This charge-neutrality level is expected to lie close to mid-gap position.

Following Heine's conjecture [39], at a metal-semiconductor interface the metal wave-functions tail into the virtual gap states of the semiconductor between E_{vi} and E_F and by this the barrier height is determined.

Electronic interface states in metal-semiconductor contacts make possible a charge flow across the junction. The resulting interface dipole Δ (see Fig. 11) then increases or decreases the interface barrier depending on the direction of the charge

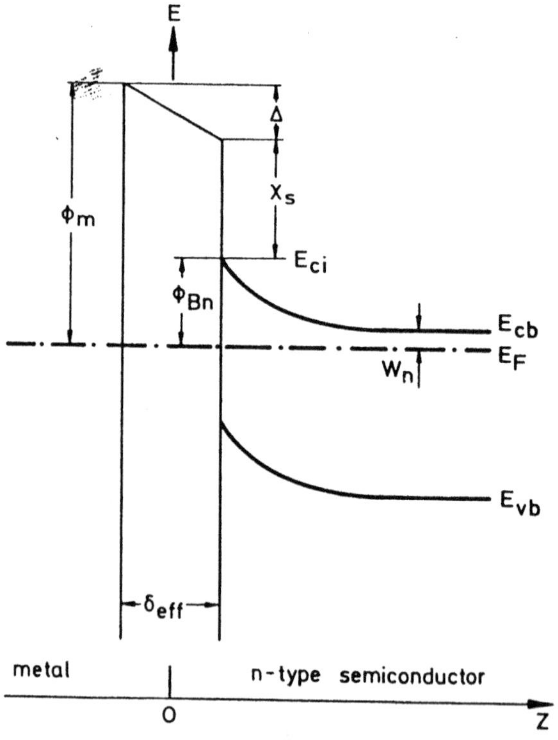

Fig. 11
Band diagram at a metal-semiconductor contact including a dipole layer at the interface.

81

transfer. Qualitatively, the sign may be easily understood by comparing the electronegativities of the metal and the semiconductor. In a single-bonded diatomic molecule AB the charge transfer is given by the difference of the electronegativities of the two atoms by

$$\Delta q_1 = \frac{0.16}{eV} |X_A - X_B| + \frac{0.035}{(eV)^2} |X_A - X_B|^2 \tag{11}$$

which is a revised version [56] of Pauling's original one [57]. The electronegativities of the group IV elements Si and Ge are 1.9 eV and 2.01 eV, respectively. The mean electronegativities $\bar{X} = 1/2 (X_A + X_B)$ of all the III–V and also the II–VI compound semiconductors are close to their average $\langle \bar{X} \rangle = (2.0 \pm 0.1)$ eV. For metal-semiconductor junctions, where the metal electronegativity is not too far from 2 eV, the Fermi level should thus coincide with the charge neutrality level of the virtual gap states of the semiconductors.

The electronegativity of, for example, gold amounts to 2.54 eV. Thus, a small charge transfer from semiconductors to gold is expected from Eq. (11). This charge flow is, however, further reduced to $0.4 \cdot \Delta q_1$ since the dipoles of individual gold-semiconductor "interface-molecules" mutually interact which causes a depolarization first evaluated by Topping [58]. In Fig. 12 the barrier heights Φ_{Bp} as measured with gold Schottky contacts are plotted versus the positions $E_0 - E_{vi}$ of the charge neutrality levels of the virtual gap states which have been calculated by Tersoff [27, 43, 59]. Obviously, both quantities are in close agreement.

When the electronegativity of the metal ranges below 2 eV then charge will be transferred across the Schottky contact from the metal to the semiconductor.

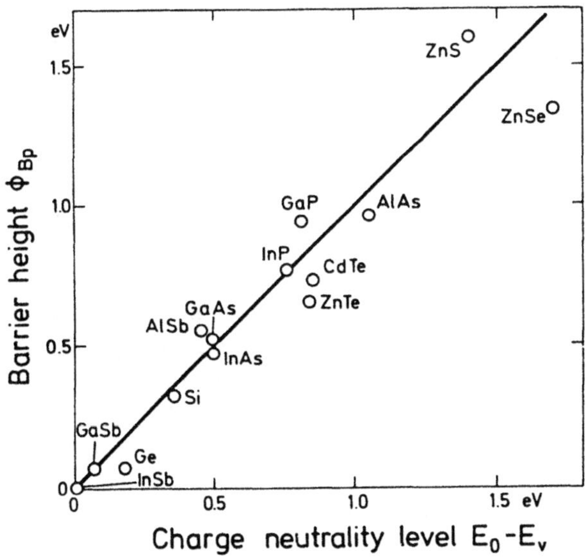

Fig. 12
Barrier heights of gold-semiconductor contacts plotted versus the position of the charge-neutrality level of the virtual gap states relative to the top of the valence band of the semiconductor.

82

This excess charge then flows into the tails of the metal wave-functions reaching into the semiconductor or with other words into the virtual gap states and, consequently, their charge-neutrality level has to move to below the Fermi level. This means that the barrier height $\Phi_{Bn} = E_{ci} - E_F$ is expected to decrease with decreasing electronegativity or, what is the same according to Eq. (7), work function of the metal. This exactly is, as shown, for example, in Fig. 4, what is found experimentally.

The density of gap states at the interface determines the slope S_Φ in the Φ_{Bn}-versus-Φ_m plots. Cowley and Sze [60] and Tejedor et al. [42] have analysed the charge balance at such interfaces and have obtained

$$\Phi_{Bn} = S_\Phi (\Phi_m - \chi_s + D_j) + (1 - S_\Phi) \cdot \Phi^0_{Bn}. \tag{12}$$

Here, $\Phi^0_{Bn} = E_{ci} - E_0$ is the barrier height when the Fermi level coincides with the charge-neutrality level of the virtual gap states, and D_j is a dipole term which depends on the metal only and accounts for the difference of the internal work function and the one measured with the bare surface [42]. The slope S_Φ is given by

$$S_\Phi = (1 + e_0^2 D_{vs} \delta_{eff}/\epsilon_0)^{-1} \tag{13}$$

where D_{vs} is the density of virtual gap states at the charge-neutrality level and

$$\delta_{eff} = \frac{t_m}{\epsilon_m} + \frac{t_s}{\epsilon_s} \tag{14}$$

is the effective width of the dipole layer at the interface. Here, t_m may be taken as the covalent radius of the metal atoms and $\epsilon_1 = 1$, since the core electrons are tightly bound, and t_s may be approximated by the decay length of the virtual gap states, and ϵ_s is the dielectric constant of the semiconductor.

Eq. (12) has two limiting cases for small and for large densities of gap states at the interface:

$$\Phi_{Bn} \approx \Phi_m - \chi_s + D_j \qquad \text{for } S_\Phi \approx 1, \tag{15a}$$
$$\Phi_{Bn} \approx \Phi^0_{Bn} \qquad \text{for } S_\Phi \approx 0. \tag{15b}$$

Relation (15a) gives the Schottky-Mott limit which is approached for $D_{vs} \to 0$ while relation (15b) represents Bardeen's model where the Fermi level is pinned at the charge-neutrality level of the interface states for $D_{vs} \to \infty$.

From the slopes S_X shown in Fig. 5 the densities of the virtual gap states may be evaluated by using Eqs. (13) and (14). With reasonable input data for δ_{eff} one obtains some 10^{14} states per eV and cm^2. The chemical trend visualized by Fig. 4 indicates that the density of states D_{vs} decreases with increasing ionizity of the semiconductor. This behaviour is plausible since the total density of virtual gap states is the same for all semiconductors, but the energy gaps are wider for the more ionic compounds compared to the more covalently bonded ones. Here, both the gaps in the middle and at the boundary of the Brillouin zone have to be considered [42, 61]. The importance of the indirect band gaps already showed up in Fig. 6 where the barrier heights measured with Au Schottky contacts were found to correlate with the indirect rather than the fundamental band-gap energies.

The chemical trends observed with the slope $S_\Phi = d\Phi_{Bn}/d\Phi_m$ can be better understood when the two quantities entering in the denominator of Eq. (13), which are the density of the virtual gap states at their charge-neutrality level and the effective width of the interface dipole-layer, are analysed in some more detail. For this purpose, the one-dimensional model shall be considered again. At their charge-neutrality level, i.e. close to mid-gap position E_1, the density of the virtual gap states is found to vary as

$$D_{vs}(E_0) \approx D_{vs}(E_1) \sim |V_1|^{-1}, \tag{16}$$

and the imaginary part of the wave vector assumes its maximum value

$$q(E_0) \approx q_{max} = |V_1| \frac{g_1}{E_1}, \tag{17}$$

where $g_1 = 2\pi/a$ is a vector of the reciprocal lattice. Since according to Eq. (14) the effective width of the interface dipole-layer varies as $\approx 1/q(E_0)$ it results

$$D_{vs} \cdot \delta_{eff} \sim |V_1|^{-2}. \tag{18}$$

Considering Eq. (13) this result means that $(1/S_\Phi - 1)$ is expected to obey a power law of the band-gap energy. This conclusion, which was derived from a one-dimensional model, should also apply to three-dimensional semiconductors.

The "effective" band gap $\langle E_g \rangle$ of a semiconductor, on the other hand, depends on the electronic contribution ϵ_∞ to the dielectric constant as ([62] or other text books on solid state theory)

$$\epsilon_\infty = 1 + \left(\frac{\hbar\omega_p}{\langle E_g \rangle}\right)^2, \tag{19}$$

where $\hbar\omega_p$ is the energy of the bulk plasmon of the valence electrons. For the semiconductors considered here, the mean energy of the bulk plasmon amounts to (16.5 ± 0.5) eV and is thus approximately constant. This means that the effective band gaps are expected to vary proportional to $(\epsilon_\infty - 1)^{1/2}$. Using this result and Eqs. (13), (18), and (19) one obtains

$$\left(\frac{1}{S_\Phi} - 1\right) = \left(\frac{A}{S_X} - 1\right) \sim (\epsilon_\infty - 1)^n. \tag{20}$$

Considering the experimental relation (7) between the work functions of metals and their electronegativities one finds $A = 1.79$. Following Eq. (20), $(A/S_X - 1)$ is plotted versus $(\epsilon_\infty - 1)$ in Fig. 13. Here, the experimental S_X values as evaluated

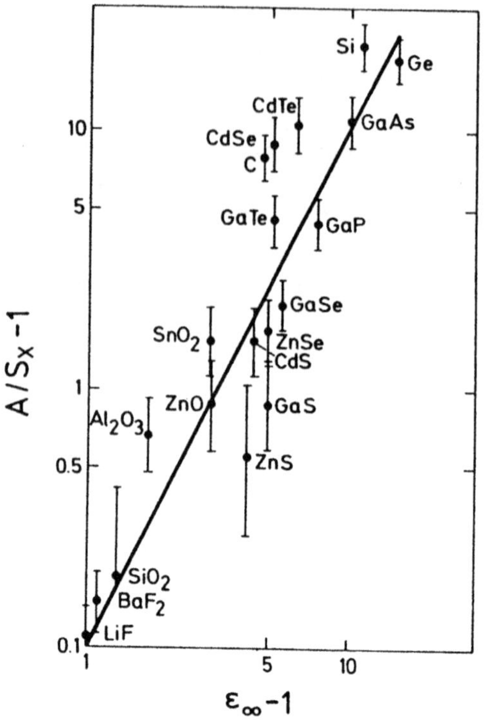

Fig. 13
Slopes $S_X = d\Phi_{Bn}/dX_m$ plotted versus the electronic contribution ϵ_∞ to the dielectric constant of the semiconductor.

by Schlüter [26] have been used. Fig. 13 clearly shows that the data indeed follow a power law and a least square fit gives

$$\left(\frac{A}{S_X} - 1\right) = 0.1\,(\epsilon_\infty - 1)^{2.0}. \tag{21}$$

The regression coefficient is 0.91. This good correlation is not obvious since the experimental barrier heights show a large amount of scatter which, at least partly, has to be attributed to the experimental conditions and/or the evaluation of the data. Threfore, straight-line fits to evaluate a slope in Φ_{Bn}-versus-Φ_m (or X_m) plots have large margins of error and sometimes become questionable.

Nevertheless, the good correlation of the data plotted in Fig. 13, which are to be described by relation (21), means that the experimentally observed slopes $S_\Phi = d\Phi_{Bn}/d\Phi_m$ are to be understood within the model that at a metal-semiconductor contact the metal wave-functions tail into the virtual gap states of the semiconductor. By this these virtual gap states, which are an intrinsic property of the semiconductor, are responsible for the determination of the barrier heights observed in Schottky contacts.

Finally, it has to be mentioned that plots similar to the one in Fig. 13 have been discussed earlier [26, 61, 63]. However, in the first publication the static dielectric constant ϵ_0 instead of the electronic contribution ϵ_∞ was taken and no power-law analysis but a correlation between S_X and $(\epsilon_0 - 1)$ was looked for. In another paper, [63], $(1 - S_\Phi)$ was plotted versus $(\epsilon_\infty - 1)$. This resulted in a diagram similar to the one shown in Fig. 5, and no correlation of the data was obtained. A very recent paper [61] assumed different power laws between S_Φ and ϵ_∞ and found limited correlation only.

6 Concluding Remarks

The concept that the virtual gap states play an important role in determining the barrier heights in metal-semiconductor contacts provides us with an understanding of the chemical trends observed with the barrier heights in such Schottky contacts. The quantitative aspects, however, are less clear since the large scatter frequently found with the experimental data makes some of the "linear approaches", such as the determination and then evaluation of the slope in barrier height-versus-metal work-function plots, questionable. On the other hand, the barrier heights measured with gold Schottky contacts exhibit a good correlation with the charge-neutrality levels calculated for the virtual gap states of various semiconductors.

The virtual gap states have also been discussed in explaining the observed band discontinuities in semiconductor heterostructures [64, 59]. It has been proposed that the bands line up such that the charge-neutrality levels of the semiconductors in contact coincide since this gives zero interface dipole. A very good correlation has been obtained between predicted and experimentally observed band offsets.

Furthermore, the pinning positions of the Fermi level observed with adsorbates on (110) surfaces of a number of III–V compound semiconductors have also been found to be close to the charge-neutrality levels of the virtual gap states of these semiconductors [65]. Only with the phosphides some deviations have been found which were attributed to adsorption-induced defects.

The experimental data on the Fermi-level position at adsorbate-covered semiconductor surfaces, the band line-ups in semiconductor heterostructures, and, last but not least, the barrier heights in metal-semiconductor contacts are strongly supporting the same concept: "adatom states", in its broadest sense, tail into the virtual gap states of the semiconductor and by this determine the line-up of the semiconductor bands and the "adatom states" at surfaces and in metal-semiconductor as well as semiconductor-semiconductor interfaces.

References

[1] *F. Braun*, Pogg. Ann. **153**, 556 (1874)

[2] *W. G. Adams* and *R. E. Day*, Proc. Roy. Soc. **25**, 113 (1876)

[3] *F. Braun*, Sitzungsber. Naturforsch. Gesell. (Leipzig), p. 49 (1876)

[4] *L. D. Grondahl*, U.S. Patent 1640335 issued Jan. 1, 1925

[5] *W. Schottky* and *W. Deutschmann*, Phys. Z. **30**, 839 (1929)

[6] *A. H. Wilson*, Proc. Roy. Soc. **A133**, 458 (1936) and **134**, 277 (1931)

[7] *O. Fritsch*, Ann. Phys. **22**, 375 (1935)
[8] *W. Schottky*, Naturwissenschaften **26**, 843 (1938)
[9] *N. F. Mott*, Proc. Cambridge Philos. Soc. **34**, 568 (1938)
[10] *W. Schottky*, Z. Phys. **113**, 367 (1939)
[11] *W. Schottky* and *E. Spenke*, Wiss. Veröff. Siemens-Werke **18**, 225 (1939)
[12] *E. Spenke* and *W. Schottky*, Wiss. Veröff. Siemens-Werke **20**, 40 (1940)
[13] *W. Schottky*, Z. Phys. **118**, 539 (1942)
[14] *H. Schweikert*, Verh. Phys. Ges. **3**, 99 (1939)
[15] *W. Schottky*, Phys. Z. **41**, 570 (1940)
[16] *S. Poganski*, Z. Physik **134**, 469 (1953)
[17] *S. Poganski*, Z. Elektrochem. **56**, 193 (1952)
[18] *E. H. Rhoderick*, Metal Semiconductor Contacts (Clarendon, Oxford 1980)
[19] *L. J. Brillson*, Surf. Sci. Repts. **2**, 123 (1982)
[20] *B. L. Sherma* (ed.), Metal-Semiconductor Schottky Barrier Junctions and Their Applications (Plenum, New York 1984)
[21] *W. Mönch*, Surf. Sci. **21**, 443 (1970)
[22] *S. Kurtin, T. C. McGill*, and *C. A. Mead*, Phys. Rev. Letters **22**, 1433 (1970)
[23] *W. Gordy* and *W. J. O. Thomas*, Phys. Rev. **24**, 439 (1956)
[24] *K. W. Frese, jr.*, J. Vac. Sci. Technol. **16**, 1042 (1979)
[25] *H. B. Michaelson*, J. Appl. Phys. **48**, 4729 (1977)
[26] *M. Schlüter*, Phys. Rev. **B17**, 5044 (1978)
[27] *J. Tersoff*, Surf. Sci. **168**, 275 (1986)
[28] *Landolt-Börnstein*: Numerical Data and Functional Relationships in Science and Technology, ed. by *O. Madelung* (Springer, Berlin 1982), Group 3, Vol. 17
[29] *C. A. Mead*, Solid-State Electron. **9**, 1023 (1966)
[30] *J. O. McCaldin, T. C. McGill*, and *C. A. Mead*, J. Vac. Sci. Technol. **13**, 802 (1976)
[31] *W. Mönch*, phys. stat. sol. **40**, 257 (1970)
[32] *K. K. Chin, S. H. Pan, D. Mo, P. Mahowald, N. Newman, I. Lindau*, and *W. E. Spicer*, Phys. Rev. **B32**, 918 (1985)
[33] *W. Mönch*, in: Festkörperprobleme: Advances in Solid State Physics, ed. by *P. Grosse* (Vieweg, Braunschweig 1984), Vol. XXIV, p. 229
[34] *R. Ludeke, T.-C. Chiang*, and *D. E. Eastman*, J. Vac. Sci. Technol. **21**, 599 (1982)
[35] *D. Bolmont, P. Chen, F. Proix*, and *C. A. Sebenne*, J. Phys. C: Solid State Phys. **15**, 3639 (1982)
[36] *R. Z. Bachrach* and *R. S. Bauer*, J. Vac. Sci. Technol. **16**, 1149 (1979)
[37] *W. Mönch*, Solid State Commun. **58**, 215 (1986)
[38] *J. Bardeen*, Phys. Rev. **71**, 717 (1947)
[39] *V. Heine*, Phys. Rev. **138**, A 1689 (1965)
[40] *F. Yndurain*, J. Phys. C: Solid State Phys. **4**, 2849 (1971)
[41] *H. Flietner*, phys. stat. sol. (b) **54**, 201 (1972)
[42] *C. Tejedor, F. Flores*, and *E. Louis*, J. Phys. C: Solid State Phys. **10**, 2163 (1977)
[43] *J. Tersoff*, Phys. Rev. Letters **52**, 465 (1984)
[44] *A. J. Bennett* and *C. B. Duke*, Phys. Rev. **162**, 578 (1967)
[45] *St. G. Louie* and *M. L. Cohen*, Phys. Rev. **B13**, 2461 (1976)

[46] *J. L. Freeouf* and *J. M. Woodall*, Appl. Phys. Lett. **39**, 727 (1986)

[47] *W. E. Spicer, P. W. Chye, P. R. Skeath*, and *I. Lindau*, J. Vac. Sci. Technol. **16**, 1422 (1979)

[48] *H. H. Wieder*, J. Vac. Sci. Technol. **15**, 1498 (1978)

[49] *R. H. Williams, R. R. Varma*, and *V. Montgomery*, J. Vac. Sci. Technol. **16**, 1418 (1979)

[50] *W. Mönch*, Surf. Sci. **132**, 92 (1983)

[51] *A. Zur, T. C. McGill*, and *D. L. Smith*, Phys. Rev. **B28**, 2060 (1983)

[52] *C. B. Duke* and *C. Mailhiot*, J. Vac. Sci. Technol. **B3**, 1170 (1985)

[53] *L. Ley, R. A. Pollak, S. P. Kowalczyk, R. McFeely*, and *A. Shirley*, Phys. Rev. **B8**, 641 (1973)

[54] *E. A. Kraut, R. W. Grant, J. R. Waldrop*, and *S. P. Kowalczyk*, Phys. Rev. Lett. **44**, 1620 (1980)

[55] *N. W. Ashcroft* and *N. D. Mermin*, Solid State Physics (Holt, Rinehart and Winston, New York 1976), p. 369

[56] *N. B. Hanney* and *C. P. Smith*, J. Am. Chem. Soc. **68**, 171 (1946)

[57] *L. Pauling*, The Nature of the Chemical Bond (Cornell University, Ithaca, N.Y., 1960)

[58] *J. Topping*, Proc. Roy. Soc. (London) **A114**, 67 (1927)

[59] *J. Tersoff*, Phys. Rev. **B30**, 4874 (1984)

[60] *A. M. Cowley* and *S. M. Sze*, J. Appl. Phys. **36**, 3212 (1965)

[61] *J. Tersoff*, Phys. Rev. **B32**, 6968 (1985)

[62] *W. Ludwig*, Festkörperphysik (Akadem. Verlagsges., Wiesbaden 1978), p. 282

[63] *J. C. Phillips*, Solid State Commun. **12**, 861 (1973)

[64] *C. Tejedor* and *F. Flores*, J. Phys. C: Solid State Phys. **11**, L19 (1978)

[65] *W. Mönch*, Proc. 13th Annual Conf. on the Phys. and Chem. of Semicond. Interfaces, Pasadena, CA (USA), 1986; J. Vac. Sci. Technol. **B4**, in print (1986)

Role of Virtual Gap States and Defects in Metal-Semiconductor Contacts

W. Mönch
Laboratorium für Festkörperphysik, Universität Duisburg, D-4100 Duisburg, Federal Republic of Germany
(Received 20 November 1986)

Chemical trends of barrier heights reported for metal- and silicide-silicon contacts are analyzed. The data are easily explained when both virtual gap states of the complex band structure of the semiconductor and electronic levels of defects created in the semiconductor close to the interface during its formation are considered. The virtual gap states determine the barrier heights when either the defect density is low or the defects are completely charged or all neutral.

PACS numbers: 73.30.+y, 73.20.−r, 73.40.Ns

The rectifying properties of metal-semiconductor contacts, which were discovered by Braun,[1] are caused by depletion layers on the semiconductor side of the interface, as was first shown by Schottky.[2] The fundamental parameter which characterizes such a junction is its barrier height, i.e., the energy distance from the Fermi level to the bottom of the conduction band at the interface when the semiconductor is doped *n* type. A basic understanding of Schottky contacts thus needs a model which explains the chemical trends of the barrier heights observed with different metal-semiconductor pairs.

Schottky[3] and Mott[4] proposed the barrier height to equal the difference of the work function of the metal and the electron affinity of the semiconductor. Although for a given semiconductor the barrier heights are generally found to increase when the work function of the metal in contact becomes larger, the simple Schottky-Mott rule is not obeyed by the experimental data. Bardeen[5] attributed this discrepancy to the presence of interface states. They could accommodate charge which is transferred between the metal and the semiconductor because of their generally different electronegativities. This means that a dipole layer exists at the interface. Since the work function of metals and their electronegativities were found to be linearly related, such interface states intuitively explain that the barrier heights are increased by metals with larger work functions but do not follow the Schottky-Mott rule. Two basically different models on the physical nature of such interface states have been suggested. In the following, they will be briefly reviewed.

The first model, which was introduced by Heine,[6] assumes that within the band gap of the semiconductor the wave functions of the metal electrons are tailing into the virtual gap states (VGS) of the complex band structure of the semiconductor. Since the virtual gap states are split off from the valence and the conduction band, their character varies across the gap from mostly donor type close to the top of the valence band to mostly acceptor type close to the bottom of the conduction band. The charge transferred between the metal and the semiconductor then pins the Fermi level above, at, or below the charge-neutrality level E_0 of the virtual gap states when the electronegativity of the metal is smaller, equal to, and larger than, respectively, the one of the semiconductor. In the following, three different and independent results will be presented which support the VGS model of Schottky contacts.

For the column-IV elemental and the III-V compound semiconductors Tersoff[7] has calculated the charge-neutrality levels of the VGS. He has obtained good agreement between $(E_{cs}-E_0)$ and the barrier heights Φ_{Bn} experimentally determined with gold Schottky contacts on samples doped *n* type. This finding is supporting the VGS model since the electronegativities of gold and of the semiconductors only differ slightly. Second, the adsorption of cesium[8] and of chlorine[9] was found to pin the Fermi level above and below, respectively, the charge-neutrality level of the VGS at cleaved GaAs(110) surfaces. Since the electronegativities of cesium and of chlorine are smaller and larger, respectively, by almost the same amount than the value of gallium arsenide the results mentioned are again in support of the VGS model.

The third indication is represented by the data plotted in Fig. 1. When interface states are assumed to be present in a metal-semiconductor junction the barrier height Φ_{Bn} increases proportionally to the work function Φ_M of the metal.[13,14] The slope parameter $S = d\Phi_{Bn}/d\Phi_M$ only depends on the product of the density of states $D_{vs}(E_0)$ around the charge-neutrality level of the interface states and the width δ of the related dipole layer as

$$S = [1 + e_0^2 D_{vs}(E_0)\delta/\varepsilon_0]^{-1}. \quad (1)$$

In the VGS model, this product $D_{vs}\delta$ is determined by the average band-gap energy of the semiconductor which, on the other hand, is related to the electronic polarizability $(\varepsilon_\infty - 1)$ of the semiconductor.[12] Although in some cases the experimental slope parameters are not well defined,[10,11] the S values of nineteen different semiconductors follow a pronounced chemical trend[12] when $(1/S-1)$ is plotted over $(\varepsilon_\infty - 1)$ as shown in Fig. 1. A least-squares fit to the data yields

$$(1/S - 1) = 0.1(\varepsilon_\infty - 1)^2 \quad (2)$$

and a regression coefficient $r = 0.91$. This result again

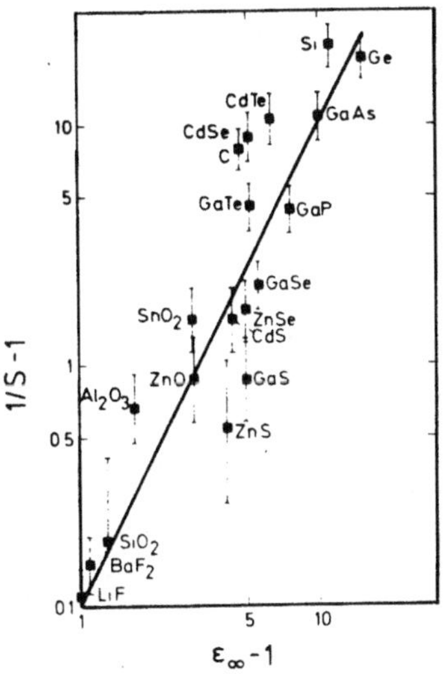

FIG. 1. Slopes $S = d\Phi_{Bn}/d\Phi_M$ plotted vs the electronic contribution ε_∞ of the dielectric constant of the semiconductor. The data were taken from Refs. 10 and 11 in the manner of Ref. 12.

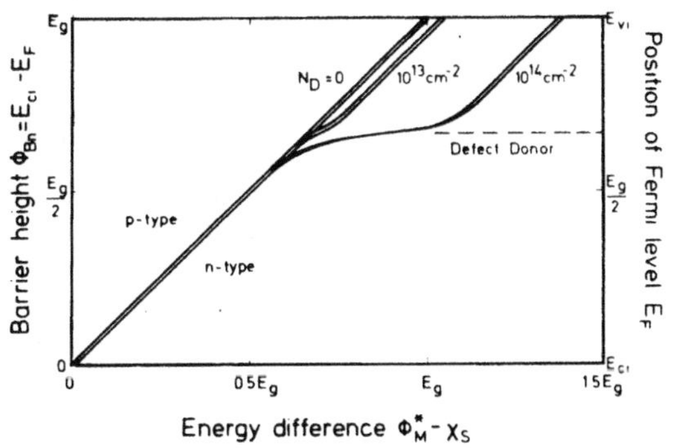

FIG. 2. Barrier height as a function of the difference between the internal part of the metal work function Φ_M^* and the electron affinity X_s of the semiconductor for three different densities of defects (shown schematically in the manner of Refs. 21 and 22).

strongly supports the VGS model of metal-semiconductor contacts.

The second model which was proposed by Wieder[15] and Spicer *et al.*[16] identifies the interface states in Schottky contacts as electronic states of native defects which are created during the formation of the junctions (see Mönch[17] for a review). The defect model was motivated by the observations that Schottky barriers on III-V compound semiconductors were found to be insensitive to within 0.2 eV to the metals used and to follow no apparent chemical trend. Up until now, no spectroscopic evidence has emerged for any defect such as vacancies or antisite defects which have been discussed theoretically.[18,19] This might indeed prove to be difficult since in many cases chemical reactions and intermixing were observed.[20] Therefore, interfaces between most metals and III-V compound semiconductors are difficult to characterize with respect to local variations in chemical composition.

The influence of defect levels on the barrier height of metal-semiconductor junctions was theoretically studied by Zur, McGill, and Smith[21] and by Duke and Mailhiot.[22] They placed defects 5–10 Å apart from the interface into the semiconductor. The metals were described by a jellium model. The results of these calculations are schematically explained in Fig. 2. When the area density N_D of defects is kept below 10^{13} cm^{-2} the position of the Fermi level moves across the energy gap of the semiconductor to the same extent as the internal work function Φ_M^* of the jellium metal is increased. For larger defect densities, the Fermi level gets transitionally pinned at the defect level, until the defects are all charged, and then further moves towards the top of the valence band at the interface as a function of metal work function.

These models have been developed in parallel to many experimental studies of metal-semiconductor junctions which have provided a huge body of data on electronic, structural, and chemical properties of such interfaces. In the following, the analysis of experimental results and the search for chemical trends shall be restricted to metal-silicon contacts for the following reason. The interfaces of Schottky contacts on III-V compound semiconductors were found to be intermixed in many cases[20] and they are thus difficult to characterize chemically. Metal-silicon junction, on the other hand, can be prepared with quite abrupt interfaces since the controlled formation of silicides, which are mostly metallic, is a well-established technique.[23] Cross-sectional pictures obtained with high-resolution transmission-electron microscopy have proven that, for example, epitaxial films of $NiSi_2$, NiSi, and Pd_2Si may be grown on silicon substrates (see, e.g., the work of Liehr *et al.*[24] and Ho[25]).

In searching for chemical trends of the barrier heights measured now with metal-silicon contacts, the main difficulty arises with the ordering of the metals. The first choice, which was motivated by the early Schottky-Mott rule, has still remained the metal work function which, however, contains an internal part plus a surface dipole contribution. The internal part of the work function, which is of interest in interfaces, may be approximated by the electronegativity of the metal. Here, the most popular scale has been the one developed by Pauling.[26] He has designed his set of values to describe the partly ionic character of covalent bonds, and it is this field where Pauling's electronegativities have their merits in semiconductor bulk and surface physics, too (see, e.g., the work of Mönch[27]). In metal-semiconductor contacts, on the other hand, the ionicity of metallic bonds comes into play. Chemical trends in the properties of metal alloys and intermetallic compounds have been successfully described by another set of electronegativities which were derived by Miedema, el Châtel, and de Boer.[28] In the present paper, the further analysis will use the electronegativities proposed by Miedema.

In Fig. 3 barrier heights measured with metal- and with silicide-silicon junctions are plotted over electronegativities based on the Miedema scale. For silicides M_mSi_n, the geometric mean $(X_M^m X_{Si}^n)^{1/(m+n)}$ of the metal and the silicon electronegativities were taken.[26,31] In this respect the plot differs from a similar one by Schmid[29] who has introduced Miedema's electronegativities in the discussion of metal-silicon junctions but has plotted Φ_{Bn} vs X_M only. In Fig. 3, the data points are obviously arranged in two groups. The straight line drawn in full is a least-squares fit to fifteen data points and is given by

$$\Phi_{Bn} = 0.17\langle X_M\rangle - 0.04 \text{ eV},$$

with a regression coefficient $r = 0.98$. The marked data point labeled CNL represents the charge-neutrality level of the virtual gap states of the complex band structure in silicon ($X_{Si} = 4.7$ eV) as calculated by Tersoff.[7] Obviously, the charge-neutrality level of the VGS fits exactly into that straight line. This finding implies that those barrier heights, which define that straight line in Fig. 3, are determined by the VGS of silicon. The broken line connecting another eighteen data points resembles the shape of the curve shown in Fig. 2 which was obtained for heavily defected metal-semiconductor contacts.

Considering the VGS and the defect model of metal-semiconductor junctions as outlined above, the data plotted in Fig. 3 suggest the following explanation. Those metal-silicon junctions, the barrier heights of which are found close to the straight-line fit, are exhibiting a density of defects below approximately 10^{13} cm^{-2} and their barrier heights are thus determined by the tailing of the metal electron wave functions into the virtual gap states of the silicon bond structure. The other silicon Schottky contacts contain a large defect density of approximately 10^{14} cm^{-2}. As the inflection of the dashed line indicates, one defect level is located at approximately 0.62 eV below the bottom of the conduction band. That defect level was already concluded by Schmid[29] from his Φ_{Bn}-vs-X_M plot for the silicide-silicon junctions. For barrier heights less than about 0.6 eV the dashed curve is running in parallel to the straight line, which is determined by the VGS of silicon, but is shifted by 0.1 eV to lower values. This indicates the presence of another defect level at or above 1.12−0.37 eV = 0.75 eV above the top of the valence band. The explanations just given for the data plotted in Fig. 3 are strongly supported by results of a study on nickel-silicide-silicon interfaces recently published by Liehr *et al.*,[24] which will be discussed in the following.

Both groups of data points in Fig. 3 contain results

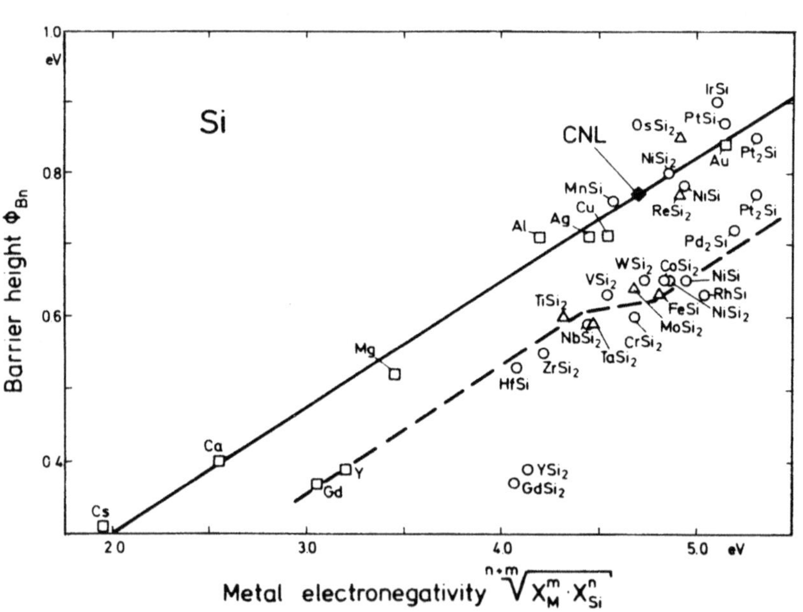

FIG. 3. Barrier heights of metal- and silicide-silicon contacts vs the effective Miedema electronegativities. The barrier heights were adopted from Refs. 24, 29, and 30.

from the study of Liehr *et al.* for $NiSi_2$ and NiSi which differ by $\Delta\Phi_{Bn} \approx 0.15$ eV. With both nickel silicides the larger values were found when epitaxial silicides were grown, which in the case of the disilicide consisted of either pure type-*A* or type-*B* interfaces. With a mixture of both types, which only differ in the stacking sequence when passing the interface, the lower value of the barrier height was observed regularly. Cross-sectional TEM pictures always revealed such interfaces exhibiting the lower Φ_{Bn} values to be less perfect, containing phase domain boundaries with faceted or stepped structures.[24] With the nickel monosilicide, large barrier heights were also observed only when the interfaces were of the same high degree of perfection as found with single-type disilicide. "The degree of perfection of the interfacial structure is more important than specific epitaxy in determining the barrier height," as was pointed out by Schmid *et al.*[32] From preliminary evaluations of further capacitance spectroscopy studies they also computed approximately 10^{12} to 10^{13} interface states per square centimeter for the single-type $NiSi_2$-Si interfaces but about 1 order of magnitude more for the mixed-type interfaces. These experimental findings by Schmid *et al.* are in excellent agreement with the explanations of the data plotted in Fig. 3 which were given above.

The results of the present paper may be summed up as follows. The analysis of the chemical trend of the barrier heights reported for 31 different metal- and silicide-silicon interfaces has revealed that both VGS and defects are needed to explain the complete set of experimental data. When, however, the experimental conditions during the preparation of Schottky contacts can be controlled such as to reduce the defect density to below approximately 10^{13} per square centimeter then the barrier height is determined by the virtual gap states of the semiconductor band structure only. In highly defected Schottky barriers the virtual gap states also determine the barrier heights when all the defects are either charged or neutral. A preliminary analysis of barrier heights observed experimentally with metal-GaAs(110) contacts show that the same concepts apply to these interfaces, too.

[1]F. Braun, Pogg. Ann. **153**, 556 (1874).

[2]W. Schottky, Naturwissenschaften **26**, 843 (1938).

[3]W. Schottky, Phys. Z **41**, 570 (1940).

[4]N. F. Mott, Proc. Cambridge Philos. Soc. **34**, 568 (1938).

[5]J. Bardeen, Phys. Rev. **71**, 717 (1947).

[6]V. Heine, Phys. Rev. **138**, A1689 (1965).

[7]J. Tersoff, Phys. Rev. Lett. **52**, 465 (1984), and Phys. Rev. B **30**, 4874 (1984).

[8]W. E. Spicer, P. E. Gregory, P. W. Chye, I. A. Babalola, and T. Sukegawa, Appl. Phys. Lett. **27**, 617 (1975).

[9]D. Troost, L. Koenders, Y.-L. Fan, and W. Mönch, to be published.

[10]S. Kurtin, T. C. McGill, and C. A. Mead, Phys. Rev. Lett. **22**, 1433 (1970).

[11]M. Schlüter, Phys. Rev. B **17**, 5044 (1978).

[12]W. Mönch, in *Festkörperprobleme: Advances in Solid State Physics*, edited by P. Grosse (Vieweg, Braunschweig, 1986), Vol. 26, p. 67.

[13]A. M. Cowley and S. M. Sze, J. Appl. Phys. **36**, 3212 (1965).

[14]C. Tejedor, F. Flores, and E. Louis, J. Phys. C **10**, 2163 (1977).

[15]H. Wieder, J. Vac. Sci. Technol. **15**, 1498 (1978).

[16]W. E. Spicer, P. W. Chye, P. R Skeath, and I. Lindau, J. Vac. Sci. Technol. **16**, 1422 (1979).

[17]W. Mönch, Surf. Sci. **132**, 92 (1983).

[18]M. S. Daw and D. L. Smith, Phys. Rev. B **20**, 5150 (1979).

[19]R. E. Allen and J. D. Dow, J. Vac. Sci. Technol. **19**, 383 (1981).

[20]L. N. Brillson, Surf. Sci. Rep. **2**, 123 (1982).

[21]A. Zur, T. C. McGill, and D. L. Smith, Phys. Rev. B **28**, 2060 (1983).

[22]C. B. Duke and C. Mailhiot, J. Vac. Sci. Technol. B **3**, 1170 (1985).

[23]S. P. Murarka, J. Vac. Sci. Technol. **17**, 775 (1980).

[24]M. Liehr, P. E. Schmid, F. K. Le Goues, and P. S. Ho, Phys. Rev. Lett. **54**, 2139 (1985).

[25]P. S. Ho, J. Vac. Sci. Technol. A **1**, 745 (1983).

[26]L. Pauling, *The Nature of the Chemical Bond* (Cornell Univ. Press, Ithaca, New York, 1960).

[27]W. Mönch, Solid State Commun. **58**, 215 (1986).

[28]A. R. Miedema, P. F. el Châtel, and F. R. de Boer, Physica (Amsterdam) **100B**, 1 (1980).

[29]P. E. Schmid, Helv. Phys. Acta **58**, 371 (1985).

[30]W. Mönch, Surf. Sci. **21**, 443 (1970).

[31]J. L. Freeouf, Solid State Commun. **33**, 1059 (1980).

[32]P. E. Schmid, M. Liehr, F. K. Legoues, and P. S. Ho, in *Thin Films—Interfaces and Phenomena*, edited by R. J. Nemanich, P. S. Ho, and S. S. Lau (Materials Research Society, Pittsburgh, 1986), p. 469.

Initial stages of Schottky barrier formation: Temperature effects

K. Stiles and A. Kahn
Department of Electrical Engineering, Princeton University, Princeton, New Jersey 08544
D. G. Kilday and G. Margaritondo
Department of Physics and Synchrotron Radiation Center, University of Wisconsin, Madison, Wisconsin 53706

(Received 10 February 1987; accepted 2 April 1987)

Recent experiments on the formation of Schottky barriers reveal a temperature-dependent pinning of the Fermi level (E_F) which cannot be explained in terms of the current published models. Al, Au, Ag, In, and Sn are evaporated on room temperature (RT) and low temperature (LT = 80 K) cleaved *n*- and *p*-GaAs(110). We find a substantial decrease in surface metal clustering at LT, leading to more homogeneous films, especially with Al, In, and Sn. Interface mixing and chemical reaction are also partly inhibited. Cancellation of the semiconductor surface relaxation is observed for Al and Sn. From the electronic point of view, low temperature has opposite effects on the pinning rate of E_F on *n*- and *p*-GaAs. Pinning on *n*-GaAs as a function of coverage is dramatically retarded. Pinning on *p*-GaAs remains faster than on *n*-GaAs, as fast as at RT, faster for Al and In. These asymmetric rates make it impossible to explain the initial stages of Schottky barrier formation for all substrates with a single mechanism. They suggest that several independent but concomitant mechanisms might be at work in the pinning process.

I. INTRODUCTION

There has been an intensive effort in recent years to characterize interfaces between metals and III–V compound semiconductors, with a strong emphasis placed on GaAs. It was established early on that, despite a variety of interface structural properties and chemical interactions with metals, the Fermi level (E_F) is pinned within a small range in the band gap of GaAs, giving a barrier height nearly independent of the metal work function. Models based on native defects,[1] interface chemistry,[2] microphase work functions,[3] and metal-induced gap states (MIGS)[4] have been proposed to explain this behavior. There remains, however, considerable controversy about this issue. The morphology, chemistry and atomic structure of metal–III–V interfaces are complex and diverse, and there is no consensus at this point on the mechanism(s) which dominate pinning during the early stages nor the final stages of interface formation.

Complementary structural and chemical measurements have shown that a variety of interactions, such as metal clustering,[5] dissociation of the semiconductor and interdiffusion in the metal overlayer,[6] chemical reaction,[7] and change of the substrate surface atomic structure,[8] take place at metal–GaAs interfaces during their formation. Two major difficulties arise, in particular in the very low coverage regime (10^{-2}–10^{-1} ML) where most of the E_F pinning occurs. First, it is hard to detect and quantify these interactions. Second, given that several interactions occur simultaneously, it is difficult to establish their individual role in the pinning process. We attempt here to simplify the problem by modifying the experimental conditions. We form metal–GaAs interfaces at low temperature (70–100 K) to inhibit some of these interactions, i.e., clustering, interdiffusion, and chemical reactions. We can then relate the changes observed in the pinning behavior to the temperature-induced changes in the morphology and chemistry of the growing interface. Under the assumption that low temperature does not introduce new elements in the interface formation process, but only provides us with a means to slow it down, we can attempt to identify more easily the important mechanisms of E_F pinning in the low coverage regime. The main result of our study is that low temperature affects the initial rate of pinning *differently* on *n* and *p* substrates, thus pointing to an interplay between several pinning mechanisms.

II. EXPERIMENTS

All measurements were done on GaAs(110) cleaved in ultrahigh vacuum (UHV) systems with base pressures $< 10^{-10}$ Torr at 300 K (RT) and at 80–100 K (LT). The GaAs was purchsed from Macom laser diode and was doped $n = 6\times10^{16} - 3\times10^{18}$ cm^{-3} (Si or Te) or $p = 4\times10^{18} - 1\times10^{19}$ cm^{-3} (Zn). Al, In, Au, Ag, and Sn layers were evaporated from resistively heated W filaments in coverages ranging from 0.02 to 32 Å. Atomic order at the interfaces was investigated with low-energy electron diffraction (LEED). Auger electron spectroscopy (AES) and electron-energy loss spectroscopy (EELS) were used to evaluate the interface composition and morphology. Soft x-ray photoemission spectroscopy (SXPS) was done at the Synchrotron radiation center of the University of Wisconsin. The Ga 3*d*, As 3*d*, valence band, and various overlayer core levels were recorded at each coverage. The position of E_F was determined from the energy shift of Ga 3*d* and As 3*d* core levels. Where the shifts of these core levels differ at high coverage due to chemical shifts, i.e., the Au/GaAs interface where Ga alloys with Au,[8] they agree in the low coverage regime of interest here. In such cases, band bending was extracted from the As 3*d* core level shift alone. Additional information about morphology and chemical interaction was extracted from SXPS by monitoring the intensity of substrate core levels versus metal coverage.

III. RESULTS

The influence of the substrate temperature during metal deposition on the overlayer morphology and structure is a major consideration. We summarize the results in this section; a more complete account of the structure work can be found elsewhere.[5] At all interfaces studied, clustering and/or interdiffusion was inhibited at LT, almost entirely with the column III metals and somewhat less with Au and Ag. The slow attenuation of the As 3*d* signal at RT, shown in a plot of the substrate core level intensity on a logarithmic scale versus In coverage in Fig. 1, results from clustering. LEED and RHEED (reflection high-energy electron diffraction) observations of substrate diffraction patterns up to In coverages larger than 50 ML confirm that the RT overlayer consists of large metallic clusters.[6] The exponential attenuation of the core level amplitude versus In coverage in Fig. 1 shows that clustering is almost completely inhibited at LT. This attenuation yields a 2 Å escape depth for the photoemitted electrons, consistent with the growth of a continuous metal film. EELS measurements of the GaAs surface transitions, in particular the 20 eV surface exciton,[7,8] confirm the increased homogeneity of the overlayer at LT. This transition, which involves the Ga-derived empty surface state, is highly sensitive to surface contamination, and is eliminated when electronic state of the Ga is modified.[8,9] Similar reduction in clustering of Al at LT agrees with previous LEED, AES, and EELS experiments.[8,10]

With Au and Ag, the substrate core level intensity plots as a function of coverage indicate that clustering and/or interdiffusion is reduced at LT, although not as drastically as with In and Al. The LT effect on the attenuation of the substrate core levels is about the same for Au, which interdiffuses with GaAs,[11] as for Ag, which does not.

Sn on GaAs gives a different overlayer morphology than the other metals studied here. The attenuation of the SXPS core level and As AES peak amplitudes as a function of coverage at RT, shown in Fig. 2, yields a straight line with a

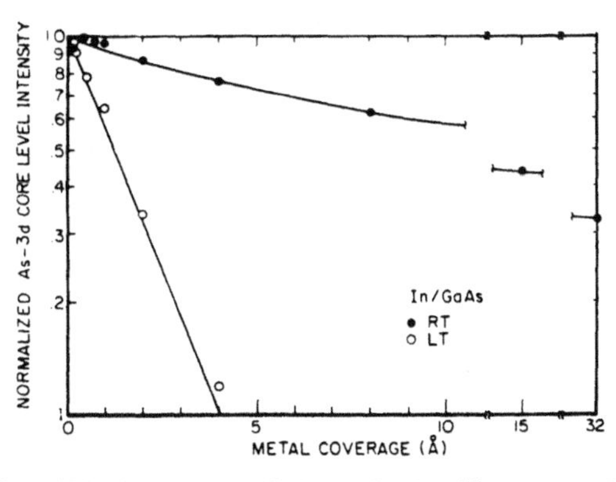

FIG. 1. PES substrate peak amplitudes as a function of In coverage on RT and LT GaAs(110). The exponential decrease at LT yields an electron escape depth of 2 Å.

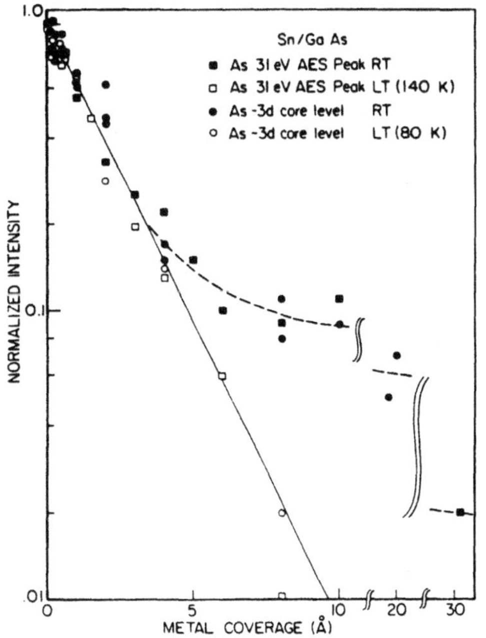

FIG. 2. AES and PES substrate peak amplitudes as a function of Sn coverage on RT and LT GaAs(110). The exponential decrease at LT yields an electron escape depth of 2 Å.

slope indicating an electron escape length of 2.0 Å up to a coverage of about 4 Å. Beyond that coverage, the core level attenuation slows abruptly as Sn follows a Stranski–Krastanov growth mode. The first 1–2 ML grow layer-by-layer and form a continuous disordered film. Clustering starts on top of this film. The plot of the core level attenuation shows that clustering beyond 4 Å is largely inhibited at LT (Fig. 2).

The changes occurring in the GaAs LEED intensity profiles as a function of Al deposition at LT show that the relaxation of the clean substrate is gradually eliminated.[10] This derelaxation can be expected since the adatoms fulfill the valency requirements of the substrate surface atoms and recreate a more bulklike environment. A regular distribution of $\frac{1}{2}$ML or 1 ML of adatoms, equivalent to one or two Al per unit cell, should complete the derelaxation. The fact that 2 ML Al are required indicates that residual clustering with cluster sizes limited to a few atoms persists at 80 K. Clustering at these temperatures probably results for a large part from surface hopping necessary to dissipate the thermal energy of the incoming adatoms. The derelaxation is observable with LEED only at LT. At RT, clustering leaves a substantial fraction of the surface uncovered and relaxed. Derelaxation under the large RT clusters also competes with the disorder induced by the Al–Ga exchange.[12] Finally, the LT derelaxation is reversible in that the relaxation reappears on areas of clean GaAs exposed by Al clustering when the sample is warmed to RT.[10]

Adatom-induced changes in the substrate relaxation are not unique to the Al interface. We have observed similar changes in the GaAs LEED intensities with the RT deposition of Au and Sn. For Au, it is still unclear whether this is due to a chemisorption-induced derelaxation, as with Al, Sn, and Sb,[13] or to bond breaking in the top GaAs layer, as was observed with the adsorption of oxygen.[14] These geometrical changes affect the electronic structure of the interface. On the clean surface, the Ga- and As-derived dangling bond surface states are swept out of the gap by the relaxation. On the derelaxed surface, however, unsaturated substrate dangling bonds could provide energy levels in the gap and participate in the initial pinning of E_F.

SXPS gives a direct measure of band bending and E_F movement from the rigid energy shifts of the substrate core levels as a function of metal coverage. This is shown for Al/GaAs in Fig. 3. At RT, band bending is rapid and approximately symmetric on *n*- and *p*-GaAs, and E_F reaches a position close to its final pinning energy with 0.5 Å Al. This rapid and symmetric movement is characteristic of many metal–GaAs interfaces.[2] There is a striking contrast between E_F movements at RT and LT. On the LT *n* substrate, the movement is considerably slowed down with respect to that at RT. The curve is nearly flat up to 2 Å, then rapidly drops toward its final RT position. On the other hand, the movement on the LT *p* substrate is more rapid than at RT and is essentially complete with 0.02 Å Al. There is also an overshoot of E_F with respect to its final position that is not observed at RT.

The movement of E_F with In coverage is similar to that obtained with Al in that it is rapid and approximately symmetric on RT *n* and *p* substrates (Fig. 4). At LT, the data of Chin *et al.*[15] on *n*-GaAs show that E_F moves down very slowly. Since this study was not extended to high coverages, it is not known whether E_F eventually approaches its RT position as it does for Al. Recent experiments do not show this happening up to coverages of 8 Å.[16] Our measurements on LT *p*-GaAs show that, as with Al, E_F moves up in the band gap faster than at RT. The overshoot with respect to the final position is also observed between 0.05 and 2 Å.

The two noble metals of our series, Au and Ag, follow the trend set by the column III metals, although some differences must be noted. The data are presented in Figs. 5 and 6. As with Al and In, E_F moves much slower on *n*-GaAs at LT

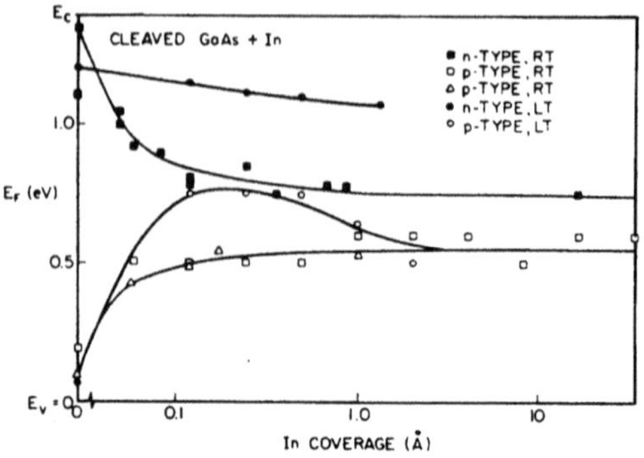

FIG. 4. Same as Fig. 3 for In/GaAs(110). (The LT data on *n*-GaAs are from Ref. 15.)

than at RT, up to 1.5–2 Å, then rapidly drops to its RT position. On the LT *p* substrate, E_F moves with both metals considerably faster than on the *n* substrate. The movement is as fast as at RT with Ag, somewhat slower with Au. We also do not observe as much overshoot as with the column III metals, and the overshoot occurs only around 2 Å with Au.

The Sn/GaAs interface is unusual in that E_F moves slower on *n*- than on *p*-GaAs at RT (Fig. 7). To our knowledge, band bending at this interface has not been previously examined. Sn is unique among the metals studied here in that it does not cluster significantly at low coverage and at RT, and does not react with GaAs. Whether one considers defects or metal-induced states as pinning mechanisms, the RT *n*-type case appears consistent with the slow formation of defect-forming or metallic clusters. The LT behavior is in accord with the trend set by the other metals. Preliminary ultraviolet photoemission spectroscopy (UPS) results indicate that E_F pinning is slowed down on *n*-GaAs. On *p*-GaAs, E_F moves somewhat slower at LT than at RT, but again significantly faster than on *n*-GaAs.

IV. DISCUSSION

These low temperature studies provide us with a snapshot of E_F pinning at low coverage with delayed interface formation. Temperature clearly affects the morphology of the met-

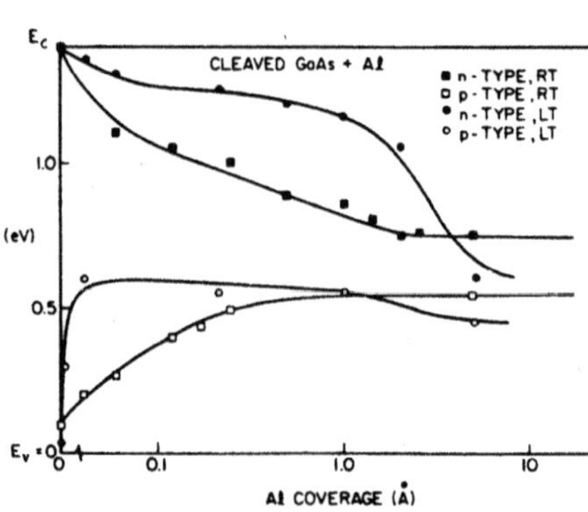

FIG. 3. Position of E_F in the gap of *n* and *p*-GaAs(110) as a function of Al coverage at RT and LT.

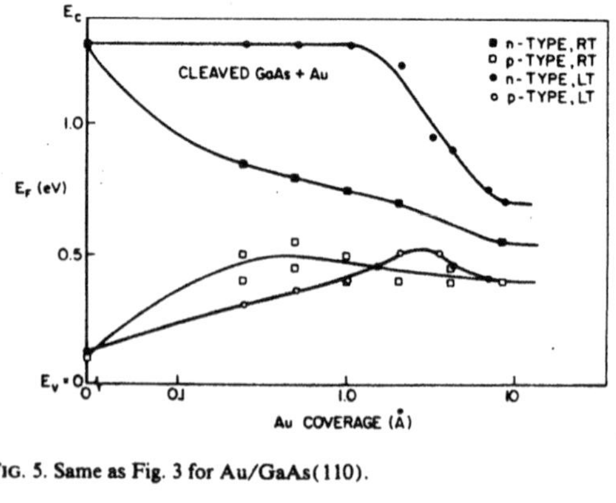

FIG. 5. Same as Fig. 3 for Au/GaAs(110).

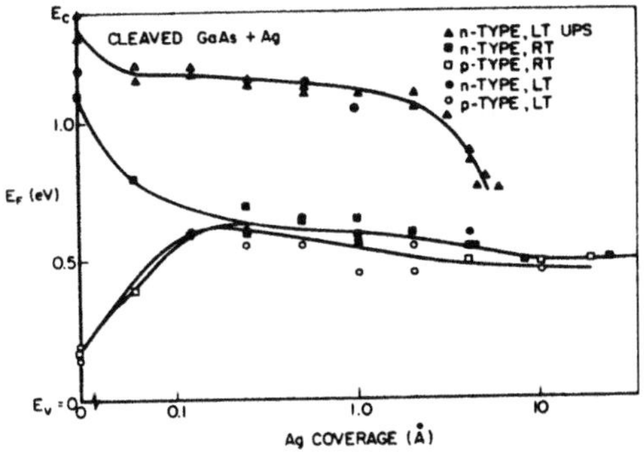

FIG. 6. Same as Fig. 3 for Ag/GaAs(110).

al–GaAs interfaces substantially. It is not clear, however, how this effect relates to surface band bending. On GaAs(110), clustering and/or interdiffusion is almost entirely eliminated at LT for Al and In, and is partially inhibited for Au and Ag. In addition, although Sn exhibits little clustering below 2–3 Å at RT, we can assume that any tendency to cluster is further inhibited at LT. There is a striking similarity in the movement of E_F on LT n-GaAs with Al, Au, and Ag. E_F moves slowly until some critical coverage, where it drops quickly toward its RT position. Preliminary Sn data show the same behavior. In might be an exception to this empirical rule, although the data are still incomplete. In the framework of defect-based models, this slowdown is consistent with a rate of defect formation that is reduced by the inhibition of clustering or chemical reaction at LT. It was suggested in an earlier study of E_F pinning with ultralow coverages of Al that clustering and condensation of metal atoms might release enough energy to break surface bonds and create defects.[17] The increase in pinning rate above 2 Å could correspond to an increase in the rate of clustering with larger concentrations of metal on top of the first homogeneous layer. The sudden drop might also correspond to the onset of metallic character in the overlayer and MIGS at the interface.

The asymmetry between the E_F movements on LT p-GaAs and n-GaAs suggests that different mechanisms are responsible for the formation of interface states that initially pin the p and n surfaces. If we examine the results of the p-type experiments within the framework of defect models, fast pinning on p-GaAs would imply that different defects with different kinetics of formation might be responsible for initial band bending, at least at LT. Pinning by multiple defects, e.g., double donor As antisite compensated by a Ga antisite with a different kinetics of formation, was recently mentioned.[18] Our data would imply that the rates of formation of these defects at LT be orders of magnitude different. This would, however, constitute a substantial increase in the level of complexity of the model that cannot be directly substantiated by experiment.

Fast pinning on LT p-GaAs suggests the role of a completely different mechanism. One can speculate that it is related to structural changes at the interface, which, if affected by the temperature, can only be more extensive at LT as the overlayer is more homogeneous and covers, for a given coverage, a larger fraction of the surface. Changes in the relaxation of the semiconductor, as observed with Al, Au, and Sn, can bring dangling bond states swept out of the gap by the clean surface relaxation back into the gap. Some of these surface bonds remain free at low coverage and, if donorlike, could participate in E_F pinning on the p substrate. Initial pinning on LT p-GaAs with Au and Ag is not as fast as with Al and In. The structural measurements discussed above, however, indicate that the elimination of clustering or reaction is not as efficient with these elements as with column III metals. A less complete derelaxation of the surface is therefore expected with these two metals.

Finally, it was proposed that in the absence of defects, which might effectively be prevented from forming at LT, the initial band bending might correspond to the Schottky limit,[16] which is to say that it becomes controlled by the metal work function. This explanation was brought about by the observation that, with Al and In, E_F was moving toward positions compatible with their respective work functions, in the upper part of the gap, producing little band bending on n-GaAs and a fast E_F movement on p-GaAs. This possibility cannot be ruled out in the cases of Al and In, although the band bending on p-GaAs at LT seems somewhat small to be consistent with this explanation. However, this model clearly cannot explain the Au and Ag results, since the large work functions of these metals would lead to large band bending on n-GaAs, which is the opposite of what is observed.

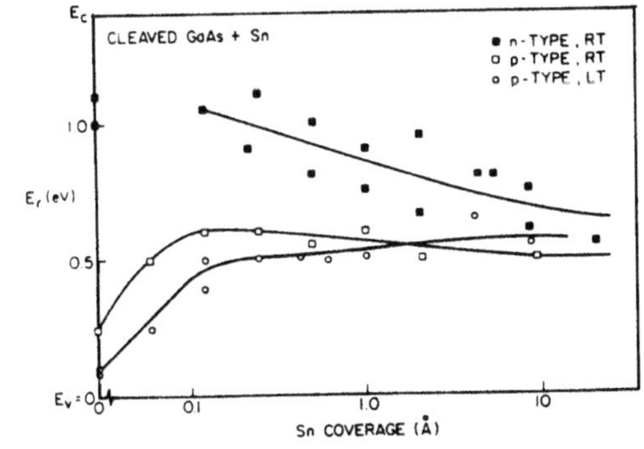

FIG. 7. Same as Fig. 4 for Sn/GaAs(110).

V. CONCLUSIONS

With the assumption that the LT experiments simplify the observation of the initial formation of the interface by slowing down all metal–semiconductor interactions, our results suggest initial pinning behaviors that cannot be explained with the current published models solely based on defects or metal-induced gap states. In particular, it appears difficult to reconcile the asymmetric E_F movements on n and p substrates with a single pinning mechanism. The qualitatively identical results obtained at interfaces between GaAs and five different metals suggest that several pinning mechanisms must now be considered. Low temperature has opposite effects on these mechanisms and, consequently, on the initial pinning rates on *n* and *p* substrates. Pinning on *p*-GaAs, in particular, calls for mechanisms that, unlike defects, are not substantially influenced by temperature. The derelaxation of the substrate by the metal adatoms could result in dangling bond states in the semiconductor gap that participate in the pinning. The existence and role of residual dangling bond states, at the completed interface, should be further investigated. Finally, the evolution of E_F above 1 Å on both *n* and *p* substrates shows that additional mechanisms come into play at higher coverages. These mechanisms are presumably responsible for the final E_F pinning position.

ACKNOWLEDGMENTS

This work was supported by the National Science Foundation Grant No. DMR-8406820 and by the Office of Naval Research. Partial support was also provided by Xerox Corporation and General Electric.

[1]W. E. Spicer, P. W. Chye, P. R. Skeath, C. Y. Su, and I. Lindau, J. Vac. Sci. Technol. **16**, 1427 (1979).
[2]L. J. Brillson, Surf. Sci. Rep. **2**, 123 (1982).
[3]J. L. Freeouf and J. M. Woodall, Appl. Phys. Lett. **39**, 727 (1981).
[4]J. Tersoff, Phys. Rev. Lett. **52**, 465 (1984), and references therein.
[5]K. Stiles, K. Young, A. Kahn, D. Kilday, N. Tache, and G. Margaritondo, Phys. Rev. B (to be published).
[6]D. E. Savage and M. G. Lagally, J. Vac. Sci. Technol. B **4**, 943 (1986).
[7]H. Lüth, M. Büchel, R. Donn, M. Liehr, and R. Matz, Phys. Rev. B **15**, 865 (1977).
[8]C. R. Bonapace, D. Tu, K. Li, and A. Kahn, J. Vac. Sci. Technol. B **3**, 1099 (1985).
[9]F. Bartel, L. Surkamp, H. J. Clemens, and W. Mönch, J. Vac. Sci. Technol. B **1**, 756 (1983).
[10]C. R. Bonapace, K. Li, and A. Kahn, J. Phys. (Paris) Colloq. **45**, 409 (1984).
[11]P. W. Chye, I. Lindau, P. Pianetta, C. M. Garner, C. Y. Su, and W. E. Spicer, Phys. Rev. B **18**, 5545 (1978).
[12]L. Smit and J. F. van der Veen, Appl. Surf. Sci. **26**, 230 (1986).
[13]C. B. Duke, A. Paton, W. K. Ford, J. Carelli, and A. Kahn, Phys. Rev. B **26**, 803 (1982).
[14]A. Kahn, D. Katnani, P. Mark, P. Chye, C. Su, I. Lindau, and W. E. Spicer, Surf. Sci. **87**, 325 (1979).
[15]K. K. Chin, T. Kendelewicz, C. McCant, R. Cao, K. Miyano, I. Lindau, and W. E. Spicer, J. Vac. Sci. Technol. A **4**, 969 (1986).
[16]T. Kendelewicz, M. D. Williams, K. K. Chin, I. Lindau, and W. E. Spicer, paper presented at the 33rd National Symposium of the American Vacuum Society, Baltimore, 1986.
[17]R. R. Daniels, A. D. Katnani, Te-Xiu Zhao, G. Margaritondo, and A. Zunger, Phys. Rev. Lett. **49**, 895 (1982).
[18]N. Newman, W. E. Spicer, T. Kendelewicz, and I. Lindau, J. Vac. Sci. Technol. B **4**, 931 (1986).

Kinetics study of initial stage band bending at metal GaAs(110) interfaces

Renyu Cao, Ken Miyano,[a] Tom Kendelewicz, Ken K. Chin,[b] Ingolf Lindau, and William E. Spicer[c]
Stanford Electronics Laboratory, Stanford, California 94305

(Received 27 January 1987; accepted 20 April 1987)

A systematic kinetics study of the initial stage band bending of metal/GaAs(110) has been done. At low temperature, independent of metals, enhanced band bending for *p*-GaAs and attenuated band bending for *n*-GaAs as a function of metal coverage has been observed. For *p*-type GaAs the band bending increases very rapidly to a value larger than that for any coverage at room temperature and then returns to the room temperature position at high coverages. It is found that this excess band bending strongly depends on the work functions of the metal deposited on GaAs. In gives 0.4 eV, Al 0.3 eV, Ag 0.2 eV, and Au only shows small excess band bending. Temperature dependence of GaAs initial stage band bending will also be discussed in this paper. The results strongly suggest that there exists another mechanism which appears to be dominant in the initial stage band bending at low temperature where the defect formation is negligible. The transition from this mechanism to the defect mechanism could be observed by increasing either system temperature or metal coverage, or more generally, disturbance to semiconductor surfaces.

I. INTRODUCTION

Schottky barrier formation at metal/semiconductor interfaces still remains as a puzzle even after more than four decades of study. By now most studies have been done at room temperature or even higher temperature. However, the real metal/semiconductor interfaces are very complicated and many things happen simultaneously, such as metal clustering, interdiffusion, chemical reaction, etc. All of these actions will affect the Schottky barrier formation. Therefore isolating some effects becomes very important in understanding the fundamental physics behind it. Last year, two groups[1–4] published some experimental results of Schottky barrier formation study at metal/GaAs(110) interfaces at low temperature. These results, which showed dramatically different band-bending behavior of metal/GaAs at low temperature compared with that at room temperature, challenge the present models about Schottky barrier formation. However, so far this kind of study has only been done for few systems. To better understand low temperature metal/semiconductor systems and discover the involved physics a systematic study has to be performed.

In this paper we will report our recent work on temperature dependence of the initial stage band-bending of In, Al, Ag, and Au on cleaved *n*- and *p*-type GaAs surfaces as a continuation of our previous work.[2] These systems have been extensively studied before with different techniques and most features have been revealed. For instance, In, Al, and Au all tend to form clusters or islands on GaAs surfaces at room temperature. The formation of clusters disturbs the semiconductor surface and triggers some reactions between metal overlayer and substrate semiconductor,[5] e.g., the Al-replacement reaction with GaAs. We consider all of these as a disturbance to the semiconductor surfaces. On the other hand, some theoretical and experimental work conclude that at low temperature metal clustering is largely inhibited and better homogeneity will be obtained.[6,7] Hence the effects induced by metal clustering will expect to be inhibited or generally the disturbance to the semiconductor surfaces will be reduced at low temperature. In our work the correlation between the semiconductor band bending and system temperature and metal coverage was systematically studied. Semiconductor-surface disturbance was reduced by decreasing the system temperature as well as the metal deposition. The evolution of semiconductor band bending and the surface Fermi level movement has been monitored. Slow band bending for *n*-GaAs and fast band bending for *p*-GaAs at low temperature has been observed as a metal independent behavior. For the first time the initial stage band bending for metal on *p*-GaAs at low temperature is found strongly depending on the work function of the metals deposited on the GaAs surface. As to be discussed later, such a phenomenon can not be explained by present models and indicates a new mechanism dominant there. It is also discovered that the transition from this low temperature mechanism to the "normal" room temperature mechanism can be observed by increasing the disturbance to the semiconductor surface (increasing temperature or the metal coverage).

II. EXPERIMENT

The experiments were performed at the 4° beam line at the Stanford Synchrotron Radiation Laboratory and a monochromatized He-discharge lamp at the Stanford Electronics Laboratories. A standard ultrahigh vacuum (UHV) chamber equipped with a double pass cylindrical mirror analyzer was used with the base pressure around 5×10^{-11} Torr. All the preparations were done *in situ* with operation pressure in the 10^{-10} Torr range. In order to perform the initial band-bending study a recently designed low-coverage evaporator[8] was used with which various metal evaporations were done starting from as low as 10^{-4} monolayers (ML) [8.94×10^{14} atoms/cm^2 on GaAs(110)] up to a few monolayers without serious heating effect on the sample surface. Surface sensitive photoelectron spectroscopy (PES) was used to track the semiconductor band bending as well as surface Fermi level position when different metals were deposited on the cleaved-clean GaAs(110) surfaces. Low temperature was obtained by feeding liquid nitrogen directly through a reservoir attached to the back of the sample and the temperature was monitored by a Chromel–Alumel thermocouple. Liquid-nitrogen temperature was obtained. Intermediate low temperatures could be obtained by controlling the liquid-nitrogen flow. The detail information about the experiment can be found in the Refs. 2 and 9.

III. RESULTS

Figures 1(a) and 1(b) present Ga 3*d* and As 3*d* core level photoemission spectra from *p*-GaAs(110) as a function of Al coverage at room temperature. (RT). When 0.001 ML of Al is deposited on the surface of GaAs(110) both Ga 3*d* and As 3*d* peaks start shifting towards the low-kinetic energy or high binding energy direction and they continue shifting as more Al is deposited. Note that with low Al coverage both Ga 3*d* and As 3*d* only show rigid shift without any change of the line shape. Also at a given Al coverage both Ga 3*d* and As 3*d* peaks (as well as valence-band maximum) shift in the same direction by the same amount. These facts indicate that the shifts represent the semiconductor band bending at the surface. At high-Al coverage a second peak appears on the high-kinetic energy side of the Ga 3*d* spectra. It has been identified as free Ga due to Al-replacement reaction with the substrate GaAs triggered by Al cluster formation.[5] Figures 1(c) and 1(d) show the photoemission spectra of Ga 3*d* and As 3*d* of *p*-GaAs(110) at different Al coverage at low temperature (−90 °C). Rigid shifts of both Ga 3*d* and As 3*d* are observed again. However, instead of continuing shifting monotonically towards the low kinetic energy direction,

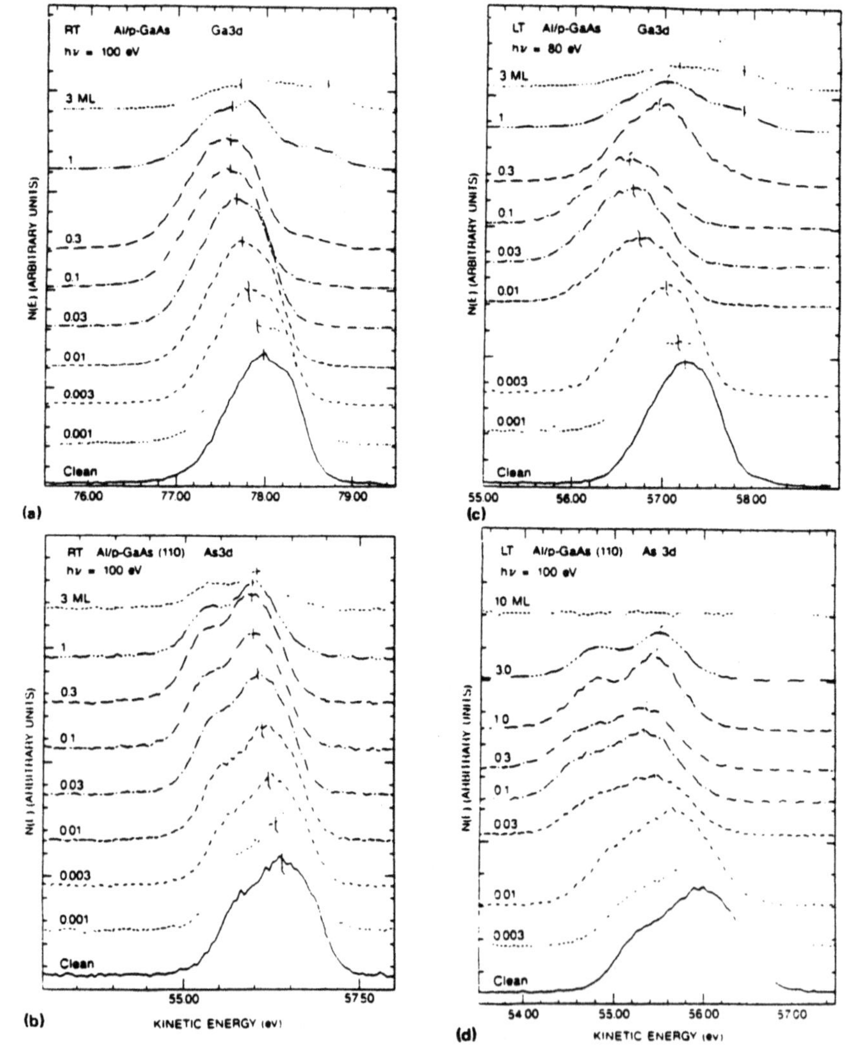

FIG. 1. PES core level spectra of *p*-GaAs as a function of Al coverage at room temperature and low temperature (−900 °C). (a) Ga 3*d*, RT; (b) As 3*d*, RT; (c) Ga 3*d*, LT; (d) As 3*d*, LT.

both Ga 3*d* and As 3*d* peak shifts reach a maximum around 0.01 to 0.1-ML-Al coverage and then drop back with further Al deposition.

The surface Fermi level positions versus Al coverage for Al/*p*-GaAs(110) at room temperature and low temperature (LT) are presented in a band-bending diagram in Fig. 2(a), including data from Al/*n*-GaAs(110). Right after the cleave flat bands are obtained and the surface Fermi levels are close to conduction-band minimum (CBM) for *n*-GaAs and valence-band maximum (VBM) for *p*-GaAs. Nonperfect cleave might induce some initial band bending for the semiconductor, which has been discussed in our report before.[10] For room temperature Al/GaAs, surface Fermi level positions (circles) monotonically change with the Al coverage and towards their final pinning positions around midgap. The band-bending curves for *n*-GaAs and *p*-GaAs are fairly symmetric. The band-bending curves for Al/GaAs at low temperature (squares) show some interesting features. In the low coverage regime (< 0.1 ML), at a given Al coverage, *n*-GaAs gives little band bending while *p*-GaAs gives large band bending, as a result, instead of symmetric band bending, the band bending for *n*- and *p*-GaAs appears very asymmetric. Additionally, *p*-GaAs shows a maximum band bending which is larger than its final value, around 0.01 to 0.1 ML of Al coverage, the excess band bending is about 0.2 eV. This kind of overshooting of the band bending is rarely observed for room temperature systems. With thick-Al coverage *p*-GaAs surface Fermi level positions merge to their corresponding room temperature final pinning positions within the experimental limit. *n*-GaAs shows the similar tendency as seen from the figure.

In addition to Al/GaAs, we also performed the kinetic study of the initial-stage band bending for In/GaAs(110), Ag/GaAs(110), and Au/GaAs(110) interfaces. Similar PES core-level spectra, which are not shown in this paper, were taken. The band-bending diagrams for these systems are also given in Figs. 2(b)–2(d). It is interesting to note that Al/GaAs is not the only case showing the type of temperature dependent band-bending behavior discussed above. All these systems show the similar behavior. At room temperature In/GaAs band bending already shows some asymmetry, which will be discussed somewhere else,[11] and at low temperature this asymmetric band bending becomes stronger, especially the overshoot for *p*-GaAs. We would also like to point out that our results are generally consistent with previous reports by Kahn *et al.*[3,4] although their low temperature Ag/GaAs data do not show the same trend as other metal/GaAs systems and our Au/*p*-GaAs(110) shows that at low temperature GaAs initial-band bending is not slower than that at room temperature.

Furthermore, in order to see the clear temperature dependence of the initial-stage band bending of metal/GaAs similar studies for some of our systems were performed at several intermediate temperatures. The band-bending diagrams of In and Al on *p*-GaAs(110) as a function of metal coverages as well as system temperatures are shown in Fig. 3. For a given system larger-semiconductor band bending and overshoot are observed at lower temperature. Our experimental data show that for In/*p*-GaAs(110), at −190 °C the excess band bending is 0.4 eV, while at −130 °C it gives 0.15 eV, and for Al/*p*-GaAs(110), at −190 °C the excess band bending is 0.3 eV, while at −90 °C it gives 0.2 eV. It is worthwhile to point out that with thick-metal coverage the

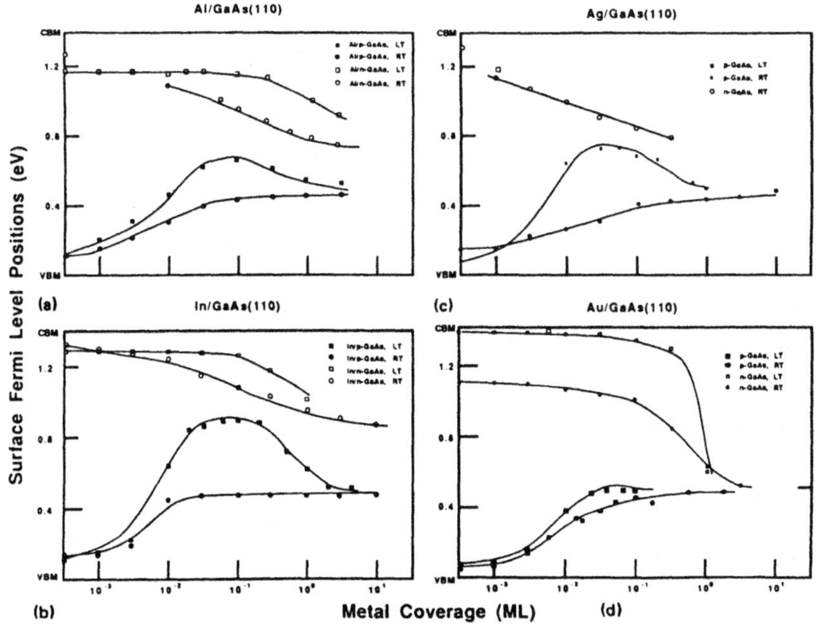

FIG. 2. Band-bending diagrams of GaAs as a function of metal coverages at RT and LT (−190 °C). (a) Al/GaAs (−90 °C); (b) In/GaAs; (c) Ag/GaAs; (d) Au/GaAs.

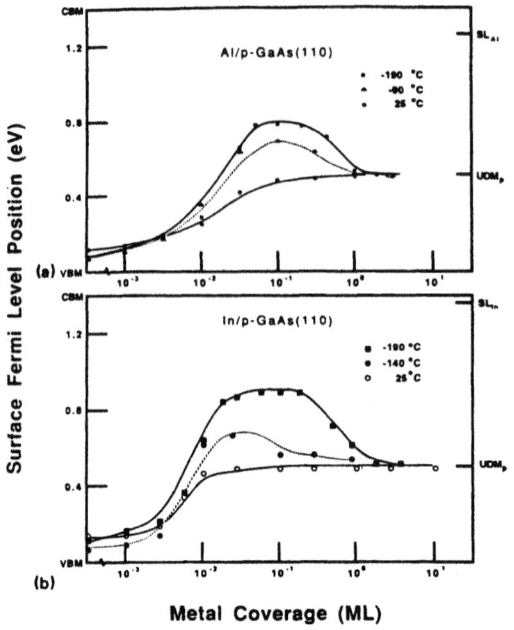

FIG. 3. Band-bending diagrams of GaAs as a function of metal coverages at different temperatures (a) In/GaAs; (b) Al/GaAs.

final surface Fermi levels are all at the same position, independent of temperature.

IV. DISCUSSIONS

From our experimental data of the kinetic study on initial-band bending of metal/GaAs(110) the systematical features have been disclosed: the more or less symmetric band bending behavior at room temperature and very asymmetric-band-bending behavior at low temperature. These results along with the overshooting of *p*-GaAs at low temperature really challenge the present models of Schottky barrier formation at metal/semiconductor interfaces[12] because there are two serious problems for most of them: first they can not explain the asymmetric band-bending behavior of *n*- and *p*-type semiconductors, even for some systems at room temperature (e.g., In/GaAs), and second they can not explain the difference of band bending upon temperature.

The unified defect model (UDM) proposed by Spicer and co-workers[13] seems to explain the experimental data about room temperature Schottky barrier formation at metal/semiconductor interfaces pretty well. As stated in the Introduction, for most systems we are studying, the deposited metal forms clusters at room temperature and the energy released by the cluster formation triggers some reaction between the adatoms and the substrate and then creates defects, in turn, donor or acceptor levels inside the semiconductor-band gap. It is these defect levels that are responsible for the semiconductor-band bending. At low temperature, however, the more uniform metal overlayer and less metal clusters as well as less disturbance to the semiconductors surface imply that less defects are formed. Hence under the framework of UDM less band bending for both *n*- and *p*-type semiconductors might be expected at low metal coverage regime, which is in contradiction to our data and those of Kahn *et al.*[3,4] Therefore if we accept the UDM explanation for room temperature systems, we have to agree that at low temperature there exists another mechanism which is quite different from that (or those) at room temperature and plays an important role in determining the initial stage band bending of the semiconductor and the Schottky barrier formation.

To find some features about this "second mechanism" let us see the stacked band-bending curves for different metals on *p*-GaAs(110) presented in Fig. 4. Even though all of the curves show the same trend (fast band bending compared with room temperature curve and excess band bending over the finally stablized value) the values of the excess band bending of the semiconductor are very different from metal to metal. In shows largest excess band bending by 0.4 eV, Al by 0.3 eV, Ag by 0.2 eV, and Au shows a small excess band bending. That is, the initial stage band bending of the semiconductor is dependent on some properties of the metals deposited on the surface of the semiconductor. When we look at the work function of the metals used in our experiment, In has 4.1 eV, Al has 4.24 eV, Ag has 4.7 eV, and Au has 5.1 eV. Although these values may vary with different crystal faces of metals the general trend will hold. The electron affinity for metal atoms has the similar trend. Then it is clear that the lower the metal-work function, the higher the excess band bending. This simple relationship between the initial-stage band bending and the metal-work function tempts us to consider a Schottky-like mechanism[14] is dominant there. In Fig. 4, we also mark the surface Fermi level positions calculated by the original Schottky model (electron affinity of GaAs is 4.07 eV). It looks that in the ultralow coverage regime, there is a tendency that the surface Fermi levels are moving towards the final positions predicted by Schottky. Schottky mechanism has been observed before at

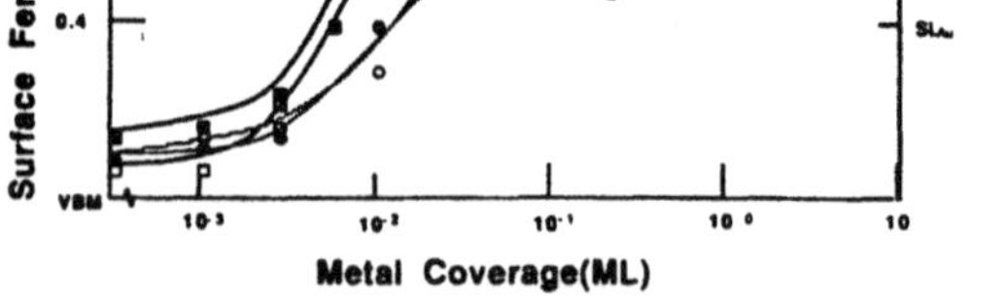

FIG. 4. Comparison of *p*-GaAs band bending diagrams with different metals (In, Al, Ag, Au) at low temperature (−190 °C) These metals have different work functions and the surface Fermi level positions in ideal Schottky cases are marked by SL. The surface Fermi level position predicted by UDM is marked by UDM.

metal/ionic semiconductor interface, such as CaS, ZnS, where the surface state density is believed very low.[15] In our experiment low temperature and low-metal deposition reduce the disturbance to the semiconductor surface and the defect formation is largely inhibited. This fact again might make us think the Schottky-like mechanism is dominant in our case. However, a precaution has to be made if one tries to use simple Schottky explanation. The band-bending diagram of metal/n-GaAs at low temperature shows a very slow initial band bending, even for Au, which has a high-work function and would be supposed to show a fast initial-band bending if the original Schottky picture were applicable. Nevertheless after we evaporate more metal on the semiconductor surfaces, the surface disturbance is increased and the number of defect states is increased too. Meanwhile, due to this disturbance this low temperature mechanism seems breakdown and defect or other mechanisms play a more and more crucial role in determining the surface Fermi level position. That is why with thick-metal coverage, the surface Fermi level position goes back to the final pinning position predicted by the UDM, as their corresponding room temperature systems.

Therefore the overall picture is that there are two mechanisms competing each other, one is dominant at low temperature, another is dominant at room temperature, such as defect mechanism. At low temperature with low metal coverage, less defects are formed, less disturbance makes low temperature mechanism dominant for the initial-band bending, while with higher temperature or more metal atoms, surface disturbance increases, defect states increase too, this mechanism becomes weak and defect mechanism becomes stronger and finally takes over. The competence can be clearly seen from Fig. 3, where temperature dependence of the semiconductor-band bending is shown. At the initial band-bending stage the lower system temperature results in the less number of defect states in the surface region and the larger excess band bending value, indicating in this case the low temperature mechanism has stronger effect. At intermediate low temperatures, it shows weak effect and the overshooting becomes small. At room temperature defects are completely dominant from the beginning and low-temperature mechanism does not show any effect at all, as a result the "normal" band-bending behavior appears. In the temperature range where the experiments were done more metal will introduce more disturbance to the semiconductor surface and may cause the existing low temperature mechanism breakdown, that is probably why with thick metal coverage all the surface Fermi levels merger to the room temperature finally stablized value indicating that the defect mechanism is totally dominant. Above all, our experimental data have shown the transition between these two mechanisms by increasing the surface disturbance.

V. CONCLUSIONS

We have done systematic kinetic study of initial-stage band bending at metal/GaAs(110) interfaces. The dependence of metal work function as well as temperature of semiconductor-band bending has been observed for the first time. We propose that at low temperature, a new mechanism which is quite different from those at room temperature may play a dominant role in determining the initial stage band bending and Schottky barrier formation at metal/semiconductor interfaces. To fully understand this behavior and get an overall picture of the Schottky barrier formation at metal/semiconductor, more experimental and theoretical work is needed.

ACKNOWLEDGMENTS

This work was supported by ONR under Contract No. N00014-85-K-0736. The experiment was partially done at Stanford Synchrotron Radiation Laboratory which is supported by the Department of Energy, Office of Basic Science Research, and the National Science Foundation, Division of Materials Research.

[a] AT&T Bell Laboratories Scholar.
[b] Present address: Department of Physics, University of Notre Dame, Notre, Dame, IN 46556.
[c] Stanford Ascherman Professor of Engineering.

[1] T. Kendelewicz, M. D. Williams, K. K. Chin, C. E. McCants, R. S. List, I. Lindau, and W. E. Spicer, Appl. Phys. Lett. **48**, 919 (1986).
[2] K. K. Chin, R. Cao, T. Kendelewicz, K. Miyano, M. D. Williams, S. Doniach, I. Lindau, and W. E. Spicer, Proceedings of the Materials Research Society, 1986.
[3] K. Stiles, A. Kahn, D. G. Kilday, and G. Margaritondo, J. Vac. Sci. Technol. A **5**, 1527 (1987).
[4] K. Stiles, A. Kahn, D. G. Kilday, G. Margaritondo, Workshop on III–V Semiconductors, Stanford, 1986.
[5] A. Zunger, Phys. Rev. B **24**, 4372 (1981).
[6] J. Ihm and J. D. Joannopoulos, Phys. Rev. B **26**, 4429 (1982).
[7] C. R. Bonapace, D. W. Tu, K. Li, and A. Kahn, J. Vac. Sci. Technol. B 3, 1099 (1985).
[8] K. K. Chin, P. McKernan, and I. Linau, J. Vac. Sci. Technol. A **4**, 1949 (1986).
[9] K. K. Chin, T. Kendelewicz, C. McCants, R. Cao, K. Miyano, I. Lindau, and W. E. Spicer, J. Vac. Sci. Technol. A **4**, 969 (1986).
[10] K. K. Chin, R. Cao, K. Miyano, C. E. McCants, I. Lindau, and W. E. Spicer, Mater. Res. Soc. Symp. Proc. **54**, 341 (1986).
[11] R. Cao, K. Miyano, I. Lindau, and W. E. Spicer (to be published).
[12] See, e.g. J. Tersoff, Phys. Rev. Lett. **52**, 465 (1984).
[13] W. E. Spicer, P. W. Chye, P. R. Skeath, C. Y. Su, and I. Lindau, J. Vac. Sci. Technol. **16**, 1422 (1979); W. E. Spicer, I. Lindau, P. Skeath, C. Y. Su, and P. Chye, Phys. Rev. Lett. **44**, 420 (1980).
[14] W. Schottky, Z. Phys. **113**, 367 (1939).
[15] S. Kurtin, T. C. McGill, and C. A. Mead, Phys. Rev. Lett. **22**, 1433 (1970).

EUROPHYSICS LETTERS 1 December 1987

Europhys. Lett., 4 (11), pp. 1303-1308 (1987)

The Schottky Contact in a Xe/Metal Interface Probed by Inverse Photoemission.

W. JACOB, E. BERTEL and V. DOSE

Max-Planck-Institut für Plasmaphysik, EURATOM Assoziation
D-8046 Garching/München, FRG

(received 15 June 1987; accepted in final form 7 September 1987)

PACS. 73.30. – Surface double layers, Schottky barriers and work functions.
PACS. 71.25T – Band structure of crystalline semiconductor compounds and insulators.
PACS. 79.20K – Other electron impact phenomena.

Abstract. – Inverse photoemission spectra of Xe on Ag(110) have been measured as a function of coverage. Work function changes have been induced by oxygen and potassium exposure of Ag(110) and the corresponding peak shifts in inverse-photoemission spectra of adsorbed xenon have been determined. The data are consistently interpreted in a Schottky-barrier model of the Xe/Ag(110) interface. No screening-dependent final-state energy shifts could be observed.

The adsorption of rare gases on metal surfaces has been studied frequently in recent years. Due to the absence of strong chemical interactions, it should provide insight into the nature of the dispersion-type forces responsible for the bonding of rare-gas atoms to the substrate [1]. Another interesting feature of such rare gas/metal systems is the interaction of excited states of the adsorbate with the metal substrate. This interaction determines, for instance, whether the energy shifts in core level photoemission are dominated by image charge screening effects [2, 3] or by work function changes [4-6]. The latter idea, *i.e.* the hypothesis that xenon core level binding energies are invariably «pinned» to the vacuum level, constitutes an essential ingredient to the so-called «PAX»-technique for the characterization of heterogeneous surfaces [6].

Very recently the body of experimental data on rare-gas/metal interfaces was augmented by two groups using inverse photoemission (IPE) to probe layer-dependent energy shifts of the unoccupied Xe levels on different metals [7, 8]. Unfortunately conflicting interpretations emerged from their results, which made additional experiments urgent. In this paper we present such additional data and explain them in the framework of a Schottky barrier being formed between the metal substrate and the growing Xe adlayer. The Schottky barrier in the Xe/metal interface provides an interesting example of a nearly pure Schottky behaviour, *i.e.* a direct correlation between barrier height and metal work function. The deviation from the Schottky limit, however, is measurable and indicates a nonnegligible density of states within the Xe band gap at the Xe/metal interface.

The experiments were performed in a UHV-chamber with base pressure below 10^{-8} Pa pumped by a combination of turbomolecular, ion-getter and liquid-nitrogen–cooled Ti-sublimation pumps. During xenon and oxygen exposures the ion-getter pump was always switched off in order to avoid regurgitation. The Ag(110) surface serving as the substrate was cleaned by standard procedures until good $O(nxl)$ LEED patterns could be generated. Potassium exposures were made by means of SAES alkali-dispensers mounted in a heated glass collimator tube. The lowest crystal temperature obtained by liquid-He cooling was 50 K. The instrumentation for recording the IPE spectra consists basically of an electron gun and a Geiger-Müller counter with CaF window and iodine gas filling at a fixed position relative to each other. Details of the photon detection system are described elsewhere [9]. Work function differences were measured using the diode method. For the work function of the clean Ag(110) surface a value of 4.52 eV [10, 11] was adopted, which is consistent with the image potential state energy of 4.1 eV measured in the present experiments. In fig. 1 a series of inverse-photoemission spectra (IPE) of Xe/Ag(110) is presented as a function of Xe-exposure ($1\,L = 10^{-6}$ Torr s, uncorrected pressure gauge reading). Comparison with the IPE spectrum of a Xe monolayer obtained by heating the sample to slightly above multilayer desorption temperature showed that the monolayer is completed at a Xe exposure of about 15 L.

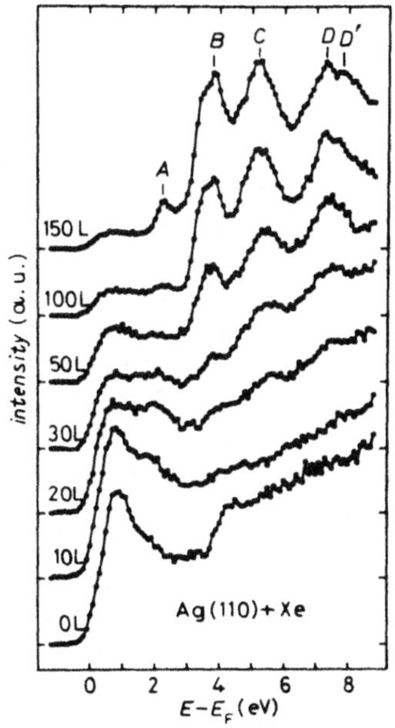

Fig. 1. – Inverse-photoemission spectra of xenon on Ag(110) as a function of exposure. (Angle of incidence $\theta = 5°$, Azimuth ΓKLU).

The IPE spectrum of clean Ag(110) shows a bulk transition just above the Fermi level, a feature at about 2 eV, and the image potential state at 4.1 eV [11]. The feature at 2 eV was previously not resolved in normal incidence IPE spectra of Ag(110) [11] and is probably related to the surface state labelled S_1 in ref. [11], which becomes quite distinct at Y. With increasing Xe-coverage the Ag(110) bulk and surface emission is increasingly suppressed and 4 Xe-related features evolve at about 2.3 eV (A), 3.7 eV (B), 5.3 eV (C) and 7.3-8.1 eV (D, D'), the latter being broad and somewhat structured. Peaks B and D show pronounced dispersion effects as the electron angle of incidence is varied [12]. This k-dependence clearly proves that the Xe-overlayer is ordered. The peak can be classified within the framework of the Xe bulk band structure [13] as follows: Peak B is dominated at normal incidence by a direct transition to the bottom of the conduction band Γ_6^+. Peak C shows little variation with $k_{\parallel}$ and is, therefore, attributed to density-of-states emission (DOS) from the s-band at the Brillouin-zone edges (L_6^+, X_6^+). Peaks D and D' behave differently, if the angle of incidence is varied and are consequently assigned to a direct transition (D) and DOS-emission (D'), respectively, both involving the flat d-bands between Γ_8^+, Γ_7^+ and L_{45}^+, L_6^+. No such straightforward explanation is at hand for peak A, which, according to the assignment given above, is located 1.5 eV below the bottom of the conduction band in the Xe band gap. It is attributed to a defect state occurring on Xe cristallite surfaces, as will be discussed elsewhere [12, 14].

As to the question of layer-by-layer peak shifts, the present results are inconclusive. Both, peak B and C, seem to shift somewhat towards the Fermi-level as the Xe-layer thickens. Peak B, however, is superimposed on the image-potential state induced feature and, therefore, cannot be located unambiguously in the low coverage spectra. The apparent shift of peak C from 5.6 eV at 20 L to 5.3 eV at 150 L exposure parallels the result of Wandelt *et al.* [8] for Xe on Ru(001). The observed shift is roughly consistent with the measured saturation work function change of 500 meV for Xe on Ag(110). However, peak C and peak B can be clearly related to the Xe bulk band structure. Accordingly, an explanation of the shift in terms of band formation during the growth of the second and higher Xe-layers cannot be safely ruled out. The present results are clearly at variance with the findings of Horn *et al.* [7]. From their results these authors deduced an 0.8 eV shift of the Xe 6s level towards higher energy as the Xe film grows from monolayer to multilayer coverages. The discrepancy seems to arise from an erroneous peak assignment in ref. [7], which ascribes the features of an essentially bandlike IPE spectrum to atomic levels.

In order to obtain further insight into work-function–induced level shifts, we prepared Ag(110) surfaces with different work functions in analogy to the experiments performed by Jacobi and Rotermund [15]. A work function increase of about 500 meV was achieved by dosing 500 L O_2 at low temperature onto the silver crystal. At 50 K, this results in the formation of an associatively chemisorbed O_2^{-2} layer [16]. The O_2^{-2} ions are closed-shell systems and, therefore, should be of minor influence on the Ag/Xe interface properties, apart from the additional surface dipole moment. Ag(110) surfaces with reduced work function were prepared by dosing K onto the silver surface. Depending on the K dose, work function changes as large as -2.4 eV can be attained ([1]).

Figure 2 presents a selection of IPE spectra of thick Xe-layers on surfaces with different work functions prepared in the way described above. Obviously, the peak positions in the Xe-IPE spectra are closely correlated to the work function. Table I summarizes peak positions with respect to the Fermi level ($E_B^F = E_B - E_F$) and the vacuum level ($E_B^V = E_B - E_V$), as measured in the present set of experiments, together with measured work function values. The data indicate a constant energy of the unoccupied Xe-levels with reference to the vacuum level independent of the dipole layer in the Xe/metal interface. This is in striking agreement with the complementary results of Jacobi and Rotermund [15] for the occupied Xe-levels obtained by UPS.

All these results can be easily understood by applying the theory of Schottky barriers to the Xe/metal interface [18, 19]. Xe can be considered as a semiconductor with a 9.3 eV wide

([1]) K-adsortpion on Ag(110) results in a surface reconstruction as shown by Hayden *et al.* [17]. Additionally, in the presence of an ionic or metallic K-layer on the Ag(110) surface considerable modifications on the Xe/metal interface may be anticipated. Such effects have actually been observed, but only for K-coverages beyond the work function minimum and will be discussed elsewhere [12].

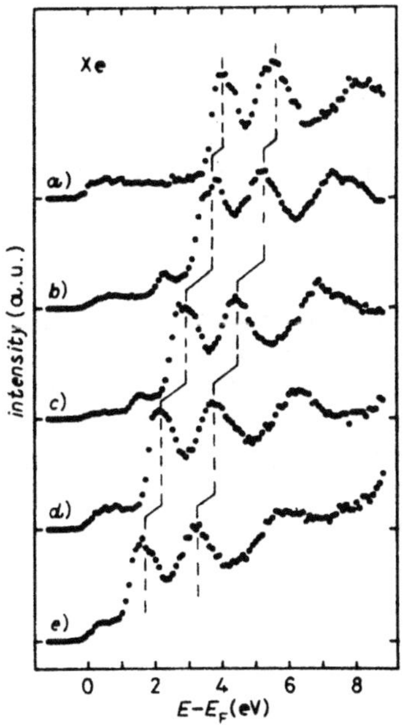

Fig. 2. – Inverse-photoemission spectra of thick Xe-films on differently prepared Ag(110) surfaces: Ag(110) + 500 L O_2(*a*)); Ag(110) (*b*)); Ag(110) + 0.16 ML K (*c*)); Ag(110) + 0.30 ML K (*d*)); Ag(110) + + 0.42 ML K (*e*)).

TABLE I. – *Energy positions of inverse-photoemission features relative to the Fermi level* (E_B^F), *to the vacuum level* (E_B^V) *and measured work functions for the* Xe-*films on differently precovered* Ag(110) *surfaces.* φ_B *is the barrier height of the Schottky contact, i.e* $\varphi_B = E(\Gamma_6^+) - E_F$, χ_S *denotes the electron affinity of* Xe. *All energies are given in* eV.

	work function φ		E_B^F			E_B^V		
	bare and precovered Ag(110) surfaces, respectively	including Xe overlayer	$B(\varphi_B)$	C	D	$B(-\chi_S)$	C	D
Ag(110)	4.5	4.0	3.7	5.3	7.3	−0.3	1.3	3.3
Ag(110) + 500 L O_2	5.0	4.4	4.0	5.5	8.1	−0.4	1.1	3.7
Ag(110) + 0.16 ML K	3.5	3.3	2.8	4.4	6.8	−0.5	1.1	3.5
Ag(110) + 0.30 ML K	2.7	2.5	2.1	3.8	6.2	−0.4	1.3	3.7
Ag(110) + 0.42 ML K	2.2	2.0	1.6	3.2	5.8	−0.4	1.2	3.8

band gap [20]. The expression for the barrier height φ_B at the Xe/metal interface is given by $\varphi_B = \varphi_M - \chi_S + \Delta\varphi$, where φ_M is the metal work function, χ_S is the energy difference between the lowest level of the conduction band Γ_6^+ in the semiconductor and the vacuum level (*i.e.* the electron affinity of the semiconductor), and $\Delta\varphi$ is the potential step in the Xe/metal interface induced by charge transfer upon Xe adsorption. The Xe films deposited in the present experiments are only a few monolayers thick, and, therefore, band bending within the Xe-films is negligible. As the barrier height φ_B is determined directly from the IPE spectra (difference between Fermi level and Γ_6^+), χ_S can be determined and is given in table I together with the relevant work function data and peak positions in IPE. For χ_S an average value of 0.4 eV is obtained in perfect agreement with optical measurements ($\chi_S = 0.5$ eV [20]). In IPE a negative-ion state is formed. In the optical measurements neutral-state properties are probed. The coincidence within 0.1 eV of the two χ_S values determined independently of these different techniques proves that the inverse-photoemission electron in the 6*s* band is completely screened in the thick Xe layer. Accordingly we do not expect to observe a final-state relaxation shift in IPE which depends on the distance to the Xe/metal interface.

The present data should be compared to the expression for the barrier height derived from a (linearized) theory of Schottky barriers: $\varphi_B = S(\varphi_M - \chi_S) + (1 - S)\varphi_B^0$ [18]. Here φ_B^0 is the barrier height if the Fermi level coincides with the charge neutrality level, and S is a proportionality factor governing the transition from Schottky behaviour ($S = 1$), with φ_B being determined by the metal work function, to the Bardeen limit ($S = 0$), with the Fermi level being pinned at the charge neutrality level. The actual value of S depends on the density of states (DOS) in the semiconductor band gap. From the present data a value of $S \approx 0.85$ can be derived. Accordingly the Xe/metal barrier is close to, but not exactly in the Schottky limit, the difference being due to a small but nonnegligible DOS in the Xe band gap. The rather limited accuracy of our work function measurements (± 0.15 eV) prevents an exact determination of the charge neutrality level, but a position considerably above the middle of the optical band gap can be inferred from the present data. This is consistent with observations on other metal/semiconductor contacts [18].

In conclusion we state that our results on unoccupied Xe levels in the Xe/Ag(110) interface closely parallel the results of Jacobi and Rotermund on occupied Xe levels in the Xe/Ni(110) interface. According to the present inverse-photoemission experiments the Xe/Ag(110) interface behaves as a Schottky barrier with a small density of states within the band gap. Consequently, the barrier height scales approximately with the work function of the metal surface, but the deviation from the exact Schottky limit is significant. Screening-dependent energy shifts are negligible in the present case. The structure of the unoccupied levels remain constant within a range of work functions of the silver surface from 4.5 to 2.2 eV.

* * *

It is a pleasure to acknowledge enlightening discussion with K. WANDELT. The work was financially supported by the Deutsche Forschungsgemeinschaft.

REFERENCES

[1] LANG N. D., *Phys. Rev. Lett.*, **46** (1981) 842.
[2] KAINDL G., CHIANG T. C., EASTMAN D. E. and HIMPSEL F. J., *Phys. Rev. Lett.*, **45** (1980) 1808.
[3] CHIANG T. C., KAINDL G. and MANDEL T., *Phys. Rev. B*, **33** (1986) 695.
[4] HULSE J., KÜPPERS J., WANDELT K. and ERTL G., *Appl. Surf. Sci.*, **6** (1980) 453.
[5] ROTERMUND H. H. and JACOBI K., *Surf. Sci.*, **126** (1983) 32.
[6] WANDELT K., *J. Vac. Sci. Technol. A*, **2** (1984) 802, and references cited therein.
[7] HORN K., FRANK K. H., WILDER A. J. and REIHL B., *Phys. Rev. Lett.*, **57** (1986) 1064.
[8] WANDELT K., JACOB W., MEMMEL N. and DOSE V., *Phys. Rev. Lett.*, **57** (1986) 1643.
[9] DOSE V., *Prog. Surf. Sci.*, **13** (1983) 225.
[10] COURTHS R., WERN H., HAU U., CORD B., BACHELIER V. and HÜFNER S., *J. Phys. F*, **14** (1984) 1559.
[11] ALTMANN W., DOSE V. and GOLDMANN A., *Z. Phys. B*, **65** (1986) 171.

[12] BERTEL E., JACOB W. and DOSE V., *Phys. Rev. B*, to be published (1987).
[13] RÖSSLER U., in: *Rare Gas Solids*, edited by KLEIN M. L. and VENABLES J. A. Vol. 1, (Academic Press, London) 1976.
[14] FRANK K. H., HORN K., WILDER J. and KOCK E. E., *Appl. Phys. A*, **44** (1987) 97.
[15] JACOBI K. and ROTERMUND H. H., *Surf. Sci.*, **116** (1982) 435.
[16] PRINCE K. C., PAOLUCCI G. and BRADSHAW A. M., *Surf. Sci.*, **175** (1986) 101.
[17] HAYDEN B. E., PRINCE K. C., DAVIE P. J., PAOLUCCI G. and BRADSHAW A. M., *Solid. State Commun.*, **48** (1983) 325.
[18] MÖNCH W., in: *Festkörperprobleme*, edited by GROSSE P., Vol. **26** (Vieweg, Braunschweig) 1986.
[19] SCHLÜTER M., *Thin Solid Films*, **93** (1982) 3.
[20] SONNTAG B., in: *Rare Gas Solids*, edited by KLEIN M. L. and VENABLES J. A., Vol. 2 (Academic Press, London) 1977.

Origin of the Excess Capacitance at Intimate Schottky Contacts

J. Werner
Max-Planck-Institut für Festkörperforschung, 7000 Stuttgart 80, Federal Republic of Germany

and

A. F. J. Levi, R. T. Tung, M. Anzlowar, and M. Pinto
AT&T Bell Laboratories, Murray Hill, New Jersey 07974
(Received 11 May 1987)

We identify the physical origin of the excess capacitance at Schottky diodes without an interfacial layer, i.e., intimate Schottky contacts. Measured capacitance in excess of the space-charge capacitance is shown to be caused by the injection of minority carriers into the bulk semiconductor, rather than by the presence of interface states, as previously thought. Minority-carrier injection depends sensitively on the properties of the Ohmic back-contact.

PACS numbers: 73.30.+y, 73.40.Cg, 73.40.Lq

The enigma of Schottky barriers at metal/semiconductor contacts still challenges solid-state research. Schottky[1] explained the rectifying properties by a barrier

$$\Sigma = \Phi_m - \chi, \tag{1}$$

where Φ_m is the work function of the metal and χ is the electron affinity of the semiconductor. Deviations from Schottky's model[1] were explained[2] with interface states. The solution of the Schottky-barrier problem was therefore seemingly reducible to interface states.

Significant progress was recently achieved in the theory[3] and in the epitaxial growth[4-6] of Schottky contacts. Understanding of carrier transport was believed to be complete: Schottky contacts are usually treated as pure majority-carrier devices. Minority-carrier injection, a fundamental prerequisite for the bipolar transistor and the birth of solid-state electronics,[7] is usually not discussed in the context of Schottky contacts. However, bipolar transport of electrons *and* holes must also occur in Schottky contacts and, in particular, the diffusing minority carriers should influence the low-frequency capacitance.

For about fifteen years, low-frequency capacitances in excess of the capacitance C_{sc} of the space-charge region have indeed been observed at forward-biased Schottky diodes. These excess capacitances were, however, not ascribed to minority carriers but to interface states.[8-10] Bardeen's[2] interface states were thought to be measurable with the help of such excess capacitances. The so-called[8] Schottky capacitance spectroscopy and accurate phase capacitance spectroscopy[9,10] are standard characterization methods. We recently[11] demonstrated a fundamental inconsistency in these techniques which violate Gauss's law when applied to Schottky contacts with an interfacial layer. Low-frequency capacitances alone are *not* sufficient to characterize such interfaces.[11] Instead, the complete frequency-dependent admittance has to be analyzed within a self-consistent model such as was used to characterize traps at Au/oxide/GaAs diodes.[11]

Excess low-frequency capacitances at contacts *without* an interfacial layer, i.e., intimate Schottky contacts, were also ascribed to interface states.[8-10] Unfortunately the interpretation is valid for the reasons outlined above. Particularly interesting are epitaxial $NiSi_2$/Si diodes: $NiSi_2$ grows in two distinct orientations (termed types A and B) on silicon (111) surfaces,[4] with a difference of at least 140 meV for the two Schottky barriers.[5] This difference, originally questioned by Liehr *et al.*,[6] was recently challenged: Ho *et al.*[9] measured excess capacitances, interpreted them in terms of interface traps, and claimed to prove *equal* barriers for types A and B due to similar densities and energies of interface states.

The present Letter examines forward-bias low-frequency *C/V* curves on intimate Schottky contacts. We here furnish the answer to the fundamental question of the real nature of the excess capacitances in the aforementioned[8-10] experiments. In contrast to the current dogma,[9] we find that these spectacular capacitances[8-10] are not caused by interface states but instead by *minority carriers* injected into the bulk semiconductor. This bulk capacitance is in turn controlled by the back-contact. The strong effect of minority carriers and, in particular, the dramatic influence of the *Ohmic* back-contact on admittances of Schottky diodes are here elucidated for the first time.

Our study examines nonepitaxial $NiSi_2$/Si, PtSi/Si, Pd_2Si/Si, as well as epitaxial $NiSi_2$/Si Schottky contacts[4,5] of types A and B with an area of 1×10^{-2} cm^2. Special attention is devoted to the Ohmic back-contacts of our 500-μm-thick wafers. We test implanted P contacts, laser-annealed Sb contacts, Al contacts, and InGa that is mechanically applied to the back side. Sb contacts are fabricated by evaporation and laser annealing of the donor Sb to yield high-quality n^+ contacts.[5] We measure capacitances at 296 K with a Hewlett-Packard model 4192A *LCR* meter and a current source to apply the bias. The precision of our setup is verified with test capacitors and resistors as well as with independent measurements using a lock-in amplifier.

Figures 1(a) and 1(b) show capacitances from two $NiSi_2$ samples of type A and two of type B that are epitaxially[4,5] grown on 4-Ω-cm, *n*-type (111) silicon. The two A-type samples of Fig. 1(a) are — as are the B-type samples of Fig. 1(b) — completely identical apart from their back-contacts. *I*/*V* curves yield $\Sigma = 0.63$ eV and an ideality[11] $n = 1.02$ for type A, and $\Sigma = 0.80$ eV, $n = 1.05$ for type B. Our measurements show no capacitance in excess of the capacitance C_{sc} of the space-charge region for Sb contacts. On the contrary, we find inductances. Such inductances are well known from *pn* junctions[12] where they are caused by injected minority carriers that modulate the bulk resistance. The same effect was predicted[13] for Schottky contacts and is here experimentally verified.

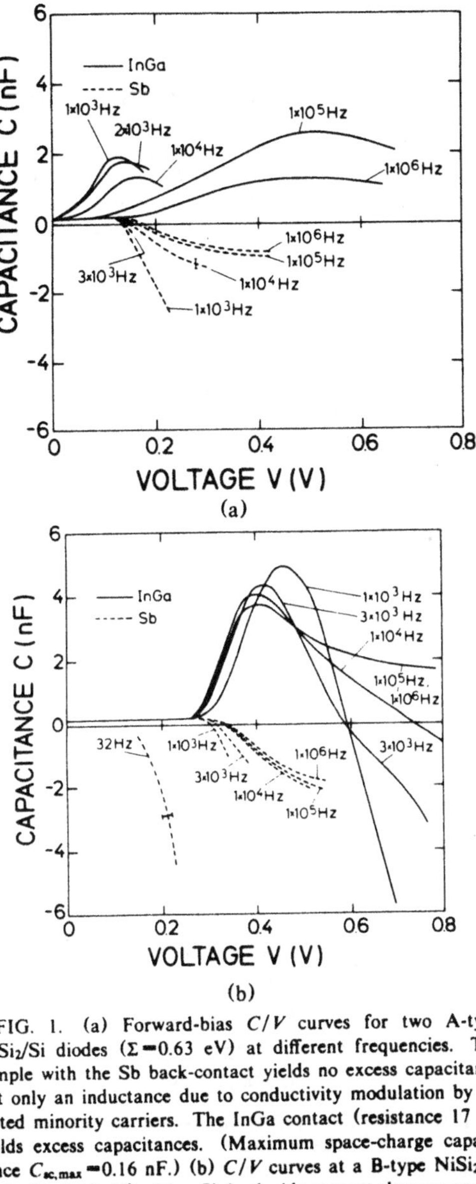

FIG. 1. (a) Forward-bias *C*/*V* curves for two A-type $NiSi_2/Si$ diodes ($\Sigma = 0.63$ eV) at different frequencies. The sample with the Sb back-contact yields no excess capacitance but only an inductance due to conductivity modulation by injected minority carriers. The InGa contact (resistance 17 Ω) yields excess capacitances. (Maximum space-charge capacitance $C_{sc,max} = 0.16$ nF.) (b) *C*/*V* curves at a B-type $NiSi_2/Si$ diode ($\Sigma = 0.80$ eV) with a Sb back-side contact show no excess capacitance. The InGa contact with 39 Ω contact resistance yields excess capacitances/inductances. ($C_{sc,max} = 0.18$ nF.)

The inductances in Figs. 1(a) and 1(b) arise when the voltage drop at the series resistance across the Si substrate is comparable to the voltage at the Schottky barrier. A comparison of the dashed curves in Figs. 1(a) and 1(b) at identical frequencies reveals that the voltages where the capacitances turn negative are higher for the B-type sample. The voltage difference (0.15 V) correlates with the barrier difference because B-type samples with their high barrier require a higher voltage to reach a regime in the *C*/*V* curve where the bulk resistance of the Si (and conductivity modulation) become comparable to the resistance of the Schottky diode.

The curve at 32 Hz in Fig. 1(b) demonstrates that even at very low frequencies there are no excess capacitances. We do not detect any excess capacitance between 5 Hz and 10 MHz that might be ascribable to interface states as long as we use implanted or laser-annealed n^+ contacts. Instead, we always find *inductances* due to minority carriers.

We find excess capacitances only if we deliberately use imperfect back-side contacts with noticeable contact resistances: InGa, Al, or unetched oxides on the back side before evaporation of Sb. The solid curves in Figs. 1(a) and 1(b) show results for InGa contacts: Large capacitance peaks occur. If we correct for the voltage drop across the InGa contact, then the voltage difference between capacitance maxima correlates again roughly with the barrier difference. Ho *et al.*[9] using Al back-contacts, interpreted this correlation as a proof of traps at the Schottky contact on the front side. We repeat their[9] experiments and find also excess capacitances which are, however, traceable to Al and its contact resistance; the acceptor Al is not expected to yield low-resistance contacts to *n*-type Si. Only a precise analysis[14] of *I*/*V* curves yields such contact resistances in addition to the bulk resistance of Si. We find total series resistances (bulk plus contact resistance) of $R_{Sb} = 19$ Ω and $R_{InGa} = 36$ Ω for the Sb and InGa contacts in Fig. 1(a) as well as $R_{Sb} = 13$ Ω, $R_{InGa} = 52$ Ω for Fig. 1(b).

We ascribe the excess capacitances to minority carriers which are injected by the Schottky contact into the bulk Si and cause there excess *diffusion* capacitances. The first hint of excess injection is the concomitance of the excess capacitance for InGa contacts with an inductance at higher voltages: Conductivity modulation by minority carriers is practically the only inductive effect in semiconductors. Inductances—also observed by the advocates of Schottky capacitance spectroscopy[15] and accurate phase capacitance spectroscopy[16]—are hard to explain with interface states.

Secondly, evidence is supplied by the dependence on the resistance of the back-contact. An interface-state capacitance at the front Schottky diode should not depend on back-contact. Minority carriers explain this dependence: Scharfetter[17] predicted minority-carrier injection in Schottky diodes under dc conditions to depend critically on recombination at the back-contact; the only theoretical treatment of ac conditions[13] unfortunately neglects recombination; no systematic experimental study has been performed. Our data prove back-contact-controlled injection: Injected minority carriers either recombine in the bulk Si or are extracted at the back side. Low-resistance n^+-back-side contacts to n-type Si form a minority-carrier reflecting barrier in the valence band of the nn^+ junction. Injection is therefore suppressed and inductances, as verified in Figs. 1(a) and 1(b) for the Sb contracts, are observable. High-resistance contacts such as Al,[9,10] or InGa suggest, on the other hand, a nn^- junction at the back side with a minority-carrier sink in the valence-band edge. Such extracting contacts support strong injection at the front Schottky diode. Excess diffusion capacitances arise as long as the diffusion length is comparable to sample thickness. Indeed, most excess capacitances were previously measured[8-10] in low-doped silicon with typical diffusion lengths of 100–500 μm.

Thirdly, spatial constancy of the total current j_{tot} requires that the excess capacitive current flows also within the space-charge region of the Schottky diode where the ac voltage $\delta\Phi$ drops. In order to obtain *excess* capacitances, this additional capacitance must be in *parallel* to the space-charge capacitance C_{sc}. Ampère's law for a semiconductor,

$$j_{\text{tot}} = \nabla H = j_n + j_p + \dot{D}, \qquad (2)$$

states that the total current j_{tot} consists of the majority-carrier current j_n, the minority-carrier current j_p, and the displacement current $\dot{D}$. There are no other currents which can be made responsible for excess capacitances within any model which is in agreement with Maxwell's electrodynamic equations.

Without an interfacial layer and with an ideality[11] $n \approx 1$ there is no possibility for an out-of-phase modulation of the band bending and therefore no excess capacitance due to a thermionically emitted majority-carrier current j_n.[11] At frequency ω, the displacement current

$$\dot{D} = i\omega C_{\text{sc}}\delta\Phi \qquad (3)$$

yields also no excess capacitance; the resistance of the bulk Si makes the band bending change $\delta\Phi$ always ***smaller*** than the applied ac voltage. The upper Maxwell equation, Eq.(2), leaves therefore only the minority-carrier current j_p for excess capacitances. Indeed, minority carriers in the space-charge region are phase shifted with respect to the band bending change $\delta\Phi$ since their transport through the bulk semiconductor is mainly controlled by diffusion. They can therefore cause a capacitive current

$$j_p^{90} = i\omega C_{\text{diff}}\delta\Phi, \qquad (4)$$

with the diffusion capacitance C_{diff} in parallel to the space-charge capacitance C_{sc}.

In addition to these arguments based on the Maxwell equations [Ampère's law, Eq.(2)] our preliminary results[18] obtained from one- and two-dimensional device simulation demonstrate also that the excess capacitance is caused by minority carriers. The calculations, which will be presented in a separate publication,[18] show the capacitances or inductances to depend sensitively on the boundary conditions of the back-contact. Our simulations qualitatively reproduce the observed inductances at $NiSi_2$ Schottky diodes with a good Ohmic contact. This inductance is influenced by spatial inhomogeneities in the Schottky barrier. The excess capacitance and the observed frequency dependence with a poor Ohmic contact are also reproduced by simulation: We obtain large excess capacitances by including regions of p^+ doping along an inhomogeneous back side as expected when one uses InGa or Al contacts to n-type silicon. In summary,

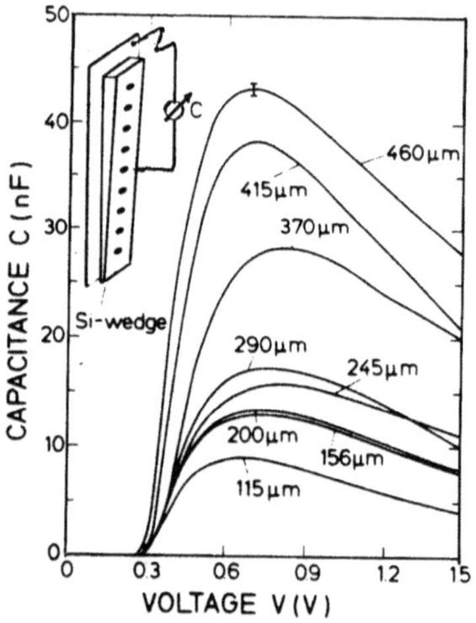

FIG. 2. Different PtSi/Si diodes with $\Sigma = 0.84$ eV on the front of the Si wedge are measured at 1 kHz against the same, single, large-area back-contact. The thickness dependence proves that the excess capacitance originates from the bulk Si and not from the interface of the PtSi/Si diode. ($C_{\text{sc,max}} = 0.08$ nF.)

all the observed capacitance and inductance behavior can be understood by inclusion of minority carriers in the description of the current transport.

An experiment utilizing a wedge as shown in Fig. 2 yields a striking *experimental* proof that the excess capacitance truly originates in the bulk Si: We evaporate Pt dots (area 5.6×10^{-3} cm^2) onto a 11-Ω-cm, *n*-type silicon wedge and form PtSi/Si diodes with $\Sigma=0.84$ eV, $n=1.07$ by heating at 400 °C. The wedge has a single, large-area, evaporated and laser-annealed Sb back-contact. The I/V curves of all Schottky contacts yield a series resistance around 250 Ω. This value is almost totally due to the contact resistance of the common back-side contact being of high resistance because we did not etch the native SiO_2 before evaporating the Sb. The thickness dependence of the maxima in Fig. 2 is due to the different volume of stored minority carriers and demonstrates that the excess capacitance is a bulk property. The Schottky diodes inject minority carriers and the back-side contact extracts them. Excess diffusion capacitance and inductive conductivity modulation add to give C/V maxima.

We conclude the following:

(i) Intimate Schottky contacts (without interfacial layers) with low-resistance, Ohmic, minority-carrier-reflecting n^+-back-side contacts show only inductive conductivity modulation. We find no excess interface-state capacitance.

(ii) Imperfect back-contacts cause the excess[8-10] capacitances. We ascribe this effect to diffusing minority carriers. Minority-carrier *extracting* back-contacts facilitate *injection* at the front Schottky diode. Minority carriers flow out of phase with the applied voltage since their transport through the bulk is controlled by diffusion.

(iii) Previously reported excess capacitances[8-10] at intimate Schottky contacts were probably caused not by interface states but by imperfect back-contacts and minority-carrier injection. Therefore, such measurements[9] constitute no proof for similar interface states at $NiSi_2$/Si diodes[4-6] of types A and B but are rather a sensitive tool for the characterization of *back-side* contacts.

(iv) Care should be taken to ensure that excess capacitance ascribed to interface states at intimate Schottky contacts with idealities $n\approx1$ are not, in fact, caused by minority carriers. In addition, any such interpretation should be shown to be consistent with Maxwell's equations [e.g., Ampère's law, Eq. (2)].

(v) The measurability of interface states at Schottky contacts requires an interfacial layer, which separates the traps from the metal and allows for an out-of-phase modulation of the semiconductor's band bending due to trapping. The traps are then detectable via capacitive majority-carrier currents j_n and displacement currents $\dot{D}$ in the interfacial layer.[11] I/V curves and frequency-dependent admittances as well as a model[11] obeying Gauss's and Ampère's law are, however, a minimum requirement for a self-consistent analysis.

We thank M. Liehr for samples and acknowledge the support of L. C. Feldman, H. Stömer, and H. J. Queisser. It is a pleasure to thank A. B. Fowler, J. L. Freeouf, S. Laux, and L. Schultheis for valuable discussions. Only the expert technical assistance of E. Kisela made our wedge experiment possible.

[1]W. Schottky, Z. Phys. **118**, 539 (1942).

[2]J. Bardeen, Phys. Rev. **71**, 717 (1947).

[3]J. Tersoff, Phys. Rev. Lett. **52**, 465 (1984).

[4]R. T. Tung, J. M. Gibson, and J. M. Poate, Phys. Rev. Lett. **50**, 429 (1983).

[5]R. T. Tung, Phys. Rev. Lett. **52**, 461 (1984).

[6]M. Liehr, P. E. Schmid, F. K. LeGoues, and P. S. Ho, Phys. Rev. Lett. **54**, 2139 (1985).

[7]W. Shockley, Bell Syst. Tech. J. **28**, 435 (1949).

[8]Parts of the extensive work on Schottky capacitance spectroscopy is reviewed by C. Barret, F. Chekir, and A. Vapaille, J. Phys. C **16**, 2421 (1983).

[9]P. S. Ho, E. S. Yang, H. L. Evans, and X. Wu, Phys. Rev. Lett. **56**, 177 (1986).

[10]H. L. Evans, X. Wu, E. S. Yang, and P. S. Ho, J. Appl. Phys. **60**, 3611 (1986).

[11]J. Werner, K. Ploog, and H. J. Queisser, Phys. Rev. Lett. **57**, 1080 (1986).

[12]T. Misawa, J. Phys. Soc. Jpn. **12**, 882 (1957).

[13]M. A. Green and J. Shewchun, Solid State Electron. **16**, 1141 (1973).

[14]J. Werner, unpublished

[15]P. Muret and A. Deneuville, Surf. Sci. **168**, 830 (1986).

[16]E. S. Yang and P. S. Ho, private communication.

[17]D. L. Scharfetter, Solid State Electron. **8**, 299 (1965).

[18]M. Pinto, unpublished.

Correlation between E_F Pinning and Development of Metallic Character in Ag Overlayers on GaAs(110)

K. Stiles and A. Kahn
Department of Electrical Engineering, Princeton University, Princeton, New Jersey 08544
(Received 11 May 1987)

Ag is deposited on room-temperature and low-temperature GaAs(110). Correlation is found between the appearance of metallicity in the Ag layer, as inferred from the width of the Ag-4*d* band, and the pinning of the Fermi level at the GaAs surface. These results suggest the importance of metal-induced gap states at high coverage (> 2 Å) in the formation of the Schottky barrier.

PACS numbers: 73.30.+y, 73.40.Ns

The models proposed to date to explain the formation of Schottky barriers at metal-semiconductor interfaces still provoke much controversy.[1-5] The complexity of the interface morphology,[6] microscopic structure,[7] and chemistry[2] has prevented unambiguous identification of the nature and origin of the states that pin the Fermi level (E_F), near midgap for GaAs. We have recently followed an approach that consists of depositing the metal on GaAs surfaces cooled to 70–100 K.[6,8-10] Metal clustering and interdiffusion are substantially inhibited at low temperature (LT), as inferred from low-energy electron diffraction (LEED) and Auger-electron spectroscopy. Chemical reactions are somewhat slowed down, but are not eliminated even at $T < 100$ K. We attempt to correlate these changes with changes observed in the formation of the Schottky barrier, in particular with E_F pinning during the initial stages of overlayer growth. As a result, we suggest the role of metal-induced gap states at the Ag/GaAs(110) interface in the high-coverage regime.

Experiments performed with Al, In, Au, and Ag evaporated on LT GaAs(110) give concordant results.[8-12] All these metals cluster to various degrees on room-temperature (RT) GaAs. Particularly significant are the soft x-ray and ultraviolet photoemission spectroscopy (SXPS and UPS) measurements which show that the coverage-dependent surface-band bending is affected differently by temperature on *n*- and *p*-GaAs.[8-12] After a small initial band bending at ultralow coverage, final pinning is *substantially* retarded on *n*-GaAs with respect to the RT case. Band bending occurs as fast on *p*-GaAs as in the RT case, faster with Al and In.[10-12] The Ag data are shown in Fig. 1. The asymmetry in the E_F movement on LT *n* and *p* samples and the LT transition around 2 Å on *n*-GaAs suggest that several mechanisms responsible for band bending at the semiconductor surface are active during the formation of the interface, in opposition to the traditional picture of a single mechanism inferred from the rapid and symmetric band bending observed at RT (Fig. 1). Our assumption is that the differences between the LT and RT data reflect a slowdown, or decomposition, of the overlayer formation process at temperatures below 100 K.

The experiments presented here were done with $n = 3\times10^{17}-3\times10^{18}$ cm^{-3} (Si or Te) and $p = 4\times10^{18}-1\times10^{19}$ cm^{-3} (Zn) doped GaAs surfaces cleaved in ultrahigh vacuum. The sample temperature was maintained at 300 K (RT) or 70–100 K (LT) during evaporation and measurements. Ag was evaporated from a W filament in equivalent coverages ranging from less than 0.01 to 32 Å. Measurements of overlayer morphology and structure with use of LEED and Auger-electron spectroscopy have been reported elsewhere.[10] In this paper, we emphasize UPS measurements of the GaAs valence band and Ag-4*d* band as functions of Ag coverage, using the He II (40.8 eV) radiation line. In the text and figures below, these measurements are correlated with surface-band bending obtained from SXPS and UPS measurements of Ga-3*d* and As-3*d* core-level energy shifts.[10]

Although this paper deals mostly with the high-coverage pinning mechanism, comments concerning the difference between the low-coverage E_F movements on *n* and *p* samples should be made. This asymmetry is difficult to reconcile with current Schottky-barrier mod-

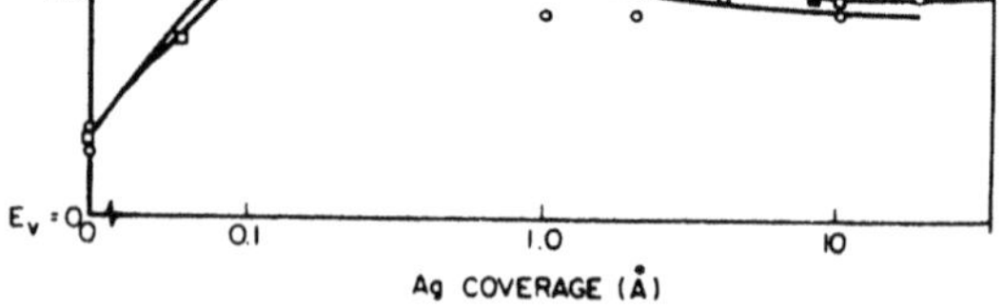

FIG. 1. Position of E_F at the (110) surface of *n*- and *p*-GaAs as a function of Ag coverage at RT and LT.

els based on near-midgap acceptor and donor defects,[1] or on metal-induced gap states.[4] In particular, the former model would require a ratio of 2 orders of magnitude between the rates of LT formation of donor and acceptor defects, e.g., As and Ga antisites, whereas these same defects would form at the same rate at RT. This is highly unlikely in the case of GaAs. Several factors could, however, contribute to this asymmetry. First, the fast initial rise of E_F on LT *p*-GaAs has been modeled, for the similar case of Al, in terms of small adatom charged clusters and their size-dependent Coulomb energy and Fermi level.[13] A good fit between theory and experiment was obtained at ultralow coverage (< 0.1 Å). Whether this model can reproduce the *n* data up to the 1-2-Å transition coverage is unknown, however. Second, it has been observed (at LT) that metals such as Al[6] and Ag[10] unrelax the (110) GaAs surface. Dangling bonds at unrelaxed sites could generate gap states, donors in the present case, which are otherwise swept out of the gap by the surface relaxation.[7] Finally, and perhaps most important, the fast low-coverage movement of E_F on LT *p*-GaAs generally exhibits an overshoot above its final pinning position (Fig. 1). The magnitude of this overshoot depends on the metal[12]: It is large with low-electronegativity metals (Al and In), smaller with Ag and Au,[10] and nonexistent with Pd.[14] In the low-coverage regime (< 0.5 Å) at LT, the overlayer consists of isolated atoms or very small clusters, as we discuss below. The low-coverage asymmetry could therefore be the manifestation of the real nature of the surface states induced by the very dilute overlayer.[15] Yet knowledge of the atomic and electronic structures (adsorption sites, substrate relaxation, bonding, and antibonding states) of atoms or small clusters on the surface must be determined before a quantitative justification of the asymmetry can be presented.

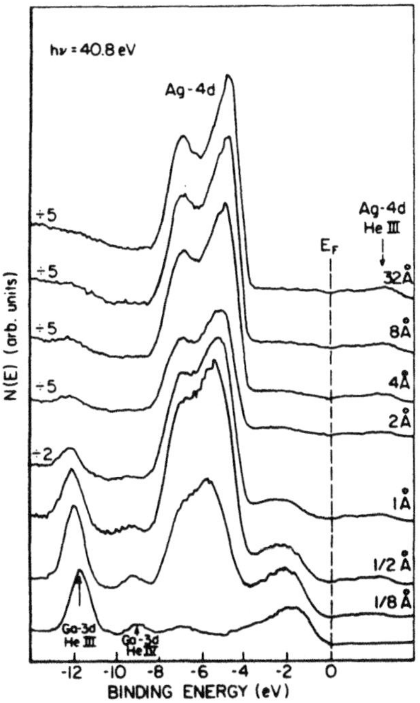

FIG. 2. GaAs valence band as a function of Ag coverage on the RT (110) surface. Extra Ga and Ag peaks are due to He III (48.4 eV) and He IV (51.0 eV) lines.

Beyond these ultralow-coverage results, the LT data indicate that another mechanism comes into play around 2 Å (Fig. 1) and pins the Fermi level on *n*- and *p*-GaAs in the high-coverage limit. Figure 2 shows the GaAs valence band and the Ag-4*d* level as a function of Ag coverage on a *p*-type RT surface. The large cross section of the Ag-4*d* level produces a peak that dominates the valence-band spectrum even at coverages below 0.5 Å. Contributions to the width of the 4*d* band come from the spin-orbit split (0.4 eV), inhomogeneities in the Ag layer (especially at LT), and the size of the Ag clusters. The evolution of the shape and width of the Ag-4*d* band has been studied as a function of Ag deposition on RT amorphous carbon (*a*-C).[16] Band narrowing, due to loss in long-range periodicity, is observed in small clusters. Other parameters such as alloy composition can affect the bandwidth equally. The Ag-GaAs interface is abrupt, however, and little if any alloying is expected, especially at LT. We can therefore attribute bandwidth variations to inhomogeneities and mostly to changes in adatom cluster size.

Two parameters become important in the evaluation of the size of the clusters in our Ag deposition: the density of nucleation sites on the GaAs surface, and the temperature-dependent surface diffusion of Ag. The typical density of nucleation sites on *a*-C substrates is $d_c = 3\times10^{12}$ cm^{-2}, much larger than on the well cleaved surfaces used here. As Ag is highly mobile on RT GaAs, clustering is not limited by surface diffusion, and we can expect cluster sizes larger than those obtained on *a*-C at equivalent coverage. A comparison of the shape and width of the 4*d* bands of Fig. 2 with those of Ref. 16 confirms this expectation. The full width at half maximum (FWHM) of our $\frac{1}{8}$-Å (7.4×10^{13} cm^{-2} Ag atoms) band is 2.85 eV, very close to the 3.0-eV value obtained with a much larger coverage of *a*-C (10^{15} cm^{-2} Ag atoms). A lower bound for the mean number of atoms per cluster, n_{Ag}, for $\frac{1}{8}$ Å can therefore be estimated with the nucleation site density on *a*-C: $n_{Ag} > 10^{15}$ cm$^{-2}/d_c \simeq 300$, well above the number of Ag atoms necessary to produce metallic behavior.[16] Thus, Ag clusters on RT GaAs are metallic early on, at coverages equal to or lower than $\frac{1}{8}$ Å. This estimation is probably more reliable than that obtained from the measurement of the Fermi edge. The *d* band has a very large cross section in UPS, and its width can be measured accurately. The Fermi edge, on the other hand, is weak at low coverage. Final-state charging effects in small (although metallic) clusters[16] can shift it to higher binding energy by a substantial amount (0.3–0.5 eV), away from the bulk Ag Fermi level and toward the top of the GaAs valence band, thus making its observation and correlation with metallicity more difficult.

The high-binding-energy peak of the Ag-4*d* doublet shows little dispersion with respect to coverage, as was shown by Chin *et al.*,[17] and the low-binding-energy peak shifts to lower binding energy, in agreement with the Ag/*a*-C data. The FWHM of the band (Fig. 3) increases smoothly as a function of coverage, indicating an increase in the long-range periodicity in the Ag clusters, with results from LEED and SXPS.[10]

The evolution of the FWHM during deposition on the LT GaAs surface is different (Figs. 3 and 4). The FWHM at $\frac{1}{8}$ Å is only 1.6 eV, about half of its high-coverage value (3.25 eV). This severe band narrowing corresponds to a small effective size of the Ag clusters. A comparison with the data of Ref. 18 indicates that the LT $\frac{1}{8}$-Å clusters are substantially smaller than those corresponding to a deposition of 2.0×10^{14}-cm^{-2} Ag atoms on *a*-C. Using again the density of nucleation sites on *a*-C, we can estimate the upper bound for n_{Ag} to be substantially less than sixty Ag atoms at $\frac{1}{8}$ Å. The clusters are therefore mostly *nonmetallic*. This corresponds to the predominance of very small AG clusters, or perhaps isolated Ag atoms, dispersed on the surface, as expected from reduced adatom diffusion at 100 K. The FWHM increases suddenly around 1 Å toward the final 3.0-eV RT value. From the considerations given above, we conclude that, at LT, the average cluster becomes metallic only with a coverage between 1 and 2 Å. Instead of the RT Volmer-Weber growth mode, the Ag overlayer grows in a quasi-2D first layer followed by 3-D nucleation and formation of metallic clusters at higher coverage. The 2D character of the first layer is compatible with LEED measurements showing *complete unrelaxation* of the GaAs surface with 1.5 Å (≈1 monolayer) of Ag deposited at LT.[10]

We propose that the 2-Å drop of E_F on LT *n*-GaAs is due to the appearance of metallicity in the overlayer. The onset of this transition occurs when the FWHM of the *d* band indicates metallic character in the clusters, i.e., FWHM > 2.6–2.8 eV between 1 and 2 Å, suggesting that the two are correlated (Figs. 1 and 3). The movement of E_F continues, however, up to a coverage of 6–8 Å. This is due to several factors. First, there is a broad distribution of cluster sizes, especially at LT where equilibrium is more difficult to achieve. Although the broad line shape of the 4*d* band of the larger clusters precludes the determination of this distribution, some clusters must remain nonmetallic up to high coverages (> 6–8 Å). Though E_F may be pinned locally under metallic clusters, the lateral extent of the depletion region is limited by the doping. This leads to inhomogeneous band bending.[18] Second, inhomogeneities in the Ag layer can also lead to bandwidth broadening unrelated to the metallic character of the overlayer.

A more fundamental question concerns the rapid movement of E_F at ultralow coverage on LT *p*-GaAs. As mentioned previously, this movement is due to an independent mechanism, i.e., charged clusters,[13] small density of defects, interface structural changes,[10] or metal-adsorption–induced states, which brings E_F close to its final position and precludes a drop or rise similar to

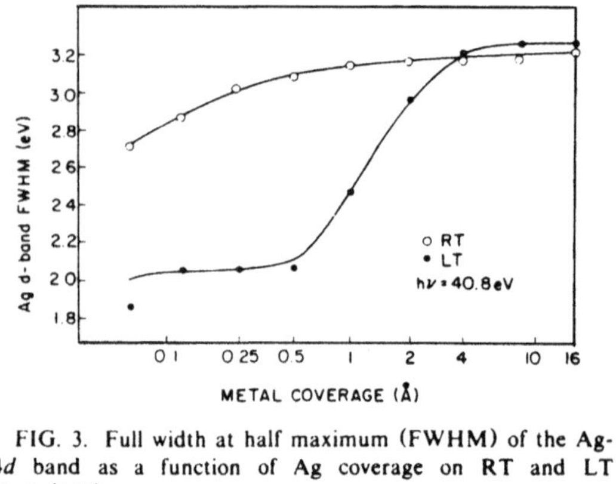

FIG. 3. Full width at half maximum (FWHM) of the Ag-4*d* band as a function of Ag coverage on RT and LT GaAs(110).

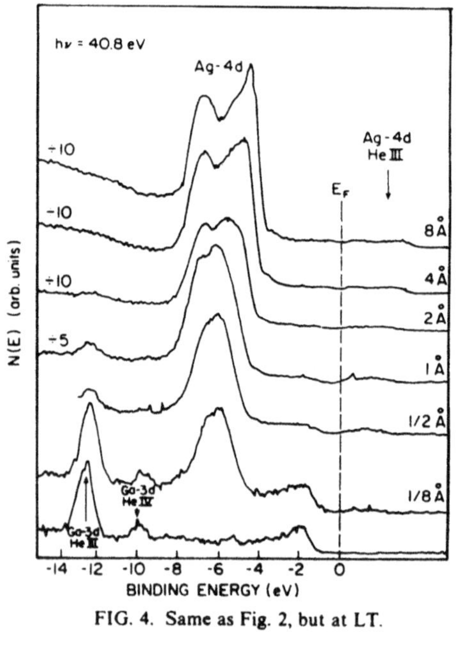

FIG. 4. Same as Fig. 2, but at LT.

that observed on n-GaAs. Yet a movement down from the overshoot position is always observed around the transition coverage. Similar movements are observed with other metals, i.e., Al, In, and Au, and by another group.[8-12] It suggests that a common pinning mechanism is active on both n- and p-type samples beyond the ultralow-coverage regime, although its manifestation is not as dramatic on LT p- as it is one n-GaAs.

Our conclusions concerning the possible pinning mechanisms are therefore the following. Mechanisms put forward in models such as the unified-defect model[1] or the effective-work-function model[3] are not likely to play a dominant role in the abrupt transition described above and the resulting E_F position at this interface. The energy necessary for defect formation, i.e., heat of condensation of the metal or energy released upon formation of small clusters (even below 1 Å at LT) should be available and produce pinning at much lower coverage than the observed transition coverage. Pinning beyond 2 Å seems to be more compatible with the appearance of metal-induced gap states which develop when delocalization of the cluster conduction electrons occurs. In other words, LT pinning is delayed until the overlayer acquires a metallic character, which does not happen until the coverage reaches 1–2 Å. At that coverage, the density of metal-induced gap states becomes sufficient to screen any interface states formed at low coverage, and to move E_F close to the GaAs charge-neutrality level. The abrupt transition is not observed at RT because of the early formation of large metallic clusters, as demonstrated from the FWHM data in Fig. 3, and RT pinning is expected to be rapid on n and p samples, as is indeed observed. We therefore suggest a correlation between E_F pinning and appearance of overlayer metallic character at the Ag/GaAs interface.

Helpful discussions with W. Mönch and G. K. Wertheim are gratefully acknowledged. This work was supported by a grant of the National Science Foundation (DMR-8406820). Partial support was also provided by General Electric and General Motors Corporation.

[1]W. E. Spicer, P. W. Chye, P. R. Skeath, C. Y. Su, and I. Lindau, J. Vac. Sci. Technol. **16**, 1427 (1979).

[2]L. J. Brillson, Surf. Sci. Rep. **2**, 123 (1982).

[3]J. L. Freeouf and J. M. Woodall, Appl. Phys. Lett. **39**, 727 (1981).

[4]J. Tersoff, Phys. Rev. Lett. **52**, 465 (1984), and references therein.

[5]W. Mönch, Phys. Rev. Lett. **58**, 1260 (1987).

[6]C. R. Bonapace, K. Li, and A. Kahn, J. Phys. (Paris), Colloq. **45**, C5-409–C5-418 (1984).

[7]A. Kahn, Surf. Sci. Rep. **3**, 193 (1983).

[8]M. K. Kelly, A. Kahn, N. Tache, E. Colavita, and G. Margaritondo, Solid State Commun. **58**, 429 (1986).

[9]K. Stiles, A. Kahn, D. Kilday, N. Tache, and G. Margaritondo, in *Proceedings of the Eighteenth International Conference on the Physics of Semiconductors, Stockholm, 1986*, edited by O. Engström (World Scientific, Singapore, 1987), p. 121.

[10]K. Stiles, A. Kahn, D. Kilday, and G. Margaritondo, J. Vac. Sci. Technol. B **5**, 987 (1987).

[11]K. K. Chin, T. Kendelewicz, C. McCant, R. Cao, K. Miyano, I. Lindau, and W. E. Spicer, J. Vac. Sci. Technol. A **4**, 969 (1986).

[12]R. Cao, K. Miyano, T. Kendelewicz, K. K. Chen, I. Lindau, and W. E. Spicer, J. Vac. Sci. Technol. B **5**, 998 (1987).

[13]S. Doniach, K. K. Chin, I. Lindau, and W. E. Spicer, Phys. Rev. Lett. **58**, 591 (1987).

[14]K. Stiles and A. Kahn, unpublished.

[15]The n vs p asymmetry can result from adsorbate-induced states if a gap opens in the density of states of the overlayer for isolated adsorbates or very small clusters (donor states in the semiconductor gap, acceptor states near or above the conduction-band minimum). The gap vanishes when the adatom clusters grow and become metallic.

[16]G. K. Wertheim, S. B. DiCenzo, and D. N. E. Buchanan, Phys. Rev. B **33**, 5384 (1986).

[17]K. K. Chin, S. H. Pan, D. Mo, P. Mahowald, N. Newman, I. Lindau, and W. E. Spicer, Phys. Rev. B **32**, 918 (1985).

[18]J. Y. F. Tang and J. L. Freeouf, J. Vac. Sci. Technol. B **2**, 459 (1984).

Direct Investigation of Subsurface Interface Electronic Structure by Ballistic-Electron-Emission Microscopy

W. J. Kaiser and L. D. Bell
Jet Propulsion Laboratory, California Institute of Technology, Pasadena, California 91109
(Received 10 December 1987)

A new technique for spectroscopic investigation of subsurface interface electronic structure has been developed. The method, ballistic-electron-emission microscopy (BEEM), is based on scanning tunneling microscopy. BEEM makes possible, for the first time, direct imaging of subsurface interface properties with nanometer spatial resolution. We report on the first application of BEEM to subsurface Schottky-barrier interfaces.

PACS numbers: 61.16.Di, 68.35.−p, 73.20.−r, 73.30.+y

The formation and properties of semiconductor interfaces have been of central importance in solid-state physics for many decades. However, a complete understanding of the fundamental characteristics of semiconductor surfaces, the influence of thin-film deposition on semiconductor surfaces, and the properties of the resulting interface system is still lacking. Metal-semiconductor Schottky-barrier (SB) interface systems have attracted the greatest attention.[1] The complexity of SB formation phenomena, including the role of interface defect formation,[2,3] electrode interdiffusion, and chemical reaction, is expected to induce inhomogeneity in interface structure and electronic properties.[1] Subsurface interface electronic properties are not directly accessible to conventional surface analytical techniques. Therefore, experimental investigation of interface phenomena is complicated by the necessity to probe interfaces buried below a heterojunction surface. In addition, conventional SB characterization methods, including photoemission, photoresponse, current-voltage, and other techniques, are limited by their lack of spatial resolution for probing the variation of SB properties over the interface plane. The conventional methods yield only a complicated spatial average of SB properties. The recently developed scanning tunneling microscopy (STM) method[4] makes possible high-resolution imaging of surface geometric and electronic structure.[5-7] However, previously developed STM techniques are insensitive to subsurface properties.

We report the first method for direct spectroscopic investigation and imaging of *subsurface* interface electronic structure with high spatial resolution. The method, ballistic-electron-emission microscopy (BEEM), employs STM for investigation of SB systems, semiconductor heterojunctions, and other interfaces. In this Letter, we describe BEEM and its first application to important SB interfaces.

Figure 1 shows a three-terminal configuration for BEEM investigation of a metal-semiconductor SB heterojunction. An STM tunnel tip is positioned near the surface of the heterojunction to permit transmission of tunnel current between tip and base (metal) electrodes. Electron tunneling from the tip to the base electrode injects ballistic electrons into the base. With typical attenuation lengths greater than 100 Å,[8] the injected ballistic electrons may propagate through a surface layer to probe a subsurface interface region. For base-tip tunnel bias less than the base-collector barrier height V_b, there is no ballistic-electron current into the collector. But, as base-tip bias V is increased above V_b [see Fig. 1(b)], a dramatic increase in base-collector current I_c occurs. The I_c-V spectrum provides a direct probe of interface electronic structure, including the important SB height, quantum-mechanical reflection of electrons at the interface, and ballistic-electron transport properties of the base film. In addition, scanning of the tunnel tip over the heterostructure during simultaneous measure-

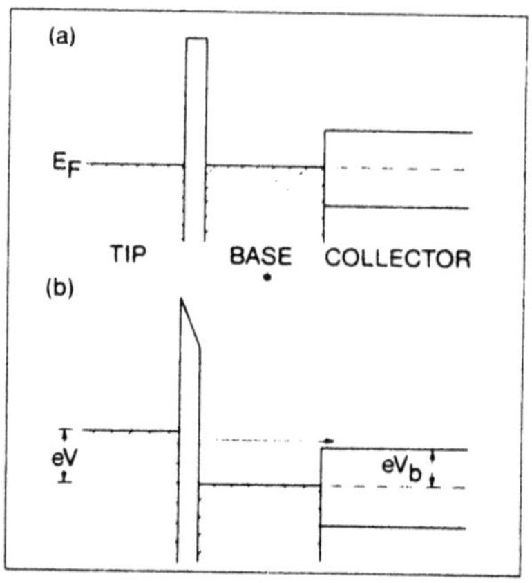

FIG. 1. Schematic energy-band diagram for the three-terminal BEEM experiment. The tunnel tip is separated by a vacuum barrier from the base metal. Terminals are applied to the tunnel tip, metal base, and semiconductor collector. Collector current I_c is measured between the base and collector. (a) The energy-band diagram for zero tunnel bias, $V=0$. (b) The energy-band diagram for tunnel bias greater than the barrier voltage, $eV > eV_b$.

ment of I_c (or I_c-V spectra) makes possible direct spatial imaging of subsurface interface structure.

Two important SB interfaces were selected for the first application of BEEM: Au-Si and Au-GaAs SB heterojunctions fabricated by standard methods. Both SB systems have been widely studied by many techniques. The Au-GaAs SB interface shows properties which are strongly affected by interface defect formation[2] and pronounced interdiffusion and alloy-formation phenomena between the Au and GaAs electrodes.[9] A wide range of SB-height values have been reported for the Au-GaAs interface.[1] In contrast to Au-GaAs, the Au-Si SB shows simple, reproducible SB characteristics.[1] The SB heterojunctions used for this study were prepared by evaporation of 1.6-mm-diam Au disk electrodes, 100 ± 5 Å thick, on chemically etched[10] n-type Si(100) ($n = 2 \times 10^{15}$ cm^{-3}) and chemically etched[11] n-type GaAs(100) ($n = 3 \times 10^{16}$ cm^{-3}) wafers. Au evaporation was performed in an ultrahigh-vacuum chamber at a pressure of 1×10^{-9} Torr with quartz-crystal microbalance control of film thickness. Current-voltage (I-V) measurements yielded SB heights of 0.85 and 0.80 eV for the Au-Si and Au-GaAs SB heterojunctions, respectively.[12]

The STM apparatus, described elsewhere,[13] is enclosed in a chamber purged with dry N_2 gas at atmospheric pressure. Au tunnel tips were employed for this work. Stable and reproducible images of the Au electrode surface were routinely obtained for all samples studied. The collector current was measured with zero applied bias between base and collector with use of a high-sensitivity (gain $= 10^{11}$ V/A), low input-impedance (10 Ω) current amplifier. From ten to forty BEEM spectra were averaged at each surface location to improve the spectral signal-to-noise ratio. The measurements reported here were made at room temperature.

Figure 2 shows typical BEEM spectra obtained at a single location on the Au electrode surface of a Au-Si SB heterojunction. The spectra were obtained by our sweeping V while holding tunnel current I_t constant under feedback control of tip-sample separation. During the sweep of V, I_c is measured. For negative V (tip positive with respect to the Au electrode) electron tunnel current originates from filled states in the Au electrode. As expected, no current has been observed between base and collector for negative V over the voltage range studied ($-2\text{ V} < V < 0$ V). Also, as expected, I_c is zero for V less than a threshold voltage. However, for V greater than $\simeq 0.8$ V, the spectra of Fig. 2(a) exhibit an abrupt threshold and a rapid increase in I_c.

A simple, one-dimensional theory explains the fundamental characteristics of BEEM. Following the treatment of photoresponse,[8] the ballistic-electron mean free path in Au is assumed to be independent of energy E over the narrow range from 0 to 2 eV above the Fermi energy, E_F. Further, the density of states in the tip and

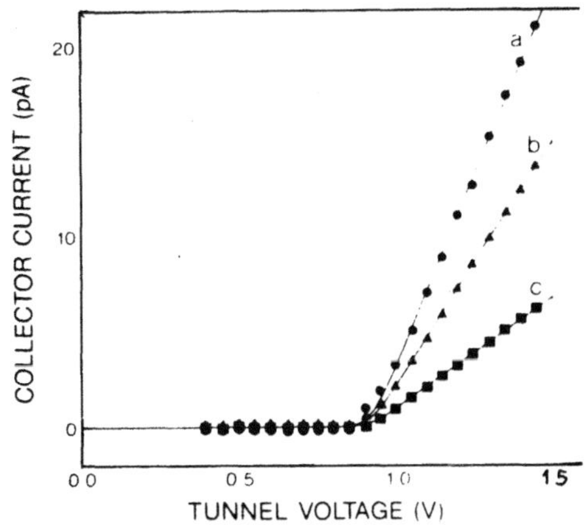

FIG. 2. BEEM spectroscopy results for the Au-Si heterojunction. BEEM spectra of collector current I_c vs tunnel voltage V are shown. Spectra a (dots), b (triangles), and c (squares) were measured at tunnel currents of 0.87, 0.57, and 0.27 nA, respectively. The calculated spectra (solid lines) correspond to barrier height values, eV_b, of 0.92 eV, and R values of 0.045, 0.045, and 0.043 eV^{-1} for spectra a, b, and c, respectively.

base electrodes is taken to be constant. Also, the energy-dependent transmission probability of ballistic-electron current through the interface with barrier height eV_b is approximated by a step function. Thus the collector current as a function of tunnel bias voltage at constant tunnel current is

$$I_c(V) = RI_t \int dE[f(E) - f(E - eV)] \times \theta(E - (E_F - eV + eV_b)), \quad (1)$$

where R is bias independent and the Fermi function, $f(E)$, is defined as $f(E) = \{1 + \exp[(E - E_F)/kT]\}^{-1}$.

A comparison of the experimental BEEM spectra with spectra calculated with use of Eq. (1) is shown in Fig. 2. Equation (1) was fitted by least squares to the experimental spectra by adjustment of R and V_b. Agreement between the simple theory and the experimental BEEM spectra is excellent. The BEEM spectra of Fig. 2 were reproduced at each location probed on the surfaces of three separate Au-Si SB heterostructures. The variation in the value of eV_b for spectra obtained at different I_t values is less than 0.01 eV at a single location on the heterostructure surface. However, experimentally derived values of eV_b varied over the narrow range from 0.86 to 0.92 eV for different regions of the three Au-Si heterostructures. The observed variation in eV_b is a direct observation of spatial variation in interface structure. It is important to note that for the Au-Si SB heterojunctions the barrier-height values determined by BEEM, 0.86–0.92 eV, are in close agreement with the spatially aver-

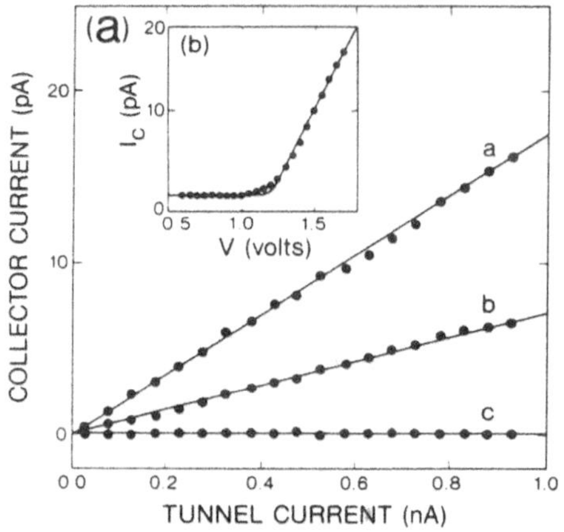

FIG. 3. (a) BEEM spectra of collector current I_c vs tunnel current I_t for the Au-Si SB. Spectra *a*, *b*, and *c* (dots) are measured at tunnel voltage, *V*, values of 1.4, 1.1, and 0.8 V, respectively. Included for comparison are straight-line fits to the spectra. (b) BEEM I_c-V spectrum (dots) for the Au-GaAs SB heterojunction obtained at $I_t = 1.0$ nA. The calculated spectrum (solid line) corresponds to an eV_b value of 1.2 eV and an R value of 0.034 eV^{-1}.

aged value determined from I-V measurements, 0.85 eV.[14]

The linear dependence of collector current on tunnel current of Eq. (1) was tested experimentally. First, it is seen that the three spectra of Fig. 2 scale linearly with I_t. Second, spectra of I_c vs I_t at constant values of V were measured. Typical I_c-I_t spectra are shown in Fig. 3(a), where the predicted linear dependence of I_c on I_t is clearly displayed.

BEEM permits direct, spatially resolved probing of heterostructure interface properties by imaging of the variation in I_c at fixed V and I_t during scanning of the tunnel tip over the heterostructure surface. Spatial variations in interface properties are revealed as spatial variations in I_c. The BEEM-imaging capabilities have been demonstrated by an investigation of the subsurface electronic properties of the Au-Si and Au-GaAs SB heterostructures. Figure 4 shows typical STM topographic and BEEM images of the Au-Si and Au-GaAs heterostructures. The STM and BEEM images were acquired simultaneously. The STM images map the topography of the heterojunction surface, shown in Figs. 4(a) and 4(c), while the BEEM images, shown in Figs. 4(b) and 4(d), display subsurface interface structure.

The STM topographic image in Fig. 4(a) shows smooth topography at the Au electrode surface for the Au-Si heterostructure. The BEEM image for Au-Si at the same location, shown in Fig. 4(b), indicates negligible spatial variation in I_c. The BEEM image, therefore, shows that the subsurface Au-Si SB interface has homogeneous electronic properties.

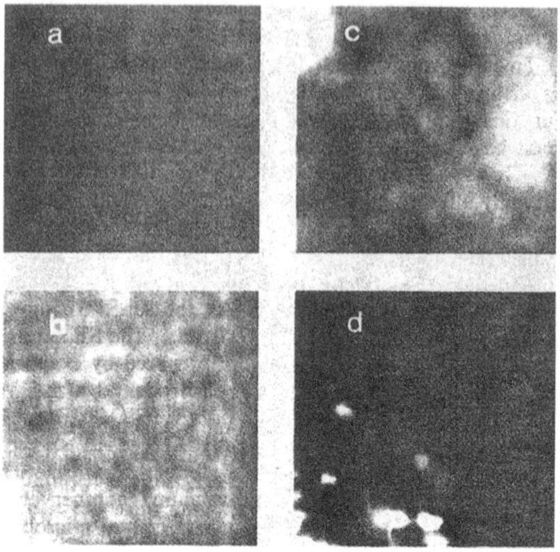

FIG. 4. STM topographic and BEEM images of the Au-Si and Au-GaAs SB heterojunctions. STM and BEEM images were acquired simultaneously. All images show a 250×250-Å^2 area. (a) STM image of the Au-Si SB. Surface height is represented by grey level, over a range from darkest (minimum height) to lightest (maximum height) of 22 Å. (b) BEEM image of the Au-Si SB obtained at $V = 1.0$ V and $I_t = 1.0$ nA. Local value of I_c is represented by grey level, over a range from darkest (minimum I_c) to lightest (maximum I_c) of 1.5 pA. The average value of I_c is 2 pA. (c) STM image of the Au-GaAs SB. Grey level extends over a range from darkest (minimum height) to lightest (maximum height) of 72 Å. (d) BEEM image of the Au-GaAs SB obtained at $V = 1.5$ V and $I_t = 1.0$ nA. Grey level extends over the range from darkest (zero I_c) to lightest (maximum I_c) of 14 pA.

Figure 3(b) contains a typical BEEM I_c-V spectrum for the Au-GaAs SB heterostructure. In contrast to the Au-Si heterostructure, indications of interface heterogeneity of the Au-GaAs system are manifested as large spatial variations in the characteristics of BEEM spectra acquired at different locations at the Au-GaAs heterojunction surface. Figures 4(c) and 4(d) compare the surface topography of a typical region of a Au-GaAs heterojunction with the subsurface interface electronic structure. The Au-GaAs BEEM images display large-amplitude spatial variation in I_c. Pronounced maxima appear in the BEEM image in Fig. 4(d) along with large regions exhibiting no detectable collector current. It is important to note that the observed spatial variations in I_c are not correlated with surface topography. The dramatic level of heterogeneity revealed by BEEM imaging may be the result of multiphase structure at the Au-GaAs SB interface.[15]

This first application of BEEM demonstrates that this new method is a unique and powerful probe of subsurface interface properties. BEEM combines STM capabilities for investigation of surface structure with high-spatial-resolution probing of subsurface layers. Experimentally measured spectra present an abrupt threshold marking the interface barrier height. This method is shown to be versatile for study of both near-ideal Au-Si and complex Au-GaAs SB systems. BEEM images indicate homogeneous subsurface interface electronic properties for Au-Si. Also, BEEM images provide direct evidence for heterogeneity at the Au-GaAs SB interface on a lateral scale as small as 20 Å.

A simple theory predicts spectra which are in excellent agreement with measured BEEM spectra. Further, it is expected that a more detailed theoretical treatment should make possible additional investigation methods for the understanding of interface physics. For instance, detailed analysis of the BEEM spectrum threshold region may yield the quantum-mechanical reflection spectrum of carriers at interface barriers. In addition, BEEM is not limited only to the investigation of SB heterojunctions, but should be applicable to the investigation of many interface systems. For example, a transmission microscopy may also be performed on free-standing thin-film materials where the collector electrode is in vacuum, thus providing a probe of the base layer and the base-vacuum interface.

The authors thank H. G. LeDuc, R. C. Jaklevic, and J. Lambe for many stimulating discussions. This work was performed by the Jet Propulsion Laboratory, California Institute of Technology, as part of its Innovative Space Technology Center, which is sponsored by the Strategic Defense Initative Organization/Innovative Science and Technology Office through an agreement with the National Aeronautics and Space Administration (NASA).

[1]For an extensive review, see L. J. Brillson, Surf. Sci. Rep. **2**, 123 (1982).

[2]W. E. Spicer, I. Lindau, P. Skeath, C. Y. Su, and P. Chye, Phys. Rev. Lett. **44**, 420 (1980).

[3]J. Tersoff, Phys. Rev. Lett. **52**, 465 (1984).

[4]G. Binnig, H. Rohrer, Ch. Gerber, and E. Weibel, Phys. Rev. Lett. **49**, 57 (1982), and **50**, 120 (1983).

[5]R. S. Becker, J. A. Golovchenko, D. R. Hamann, and B. S. Swartzentruber, Phys. Rev. Lett. **55**, 2032 (1985).

[6]R. M. Feenstra, W. A. Thompson, and A. P. Fein, Phys. Rev. Lett. **56**, 608 (1986); J. A. Stroscio, R. M. Feenstra, and A. P. Fein, Phys. Rev. Lett. **57**, 2579 (1986).

[7]R. J. Hamers, R. M. Tromp, and J. E. Demuth, Phys. Rev. Lett. **56**, 1972 (1986).

[8]C. R. Crowell, W. G. Spitzer, L. E. Howarth, and E. E. Labate, Phys. Rev. **127**, 2006 (1962).

[9]P. W. Chye, I. Lindau, P. Pianetta, C. M. Garner, C. Y. Su, and W. E. Spicer, Phys. Rev. B **18**, 5545 (1978).

[10]P. O. Hahn and M. Henzler, J. Vac. Sci. Technol. A **2**, 574 (1984).

[11]J. R. Waldrop, J. Vac. Sci. Technol. B **2**, 445 (1984).

[12]The SB heights measured by the *I-V* method for the Au-Si and Au-GaAs heterostructures are in close agreement with the SB values for Au-Si, 0.8 eV (see Ref. 1), and for Au-GaAs, 0.85 eV (see Ref. 11). The Schottky-barrier lowering of approximately 0.04 eV (included in the analysis of Ref. 11) is not included for this comparison.

[13]W. J. Kaiser and R. C. Jaklevic, Surf. Sci. **181**, 55 (1987).

[14]The commonly used *I-V* method measures a spatial average of SB height which is heavily weighted by small barrier-height regions because of the exponential dependence of heterojunction current on voltage. It is expected, therefore, that in the presence of heterogeneous interface properties, typical, single-location BEEM measurements may show larger SB heights than indicated by *I-V* measurements.

[15]J. L. Freeouf and J. M. Woodall, Appl. Phys. Lett. **39**, 727 (1981).

Chemical trends in Schottky barriers: Charge transfer into adsorbate-induced gap states and defects

Winfried Mönch
Laboratorium für Festkörperphysik, Universität Duisburg, D-4100 Duisburg, Federal Republic of Germany
(Received 25 March 1987; revised manuscript received 28 September 1987)

The chemical trends reported for barrier heights in metal-GaAs contacts are explained by a charge transfer between the metal and adsorbate-induced gap states, which are identified as the virtual gap states of the complex band structure of GaAs, as well as fabrication-induced defects of donor type. Following the concept of the ionicity of chemical bonds, the charge transfer is described by the difference in the electronegativities of overlayer and substrate atoms. The density of fabrication-induced defects varies considerably.

As early as 1938, Schottky explained the rectifying behavior of metal-semiconductor contacts by a depletion layer on the semiconductor side of the interface.[1] The electron flow across the junction is thus determined by the energy distance from the Fermi level to the bottom of the conduction band when the semiconductor is doped *n*-type. Several models have been proposed for explaining the chemical trends observed with the barrier heights in such Schottky contacts.

The first proposal of this kind dates back to Mott,[2] who suggested that the barrier height of a metal-semiconductor contact equals the difference of the metal work function and the semiconductor electron affinity. The first systematic investigation[3] indeed revealed the barrier heights of metal-selenium contacts to be linearly related to the work functions of the metals involved but to be considerably smaller than expected from the Mott-Schottky rule just mentioned. To resolve this discrepancy, Bardeen[4] proposed interface states to determine the position of the Fermi level within the band gap at the interface. Such interface states result in the formation of a double layer at the interface, which was later on included in the analysis of the charge balance at such junctions.[5,6] The Mott-Schottky rule is then obtained with no interface states present at all while very large densities of interface states are pinning the Fermi level at their charge-neutrality level and the barrier heights become independent of the metals used. In the past, two basically different models have been proposed to explain the physical nature of such interface states.

Heine[7] argued that the wave functions of the metal electrons are tailing into the semiconductor for energies where the conduction band of the metal overlaps the band gap of the semiconductor. By this a dipole layer forms as it does at the metal-vacuum interface. The properties of those tails of the metal-electron wave functions are thus determined by the virtual gap states (VGS's) of the complex band structure of the semiconductor. The character of these states is predominantly acceptorlike and donorlike closer to the conduction and the valence band, respectively. At their branch point or, in other words, their charge-neutrality level, their decay length is smallest but increases towards both band edges. These tails of the metal wave functions into the semiconductor have also been called adsorbate-induced surface states[8] or metal-induced gap states (MIGS's).[9]

Later on, Wieder,[10] Spicer *et al.*,[11] and also Williams *et al.*[12] suggested that at semiconductor interfaces the Fermi level gets pinned by energy levels of native defects such as vacancies,[11,13] or antisite defects,[14] which are created in the semiconductor during the formation of the interface. For a review the reader is referred to Refs. 15 and 16.

Quite recently, Mönch[17] has analyzed the chemical trends of barrier heights reported for metal- and silicide-silicon contacts. The formation of dipole layers at the interface was attributed to the differences in electronegativity between silicon and the metals or silicides. The large body of experimental data and their chemical trends could be explained when adsorbate- or metal-induced gap states, which are identified as the VGS's of the complex band structure of the semiconductor, and fabrication-induced defects of donor-type were considered. This new VGS-plus-defects model of the barrier heights in Schottky contacts will be applied to metal-GaAs junctions in the present paper.

In the past, the analysis of the barrier heights in metal-semiconductor contacts was often based on the charge-neutrality condition

$$Q_{SS}+Q_{SC}=0 , \tag{1}$$

i.e., the surface band bending adjusts so as to achieve a balance between the space charge Q_{SC} and the excess charge Q_{SS} in surface states. Equation (1) is valid at clean semiconductor surfaces. At interfaces, where two different materials are in contact, the charge-neutrality condition (1) needs to be modified since a charge transfer may additionally occur across the interface due to chemical bonding. This phenomenon is most easily demonstrated by considering the adsorption of atoms, which are much more electropositive or electronegative than the semiconductor substrate used.

The adsorption of cesium,[18–20] of chlorine,[21] and of sulfur[22] on cleaved GaAs(110) surfaces, for example, is known to cause adsorbate-induced surface dipoles and surface states, which are observed via changes of the ionization energy and the buildup of surface band bending, respectively, as a function of adsorbate coverage. The adsorbate-induced dipoles may be described by an electric double layer made up by adequately charged adatoms and adsorbate-induced surface states in the semiconductor. At an interface the charge-neutrality condition (1) of the clean surface has now to be replaced by

$$Q_{ad}+Q_{SS}+Q_{SC}=0\ , \tag{2}$$

where Q_{ad} is the total charge per cm^2 carried by the adatoms.

The adsorption of cesium decreases the ionization energy of the GaAs substrate while it is increased by the adsorption of chlorine or sulfur. The adsorbed atoms of cesium and chlorine or sulfur are thus charged positively and negatively, respectively. As a consequence, the adsorbate-induced surface states are carrying a negative charge with cesium and a positive one with chlorine or sulfur adsorbed. This opposite sign of the charge transfer between adsorbate and substrate results from the fact that cesium and chlorine or sulfur are much more electropositive and electronegative, respectively, than gallium arsenide. The magnitude and the sign of the charge transfer may be estimated from the differences in the electronegativities of the adsorbates and the substrate by following Pauling's concept of the partial ionic character of covalent bonds.[23] The physical origin of the adsorbate-induced surface states, which are forming the adsorbate-induced surface dipoles together with the partly charged adatoms, shall be discussed in the following.

In a one-dimensional semiconductor, the sign of the appropriate Fourier coefficient of the periodic potential in conjunction with the position of the matching plane determines whether or not a virtual gap state of the complex band structure of the chain will form a real surface state. The energy and by this the character of this surface state are given by the same Fourier coefficient of the potential and the depth of the potential well at the surface state are given by the same Fourier coefficient of the potential and the depth of the potential well at the surface.[24–26] At adsorbate-covered surfaces, those virtual gap states will then happen to become adsorbate-induced difference in electronegativities.[7]

The charge transfer between adatoms and adsorbate-induced surface states of the substrate also mainly determines the position of the Fermi level at the surface, i.e., the surface band bending. The density of virtual gap states and thus of adsorbate-induced surface states measures some 10^{14} states per eV and cm^2 (Ref. 6) and thus the space charge Q_{SC} only plays a minor role in determining the surface band bending via the charge-neutrality condition (2).

The VGS model of adsorbate-induced surface states is strongly supported by experimental and theoretical data. In Fig. 1 the positions of the Fermi level below the conduction-band bottom as measured with cesium, sulfur, and chlorine adsorbed on cleaved GaAs(110) surfaces are plotted versus the difference $\Delta X = X_{ad} - X_{sub}$ between the electronegativities of the adsorbates and the GaAs substrate. Here ΔX stands for the charge transfer between adatoms and adsorbate-induced surface states. The electronegativities given by Miedema[27] are used; they are linearly related to the Pauling scale by

$$X_{Mied}=1.79X_{Paul}-0.8\ . \tag{3}$$

The electronegativities given by Miedema are preferred since they were derived from chemical trends in the properties of metallic alloys and compounds. The conclusions reached remain unchanged when Pauling's scale is used instead of Miedema's. The linear interpolation between the three experimental data points ($r = 0.99$) results in a charge-neutrality level of the adsorbate-induced surface states which agrees to within 40 meV with the energy predicted for the branch point of the VGS of GaAs by Tersoff.[28] An earlier estimate by Tejedor and Flores[29] deviates from this value by only 50 meV. The same model shall now be adopted for metal-semiconductor contacts.

The barrier heights reported for metal-GaAs contacts[30–33] are also plotted in Fig. 1. The selection of the data is based on the following considerations. First, all the barrier heights displayed were evaluated from current-voltage characteristics of Schottky diodes and have been corrected for image-force lowering. This means that, for example, the large body of results obtained with photoemission spectroscopy is not included since the metal coverage, at which the Fermi level gets pinned, generally depends on the growth conditions of the metal film as, for example, the substrate temperature and the evaporation rate. Second, with the exception of the Ru data point,[32] results communicated by Waldrop[30,31] for (100)- and by Newman *et al.*[33] for (110)-oriented GaAs substrates were considered only. Each of the two groups prepared all their samples under the same experimental conditions although the techniques used were different for the two substrate orientations.

In Fig. 1 the data points for the Schottky diodes fabricated with rare-earth metals (RE), titanium, lead, silver,

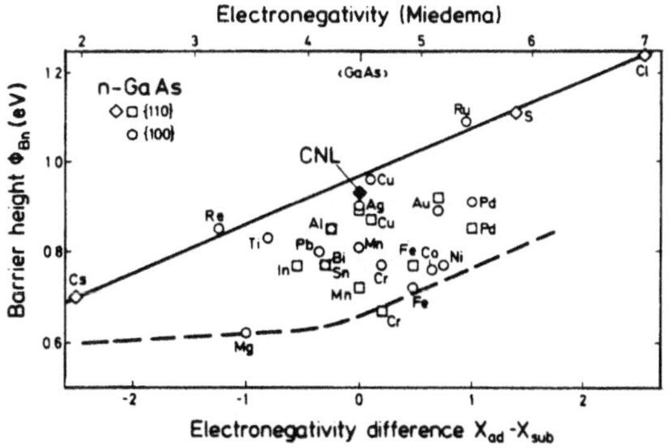

FIG. 1. Barrier heights Φ_{Bn} reported for metal-GaAs contacts and for adsorbates on GaAs cleaved surfaces plotted vs the electronegativity differences between the adatoms and the substrate with $X_{GaAs} = 4.45$. The electronegativities were taken from Miedema (Ref. 27), and the charge-neutrality level CNL of the virtual gap states of the complex band structure of GaAs was given by Tersoff (Ref. 28).

copper, and ruthenium are all following the chemical trend predicted from the VGS model of adsorbate- (Ref. 8) or, what is the same, metal-induced gap states.[9,34] Remarkably, the other metals give barrier heights which are lower than predicted by that model. This conclusion will not be altered when the pinning positions of the Fermi level as deduced from the energy-distribution curves of photoemitted electrons are included. The scatter of the data points entered in Fig. 1 will be attributed to the presence of fabrication-induced defects of donor type.

In defect-free metal-semiconductor contacts the variation of the barrier height as a function of the transfer of charge across the interface or, what is equivalent, of the electronegativity of the metal is determined by the density of states D_{VGS} of the VGS of the semiconductor and the effective width δ_{eff} of the dipole layer at the interface.[5,6] With the charge-neutrality level E_0 of the VGS lying well within the bulk band gap of a direct semiconductor, both the density of states D_{VGS} as well as the effective width δ_{eff}, which is determined by the decay length of the VGS,[7] are almost constant when the electron energy deviates by less than approximately one quarter of the gap energy from the charge-neutrality level of the VGS.[6] For a specific semiconductor, the barrier heights then vary linearly as a function of the metal electronegativity.[5,6,35]

With fabrication-induced defects present at a metal-semiconductor contact the condition of interface charge-neutrality now reads

$$Q_{ad}+Q_{SS}+Q_{def}+Q_{SC}=0 \ , \tag{4}$$

since charge Q_{def} in defects has to be considered in addition to the space charge and charged adatoms as well as adatom-induced interface states. Let us assume interface defects of donor type which exhibit an energy level in the upper half of the band gap. For very large transfer of negative charge into the adatom-induced gap states, the VGS model of adatom-induced interface states predicts low barrier heights, i.e., the Fermi level is found close to the conduction-band bottom and the interface defects assumed are neutral. With increasing electronegativity of the metal or, in other words, with decreasing negative charge in metal-induced gap states, the Fermi level moves towards the charge-neutrality level of the VGS and by this approaches the energy level of the fabrication-induced donors, which then become positively charged. The Fermi level gets pinned close to the defect level, i.e., the barrier height remains constant, until all the defects are charged and, eventually, the adatom-induced gap states will again accommodate charge. The Φ_{Bn}-versus-ΔX curve then runs parallel to that line observed with no fabrication-induced defects being present. The dashed line in Fig. 1 gives the relationship $\Phi_{Bn}(\Delta X)$ for the largest density of defects being compatible with the experimental data. The "pinned" part of the curve, which is determined by an assumed position of a defect level at 0.65 eV below the conduction-band bottom, is very ill defined since it is based on one data point only. In the framework of the model presented here, the densities of fabrication-induced defects are lowest with the rare-earth metals, Ti, Pb, Ag, Cu, and Ru, but largest with Fe, Co, and Ni.

The density of states of the adsorbate-induced gap states may be estimated from the difference in the position of the Fermi level at the interface of Schottky diodes prepared with substrates doped *n*- and *p*-type. Waldrop[30] has measured the respective barrier heights Φ_{Bn} and Φ_{Bp} for 14 different metals and has found them to add up to the band-gap energy minus $\delta\Phi_B=0.04\pm0.03$ eV on the average. Since the barrier heights amount to $E_g/2+0.1$ eV on the average depletion layers are present with substrates doped *p*- as well as *n*-type. According to the charge-neutrality condition (2), the density of adsorbate-induced gap (AIGS) states is then given by $D_{VGS}=(Q^n_{SC}-Q^p_{SC})/\delta\Phi_B$ since due to their energy position in the upper half of the band gap the fabrication-induced defects will always be charged independent of the type of substrate doping. Here, Q_{sc} is the space-charge density in the depletion layer. With the assumption that the GaAs samples used only contained shallow impurities, the density of states of the AIGS is estimated as 5×10^{13} states per cm^2 and eV. Theoretical models predicted 3×10^{14} states per cm^2 and eV.[6,34,36] The deviation is partly attributed to the large margins of experimental error with Φ_{Bp}, which is mostly due to the nonideality of the Schottky contacts on *p*-type GaAs, and to the unknown density of deep impurities in the substrates. This means that the value estimated for the density of AIGS represents a lower limit only.

The data plotted in Fig. 1 clearly show that the chemical trends observed in the barrier heights of metal-GaAs contacts may be understood by a chemically driven charge transfer across the interface into adsorbate-induced gap states, which are identified as the virtual gap states of the complex band structure of the semiconductor, and into fabrication-induced defects of donor type with energy levels in the upper half of the band gap. The same model was also successfully used to describe the chemical trends of the barrier heights reported for metal- and silicide-silicon contacts.[17] The charge transfer across the metal-semiconductor interface is modeled by the difference of the electronegativities of the metal and the semiconductor. Neither for GaAs nor for Si Schottky contacts can the energy position of the fabrication-induced donor-type defects be determined from the data available hitherto.

The recent experimental studies on the interaction of chlorine and sulfur with GaAs cleaved surfaces, which have triggered and stimulated the present work, were made possible by support from the Ministerium für Wissenschaft und Forschung des Landes Nordrhein-Westfalen and the Deutsche Forschungsgemeinschaft, respectively.

[1]W. Schottky, Naturwissenschaften **26**, 843 (1938).
[2]N. F. Mott, Proc. Cambridge Philos. Soc. **34**, 568 (1938).
[3]H. Schweikert, Verh. Dtsch. Phys. Ges. **3**, 99 (1938) as cited by W. Schottky, Phys. Z. **41**, 570 (1940).
[4]J. Bardeen, Phys. Rev. **71**, 717 (1947).
[5]A. M. Cowley and S. M. Sze, J. Appl. Phys. **36**, 3212 (1965).
[6]C. Tejedor, F. Flores, and E. Louis, J. Phys. C **10**, 2163 (1977).
[7]V. Heine, Phys. Rev. **138**, A1689 (1965).
[8]W. Mönch, Phys. Status Solidi **40**, 257 (1970).
[9]St. G. Louie and M. L. Cohen, Phys. Rev. B **13**, 2461 (1976).
[10]H. H. Wieder, J. Vac. Sci. Technol. **15**, 1498 (1978).
[11]W. E. Spicer, P. W. Chye, P. R. Skeath, and I. Lindau, J. Vac. Sci. Technol. **16**, 1422 (1979).
[12]R. H. Williams, R. R. Varma, and V. Montgomery, J. Vac. Sci. Technol. **16**, 1418 (1979).
[13]M. S. Daw and D. L. Smith, Phys. Rev. B **20**, 5150 (1979).
[14]R. E. Allen and J. D. Dow, J. Vac. Sci. Technol. **19**, 383 (1981).
[15]L. N. Brillson, Surf. Sci. Rep. **2**, 123 (1982).
[16]W. Mönch, Surf. Sci. **182**, 92 (1983).
[17]W. Mönch, Phys. Rev. Lett. **58**, 1260 (1987).
[18]T. Madey and J. Yates, J. Vac. Sci. Technol. **8**, 39 (1971).
[19]W. E. Spicer, P. E. Gregory, P. W. Chye, I. A. Babalola, and T. Sukegawa, Appl. Phys. Lett. **27**, 617 (1975).
[20]H. J. Clemens, J. von Wienskowski, and W. Mönch, Surf. Sci. **78**, 648 (1978).
[21]D. Troost, L. Koenders, L.-Y. Fan, and W. Mönch, J. Vac. Sci. Technol. B **6**, 1119 (1987).
[22]L. Koenders, M. Blömacher, and W. Mönch, J. Vac. Sci. Technol. B (to be published).
[23]L. Pauling, *The Nature of the Chemical Bond* (Cornell, Ithaca, New York, 1960).
[24]A. W. Maue, Z. Phys. **94**, 717 (1935).
[25]E. T. Goodwin, Proc. Cambridge Philos. Soc. **35**, 205 (1939).
[26]J. Zak, Phys. Rev. B **32**, 2218 (1985).
[27]A. R. Miedema, J. Phys. F **3**, 1558 (1973).
[28]J. Tersoff, Surf. Sci. **168**, 275 (1986).
[29]C. Tejedor and F. Flores, J. Phys. C **11**, L19 (1978).
[30]J. R. Waldrop, J. Vac. Sci. Technol. B **2**, 445 (1984).
[31]J. R. Waldrop, Appl. Phys. Lett. **47**, 1301 (1985).
[32]M. Ludwig and G. Heymann, J. Vac. Sci. Technol. B **4**, 485 (1986).
[33]N. Newman, W. E. Spicer, T. Kendelewicz, and I. Lindau, J. Vac. Sci. Technol. B **4**, 931 (1986).
[34]St. G. Louie, J. R. Chelikowsky, and M. L. Cohen, Phys. Rev. B **15**, 2154 (1977).
[35]W. Mönch, in *Festkörperprobleme: Advances in Solid State Physics,* edited by P. Grosse (Vieweg, Braunschweig, 1986), Vol. 26, p. 67.
[36]S. B. Zhang, M. L. Cohen, and St. G. Louie, Phys. Rev. B **34**, 768 (1986).

The advanced unified defect model for Schottky barrier formation

W. E. Spicer
Stanford Electronics Labs, Stanford University, Stanford, California 94305

Z. Liliental-Weber
Material and Molecular Research Division, Lawrence Berkeley Laboratory, Berkeley, California 94720

E. Weber
Department of Materials Science and Mineral Engineering, University of California, Berkeley, California 94720

N. Newman
Stanford Electronics Labs, Stanford University, Stanford, California and Material and Molecular Research Division, Lawrence Berkeley Laboratory, Berkeley, California 94720

T. Kendelewicz, R. Cao, C. McCants, P. Mahowald, K. Miyano, and I. Lindau
Stanford Electronics Labs, Stanford University, Stanford, California 94305

(Received 3 February 1988; accepted 10 May 1988)

The advanced unified defect model (AUDM) for GaAs proposed in this paper can be looked upon as a refinement of the unified defect model (UDM) proposed in 1979 to explain Fermi level pinning on 3–5 compounds due to metals or nonmetals. The refinement lies in identifying the defect producing pinning at 0.75 and 0.5 eV above the valence band maximum as the As_{Ga} antisite. Since the As_{Ga} antisite is a double donor, a minority compensating acceptor is necessary. This is tentatively identified as the Ga_{As} antisite. The concentration of As excess or deficiency due to processing or reactions at interfaces is particularly emphasized in this model. A wide range of experimental data is discussed in terms of this model and found to be in agreement with it. This includes the original data on which the UDM was based as well as more recent data including Fermi level pinning on the free-GaAs(100) molecular-beam epitaxy surface, Schottky barrier height for thick (~1000 Å) Ga films on GaAs, and the LaB_6 Schottky barrier height on GaAs (including thermal annealing effects). Of particular importance is the ability of this model to explain the changes in Schottky barrier height for Al and Au on GaAs due to thermal annealing and to relate these changes to interfacial chemistry.

I. THE ADVANCED UNIFIED DEFECT MODEL (AUDM)

In 1979, our group at Stanford suggested in the unified defect model (UDM) that native defects were important for pinning the Fermi level at 3–5 compound interfaces with both metal (Schottky barrier) and nonmetal overlayers.[1] Allen and Dow[2] proposed that antisites were the key native defects. Freeouf and Woodall[3] suggested in 1981 that excess As at the interface played a key role. In 1982, Weber *et al.*[4] argued that the key defect was the As_{Ga} antisite (an As atom on a Ga site in the GaAs lattice). Mönch and co-workers[5] reported in 1983 the occurrence of precipitates of excess As on surfaces of cleaved GaAs, indicating the presence of excess As in the GaAs crystals. As we will develop in this article, it appears that there is something important to the understanding of GaAs interfaces in each of these suggestions.

On the left-hand side of Fig. 1, we reproduce results from the 1979 Stanford paper. Here we present the pinning positions of n- and p-GaAs(110) surfaces for submonolayer to monolayer coverages of the indicated atoms. The horizontal lines indicate the energy levels deduced for the proposed defect levels. On the right-hand side, we show the energy levels which Weber *et al.*[4] found for the As_{Ga} antisite. As can be seen, the coincidence of the energy levels is striking (0.75 eV for the upper level and 0.5 to 0.52 for the lower level). This led Weber *et al.* to suggest that the As_{Ga} antisite was the key defect in the UDM. In recent years, the values of the energy levels by Weber *et al.* have received increasing acceptance (see, for example Ref. 6, and references therein).

The As_{Ga} antisite also seems to be the native defect central to the EL-2 center which is responsible for the compensation of undoped semi-insulating properties of GaAs.[7] However,

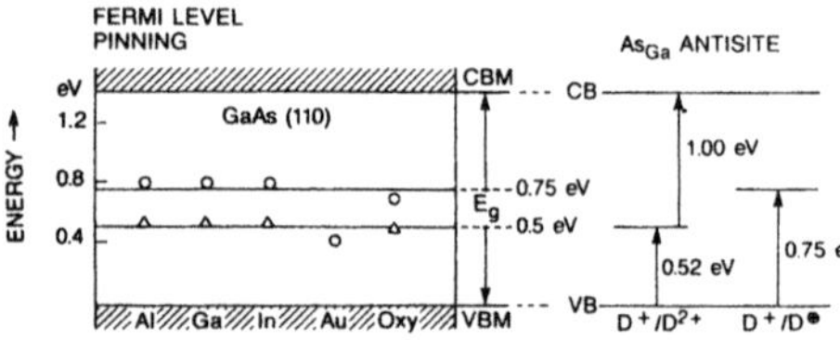

FIG. 1. The diagram to the left, PES results, is taken from a 1979 paper (Ref. 1). This energy level diagram indicates the pinning positions obtained on (○) n- and (△) p-GaAs with very low coverages (order 1 ML or less) of the indicated element measured at room temperature ($E_g = 1.42$ eV). The vertical lines through those data points indicate the defect energy levels deduced in 1979. The right diagram, photospin resonance results indicates the energy of the As_{Ga} antisite levels from the VBM measured at 8 K ($E_g = 1.52$ eV) by Weber *et al.* in 1982 (Ref. 4). As can be seen, the two sets of levels are found to agree

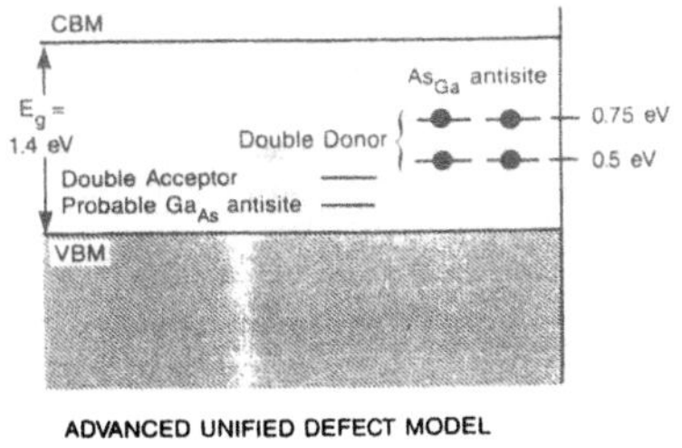

FIG. 2. This energy level diagram for the AUDM. The As_{Ga} antisite double donor with levels of 0.75 and 0.5 eV and the compensating acceptor (probably the Ga_{As} antisite) with energy levels below 0.5 eV are shown. Both defects are located in the same spatial region near the surface. The surface Fermi level position, E_{fi}, for the free surface will be determined by the relative densities of the two defects in the near surface region. In the usual case where $E_{fi} > 0.5$ eV, the density of $As_{Ga} > Ga_{As}$.

for neither EL-2 nor the interface behavior is the As_{Ga} antisite sufficient. As Weber *et al.*[4] found, the As_{Ga} antisite is a double donor. Thus, it must have a "spectator" minority acceptor defect in order that the Fermi level could be pinned at 0.75 eV for *n*-type material. For the interface, we would suggest that this spectator defect is most likely a Ga_{As} antisite (a Ga on an As site in GaAs) and that the Ga_{As} antisite provides two acceptor levels between the 0.5 eV level of the As_{Ga} antisite and the valence band maximum (VBM). For discussions of this defect see Ref. 8.

Based on this discussion, we would like to suggest the defect model for the interface shown in Fig. 2. The As_{Ga} antisite provides the pinning levels seen experimentally at 0.75 and 0.5 eV. However, the minority spectator plays an essential role. From computer modeling of the Fermi level position at the surface we come to the conclusion that the number of As_{Ga} antisites must be close to twice the number of Ga_{As} antisites and other acceptors in order to explain the *n*- and *p*-type pinning phenomena shown in Fig. 1. Simple calculations lead to a similar conclusion. Thus the near interface region must be As rich. In this model, As-rich interfaces are expected to be pinned near the midgap level, whereas less As-rich interfaces should exhibit pinning nearer the lower level.

It should be remembered that the bulk-Bridgeman-grown GaAs used to provide the cleaved (110) surfaces is grown from the As-rich side of the phase diagram, and, thus the crystals are expected to be As rich. And in fact there is ample evidence for this given by such diverse experiments as the ubiquitous presence of the EL-2 center[6] in as-grown bulk materials and the As inclusions on cleaved surfaces as reported by Bartels *et al.*[5] Recent evidence that clouds of EL2 (and therefore the As_{Ga} point defect) form around dislocations in GaAs[9] possibly due to the presence of long-range strain fields and excess As, suggests that an analgous situation may also occur at thick-film metal/GaAs interfaces.

Having defined the AUDM, we will in the following sections of this paper examine certain experimental data which may be explained in terms of it and further specify this model. There are many data in the literature which appear related to the AUDM; because of the limitation of space we can discuss only a limited set.

It should be mentioned that the AUDM was first presented at the Fourth International Conference on Solid Films and Surfaces (ICSFS) in August, 1987.[10]

II. THE GaAs-FREE SURFACES

It is important to develop perspective in terms of the free surface of GaAs. For the (110) surface, it is clear that surface reconstruction moves the intrinsic surface states out of the band gap.[11] This appears to also be the case for the other GaAs surfaces. However, only the cleaved (110) has been reported to be free of Fermi level pinning. For all other surfaces studied to date, the Fermi level at the surface has been found to be pinned near midgap. Since the pinning position cannot be explained in terms of the intrinsic surface states, it appears most likely that they are due to imperfections.

Svensson *et al.*[12] have reported a systematic study of GaAs(100) surfaces prepared by molecular-beam epitaxy (MBE). All the surfaces studied were As rich. In Fig. 3 we present the surface Fermi level (E_{fs}) positions which they found for three different surface reconstructions of *n*- and *p*-type GaAs. As Svensson *et al.* pointed out, the E_{fs} positions are very similar to those of Fig. 1, i.e., close to the two defect levels at 0.5 and 0.75 eV proposed in 1979. For sake of comparison, the 0.75 and 0.5 eV levels are represented by horizontal lines in Fig. 3. The As excess increases as one goes from the (4×6) to the *c*(4×4) reconstruction and the E_{fs} moves towards the conduction band maximum (CBM). This is what one would expect, using the model of Fig. 2, if the number of As_{Ga} antisites were increased relative to the number of Ga_{As} antisites. Our computer modeling confirms this conclusion.

Although they have not been as thoroughly studied, Ga-rich surfaces can also be formed by MBE. In Fig. 4, we pres-

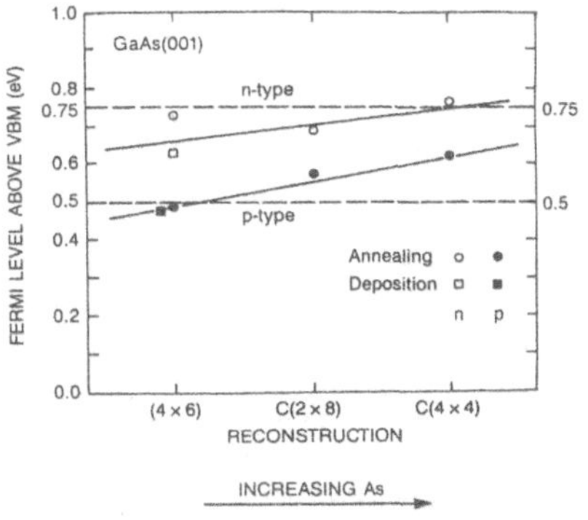

FIG. 3. The surface Fermi level positions E_{fi}, of the free (100) surface grown by MBE for *n*- and *p*-GaAs. The three surface reconstructions are indicated on the horizontal axis. All surfaces are As rich with the amount of excess As increasing from (4×6) to *c*(4×4). The energy levels of the AUDM model are shown. Note that these are the same as the pinning levels found on cleaved GaAs (110) surfaces due to deposition of metals or nonmetals (Fig. 1) (Ref. 12).

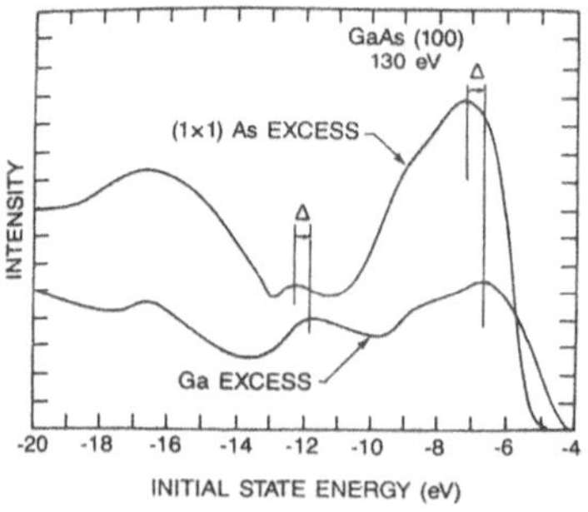

FIG. 4. Shift in the energy distribution of photoemitted electrons due to going from surfaces with As excess to those with Ga excess. The shift (Δ) of about 0.5 eV is attributed to the Ga_{As} antisite density becoming larger than the As_{Ga} antisite density when Ga excess is present (Ref. 13).

TABLE I. A summary of published results on Ga/GaAs contacts. By comparison with other metals of comparable electronegativity, one would expect a barrier height of about 1.1 eV or much higher.

Various reports of the electrical characteristics of Ga contacts on GaAs.
Ohmic contact on p-GaAs.
R. Z. Bachrach and A. Bianconi, J. Vac. Sci. Technol. **15**, 525 (1978).
New filled states extending below E_f due to Ga deposition.
D. Bolmont, P. Chen, and C. A. Sebenne, Surf. Sci. **117**, 417 (1982).
1.05 eV Schottky barrier height on MBE, As Rich, (100) *n*-GaAs for thick Ga layers.
S. P. Svensson, J. Kanski, and G. Anderson, Phys. Rev. B **30**, 6033 (1984).
1.15–1.2 eV Schottky barrier height for electrochemical metal deposition on *n*-GaAs: Relate to metallic Ga which they believe is formed.
R. Reinke and Meming, Surf. Sci. **192**, 66 (1987).
1.0–1.1 eV Schottky barrier height for thick Ga on GaAs(110).
A. B. McLean and R. H. Williams, Semicond. Sci. Technol. **21**, 654 (1987).

ent results from a study by Bachrach *et al.*[13] of the changes between As and Ga-rich MBE surfaces. As can be seen, there is a rigid shift of about 0.5 eV to a more negative initial state energy for the As-rich as compared to the Ga-rich surfaces. This is what is to be expected according to the model summarized on Fig. 2 when the Ga_{As} antisite defects become more numerous than the As_{Ga} antisites. Under this condition E_{fs} moves from the As_{Ga} levels toward the Ga_{As} acceptor levels which lie closer to the valence band maximum.

III. SCHOTTKY BARRIER HEIGHT FOR THICK Ga FILMS ON GaAs

If for Ga-rich MBE surfaces E_{fs} moves toward the VBM, should not similar behavior be expected for Ga on GaAs Schottky barriers? The early experiments with submonolayer to monolayer coverages of Ga reported no such effects (see Fig. 1 and Ref. 1). However, one must question whether deposition of a few monolayers of Ga on GaAs at room temperature, is sufficient for Ga atoms to move into the GaAs forming large numbers of Ga_{As} antisites.

The literature for thick Ga films on GaAs is quite different from that on the very thin films. As can be seen from Table I, a number of workers found E_{fs}, the Fermi level of the interface, to be surprisingly close to the VBM. Based on results for metals with similar electronegativity[13] one would expect E_{fs} for Ga to be 0.7–0.6 eV above the VBM. However, the barrier height measurements given in Table I lie 0.5–0.3 eV above the VBM, i.e., anomalously low in the forbidden gap.[14]

The low value of E_{fs} for these thick Ga films can easily be explained in terms of the AUDM. If Ga moves beneath the surface of the GaAs forming large numbers of additional Ga_{As} antisite, and these become more numerous than the As_{Ga} antisites, E_{fs} will move below the lowest As_{Ga} level 0.5 eV above VBM, as observed.

IV. LaB_6 ON (100) GaAs

Workers at the Optoelectronics Joint Research Laboratory in Japan have studied Schottky barriers of LaB_6 on GaAs.[15] The lack of stong chemical reactions of LaB_6 with GaAs makes it of interest for self-aligned gate technology.

Studies were made of LaB_6 on both MBE and chemically prepared (100) surfaces. In the case of MBE, *in situ* measurements of the Fermi level position at the surface were performed on thin layers (up to a few monolayers) of LaB_6 deposited and thermally annealed. Thick (1800 Å) films were deposited on both the MBE and chemically prepared surfaces and electrical measurements were performed to determine the Schottky barrier height as a function of annealing. For both the MBE-thin film studies and the thick studies, the effects of annealing were related to As loss. For example, the MBE samples were known to be As-rich as-grown and As was believed to be removed by the metal deposition or (if that was not sufficient) by annealing.

The experimental data are presented in Fig. 5. Two pinning positions were found in both the thin and thick film experiments. The original Schottky barrier height on *n*-GaAs for the thick films was close to 0.7 eV and this was increased to about 0.9 eV by annealing—(i.e., reduction of the As excess). In the AUDM, the two positions would be associated with the two As_{Ga} antisite levels near 0.65 and 0.9 eV below the CBM (see Fig. 5). With sufficient excess As and thus a As_{Ga}/Ga_{As} ratio close to two, one would expect pinning near the 0.65 eV As_{Ga} antisite level as observed. As the As/Ga ratio is reduced, the pinning position would be expected to move to the second As_{Ga} antisite level at 0.9 eV below the CBM minimum, as observed. Thus, this data set (which is somewhat unique in the combination of MBE-thin film studies with the thick film studies) can be explained by the AUDM.

V. ANNEALING STUDIES OF THICK UHV PREPARED AL AND AU DIODES

A. Interfacial chemistry and Scottky barrier changes

Two of the most studied Schottky barrier metals on GaAs are Au and Al. Recently a collaborative work between Stanford and Berkeley has carried out an extensive investigation of these systems. These studies have taken extreme care to understand and remove effects due to lack of cleanliness on

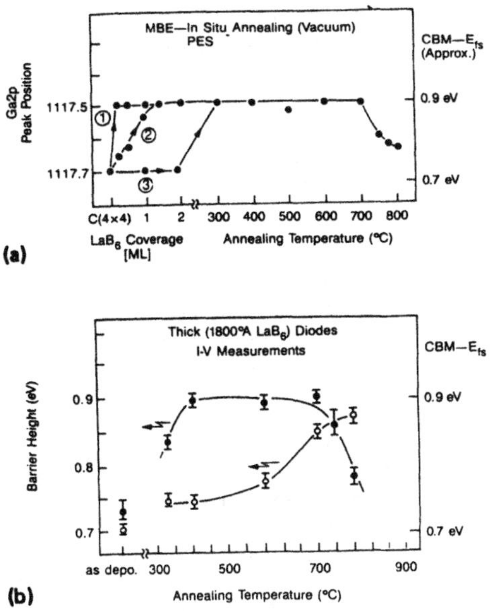

FIG. 5. Surface or interface Fermi level position E_{fi} for LaB_6 on GaAs. (Ref. 15.) (a) The Ga 2*p* core level position on MBE (100) GaAs due to LaB_6 deposition and/or annealing. The left-hand scale was obtained by assuming a 0.9 eV position after annealing. The arrows (1), (2), and (3) indicate changes due to LaB_6 deposition on different starting surface. (b) The Schottky barrier heights as a function of annealing for 1800 Å of LaB_6 on GaAs. (O) indicate MBE [$c(4\times4)$] and (●) chemically etched (100) GaAs surfaces.

the starting GaAs surfaces and identify and remove effects due to leakage currents at the perimeter of the diode.[16–21] In this way it has been possible to clearly identify and quantify changes in the actual Schottky barrier height due to annealing of the metal/GaAs interfaces. This, in turn, allows one the possibility of connecting the change in barrier height with the interfacial chemistry which goes on between the metal and GaAs. The electron microscope and other techniques were used to examine the interface region for reaction products and other phenomena. In this way, models for the change in Schottky barrier height may be tested more comprehensively.

For the AUDM, the barrier height depends on the number and position of the antisite defects and work function difference of the semiconductor and metal. For the case of as-deposited diodes formed on clean GaAs(110) surfaces, we find the large work function metals tend to pin near the 0.5 eV level and small work function metals tend to pin near the 0.75 eV level.[22] This indicates that charge exchange between the metallic overlayer and interface states can be important, as pointed out by Bardeen in 1947[23] and Zur *et al.*[24] more recently. However, because changes in barrier height are found at annealing temperatures in which little or no significant changes occur in the overlayer structure (see, for example, Refs. 16–21), the relationship between changes in barrier height upon annealing and reaction products appears quite straightforward. As discussed earlier, if the reaction produces excess As, one would expect E_{fi} to move toward the CBM; conversely, the decrease of the excess As, should move E_{fi} toward the VBM. The results will be discussed in these terms.

TABLE II. Interfacial chemistry and changes in Schottky barrier height. Relation between interfacial chemistry and changes in Schottky barrier height. For Al, excess Ga is produced by the reaction and the interfacial Fermi level E_{fi} moves toward to VBM. For Au, the opposite is the case, excess As is produced and E_{fi} moves toward the CBM.

Metal	Effect of reaction with GaAs	Change in ϕ_n due to anneal	Expectation from AUDM
Al	Excess Ga produced	+ 0.09 eV (0.08 to 0.89 eV)	Increase in ϕ_n
Au	Excess As produced	− 0.12 eV (0.92 to 0.8 eV)	Decrease in ϕ_n

As Table II indicates, careful studies have found that the barrier height for Al on *n*-type GaAs increases by about 0.1 eV (i.e., from 0.8 to 0.89 eV) on annealing; whereas, that for Au decreases by about the same amount (i.e., from 0.92 and 0.8 eV). With such a clear cut difference, it is interesting to examine the interfacial chemistry of these two metals with GaAs.

Interfacial chemistry between GaAs and metals have been the subject of considerable study recently. It would be appropriate to attempt to review that literature. From a few of the most relevant references for our discussion [e.g., Refs. 21, 25, and 26], one can obtain the essence of the interface chemistry for Al and Au. For Au/GaAs, the principal reaction can be symbolically indicated as:

$$x\text{Au} + \text{GaAs} \rightarrow \text{Au}_x\text{Ga} + \text{As}, \qquad (1)$$

i.e., a Au_xGa solution or alloy is formed with liberation of As. Energy is gained by the formation of Au_xGa; however, it appears necessary for the As to sublimate from the system in order for the reaction to proceed.[26] To put it another way, due to the presence of the tie line between Au and GaAs the thermodynamics appears to be favorable for the reaction only if the As can irreversibly leave the reacting interface. Note that As does not have a large solubility in either Au or GaAs. The photoelectron spectroscopy (PES) studies[25] indicate that a limited reaction takes place upon deposition of Au on clean GaAs, even if the liberated As does not leave. It is also clear that a similar (limited) reaction takes place between Au and GaAs for a clean surface and 1000 Å of Au when the system is annealed to 400 °C under an atmosphere of N_2.[27] Because the reaction is a source of extra As, As can be trapped at the interface.

The situation for Al is quite different from Au. Whereas, the driving force for the Au/GaAs interfacial reaction is the formation of Au_xGa, that for the Al/GaAs is the strong Al–As bond. To first order, one can neglect the effect of the formation of an $AlGa_x$ alloy. The strong AlAs bond exhibits itself strongly in the well established replacement reaction[28]:

$$x\text{Al} + \text{GaAs} \rightarrow \text{Ga}_{1-x}\text{Al}_x\text{As} + x\text{Ga}, \qquad (2)$$

i.e., Al replaces Ga *in the GaAs lattice* at the interface, and Ga metal is liberated. It was because of this reaction that we first suggested that the increase in *n*-GaAs barrier height was due to the increased band gap of the alloy formed at the interface[20]; however, recent work has established that this is not the case and the change in barrier height is due to a change in the Fermi level pinning position within the GaAs band gap.[29] Let us look more closely the Al–GaAs chemistry.

Consider the fate of any excess As, for example As_{Ga}, in the GaAs. Any such As reaching the interface will strongly react with the Al forming AlAs. Thus, the interface will act as a sink for As. In addition to the reaction of Eq. (2), a more generic symbolic chemical reaction might be indicated as

$$Al + GaAs_{1+\delta} \rightarrow AlAs_{\Delta} + GaAs_{1+(\delta-\Delta)}. \qquad (3)$$

The actual reaction would be a sum of Eqs. (2) and (3).

Let us return to the AUDM and consider how the interfacial chemistry could serve as a source or sink for excess As and/or Ga. The key thing is whether the interfacial chemistry tends to drive the interfacial Fermi level toward or away from the VBM. If a reaction at the interface tends to increase the As_{Ga} antisite density relative to the defect acceptor density, it will move the interfacial Fermi level toward the CBM. If the opposite is true, it will drive the Fermi level toward the VBM. From the discussion of interfacial chemistry described above, it seems clear that excess As will be produced at the Au/GaAs interface, whereas, excess As already present in the GaAs near the interface will be reduced by the Al/GaAs reaction. Thus, the AUDM would predict an increase in barrier height for annealing Al and a decrease for annealing Au on *n*-GaAs. The results are summarized in Table II. One should expect the barrier height to change in the opposite direction by the same amount for *p*-GaAs. Careful studies have shown this to be the case for Al[29]; comparable studies have not yet been done for Au.

Although excess As is generated at the interface in the case of Au, one must consider whether or not it can be removed from the interface due to its inherently high-vapor pressure. This depends on whether the annealing takes place under conditions which inhibit As from leaving the interface, such as an inert atmospheric (N_2 atmosphere) pressure or under conditions where the As can leave the interface, e.g., in vacuum.[30] The Au annealing described above took place under a N_2 atmosphere which prevented or reduced As leaving. However, As may leave due to annealing under other conditions. Preliminary results have been obtained with Au anneal in vacuum, i.e., *in situ* after evaporation. In this case, one would expect any excess As to evaporate during the annealing; and, therefore, one would not expect the barrier height to be significantly changed by the annealing. In fact this is what preliminary measurements find. Chemically one would expect Ti to act much like Al on annealing, and a similar increase in the Schottky barrier height of Ti/*n*-GaAs is also found.[31]

B. Direct measurements of As excess buildup or reduction due to annealing

Because of the volatility of As, one might expect it to be difficult to determine whether or not excess As is present near the surface. This indeed proves to be the case. However, it has been possible to detect excess As through two techniques discussed below.

In the course of transmission electron microscope (TEM) studies of the interfaces of GaAs before and after annealing (these studies include the samples discussed above on which the Schottky barrier changes were observed), energy dispersive x-ray spectroscopy (EDXS) has been used to examine the stoichiometry of the GaAs.[16,32] For Au, increased As was found after annealing, in agreement with the discussion of interfacial chemistry.[16,32] Recently, we have found a decrease in excess As after annealing in the Al case. Again this correlates with the increase in Schottky barrier height (SBH) on *n*-GaAs as expected for the AUDM.

Becker and his co-workers have recently developed a new technique for chemical analysis.[33] This is similar to secondary ion mass spectroscopy (SIMS) except that in SIMS the ions are provided by the sputtering process: whereas, in the new technique, the neutrals produced by sputtering are laser ionized. This technique, called surface analysis by laser ionization (SALI), has been applied to the Au/GaAs system.[34] The results are given in Fig. 6. For the annealed sample an increase in As is seen (as compared to the unannealed sam-

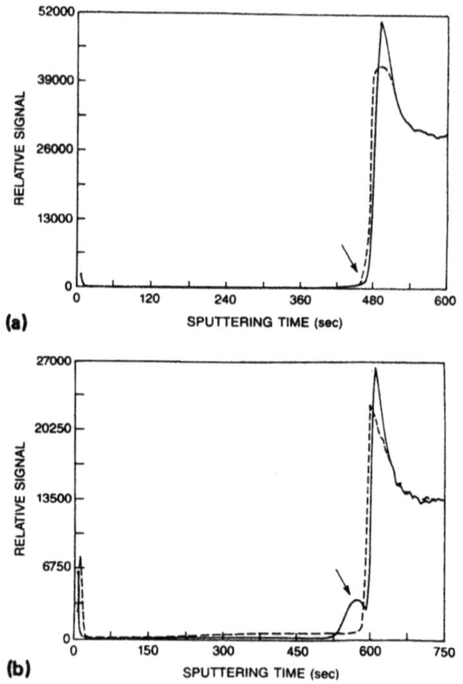

FIG. 6. (a) SALI analysis of Pallix, Becker, and Newman (Ref. 34) for 1000 Å of Au on GaAs. Note in particular the region of the Au/GaAs interface. The upper panel gives the results of an unannealed sample. There is little As—or Ga(---) in the Au. Panel (b) gives the results after thermal annealing 405 °C, 10 min in N_2. The most striking change is the large increase in As as the sputtering approaches the Au/GaAs interface (indicated by the arrow). Ga (---) and, to a lesser extent, As (—) are also seen in the Au and at the outside surface.

ple) as one approaches the interface. Thus, the decrease in SBH for Au/n-GaAs (see Table I) is confirmed to be associated with a buildup of As near the interface, in agreement with the AUDM.

VI. FERMI LEVEL MOVEMENT DUE TO DEPOSITION OF METALS AT LOW TEMPERATURES

Recently it has been found that the pattern of Fermi level pinning with metal deposition is changed by cooling the GaAs well below room temperature.[35–38] Typical data are shown in Fig. 7.[38] This is for Al deposited at − 190 °C. The pinning rate is increased for p-type GaAs but decreased for n type.

The behavior for metal coverages <0.2 monolayer (ML) coverage could be explained by the role of As_{Ga} antisite formation being much more important than that of the Ga_{As} antisite. Since As_{Ga} is a double donor, it would move the interface Fermi level on p- and not n-type GaAs. That this could be the case for these conditions may follow from the fact that the GaAs crystals have excess As due to their growth from the As-rich side of the phase diagram. This is combined with the observation that, at low temperatures, the clumping of the metal necessary to release the large amounts of energy[39] needed for formation of Ga_{As} antisites is suppressed. However, for coverages above 0.2 ML, sufficient clumping of Al can take place (due to Al atoms being statistically likely to find Al neighbors) so that formation of Ga_{As} antisites can proceed; and, thus, the Fermi level will begin to move on n-GaAs to the RT positions. Detailed studies should be made to test this suggestion.

VII. CONCLUSIONS

The original UDM presented in 1979 has been refined by associating the pinning levels at 0.75 and 0.5 eV (above the VBM) with the As_{Ga} antisite defect. A second minority acceptor defect located below 0.5 eV is also necessary. This accepter defect is believed to be the Ga antisite. This refined defect model is called for distinction the advanced unified defect model (AUDM).

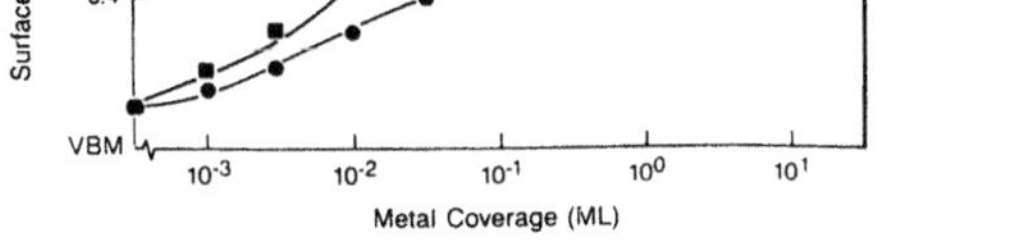

FIG. 7. Fermi level movement for Al deposited on n- and p-type GaAs at room (RT) and low (LT) temperatures. The Fermi level position (E_{fi}) is plotted on the ordinate vs the amount of Al deposited in monolayers (ML) is plotted on the abscissa. The striking change at LT is the decrease in movement of E_{fi} for n-type and the increase for p-type GaAs at LT. This can be qualitatively explained in terms of dominant As_{Ga} antisite formation for coverages $<\sim 0.1$ ML and the onset of strong Ga_{As} (as well as As_{Ga}) formation above that coverage (Ref. 38).

The experimental results on which the UDM was based are consistent with the AUDM. Also in agreement with the AUDM are more recent data on the "free" surface of MBE grown GaAs which shows pinning near the 0.75 and 0.5 eV levels and depend on As excess.

A more demanding test of the AUDM is given by examining data related to the presence of excess As or Ga at the interface and changes in the Fermi level (E_{fi}) with the change of the As to Ga ratio. Based on the AUDM, an As/Ga increase enhances the ratio of As_{Ga}/Ga_{As} antisites in the GaAs near the interface and this is expected to move the Fermi level toward the CBM; whereas, a decrease should move it toward the VBM.

Changes in E_{fi} on the MBE grown clean surface due to an increase in the As excess were shown to be in agreement with this AUDM. For the LaB_6/GaAs interface, E_{fi} has been shown to move about 0.2 eV and this movement associated with removal of excess As. Again in agreement with the model. Most available data show that Ga can move E_{fi} anomalously close to the VBM. The AUDM gives an explanation for this in terms of formation of Ga_{As} antisites.

The consideration of the preceding paragraph provides a way in which the AUDM may be used to relate interfacial chemistry to changes in E_{fi} and thus the Schottky barrier height (SBH). It is applied in the cases of careful detailed investigations of Au and Al on GaAs. From the interfacial chemistry one would expect a decrease in SBH for Au on n-GaAs due to annealing and an increase for Al. The decrease for Au would be due to the increase of As excess at the interface. For Al, the converse would be true. Independent experiments are reported in which the As buildup is increased for Au and decrease for Al have been observed.

Another popular model for Schottky barrier formation is the "metal-induced gap state" (MIGS) model.[40,41] This predicts E_{fi} to lie at a fixed energy of 0.5 eV above the VBM. Experiment finds E_{fi} to lie in a wide range of energies depending on the metal used. The AUDM is capable of explaining the range of pinning in terms of electronegativity and interfacial chemistry. As demonstrated in this paper it is also capable of explaining changes in E_{fi} and (thus SBH) due to annealing or other processes. It does not appear that the MIGS model can do this. The AUDM can explain uniquely a range of Schottky barrier and changes due to annealing. However, the history of attempts to explain Schottky barrier formation has been very complex. Thus, much more study is necessary before we can be absolutely sure as to the validity of this model and the range of conditions in which it applies.

ACKNOWLEDGMENTS

Useful discussions with Izuo Hayashi and Harry Wieder are gratefully acknowledged. Work at Stanford supported by DARPA and ONR under Contract No. N00014-83-K-0073, work at University of California, Berkeley supported by the DOE under Contract No. DE-AC03-76SF00098.

[1]W. E. Spicer, P. W. Chye, P. R. Skeath, C. Y. Su, and I. Lindau, J. Vac. Sci. Technol. **16**, 1427 (1979); **17**, 1019 (1980); Phys. Rev. Lett. **44**, 420 (1980).
[2]R. E. Allen and J. D. Dow, J. Vac. Sci. Technol. **19**, 383 (1981).
[3]J. L. Freeouf and J. M. Woodall, Appl. Phys. Lett. **39**, 727 (1981).
[4]E. R. Weber, H. Ennen, V. Kaufmann, J. Windscheif, J. Schneider, and T. Wosinski, J. Appl. Phys. **53**, 6140 (1982); E. R. Weber and J. Schneider, Physica B **116**, 398 (1983).
[5]F. Bartels, H. J. Clements, and W. Monch, Physica B **117/118**, 801 (1983).
[6]G. Baraff and M. Schluter Phys. Rev. Lett. **55**, 1327 (1987); Phys. Rev. B **35**, 6154 (1987).
[7]G. M. Martin, J. P. Farges, G. Jacob, and J. P. Hallais, J. Appl. Phys. **51**, 2840 (1980).
[8]K. R. Eliott, Appl. Phys. Lett. **42**, 274 (1983); T. Figielski, Appl. Phys. A **35**, 255 (1984); Z. G. Wang, L. A. Ledebo, and H. G. Grimmeiss, J. Phys. C **17**, 250 (1984).
[9]D. J. Stirland, M. R. Brozel, and I. Grant, Appl. Phys. Lett. **46**, 1066 (1985).
[10]W. E. Spicer, T. Kendelewicz, N. Newman, R. Cao, C. McCants, K. Miyano, I. Lindau, and E. R. Weber, Appl. Surf. Sci. (in press).
[11]S. Y. Tang, W. N. Mei, and G. Xu, J. Vac. Sci. Technol. B **2**, 393 (1984), and references therein.
[12]S. P. Svensson, J. Kanski, T. G. Anderson, and P. O. Nilsson, J. Vac. Sci. Technol. B **2**, 235 (1984).
[13]R. Z. Bachrach, R. S. Bauer, P. Chiaradia, and G. V. Hansson, J. Vac. Sci. Technol. **19**, 335 (1981).
[14]It should be noted that Woodall reported Ohmic behavior for Ga on *n*-GaAs. J. Woodall and C. Lauza, J. Vac. Sci. Technol. **15**, 1436 (1978).
[15]T. Yokotsuka, T. Narusawa, Y. Uchida, and H. Nakashima, Appl. Phys. Lett. **50**, 591 (1987); Y. Uchida, T. Yokotsuka, H. Nakashima, and S. Takatani, *ibid.* **50**, 670 (1987).
[16]Z. Liliental-Weber, R. Gronsky, J. Washburn, N. Newman, W. E. Spicer, and E. R. Weber, J. Vac. Sci. Technol. B **4**, 912 (1986). Z. Liliental-Weber, E. R. Weber, N. Newman, W. E. Spicer, R. Gronsky, and J. Washburn, *Defects in Semiconductors*, edited by H. J. von Bardeleben, Vols. 10–12 in Material Science Forum (Trans. Tech., Switzerland, 1986), pp. 1223–1228.
[17]Z. Liliental-Weber, J. Vac. Sci. Technol. B **5**, 1020 (1987).
[18]N. Newman, W. G. Petro, T. Kendelewicz, S. H. Pan, S. J. Eglash, and W. E. Spicer, J. Appl. Phys. **57**, 1247 (1985); D. Coulman, N. Newman, G. A. Reid, Z. Liliental-Weber, and W. E. Spicer, J. Vac. Sci. Technol. A **5**, 1521 (1987).
[19]Z. Liliental-Weber, J. Washburn, N. Newman, W. E. Spicer, and E. R. Weber, Appl. Phys. Lett. **49**, 1514 (1986).
[20]N. Newman, K. K. Chin, W. G. Petro, T. Kendelewicz, M. D. Williams, C. E. McCants, and W. E. Spicer, J. Vac. Sci. Technol. A **3**. 996 (1985).
[21]N. Newman, W. E. Spicer, and E. R. Weber, J. Vac. Sci. Technol. B **5**, 1020 (1987).
[22]N. Newman, W. E. Spicer, T. Kendelewicz, and I. Lindau, J. Vac. Sci. Technol. B **4**, 931 (1986), and references therein.
[23]John Bardeen, Phys. Rev. **71**, 717 (1947).
[24]A. Zur, T. C. McGill, and D. L. Smith, J. Vac. Sci. Technol. B **1**, 608 (1983).
[25]W. G. Petro, T. Kendelewicz, I. Lindau, and W. E. Spicer, Phys. Rev. B **34**, 7089 (1986).
[26]C. T. Tsai and R. S. Williams, Mater. Res. **1**, 820 (1986).
[27]Earlier we had proposed that the movement of the E_{fi} toward the CBM was due to the formation of $AuGa_x$ alloy at the interface (Ref. 20). However, more recent studies indicate that the amount of Ga in the Au when the change is first observed is insufficient to form an alloy; instead a solution of a few percent Ga is dissolved in a crystalline Au phase (Refs. 16 and 19).
[28]C. B. Duke, A. Paton, R. J. Mey, L. J. Brillson, A. Kahn, D. Kanani, J. Carelli, J. L. Yeh, G. Margaritondo, and A. D. Katnani, Phys. Rev. Lett. **46**, 440 (1981).
[29]N. Newman, W. E. Spicer, and E. R. Weber, J. Vac. Sci. Technol. B **5**, 1020 (1987).
[30]E. Kinsbron, P. K. Gallagher, and A. T. English, Solid State Electron. **22**, 517 (1979); S. Leung, L. K. Wong, D. D. L. Chung, and A. G. Milnes, J. Electrochem. Soc. **130**, 462 (1983).
[31]C. E. McCants, T. Kendelewicz, P. H. Mahowlad, K. A. Bertness, M. D. Williams, N. Newman, I. Lindau, and W. E. Spicer, J. Vac. Sci. Technol. A **6**, 1466 (1988).
[32]Z. Liliental-Weber, E. R. Weber, N. Newman, W. E. Spicer, R. Gronsky, and J. Washburn, in *Defects in Semiconductors*, edited by H. J. von Bardeleben, Vols. 10–12 in Material Science Forum (Trans Tech., Switzerland, 1986), p. 1223.
[33]C. H. Becker and K. T. Gillen, Anal. Chem. **56**, 1671 (1984); J. Vac. Sci. Technol. A **3**, 1347 (1985).
[34]J. B. Pallix, C. H. Becker, and N. Newman, Mater. Res. Bull. **12**, 52 (1987).
[35]T. Kendelewicz, M. D. Williams, K. K. Chin, C. E. McCants, R. S. List, I. Lindau, and W. E. Spicer, Appl. Phys. Lett. **48**, 919 (1986).
[36]K. Stiles, A. Kah, D. G. Kilday, and G. Margaritondo, J. Vac. Sci. Technol. B **5**, 987 (1987).
[37]M. D. Williams, K. K. Chin, C. E. McCants, P. H. Mahowald, and W. E. Spicer, J. Vac. Sci. Technol. B **4**, 939 (1986).
[38]R. Cao, K. Miyano, T. Kendelewicz, K. K. Chin, I. Lindau, and W. E. Spicer, J. Vac. Sci. Technol. B **5**, 998 (1987).
[39]A. Zunger, Phys. Rev. B **24**, 4372 (1981).
[40]V. Heine, Phys. Rev. A **138**, 689 (1965).
[41]J. Tersoff, Phys. Rev. B **32**, 6968 (1985); Phys. Rev. Lett. **52**, 465 (1984).

Screening and delocalization effects in Schottky barrier formation

R. Ludeke, G. Jezequel,[a)] and A. Taleb-Ibrahimi
IBM T. J. Watson Research Center, P. O. Box 218, Yorktown Heights, New York 10598

(Received 2 February 1988; accepted 4 May 1988)

In order to assess the role of metallic screening at the metal–semiconductor interface, we have investigated the effects of Ag deposition on the band bending of GaAs(110) surfaces previously "pinned" with vanadium atoms. Their coverages ranged from 0.05 to 1.5 monolayers (ML), which positioned the surface Fermi level (E_F) ~0.85–0.6 eV above the valence band maximum (VBM). For subsequent depositions of Ag additional changes in E_F of up to 200 meV towards the VBM were observed on both *n*- and *p*-type GaAs. The onset of these changes, which occurs for mean Ag coverages of ~0.1 ML, coincides with the appearance of metallic characteristics in the Ag clusters. These changes in E_F can be explained in terms of a new model in which the interface impurity levels in the band gap of the semiconductor become delocalized through interaction with states of the metal overlayer. A functional expression is developed which relates the Fermi level to the metal and semiconductor parameters, and to an effective charge in each delocalized level. These levels, actually broadened resonances of the impurity-metal system, accommodate the charge densities required to equalize E_F throughout the system. In analogy to a partially filled band, E_F is determined by the average effective charge residing in each resonance. The resonances are nearly neutral and can accept or donate electronic charge as needed to compensate for the conductivity type of the semiconductor. Calculated changes of E_F based on this model agree well with the experimental observations.

I. INTRODUCTION

The role of metallic character in the overlayer during the formation process of the metal–semiconductor interface has received scant attention, aside from the necessary presence of the metallic phase in the description of the metal induced gap states (MIGS) models.[1–5] In general, the position of the interface Fermi level E_F has reached a value in the band gap near its final position prior to the establishment of metallic characteristics,[6–9] the latter occurring in the coverage range of ~0.1–1 monolayer (ML) [referred here to the atomic density of $\sim 8.86\times 10^{14}$ cm^{-2} of the GaAs (110) surface]. These observations have in the past been used to argue against MIGS models.[8,9] Nevertheless, some additional motion of E_F occurs near the onset of metallic behavior,[9] which may be associated with metallic characteristics in the overlayer. A particularly unusual behavior was observed some time ago for noble metals on *p*-type GaAs,[10,11] for which E_F first rises in the band gap with metal coverage, levels off at the onset of metallic behavior, and decreases slowly to its final position for coverages exceeding several ML. Similar effects have subsequently been reported for reactive metals,[9] and very recently for metals deposited at low temperatures, for which the cresting effect in E_F with coverage is particularly pronounced.[12,13] These results cannot be explained by present Schottky barrier models, which suggests that effects which have not been considered previously seem to play an important role at the metal–semiconductor interface.

We set out to study the role of the onset of metallic behavior by first stabilizing E_F with a previously studied metal adsorbate, vanadium, and then covering the surface with a noninteracting metal, for which we chose Ag. On the GaAs(110) surface V reacts by an exchange reaction which predominantly involves the Ga surface atoms.[9,14] The reactions produce impurity states which readily "pin" E_F near midgap for ~0.1 ML of V.[9] The initial goal of the experiments was to test the notion of Zur *et al.*[15] that metallic screening required an interface charge density as large as $\sim 10^{14}|e|$cm^{-2} to stabilize or pin E_F at the energy of a midgap defect level, where $|e|$ is the magnitude of the electron charge (in contrast, a charge density of only $\sim 10^{12}$ cm^{-2} is required for the free surface). Additional considerations by Duke and Mailhiot[16] supported these findings. Thus interface defect or impurity densities substantially $< 10^{14}$ cm^{-2} should not affect the final E_F of a metal (such as Ag) with a different defect level. On the other hand, a density substantially larger should pin E_F, with its value unchanged by coverage of the aforementioned metal. As will be discussed in the next section, the latter situation is not observed experimentally, that is, changes in E_F are measured with Ag coverage even for estimated impurity densities exceeding 10^{15} cm^{-2}, an observation which again cannot be explained by existing Schottky barrier models.

The experimentally observed changes in E_F with Ag coverage can be explained by a new model which considers the interaction of the impurity states with metallic states in the limit of full metallic coverage. This model will be developed in Sec. III, where we will show that an impurity or defect level localized on the surface of the semiconductor will delocalize and become a (broadened) resonance in the presence of a continuum of states (the metal). An important property of the resonances is their ability to donate or accept incremental charge from the metal, the amount depending on the requirements for maintaining a constant Fermi level throughout the system. This property obviates the need for separate donor or acceptor levels in the defect models,[17,18] a requirement which has led to some confusion and considerable speculation.[17–20] In Sec. IV we will discuss implications of this model, including the role of broadening of the defect states. Such broadening, unless the result of the delocalization phenomena, cannot account for the experimental observations. We conclude the last section with a discussion of the relationship of our model to MIGS and defect models.

II. THE EFFECT OF METALLIZATION ON E_F: EXPERIMENTS

Changes in the positions of the Fermi level relative to the semiconductor band edges were obtained from changes in kinetic energy in the photoemission spectra of the Ga and As 3*d* core levels. Because of the strong interaction of V with the GaAs(110) surface, substantial contributions from the reaction products (chemically shifted peaks) appear in the core level spectra. These disguise the substrate contributions to the core level spectra, which have to be resolved in order to accurately determine changes in the Fermi level. Consequently we limited our maximum V coverage to about 3 Å (~3 ML). An accuracy of ±0.03 eV in the determination of the kinetic energy of the bulk components could still be achieved at this coverage by computer aided line shape analysis.[9] In the experiments, V was deposited at room temperature onto freshly cleaved *n*- and *p*-type GaAs(110) (doped $5\text{–}10\times10^{17}$ cm^{-3}). Several samples were annealed for 20 min at ~300 °C to improve the surface stoichiometry,[21] a process which also changed E_F relative to the band edges. Subsequently, Ag was stepwise deposited at room temperature to thicknesses of 30 Å. Photoemission spectra were taken after each deposition or treatment.

The measured changes in kinetic energy following the depositions were referenced to the known position of E_F for the freshly cleaved surfaces. The results are shown in Fig. 1. We will first discuss the unannealed samples (open data points).

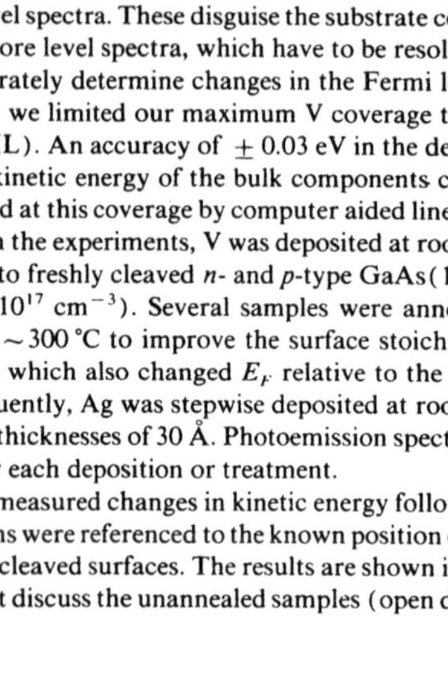

FIG. 1. Changes in the interface Fermi level with Ag coverage for (a) *n*-type GaAs(110) and (b) *p*-type GaAs(110) for indicated thicknesses of V interlayers. Open data points are for unannealed samples (○) 0.1 Å V, (□) 0.3 Å V, (△) 1 Å V; filled data points (▼) 3 Å V are for annealed samples. (---) represent the behavior for Ag without V at the interface.

The lowest vanadium coverage shown corresponds to $\sim7\times10^{13}$ cm^{-2}. For lower vanadium coverages the changes followed closely those of Ag on the bare GaAs(110) surface, shown as the dashed line in Fig. 1; these results are discussed elsewhere.[22] Prior to the Ag deposition, E_F varies according to the V coverage, as shown by the values on the left ordinate. Its value stabilizes to 0.62 eV above the valence band maximum (VBM) on both *n*- and *p*-type surfaces for V coverages exceeding 0.5 Å (~0.4 ML). With deposition of Ag a similar behavior is observed in the movement of E_F for the different V starting coverages: a nearly constant value for Ag coverages <0.5 Å, followed by a drop and a subsequent leveling off beyond 10 Å. The downturn coincides with the onset of metallic behavior in the Ag overlayer. This onset is manifested by a broadening of the Ag 4*d* bands and by the appearance of a metallic Fermi edge; features that are observable in the valence band spectra.[11] A surprising feature of the data is the close proximity, within 90 meV, of the final positions to the value for the Ag–GaAs interface (dashed line) on *n*-type surfaces, and their coincidence on *p*-type samples. The data appear to suggest that Ag almost totally screens the impurity levels generated by the V. The Ag does not remove the V,[21] although it has a tendency to concentrate (snow plow) the replaced Ga atoms in the surface region.[22] These observations are already inconsistent with both the MIGS[1–5] and the unified defect models[17,18] as both would predict invariance with the deposition of the second metal. The experiment was repeated with annealed surfaces covered with 3 Å V (solid data points and line in Fig. 1). The anneal produces a more homogeneous surface with a Fermi level position below that of the unannealed surface. An identical trend as discussed before for the unannealed surfaces is observed when Ag is deposited, with the important distinction that the final position of E_F is *below* that for Ag alone on GaAs. Clearly, it is not just the Ag that determines the final position of the Fermi level, otherwise its position would have coincided with the final positions for the unannealed samples. An adequate explanation of the observed behavior is beyond the capabilities of previously proposed Schottky barrier models, and we must look for an explanation based on concepts not applied previously to the metal–semiconductor interface. Although screening appears to be a plausible mechanism, in particular the lowering of energy of a charged impurity level due to image force effects, a detailed accounting, given in the next section, indicates that this effect is generally small, typically <10 meV. In contrast the observed changes in Fig. 1 are of the order of 140–270 meV, which are quite substantial when compared to the reported variations of ~300 meV in Shottky barrier heights for metals on GaAs.[9,18,23] The underlying effect which explains the observations is based on the profound changes occurring in a localized stationary state in the presence of a continuum of states. This phenomenon is well known in nuclear physics and has also been applied in the study of work function changes of adsorbates on metal surfaces.[24–26] We will follow a similar development here. To a first approximation the localized defect level may be though of as a suitable atomic level imbedded in a dielectric medium (the semiconductor). We will develop this model next.

III. THE DELOCALIZATION MODEL

A. Model formulation

Let us consider an impurity or defect in the surface plane of the semiconductor which generates a localized state in the band gap that can either accept or donate charge. Such a state can be adequately described by an atomiclike defect potential[27] $V_i \sim -e^2/\epsilon(r)r$, where $\epsilon(r)$ is a suitable dielectric function of the semiconductor. Let us assume that the impurity level E_i^0 is an eigenvalue of the Schrödinger equation $(\mathscr{H}_0 + V_i)\psi_i = E_i^0\psi_i$, where $\mathscr{H}_0$ is the appropriate Hamiltonian of the semiconductor, and ψ_i the eigenvector for state i, which we assume to be known. The metal will be described as free-electronlike (jellium) and satisfies the relationship $\mathscr{H}_m\psi_k = E_k\psi_k$. The boundary conditions at the metal surface ($x=0$) require that the ψ_k are propagating in the metal for $x>0$ and decaying for $x<0$, for which $\psi_k \sim e^{ik_y y + ik_z z} e^{\kappa x}$, where κ is a real positive number.

Conceptually, as the metal approaches the semiconductor surface the impurity and metal wave functions begin to overlap. Electrons will be able to tunnel between the two states in a time τ, which results in the delocalization of the former stationary impurity levels. A consequence of the finite lifetime τ is a broadening in energy of the impurity level. This broadening, represented by Γ, is related to τ through the uncertainty principle: $\Gamma = \hbar/\tau$. If in addition the level is charged, correlation effects may also change its average energy. This effect can be adequately described in the framework of the image force $V_{\rm im}$, which in the classic limit assumes the form $V_{\rm im} = e^2/4\epsilon s$, where s is the impurity–metal separation. An energy diagram for the system at close metal–semiconductor contact is shown in Fig. 2. The potential of the impurity–metal system is schematically shown by the dashed curve. For tunneling to be effective, the barrier should be of the order of an eV or less, which is realized when the metal work function ϕ_m is comparable to that of the intrinsic semiconductor. The impurity level E_i^0 is shown in Fig. 2 to have broadened and shifted downward by ΔE_i. This shift, which is due to $V_{\rm im}$, is different in sign from shifts in the metal adsorbate system because of an absence of an image force contribution from the ion core.[26] In general, ΔE_i is small (<20 meV); consequently correlation effects can only partially account for the observed changes. At intimate contact the delocalized impurity level is a resonance of the whole metal–semiconductor system, and an arbitrary amount of charge can be added or removed from the vicinity of the resonance as needed to equalize the Fermi level throughout the system. The broadened level may now be viewed as a local density of states $\rho_i(E)$, with the property $\int_{+\infty}^{-\infty} \rho_i(E)dE = 1$, that is, it can accommodate up to one electronic charge. Its functional dependence on E may be assumed to be Lorentzian,[24,25] which, for the isolated adsorbate–metal system, can be shown to be approximately correct by means of a Green's function approach due to Newns and Anderson.[28]

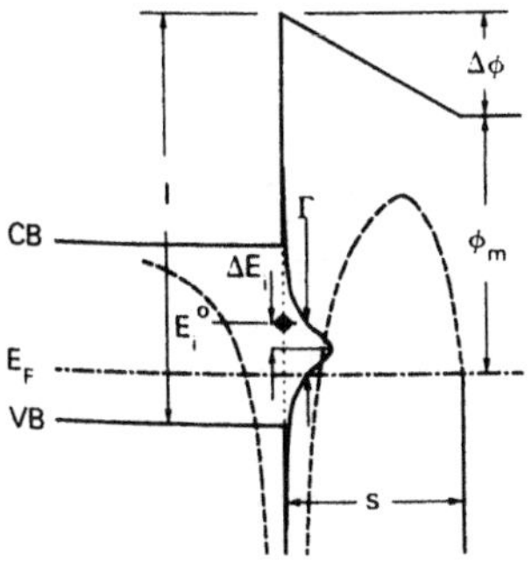

FIG. 2. Energy diagram for a semiconductor with a surface impurity level E_i^0 in close contact with a metal. The impurity level broadens by an amount Γ in the presence of the metal and shifts by ΔE_i due to screening (image force) effects. The dashed line represents the potential of the impurity–metal system.

We will adapt the Newns–Anderson formalism to the present problem of an isolated impurity level in close proximity to the metal. As for the adsorbate–metal system, we will assume that the eigenfunctions of the separate metal and impurity Hamiltonians form an orthonormal set, i.e., $\langle i|k\rangle = 0$; $\langle k|k'\rangle = \delta_{kk'}$. The complete Hamiltonian is given by $\mathscr{H} = \mathscr{H}_m + V_{\rm im} + \mathscr{H}_0 + V_i$, from which the following nonzero matrix elements are obtained:

$$\mathscr{H}_{kk'} = E_k\delta_{kk'};\ \mathscr{H}_{ik} \equiv V_{ik};\ \mathscr{H}_{ki} \equiv V_{ki},$$

$$\Delta E_i \equiv -\langle i|V_{im}|i\rangle;\ \mathscr{H}_{ii} \equiv E_{ii} = E_i^0 - \Delta E_i. \tag{1}$$

We define the Green's function of the metal–impurity system by

$$G = 1/(E - \mathscr{H} - i\alpha). \tag{2}$$

It can readily be shown,[26] that

$$\rho_i(E) = (1/\pi)\,\mathrm{Im}\{G_{ii}\} \tag{3}$$

with G_{ii} given by

$$G_{ii} = \langle i|G|i\rangle = 1/(E - E_{ii} - \Lambda - i\Gamma/2), \tag{4}$$

where

$$\Gamma(E) = 2\pi \sum_k |V_{ik}|^2\delta(E - E_k) \tag{5}$$

and

$$\Lambda(E) = P\sum_k [|V_{ik}^2|/(E - E_k)]. \tag{6}$$

Taking the imaginary part of Eq. (4) we obtain:

$$\rho_i(E) = \frac{1}{\pi}\frac{\Gamma/2}{(E - E_{ii} - \Lambda)^2 + (\Gamma/2)^2}. \tag{7}$$

In general, Γ and Λ are functions of E; however, as we will determine, Γ is much smaller than the bandwidth of the metal and the band gap of the semiconductor. Consequently we will assume it to be energy independent, with the additional consequence that $\Lambda \sim 0$. Implicit in the development of Eq. (7) is the neglect of coulombic effects on the unperturbed impurity level E_i^0. Since the average charge in the level is quite small, as will become evident later, we will continue to neglect this effect. With these assumptions, and substituting the appropriate matrix elements in Eq. (7), we obtain the desired Lorentzian form of $\rho_i(E)$:

$$\rho_i(E) = \frac{1}{\pi}\frac{\Gamma/2}{\{E - [E_i^0 - \Delta E_i(q_{\rm eff})]\}^2 + (\Gamma/2)^2}, \tag{8}$$

where $\Delta E_i(q_{\rm eff})$ reflects the image force lowering due to an effective charge $q_{\rm eff} \leqslant 1$ in the resonance level. We will now proceed to evaluate an expression to calculate $q_{\rm eff}$.

A charge redistribution will generally occur together with the delocalization of the impurity level. The redistribution is necessary to equalize the Fermi level throughout the metal–semiconductor system. Charge will transfer into the impurity level from the metal, leaving a counter charge of opposite polarity at the metal surface. This establishes a dipole potential $\Delta\phi$ at the interface of sufficient magnitude to equilibrate the Fermi level in the absence of an external field. Hence at equilibrium the Fermi level is given by:

$$E_F = I - \phi_m - \Delta\phi, \quad (9)$$

where I represents the ionization potential of the semiconductor and ϕ_m the metal work function. These two quantities represent the internal potentials, whereas in practice only the total (internal + surface) potentials are measured. Since I and ϕ_m always appear as a difference, the surface contributions, which are comparable,[16] tend to cancel when the total or experimental values are used in the calculations. For a surface density N_i of defects with identical impurity levels, the dipole can be represented by $\Delta\phi = 4\pi e^2\lambda_{\text{eff}} N_i q_{\text{eff}}$, where q_{eff} is the average charge per defect level and λ_{eff} is the effective separation of q_{eff} and its countercharge near the ion potential in the metal. Its value has been estimated as $\lambda_{\text{eff}} \sim \xi + s/\epsilon$, where ξ is the Fermi–Thomas screening length.[3] The effective charge q_{eff} is now obtained by integrating $\rho_i(E)$, Eq. (8): $q_{\text{eff}} = \int_{+\infty}^{-\infty} F(E)\rho_i\, dE$, where $F(E)$ is the Fermi function. In the low-temperature limit we obtain:

$$q_{\text{eff}} \overset{T\to 0}{=} \int_{-\infty}^{E_F} \rho_i(E)dE$$

$$= \frac{1}{\pi}\left\{\tan^{-1}\frac{E_F - [E_i^0 - \Delta E_i(q_{\text{eff}})]}{\Gamma/2} + \frac{\pi}{2}\right\}. \quad (10)$$

Since both ΔE_i and E_F, the latter now expressed as:

$$E_F = (I - \phi_m) - 4\pi e^2 \lambda_{\text{eff}} N_i q_{\text{eff}} \quad (11)$$

depend on q_{eff}, Eq. (10) must be solved self-consistently.

B. Numerical evaluations

The following general approximations were made in the calculation of the parameters: (a) the metal is treated as a jellium; (b) metal–semiconductor interactions beyond those forming the impurity/defect levels (bond formation) are neglected; and (c) the unperturbed impurity state is isolated and represented by a single impurity wave function, i.e., impurity bands are neglected. This approximation is certainly not correct for coverages approaching a ML; however the broadening due to delocalization effects would equally apply to band states and could be approximated by convoluting the impurity band density of states with an appropriate broadening function. We will readdress this point later.

We will briefly outline the calculation of Γ represented by Eq. (5). The procedure follows that of Gadzuk.[25] For the defect wave function we chose a hydrogenic 2-s wave function

$$\psi_i = (a^2/\pi)^{1/2}(1 - ar)e^{-ar},$$

where a is an adjustable parameter chosen so that $r^2|\psi_i|^2$ matches the main lobe of a model defect wave function.[29] For the present case $a = 1.24$ Å^{-1}. The relevant, decaying metal wave function (for $x \leqslant 0$) is represented by:

$$\psi_k = (2k_F/k_0)\, e^{\kappa x} e^{ik_y y + ik_z z},$$

where $k_0 = [(2m/\hbar^2)(\mu_m + \phi_m)]^{1/2}$, with μ_m the Fermi energy in the metal measured relative to the bottom of the free electron band, k_F is the Fermi wave vector, and $\kappa = (k_0^2 - k_F^2)^{1/2}$. In evaluating the overlap integral

$$V_{ik} = \int \psi_i^* [e^2/\epsilon(r)r]\psi_m\, d\tau$$

in Eq. (5) we asume that $\epsilon(r)$ is a slowly varying function over the region of wave function overlap, and can be removed from the integral. As we will see this is a good approximation. Evaluation of this integral[25] leads to the following expression of Eq. (5):

$$\Gamma = \frac{a^4 e^2 k_F s^3 e^{-2as}}{2a_0 k_0^2 \epsilon^2}\left(1 - \frac{4}{as} - \frac{4}{a^2s^2} + \frac{16}{a^3s^3} + \frac{32}{a^4s^4}\right). \quad (12)$$

Here a_0 is the Bohr radius. As stated above, ϵ is also a function of the impurity–metal surface separation s, and must be evaluated for the appropriate value of s.

The image potential of a conduction electron in a semiconductor near the interface to a metal has been evaluated by Inkson,[30] and corresponds to the potential of an electron on the vacuum side of a metal surface reduced by the dielectric constant of the semiconductor. In close vicinity of the metal surface strong deviations from the classical expression $e^2/4d$ occur. However, as Newns pointed out,[31] a good approximation is obtained by replacing the separation d by $d + \xi$, where ξ is the Thomas–Fermi screening distance. We therefore assume that for the impurity–metal interface the image potential can be represented by $V_{\text{im}} = q_{\text{eff}}^2/4\epsilon(s + \xi)$. To obtain ΔE_i, we evaluate the matrix element: $\Delta E_i = \langle i|V_{\text{im}}|i\rangle \sim q_{\text{eff}}^2/4\epsilon(s + \xi)$.

An important function we need to evaluate is the dielectric function ϵ, which is a parameter in the three relevant variables Γ, ΔE_i, and λ_{eff}. In the evaluation of Γ and ΔE_i we removed ϵ from the integrals. To recoup from this lack of rigor we use an appropriate dielectric function ϵ weighted by the impurity wave function: $\epsilon_{\text{eff}}(s) = \int_0^s \psi_i \epsilon(r)\psi_i d\tau / \int_0^s \psi_i^* \psi_i d\tau$. Using an exponential approximation to the calculated $\epsilon(r)$ for GaAs,[32] we evaluated $\epsilon_{\text{eff}}(s)$ for $1 \leqslant s \leqslant 3$ Å. We made the additional assumption that at the free surface the dielectric function $\epsilon^{\text{surf}}(s) = 0.5\epsilon^{\text{bulk}}(s)$ in the vacuum direction, reaching a maximum near 2.5 Å and subsequently decaying to the vacuum value of 1. The details of the functional behavior beyond 2.5 Å are unimportant however, since typical s values fall in the range 1.6–2.4 Å. The results are shown in Fig. 3(a), together with the evaluation of Γ for parameters corresponding to Ag and GaAs. As can be seen, Γ is extremely sensitive on the separation s, whereas ϵ_{eff} varies slowly over the same range, justifying the previous assumptions. We define $s = r_{sc} + r_m - d_m/2$, where the constants represent one-half of the GaAs bond length, the metallic radius and the metal interplanar spacing, respectively. For Ag on GaAs(110) we estimate a value of $s = 1.84$ Å. In the vicinity of this value $\epsilon_{\text{eff}} \sim 4$, which is a factor of 2 larger than the value estimated by Louie *et al.*[3] Based on the

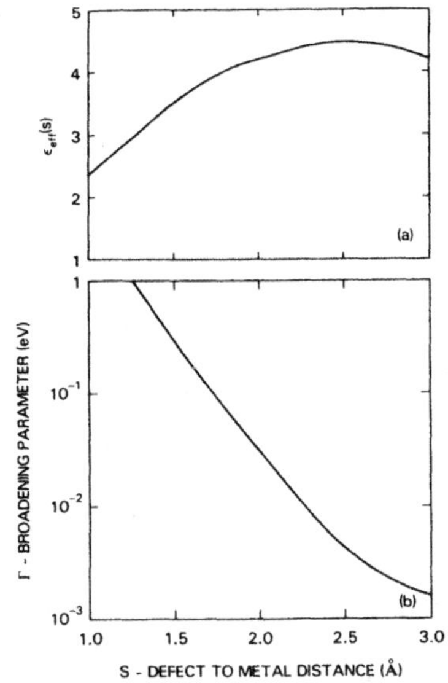

FIG. 3. Dependence of: (a) the effective dielectric function ϵ_{eff}, and (b) the broadening parameter Γ on the defect-to-metal separation s. Values are calculated for GaAs and Ag.

more recent calculations of $\epsilon(r)$,[32] the latter appears to be an underestimate, although such low value would enhance by a factor of 4 our estimates for Γ. For our parameters we calculate $\Gamma = 0.062$ eV for Ag.

Two additional quantities remain to be evaluated, namely λ_{eff} and N_i in Eq. (11). Following Louie *et al.*,[3] $\lambda_{eff} \sim \xi + s/\epsilon_{eff} \sim 1$ Å. The defect density N_i is proportional to the metal coverage. If the metal is dispersed, as the V in the present case, each atom contributes to N_i up to a coverage $\sim 1/2$ ML $\sim 5\times 10^{14}$ cm^{-2}, a value in good agreement with other estimates.[3] As we will show in Sec. IV, q_{eff}, as calculated from Eq. (10), is not very sensitive on N_i for values in the mid 10^{14} range. For our calculations we choose $N_i = 5\times 10^{14}$ cm^{-2} irrespective of the V coverage. This assumption is based on the argument that Ag itself will create additional interface states on V-free regions of the GaAs surface, thus forcing N_i to its saturation value. We are now in a position to calculate q_{eff} from Eq. (10) and obtain values for E_F through Eq. (11). For $I = 5.56$ eV[33] and $\phi_m = 4.52$ eV[34] we have calculated the changes in E_F ($\Delta E_F = E_i^0 - E_F$) following metallization for the various V coverages shown in Fig. 1. The results are summarized in Table I, which also lists the values of E_i^0 of the pinned starting surfaces. Again we must emphasize that we have assumed that the V pins the surface homogeneously, that is, for a given coverage all impurity levels exhibit the same value. We will discuss this point further in the next section. The generally good agreement exhibited in Table I between theory and experiment supports the viability of our model and its underlying premises. It should be re-emphasized that the model accounts only for the changes in E_F following the establishment of the complete, bulklike metal overlayer, for which the relevant parameters are assumed to be known. Since the formation of the metal–semiconductor interface is an inhomogeneous process, a full description in terms of the evolution of E_F with metal coverage would require detailed knowledge of the morphology and electronic structure of the developing metal islands or clusters. This information is presently unavailable. We will next assess the dependence of the solution of Eqs. (10) and (11) on the various parameters, and discuss more general trends predicted by this model.

IV. SOME MODEL PREDICTIONS AND DISCUSSION

To assess the dependence of the solutions on the parameters we must solve Eqs. (10) and (11) self-consistently and display the results graphically. Because of the many variables, we will discuss here only two sets of independent variables, the product $\lambda_{eff} N_i$ and a new variable $\Delta\phi_0 \equiv I - \phi_m - E_i^0$. The latter representation allows for an easy assessment of changes in any of the three variables, such

TABLE I. Comparison of experimental shifts in E_F (column 3) with calculated values (column 4), for Ag deposited on GaAs (110) surfaces predeposited with V. The pinning position prior to Ag deposition is given by E_i^0. q_{eff} for each case is also given. $\lambda_{eff} N_i = 5\times 10^{14}$ cm^{-2} Å, and $\Gamma = 0.062$ eV. The surface charge density ($q_{eff} N_i$) is given in column 6, assuming an $N_i = 5\times 10^{14}$ cm^{-2}.

	Experiment		Model prediction		
Sample	E_i^0 (eV)	ΔE_F (eV)	ΔE_F (eV)	q_{eff}	$q_{eff} N_i$ ($\times 10^{13}$cm^{-21})
n type					
0.1 Å V	0.86	0.28	0.222	0.044	2.22
0.3 Å V	0.73	0.19	0.181	0.054	2.72
1 Å V	0.63	0.14	0.157	0.063	3.14
3 Å V	0.56	0.13	0.143	0.069	3.44
p type					
0.1 Å V	0.74	0.26	0.184	0.054	2.68
0.3 Å V	0.64	0.16	0.159	0.062	3.09
1 Å V	0.62	0.14	0.155	0.064	3.18
3 Å V	0.52	0.14	0.136	0.072	3.62

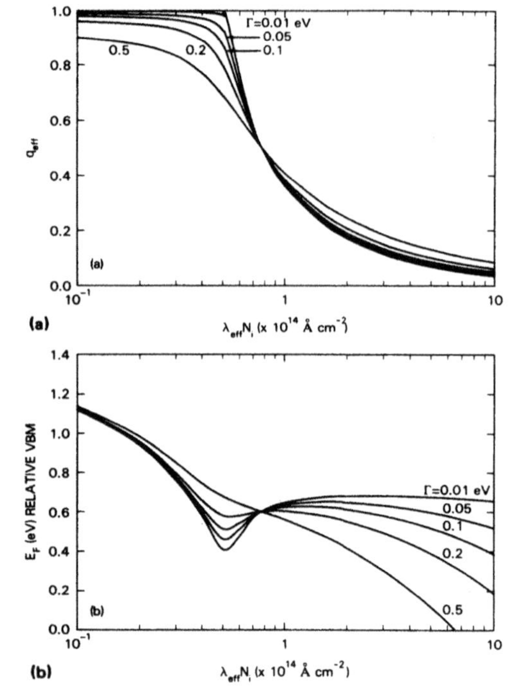

FIG. 4. Dependence of: (a) the effective charge per impurity resonance, and (b) the position of the interface Fermi level on the impurity density for values of $\Gamma = 0.01$, 0.05, 0.1, 0.2, and 0.5 eV and $\Delta\phi_0 = 0.60$ eV. $\Delta\phi_0 = I - \phi_m - E_i^0$.

as changes in metal (ϕ_m), semiconductor (I), or defect level (E_i^0).

Choosing first $\lambda_{eff} N_i$ as a variable, we show in Fig. 4(a) the solutions for q_{eff} and in Fig. 4(b) those for E_F for various values of Γ and a specific value of $\Delta\phi_0 = 0.60$ eV. For example, the latter value corresponds to $E_i^0 = 0.7$ eV for Al ($\phi_m = 4.26$ eV) on GaAs ($I = 5.56$ eV). As can be seen in Fig. 4(a), for a low-defect density q_{eff} is near unity, as the screening charge necessary to equalize the Fermi level fills every available defect site. In practice this range of low-defect density is not achievable since every known metal produces interface states in addition to the postulated impurity or defect states represented by N_i. Hypothetically, the range below $\sim 1\text{–}3 \times 10^{14}$ cm^{-2} is the jellium limit (noninteracting metal). Figure 4(b) shows the corresponding variations in the Fermi level. The decrease for low $\lambda_{eff} N_i$ is due to the increasing contribution of the screeing term ΔE_i (image potential). This contribution, which is $\propto q_{eff}^2$, becomes negligible for $q_{eff} \lesssim 0.2$. When q_{eff} is $< 1/2$, E_F is predominantly determined by the width of the resonance and how far up the resonance is "filled" with charge [Eq. (10)]. Thus delocalization dominates the position of E_F byond $\lambda_{eff} N_i \sim 10^{14}$ cm^{-2}, with E_F exhibiting the lowest values for low-charge contents in the broadest levels. It can also be ascertained that for the lower values of Γ, E_F varies slowly with $\lambda_{eff} N_i$, so that a precise determination of its value is unnecessary.

The variations of q_{eff} and E_F with $\Delta\phi_0$ are shown in Figs. 5(a) and 5(b), respectively, for several values of $\lambda_{eff} N_i$ and $\Gamma = 0.1$ eV. A large value of $\Delta\phi_0$ represents a low ϕ_m and/or an impurity level deep in the band gap. Thus a larger dipole needs to be set up which requires a larger q_{eff} per impurity site. The reverse applies for the lower values of $\Delta\phi_0$, which can also have negative values. For the lower defect densities, changes in E_F with increasing $\Delta\phi_0$ are characterized by an initial decrease due to image force effects, followed by a linear rise that represents the "Schottky" limit. Again, this limit is not achieved in practice because of the invariably higher interface state densities in all real systems. For more realistic densities of $\lambda_{eff} N_i \sim 5 \times 10^{14}$ cm^{-2}, our model would "predict" a sublinear variation of E_F with $\Delta\phi_0$, as shown by the solid curve in Fig. 5(b). Equivalently, this means that for metals with low ϕ_m, Schottky barrier heights would change slowly, but increase more rapidly for metals with larger ϕ_m. Since Γ and presumably E_i^0 are metal dependent, the smooth variation predicted in Fig. 5(b) is an idealization. Scatter in the existing experimental data does not yet allow a clear resolution of the functional dependence (linear or not) of E_F with the metal work function or related chemical quantity, such as the electronegativity. Nevertheless, seen with the simple assumption of a universal defect level, the models predict Schottky barrier variations of ~ 300 meV, which are comparable to experimental observations.[8,9,23]

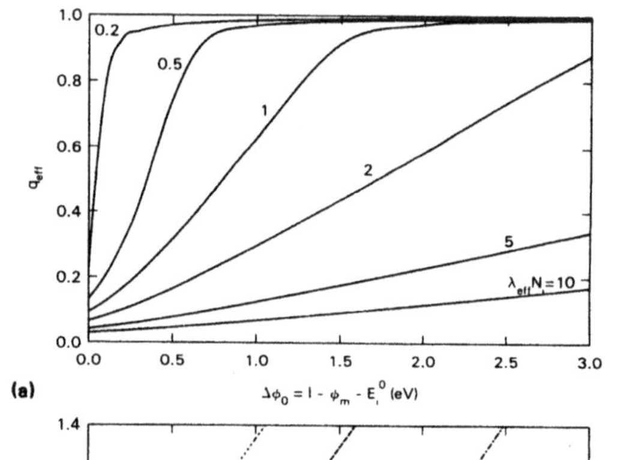

FIG. 5. Dependence of: (a) the effective charge per impurity resonance, and (b) the position of the interface Fermi level on $\Delta\phi_0$ for values of $\lambda_{eff} N_i = 0.2, 0.5, 1, 2, 5$, and 10 ($\times 10^{14}$ Å cm^{-2}), and $\Gamma = 0.1$ eV. The curve for $\lambda_{eff} N_i = 5$ corresponds approximately to GaAs with $E_i^0 = 0.7$ eV.

In the above discussions we have assumed that the impurity (or defect) levels were degenerate with a single value of E_i^0, that is, their environment was assumed to be homogeneous and interactions with neighboring impurity sites (band formation) were neglected. The latter assumption is quite reasonable since the overlap of the localized wave functions on adjacent impurity sites is negligible.[35] Since there is no evidence that the surface orders with metal adsorbates, it may be assumed that local structural and/or bonding variations exist; these would lead to variations in the energy of the defect levels. The resulting disorder-induced broadening in the distribution of impurity levels is unknown. Prior to the onset of metallic behavior, the lower levels are occupied by discreet charges (densities $\sim 10^{12}$ cm^{-2}). This broadening by itself cannot explain the changes in E_F with metallization. On the contrary, the additional charge of $\sim 10^{14}$ cm^{-2} needed to overcome metallic screening[15,16] would raise E_F slightly, instead of lowering it as observed experimentally. With onset of metallization all of the impurity levels at the interface will broaden due to the delocalization. If the latter is comparable to or greater than the disorder-induced broadening, the predictions of this model would in general still be valid. In the reverse case, the changes in E_F with metallization would be smaller, particularly for low work function metals [large $\Delta\phi_0$ in Fig. 5(b)]. In contrast, for large work function metals, for which $\delta E_F/\delta\Delta\phi_0$ is large [Fig. 5(b)], appreciable changes in E_F would still be observable because of the small average q_{eff}. This charge would fill only the portion of the total density of states that is dominated by the "tails" of the Lorentzians, which are considerably broader than just Γ.

We will discuss next the relationship between the present model and the MIGS and unified defect models. Resemblance between the delocalization model and the unified defect model ceases beyond the common requirement of an interface defect or impurity level. The delocalization model requires that the impurities/defects are distributed in or on top of the semiconductor surface layer, i.e., adjacent to the metal overlayer. The defect model, on the other hand, requires separate acceptor and donor levels in the near suface region to accommodate doping type differences in the semiconductor, a requirement that severely limits the possibilities. A recent extension of the model[17] proposed As_{Ga} (double donor) and Ga_{As} (acceptor) antisite defects as the responsible defects for pinning E_F for all metals on GaAs. This variation of the model specifies rather stringent requirements on both the relative abundance of the defects (number of donors > number of acceptors) and their location away from the immediate interface. The latter requirement is based solely on the need of identifying the pinning level with the bulk defect level, as an equivalent surface defect would have different energies. However, the notion of an adsorbate-generated defect level removed several atomic distances from the surface in order to remain "unperturbed" and bulk-like is physically counterintuitive, since any generation of defects by an adsorption process is expected to be in the immediate vicinity of the adsorbate atom. However, even a subsurface defect has a finite overlap with the encroaching metal wave functions, so that the concept of a fractionally occupied, but narrower resonance would still apply. It is thus difficult to reconcile the premises of the defect model with the physical changes that are sure to occur during metal–defect interactions. The relationship between the delocalization and the MIGS[1–6] models may be briefly stated as follows: In general, the MIGS models focus on the interaction of the tailing metallic states with the *unperturbed* semiconductor, which includes the action of MIGS on the dangling bonds of the semiconductor surface.[4,36] That is, chemistry (rebonding) at the interface has been ignored in previous MIGS models, with the exception of a recent work which included the interaction of jellium Al with the As surface atoms.[37] In contrast the delocalization model is based on bond rehybridization at the interface, which generates new gap states upon which the metallic states act. The resulting delocalized levels or resonances may be looked at as MIGS, as they are new states of the metal–impurity system, but unlike the traditional MIGS, they are of an entirely different origin.

In concluding, we will briefly summarize the key features of the delocalization model. A basic precept of the model is the presence of defect or impurity levels at the surface of the semiconductor, which are generated during the initial stages of the metallization process. As far as is known, every metal generates such states in most semiconductors, including GaAs, prior to the formation of the metallic phase. Upon formation of this phase the levels will become resonances, into which an incremental effective charge can flow to establish a dipole potential that equalizes work function differences between the metal and semiconductor. The effective charge is generally small, thus the level is nearly netural. The resonance can accommodate additional charge to compensate the large potential differences between *n*- and *p*-type semiconductors. This charge, although small ($\sim 0.1 q_{\text{eff}}$), generates differences in Fermi level positions for *n*- and *p*-type material of a few meV for most metals, but which could be as large as 30 meV for the larger work function metals.[38] The predictive power of the model depends on the knowledge of the impurity levels prior to the formation of the metallic phase. These are not isolated levels, but more likely disorder-broadened "bands" near midgap. Their determination remains a challenging problem. Nevertheless, the delocalization model estimates rather well the experimental changes discussed earlier for a metal acting on impurity levels of measurable energies. This agreement provides encouraging support to the fundamental premises of the delocalization model proposed here.

ACKNOWLEDGMENTS

The experiments were performed at National Synchrotron Light Source, Brookhaven National Laboratory, which is funded by the U. S. Department of Energy. The technical assistance of M. Prikas and A. Marx during the experimental portion of this research is gratefully acknowledged.

[a] Present address: Université de Rennes I, 35042 Rennes, France.

[1] V. Heine, Phys. Rev. A **138**, 1689 (1965).
[2] E. Louis, F. Yndurain, and F. Flores, Phys. Rev. B **13**, 4408 (1976).

[3]S. G. Louie, J. R. Chelikowsky, and M. L. Cohen, Phys. Rev. B **15**, 2154 (1977).
[4]E. J. Mele and J. D. Joannopoulos, Phys. Rev. B **17**, 1528 (1978).
[5]J. Tersoff, Phys. Rev. Lett. **52**, 465 (1984).
[6]L. J. Brillson, C. F. Brucker, A. D. Katnani, N. G. Stoffel, R. Daniels, and G. Margaritondo, J. Vac. Sci. Technol. **21**, 564 (1982).
[7]T. Kendelewicz, N. Newman, R. S. List, I. Lindau, and W. E. Spicer, J. Vac. Sci. Technol. B **3**, 1206 (1985).
[8]W. E. Spicer, T. Kendelewicz, N. Newman, K. K. Chin, and I. Lindau, Surf. Sci. **168**, 240 (1986).
[9]G. Hughes, R. Ludeke, F. Schäffler, and D. Rieger, J. Vac. Sci. Technol. B **4**, 924 (1986); R. Ludeke and G. Landgren, Phys. Rev. B **33**, 5526 (1986); F. Schäffler, G. Hughes, W. Drube, R. Ludeke, and F. J. Himpsel, Phys. Rev. B **35**, 6328 (1987).
[10]P. Skeath, C. Y. Su. I. Hino, and W. E. Spicer, Appl. Phys. Lett. **39**, 349 (1981).
[11]R. Ludeke, T.-C. Chiang, and T. Miller, J. Vac. Sci. Technol. B **1**, 581 (1983).
[12]K. Stiles, A. Kahn, D. G. Kilday, and G. Margaritondo, J. Vac. Sci. Technol. B **5**, 987 (1987).
[13]R. Cao, K. Miyano, T. Kendelewicz, K. K. Chin, I. Lindau, and W. E. Spicer, J. Vac. Sci. Technol. B **5**, 998 (1987).
[14]M. Grioni, J. J. Joyce, and J. H. Weaver, J. Vac. Sci. Technol. A **3**, 918 (1985).
[15]A. Zur, T. C. McGill, and D. L. Smith, Phys. Rev. B **28**, 2060 (1983).
[16]C. B. Duke and C. Mailhiot, J. Vac. Sci. Technol. B **3**, 1170 (1985).
[17]W. E. Spicer, P. W. Chye, P. R. Skeath, C. Y. Su, and I. Lindau, J. Vac. Sci. Technol. **16**, 1422 (1979). W. E. Spicer, T. Kendelewicz, N. Newman, R. Cao, C. McCants, K. Miyano, I. Lindau, and E. R. Weber (to be published).
[18]N. Newman, W. E. Spicer, T. Kendelewicz, and I. Lindau, J. Vac. Sci. Technol. B **4**, 931 (1986).
[19]R. E. Allen and J. D. Dow, Phys. Rev. B **25**, 1423 (1982).
[20]S.-F. Ren and R. E. Allen, Surf. Sci. **148**, L637 (1984).
[21]A. Taleb-Ibrahimi, G. Jezequel, and R. Ludeke, J. Vac. Sci. Technol. B **6**, 1427 (1988).
[22]G. Jezequel, A. Taleb-Ibrahimi, R. Ludeke, and F. Schäffler, J. Vac. Sci. Technol. A **6**, 1561 (1988).
[23]A. B. McLean, R. H. Williams, and J. F. McGilp, Solid State Commun. **65**, 1415 (1988).
[24]A. J. Bennett and L. M. Falicov, Phys. Rev. **151**, 512 (1966).
[25]J. W. Gadzuk, Surf. Sci. **6**, 133 (1967).
[26]S. K. Lyo and R. Gomer, *Topics in Applied Physics*, edited by R. Gomer (Springer, New York, 1975), Vol. 4, Chap. 2.
[27]M. Jaros, *Deep Levels in Semiconductors* (Adam Hilger Ltd, Bristol, 1982).
[28]D. M. Newns, Phys. Rev. **178**, 1123 (1969); P. W. Anderson, *ibid.* **124**, 41 (1961).
[29]S. Brand, M. Jaros, and C. O. Rodriguez, J. Phys. C **14**, 1243 (1981).
[30]J. C. Inkson, J. Phys. C **4**, 591 (1971).
[31]D. M. Newns, J. Chem. Phys. **50**, 4572 (1969).
[32]P. K. W. Vinsome and D. Richardson, J. Phys. C **4**, 2650 (1971).
[33]A. Huijser, J. Van Laar, and T. L. Van Rooy, Surf. Sci. **62**, 472 (1977).
[34]A. W. Dweydari and C. H. B. Mee, Phys. Status Solidi A **27**, 223 (1975).
[35]The analogy may be made between the maximum effective impurity density which generates the acceptor gap states and the density of dangling Ga bonds for the free surface. Thus the average separation between impurity sites is about 4 Å, which results in negligible broadening [a rough estimate can be obtained from Fig. 3(b)]. With metal coverage, the overlap with the metallic states is much larger, consequently the metal induced broadening would dominate eventually.
[36]I. Lefebvre, M. Lannoo, C. Priester, G. Allen, and C. Delerue, Phys. Rev. B **36**, 1336 (1987).
[37]S. B. Zhang, M. L. Cohen, and S. G. Louie, Phys. Rev. B **34**, 768 (1986).
[38]R. Ludeke (unpublished).

EUROPHYSICS LETTERS 1 October 1988

Europhys. Lett., 7 (3), pp. 275-279 (1988)

Tight-Binding Model of Surface Donor-States Induced by Metal Adatoms on GaAs(110) Surfaces.

W. MÖNCH

Laboratorium für Festkörperphysik and Sonderforschungsbereich 254
Universität Duisburg, D-4100 Duisburg, F.R.G.

(received 21 March 1988; accepted in final form 15 July 1988)

PACS. 73.20H – Impurity and defect levels; energy levels of adsorbed species.

Abstract. – Metals deposited on clean GaAs(110) surfaces held at low temperatures are inducing extrinsic surface states of donor type. For metals with an outer *s*-electron the energy levels of such surface donors are calculated in a tight-binding model. The approach used describes the adatom-substrate bonding as a diatomic, heteropolar «surface-molecule» in which the metal *s*-electrons are forming bonds with dangling sp^3-hybrids of surface Ga-atoms. The chemical trend of the calculated energy of the bonding states as a function of the atomic term values of the metal *s*-electrons is in agreement with experimental data.

Clean and well-cleaved GaAs(110) surfaces exhibit no intrinsic surface states within the band gap and the bands are thus flat up to the surface. The adsorption of, for example, chlorine at room temperature causes the build-up of a depletion layer on GaAs samples doped *n*-type, while the bands are remaining unbent with substrates doped *p*-type [1]. Such behaviour indicates the induction of extrinsic surface states of acceptor type. With metals (see, for example, ref. [2, 3]), oxygen [4] and also hydrogen [5] deposited or adsorbed on GaAs(110) surfaces at room temperature the formation of depletion layers was observed independent of the type of bulk doping. When, however, the substrates were held at approximately 100 K during the evaporation of metals such as Al, Ga, In, Ag, Au and Sn the bands were found to remain flat on substrates doped *n*-type but to bend upward on samples doped *p*-type [2, 3]. These findings evidence the formation of extrinsic surface states of donor type. A detailed analysis of experimental data at submonolayer coverages revealed each metal adatom to contribute one of those surface donors [6].

The different behaviour found with metals deposited at room and at low temperature is due to the reduced mobility and reactivity of the adatoms at low temperatures. At submonolayer coverages, a decreased surface mobility favours the occurrence of single adatoms and nonmetallic clusters but suppresses the formation of metallic islands which exhibit identical Fermi-level positions within the semiconductor band gap at the interface or, in other words, cause depletion layers independent of the type of doping. A lower reactivity, on the other hand, reduces the possible creation of defects the energy levels of which would modify the barrier heights at metal-semiconductor contacts.

Meanwhile, the variation of the Fermi level within the band gap has been studied by using core-level photoemission spectroscopy with the alkali-metals Na, K, Rb, Cs, the group III metals Al, In, Ga, the noble metals Ag, Au, the transition metal Mn and also Sn deposited on cleaved GaAs(110) surfaces held at the temperature of liquid nitrogen [2, 3, 7, 8]. With increasing metal coverage, the Fermi level was always found to stabilize on samples doped p-type when the coverage exceeds approximately 10^{-2} monolayers and then to move to its final position for coverages above some tenth of a monolayer. The initial, low-coverage pinning positions of the Fermi level, which are identified as the energy of the metal-induced surface states of donor type, are plotted *vs.* the first ionization energies of the free atoms [9] in fig. 1. The diagram also contains data measured with some of the transition metals, the rare-earth metal Tm and Ca deposited at room temperature [10-14]. This procedure seems to be justified since depositions of Mn at room and at low temperature are giving the same energy position of the stabilized Fermi level on p-DaAs(110) [8, 10]. This finding seems to be plausible since after evaporations at room temperature Mn atoms, as other transition metal atoms, too, were found to at least partly replace surface Ga atoms and they are thus in isolated, adatomlike positions.

Figure 1 shows the energies of the donor-type surface states induced by metal adatoms at GaAs(110) surfaces to be well correlated with the first ionization energies of the free metal atoms. This chemical trend shall be investigated in a tight-binding approach by describing the adatom-substrate bonding as a diatomic, heteropolar «surface molecule». Such a simple model seems to be justified since, for example, the decrease of the work function of Si(111) as well as GaAs(110) surfaces caused by the adsorption of Cs could be explained by localized dipoles [15, 16]. Adsorbate bond lengths were determined for Cl [17], Br [18], and K [19] adsorbed on Si and for Cl [17] adsorbed on Ge surfaces by using surface-extended X-ray absorption fine structures (SEXAFS) [17, 19] or X-ray standing waves [18]. The nearest-neighbour distances between substrate and adatoms were found to follow Pauling's rule [20] which estimates bond lengths in heteropolar molecules as the sum of the covalent radii of the involved atoms corrected for the difference of their electro-negativities. Using this rule, then the resulting dipole charge agrees with the charge transfer which is estimated for a partial ionic bond between adsorbed and substrate atoms from the difference of their electronegativities [20, 21]. The surface-molecule approach is also strongly supported by the

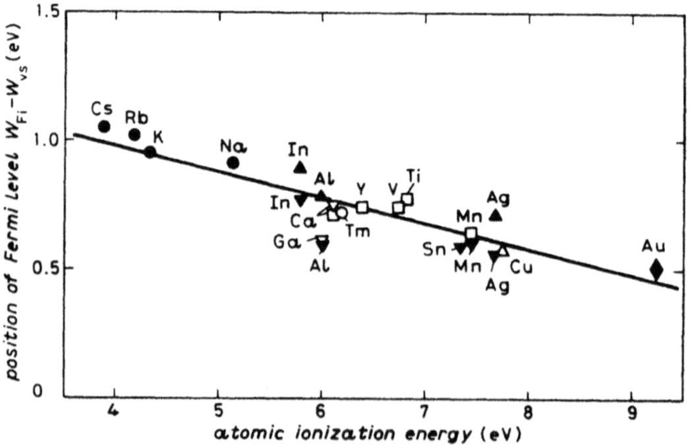

Fig. 1. – Initial pinning position of Fermi level above the valence-band top as a function of first ionization energy of metal atoms deposited on p-GaAs. Full and empty symbols indicate metal depositions on substrates held at liquid-N_2 temperatures and room temperature, respectively. ▽ data from ref. [2, 8, 14], △data from ref. [3], ○ data from ref. [7, 12], □ data from ref. [10, 11 13].

self-consistent electronic-structure calculations for Al on Ge(100) surfaces presented by Batra and Ciraci [22]. At a coverage of half a monolayer of Al, they found the adatoms not to interact with each other and to form covalent bonds with substrate atoms.

In the surface-molecule model the adatoms are considered to bond with a dangling hybrid of a surface atom of the substrate. For adatoms with an outer s-electron the energy of the bonding state is then given by [23]

$$W_b = \frac{W_h^i + W_s}{2} - \sqrt{\left(\frac{W_h^i - W_s}{2}\right)^2 + V_2^2}, \tag{1}$$

where W_s is the atomic term value of the s-level of the adatoms. The energy W_h^i of the dangling-bond sp^3 hybrid is the average

$$W_h^i = \frac{W_s^i + 3W_p^i}{4}, \tag{2}$$

of the s- and p-term value of the free atoms. The superscript i denotes whether substrate cations (c) or anions (a) are considered. The matrix element is obtained as

$$V_2 = \frac{-V_{ss\sigma} + 3V_{sp\sigma}/\sqrt{3}}{2}. \tag{3}$$

The interatomic matrix elements $V_{ss\sigma}$ and $V_{sp\sigma}$ are assumed to scale as the interatomic distances d as

$$V_{ll'm} = \eta_{ll'm} \cdot \frac{\hbar^2}{m_0} \cdot \frac{1}{d^2}, \tag{4}$$

where m_0 is the electron mass. The interatomic distances are taken as the sum of the covalent radii of the adatom and the respective substrate atom

$$d = r_{\text{cov}}^{\text{ad}} + r_{\text{cov}}^{\text{i}}. \tag{5}$$

In the present approach the same basis set of parameters is applied which was successfully used by Harrison [24] for fitting the bands of covalent solids in a tight-binding theory. These parameters are the Hartree-Fock orbital binding energies computed by Fischer [25] and the coefficients in the interatomic matrix elements (4) proposed by Harrison [24].

The charge transfer in the surface molecules may be partly screened by the substrate charge density which may be considered in (1) by introducing a surface dielectric constant. This was found to have only little effect on the binding energies, and it was, therefore, neglected in the present analysis.

In fig. 2, the energies of the binding states (1) calculated for surface molecule consisting of Na, K, Rb, Cs, Cu, Ag, and Au adatoms and surface Ga atoms are plotted *vs.* the atomic term values of the outer s-electrons of those alkali and noble metal atoms. From the data displayed in fig. 2 the following conclusions may be drawn. First of all, the theoretical model excellently confirms the linear correlation which was found between the experimental energies of the donor-type surface states induced by metals adsorbed on GaAs(110) surfaces and the first ionization energies of the free metal atoms and which is shown in fig. 1. Second, the energies of the binding states calculated for the adatom-Ga surface molecules are found to lie within the tight-binding band-gap of GaAs. The top of the valence and the bottom of

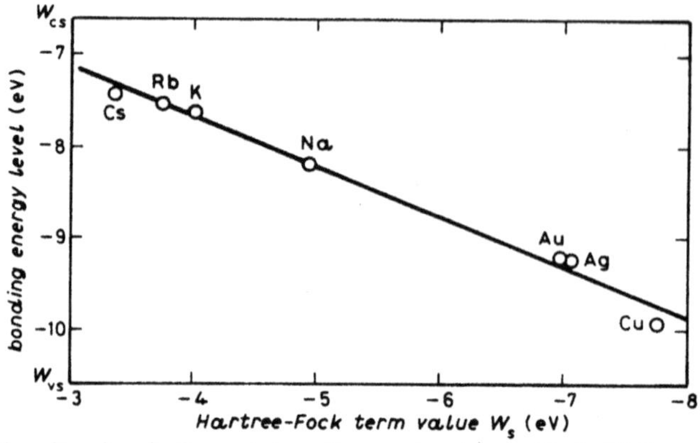

Fig. 2. – Energy of bonding level of s-metal surface-molecules on GaAs(110) surfaces as a function of the Hartree-Fock atomic-term values of the s-electrons of the metal adatoms.

the conduction band are given by [26]

$$W_v = \frac{W_p^c + W_p^a}{2} - \sqrt{\left(\frac{W_p^c - W_p^a}{2}\right)^2 + (4E_{xx})^2},$$

with the matrix element

$$E_{xx} = \frac{V_{pp\sigma} + 2V_{pp\pi}}{3}$$

and

$$W_c = \frac{W_s^c + W_s^a}{2} - \sqrt{\left(\frac{W_s^c - W_s^a}{2}\right)^2 + (4V_{ss\sigma})^2},$$

respectively. These band edges are represented by the lower and upper boundaries, respectively, of fig. 2. It shall be mentioned that in the present minimal-basis tight-binding model the band gap is by a factor 2.5 larger than the experimental value which behaviour is characteristic for such crude approaches [23].

For adatom-As surface-molecules, on the other hand, the energies of the binding states are found by 1.75 to 2.75 eV below of the valence-band maximum. In addition, the «average hybrids» as discussed by Tersoff and Harrison [27] in their studies of the dangling bonds in Ga vacancies of GaAs were considered. For such adatom-⟨Ga/As⟩ surface molecules the energies of the binding states are shifted to slightly above and below the valence-band top for the alkali and the noble metals, respectively.

The results presented here clearly show that the chemical trend, which is observed with the energy of the metal-induced donor-type surface states on GaAs(110) surfaces, is well described by a surface-molecule model using a minimal-basis tight-binding approach. The results of the calculations seem to indicate that the s-electron metal atoms are predominantly bond with surface Ga atoms. The observation that on GaAs(110) surfaces the surface excitonic energy loss of 20 eV, which results from electronic transitions from Ga($3d$) core levels into empty Ga dangling-bond surface states, has completely decayed after the adsorption of half a monolayer of cesium [28], is strongly supporting this conclusion.

The adsorption of metals such as Al, Ga, and In, which exhibit two s- and one p-electron in their outer electron shells, needs a refinement of the surface molecule approach which has to include the interaction of sp^3 substrate and sp^2 adatom hybrids. Such a model is developed for Al adsorbed on Si(111) surfaces by Klepeis and Harrison [29]. Even more elaborate models are necessary for the treatment of the acceptor-type surface states induced by the adsorption of chlorine [1] sulfur [30], and oxygen [31, 32] on GaAs(110) surfaces. As in the present case of the *s*-electron metal adatoms, a surface-molecule approach should at least give the correct chemical trends for the more complicated systems.

* * *

The author should like to thank Prof. W. A. HARRISON for fruitful and encouraging discussions. Stimulating comments by L. KOENDERS and Prof. J. POLLMANN are acknowledged. This study was partly supported by the Deutsche Forschungsgemeinschaft.

REFERENCES

[1] TROOST D., KOENDERS L., FAN L.-Y. and MÖNCH W., *J. Vac. Sci. Technol. B*, **5** (1987) 1119.
[2] STILES K., KAHN A., KILDAY D. and MARGARITONDO G., *J. Vac. Sci. Technol. B*, **5** (1987) 987.
[3] CAO R., MIYANO K., KENDELEWICZ T., CHIN K. K., LINDAU I. and SPICER W. E., *J. Vac. Sci. Technol. B*, **5** (1985) 998.
[4] MÖNCH W., in *Festkörperprobleme: Advances in Solid State Physics*, edited by P. GROSSE, Vol. **24**, (Vieweg, Braunschweig) 1984, p. 229.
[5] BARTELS F., SURKAMP L., CLEMENS H. J. and MÖNCH W., *J. Vac. Sci. Technol. B*, **1** (1983) 756.
[6] MÖNCH W., *J. Vac. Sci. Technol. B*, **6** (1988) in press (Proc. PCSI-15).
[7] PRIETSCH M., *Dissertation* (FU, Berlin) 1987.
[8] STILES K., HORNG S. F., KAHN A., MCKINLEY J., KILDAY D. G. and MARGARITONDO G., *J. Vac. Sci. Technol. B*, **6** (1988) in press.
[9] WEAR R. C. (Editor), *Handbook of Chemistry and Physics* (CRC Press, Boca Raton, FL) 1987.
[10] HUGHES G., LUDEKE R., SCHÄFFLER F. and RIEGER D., *J. Vac. Sci. Technol. B*, **4** (1986) 924.
[11] LUDEKE R. and LANDGREN G., *Phys. Rev. B*, **33** (1986) 5526.
[12] PRIETSCH M., DOMKE M., LAUBSCHAT C. and KAINDL G., *Phys. Rev. Lett.*, **60** (1988) 436.
[13] LUDEKE R. and JEZEQUEL G., private communication.
[14] MAO D., YOUNG K., STILES K. and KAHN A., private communication.
[15] MÖNCH W., *Phys. Status Solidi*, **40** (1970) 257.
[16] CLEMENS H. J., v. WIENSKOWSKI J. and MÖNCH W., *Surf. Sci.*, **78** (1978) 648.
[17] CITRIN P. H., ROWE J. E. and EISENBERGER P., *Phys. Rev. B*, **28** (1983) 2299.
[18] FUNKE B. and MATERLIK G., *Surf. Sci.*, **188** (1987) 378.
[19] KENDELEWICZ T., SOUKIASSIAN P., LIST S., WOICIK J. C., PIANETTA P., LINDAU I. and SPICER W. E., *Phys. Rev. B*, **37** (1988) 7115.
[20] PAULING L., *The Nature of the Chamical Bond* (Cornell University, Ithaca, N.Y.) 1960.
[21] HANNEY N. B. and SMITH C. P., *J. Am. Chem. Soc.*, **68** (1946) 171.
[22] BATRA I. P. and CIRACI S., *Phys. Rev. B*, **33** (1986) 4312.
[23] HARRISON W. A., *Electronic Structure and the Properties of Solids* (Freeman, San Francisco, CAL) 1980.
[24] HARRISON W. A., *Phys. Rev. B*, **24** (1981) 5835.
[25] FISCHER C., *Atomic Data*, **4** (1972) 301.
[26] CHADI D. J. and COHEN M. L., *Phys. Status Solidi B*, **68** (1975) 405.
[27] TERSOFF J. and HARRISON W. A., *Phys. Rev. Lett.*, **58** (1987) 2367.
[28] GRUNWALD F., Diplomarbeit, Universität Duisburg (1987).
[29] KLEPEIS J. E. and HARRISON W. A., *J. Vac. Sci. Technol. B*, **6** (1988) in press (Proc. PCSI-15).
[30] KOENDERS L., BLÖMACHER M. and MÖNCH W., *J. Vac. Sci. Technol. B*, **6** (1988) in press (Proc. PCSI-15).
[31] STROSCIO J. A., FEENSTRA R. M. and FEIN A., *Phys. Rev. B*, **36** (1987) 7718.
[32] STILES K., MAO D. and KAHN A., *J. Vac. Sci. Technol. B*, **6** (1988) in press (Proc. PCSI-15).

Switching of Band Bending at the Nonreactive CsO_x/GaAs(110) Interface

C. Laubschat, M. Prietsch, M. Domke, E. Weschke, G. Remmers, T. Mandel, J. E. Ortega,[a] and G. Kaindl
Institut für Atom und Festkörperphysik, Freie Universität Berlin, Arnimallee 14, D-1000 Berlin 33, Germany
(Received 9 January 1989)

The behavior of band bending in nonreactive interfaces between thin CsO_x films and GaAs(110) is studied by photoemission. Successive cesium and oxygen exposures change the CsO_x stoichiometry in a way that leads to a back-and-forth switching of overlayer metallicity and band bending. For both *n*- and *p*-type GaAs(110), oxygen-rich nonmetallic overlayers create almost flat-band conditions, while cesium-rich metallic overlayers result in Fermi-level positions close to midgap. These observations demonstrate the dominant influence of metal-induced gap states in nonreactive metal-semiconductor interfaces.

PACS numbers: 73.30.+y, 73.20.Hb, 79.60.Gs

The investigation of metal-semiconductor interfaces has raised increasing interest in the recent past mainly due to technological aspects.[1,2] Despite widespread applications of Schottky diodes, the basic mechanisms of Schottky-barrier formation are still not fully understood and are being discussed in a controversial way. It is by now well accepted that interfacial states in the semiconductor band gap play a crucial role in determining band bending and hence the Schottky-barrier height. Essentially two types of such interfacial states are discussed: (i) defect states that are connected with vacancies or antisite defects in the interfacial layer of the semiconductor,[3,4] and (ii) virtual gap states of the semiconductor coupling to itinerant electronic states of the overlayer that penetrate a few Å into the semiconductor.[5-9] In the case of metallic overlayers, these states are denoted as metal-induced gap states (MIGS). Both types of interfacial states can play a role at metal-semiconductor interfaces, raising the question of their relative importance in determining the Fermi-level position. In the case of GaAs(110), both the defect states and the MIGS have been claimed to result in a Fermi-level position close to midgap.[4,8,9]

In this context, core-level photoemission (PE) from interfaces formed by depositing thin metal films on single-crystalline semiconductor substrates has proven to be a powerful tool for gaining detailed information on band bending from the binding energies (BE) of semiconductor core levels. In addition, the PE line shapes contain information on interfacial reactivity and overlayer metallicity, allowing an analysis of their influence on band bending during interface formation. Recently, an abrupt change in Fermi-level position upon metallization of the overlayer was observed for the reactive Tm/GaAs(110) interface; it was attributed to a change in the dominant band-bending mechanism from defect-state to MIGS pinning.[10] A dominance of MIGS relative to defect states was also concluded from a PE study of the Na/GaAs(110) interface,[11] where the same Fermi-level positions were obtained for reactive and nonreactive interfaces in the case of thick metallic overlayers.

In the present Letter, we report on the first observation of an adjustable band bending as a function of overlayer metallicity for nonreactive interfaces of Cs oxides with GaAs(110). While approximately 2 monolayers (ML) of Cs on both *n*- and *p*-type GaAs(110) lead to interfacial Fermi-level positions close to midgap, additional deposition of oxygen causes a drastic decrease in band bending, with Fermi-level positions close to those of the uncovered substrates. This behavior is explained by oxidation of the Cs overlayer, i.e., loss of its metallic properties, while the GaAs substrates are chemically unaffected. Consecutive deposition of Cs leads again to a metallization of the overlayer and hence to Fermi-level positions close to midgap, as in the original cesium-metal-exposed case, while subsequent oxygen exposure results in a further strong decrease in band bending. These observations strongly support the applicability of the MIGS model in the case of nonreactive interfaces.

The PE measurements were performed with synchrotron radiation from the HE-PGM-2 beam line at the Berliner Elektronenspeicherring für Synchrotronstrahlung (BESSY) employing a hemispherical electron-energy analyzer operated in normal-emission geometry. The overall-system resolution (FWHM) was $\simeq 0.2$ eV (at $h\nu = 84$ eV), and BE changes could be determined to an accuracy of ± 0.1 eV. Mirrorlike (110) surfaces were obtained by cleavage of Si-doped *n*-type GaAs or Zn-doped *p*-type GaAs (with dopant concentrations of $\simeq 3\times 10^{17}$ atoms/cm^3), that were kept at temperatures of 140 ± 20 K. Cs films were deposited from commercial chromate dispensers with deposition rates of $\simeq 0.5$ ML/min. Relative coverages were calibrated via deposition time (estimated accuracy of $\pm 10\%$), while absolute coverages were obtained from the coverage dependence of the plasmon energy (estimated accuracy of $\pm 30\%$).[12] During oxygen exposure the ion gauge was kept operating, and a small amount of molecular oxygen may be activated.[13]

Figure 1 shows representative wide-scan PE spectra for a cleaved *n*-type GaAs(110) surface, and (b)-(d) after successive cesium and oxygen exposures at $\simeq 140$ K, as well as (e) after annealing at 800 K. The spectrum from freshly cleaved *n*-type GaAs(110) [Fig. 1(a)] is dominated by intense emission from the As-3*d* and Ga-3*d* core levels, while the valence-band emission is very weak. Deposition of 2 ML of Cs [Fig. 1(b)] results in additional emission lines from the Cs-5*s* and Cs-5*p* core levels as well as in strong satellites for all PE lines at $\simeq 2$ eV higher BE (indicated by arrows) that are assigned to extrinsic plasmon-energy losses in the Cs layer. Additional adsorption of 12 L [1 Langmuir (L) $= 10^{-6}$ Torr Sec] oxygen [Fig. 1(c)] leads to a quenching of these satellites as well as to the appearance of a multipeaked, O-2*p*-derived valence-band emission that is assigned mainly to Cs_2O_2 (Refs. 14 and 15). Upon subsequent deposition of 2 ML of Cs [Fig. 1(d)], this oxide layer is chemically reduced mainly to Cs_2O, leading again to the appearance of plasmon-loss satellites.

The absence of chemically shifted components in the As-3*d* and Ga-3*d* core-level PE lines [Figs. 1(a)-1(d)] clearly shows that the oxidation of the Cs adsorbate layer is not affecting the chemical composition of the substrate at the interface. On the other hand, annealing of a CsO_x/GaAs(110) interface at $\simeq 800$ K [Fig. 1(e)] leads to additional components shifted to higher BE by $\simeq 3.5$ eV (As 3*d*) and $\simeq 1$ eV (Ga 3*d*), respectively, due to substrate oxidation.[13] These observations show that the CsO_x/GaAs(110) interface formed at $\simeq 140$ K is nonreactive and abrupt.

In Fig. 2, the Ga-3*d* PE core-level lines are shown in more detail for both *n*- and *p*-type substrates, exposed successively to cesium, oxygen, cesium, and oxygen. The striking information contained in Fig. 2 is the fact that even repeated exposures with cesium and oxygen, respectively, shift the Ga-3*d* lines in opposite directions for

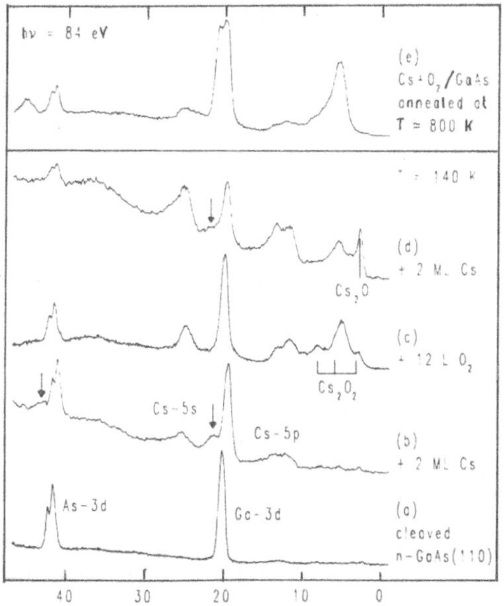

FIG. 1. Wide-scan PE spectra of (a) a cleaved *n*-type GaAs(110) surface at $T \simeq 140$ K, and (b)-(d) after successive depositions of cesium and oxygen. Plasmon-loss satellites are marked by arrows. For comparison, the spectrum of a CsO_x/GaAs(110) interface after annealing at $\simeq 800$ K is shown in (e). The spectra are normalized to equal height.

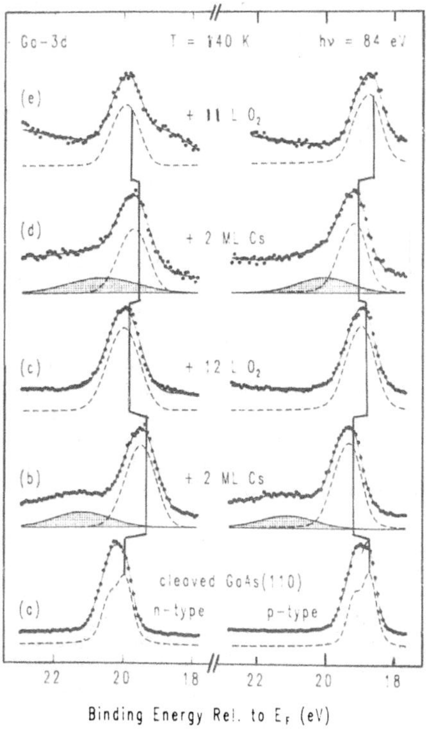

FIG. 2. Ga-3*d* PE spectra of cleaved *n*- and *p*-type GaAs(110) surfaces and after successive cesium and oxygen exposures ($T \simeq 140$ K). Dashed subspectra represent the substrate bulk, dotted subspectra represent the substrate surface, and the shaded subspectra correspond to plasmon-loss satellites that are assigned to overlayer metallicity. The shoulder at $\simeq 19$ eV BE in (e) is attributed to O-2*s* emission. The vertical solid line traces the positions of the Ga-$3d_{5/2}$ components. Note the obvious correlation of band-bending-induced BE shifts and overlayer metallicity monitored by plasmon-loss satellites. The spectra are normalized to equal height.

both n- and p-type substrates. Spectra (a) and (b) at the bottom of Fig. 2 reveal that the initial deposition of 2 ML of Cs causes BE shifts of ≃0.6 eV in opposite directions for n- and p-type GaAs, which are attributed to band bending towards a common Fermi-level position close to midgap typical for alkali-metal/GaAs(110) interfaces.[11,12] The appearance of a plasmon-energy-loss satellite (shaded subspectrum) at ≃2 eV higher BE [Fig. 2(b)] reflects a delocalization of valence electrons in the Cs layer, indicating overlayer metallization. It was previously found that a plasmon-loss energy of ≃2 eV is reached when the Cs overlayer metallizes at coverages between 1.5 and 2 ML.[12] In this way we use the plasmon-loss satellites as a monitor for metallization. Upon additional adsorption of 12 L oxygen, the plasmon-loss structures disappear [Fig. 2(c)], signaling a demetallization of the overlayer due to oxidation. This causes the backshifts of the Fermi-level positions to almost the original values observed for freshly cleaved n- and p-type GaAs(110). By additional deposition of 2 ML of Cs, the metallicity of the overlayer as well as the Fermi-level positions close to midgap are reestablished [Fig. 2(d)], while another exposure to 11 L oxygen [Fig. 2(e)] results again in a nonmetallic overlayer accompanied by a strong decrease in band bending.

The spectra in Fig. 2 were least-squares fitted with a superposition of Gaussian lines for instrumental and inhomogeneous broadening convoluted with a Lorentzian containing lifetime effects (FWHM = 0.2 eV). They consist of spin-orbit-split doublets with intensity ratios close to 3:2 and a splitting of 0.46 ± 0.02 eV,[11,12] plus a background that was approximated in all cases by straight lines. This curve-fitting procedure is most clearly visible in the spectra of the freshly cleaved GaAs substrates presented in Fig. 2(a), which are deconvoluted into bulk (dashed) and surface (dotted) components. Note also the Ga-3d BE difference of ≃1.3 eV for n- and p-type GaAs that reflects the different Fermi-level positions in the GaAs band gap close to the conduction-band minimum (CBM) in case of n-type and the valence-band maximum (VBM) in case of p-type GaAs(110), respectively.

The variation in band bending upon successive cesium and oxygen exposure on GaAs(110) kept at ≃140 K is shown in more detail in Fig. 3. This data plot clearly shows that a common Fermi-level position is obtained for n- and p-type GaAs(110) for a Cs coverage of 2 ML,[12] while overlayer oxidation leads to a decrease in band bending for both doping types, approaching flat-band conditions. The second Cs exposure results again in an increase in band bending, but a common Fermi-level position is no longer obtained. The second oxygen exposure flattens the bands again.

In this way it becomes clear that the switching of overlayer metallicity is directly related to the changes in band bending. Such a behavior unambiguously rules out defect states near midgap as a possible cause for band bending at these interfaces, since defects can obviously not be removed by oxidation. On the other hand, the initial Fermi-level positions for n- and p-type GaAs(110) are not completely reproduced by oxidation of the metallic overlayer, reflecting the fact that the CsO_x/GaAs(110) interface represents some kind of a semiconductor heterojunction.[16] Therefore, the Fermi-level positions for n- and p-type GaAs(110) are expected to be governed by the energetic positions of the CBM and VBM, respectively, of the cesium-oxide overlayer, which in turn may be expected to change with overlayer stoichiometry. Such a dependence on stoichiometry could be the reason for our observation of slightly different Fermi-level positions after the first and second oxygen exposures (see Fig. 3).

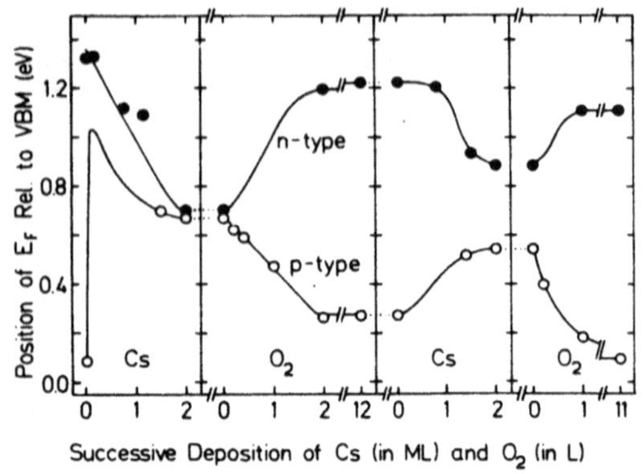

FIG. 3. Variation in Fermi-level position with respect to the VBM (accuracy of ±0.1 eV) for n- and p-type GaAs(110), kept at ≃140 K, as a function of successive depositions of cesium and oxygen. The data are obtained from least-squares fits of As-3d and Ga-3d PE spectra (see Fig. 2) and contain a correction for finite sampling depth of the PE measurements (Refs. 11 and 12). The initial band-bending curve for Cs/p-GaAs(110) is taken from Ref. 12.

Differences in overlayer stoichiometry may also be responsible for the fact that the original metal-induced band bending is not fully reproduced upon the second deposition of 2 ML of Cs on the oxidized interface. The chemical reduction of Cs_2O_2 during the second Cs deposition [see Fig. 1(d)] may produce a mixture of Cs_2O and metallic Cs that is characterized by a lower density of conduction electrons than is effective in an elemental Cs overlayer. Such a mechanism is actually supported by the lower plasmon energy of ≃1.5 eV observed in this case [see Fig. 2(d)]. Furthermore, inhomogeneities in the overlayer may result in some kind of a Cs/Cs_2O/GaAs(110) multilayer structure, where the influence of the cesium-metal layer on the semiconductor is screened by the oxide layer. Both effects tend to decrease the density of MIGS, which may then not be sufficiently high for a complete band bending to a common Fermi-level position for both *n*- and *p*-doped substrates, as observed (see Fig. 3). Such a common Fermi-level position is theoretically expected for band bending through MIGS[5–9,17] and has been observed recently upon metallization.[11,12]

In summary, the present work clearly shows that the Fermi-level position at nonreactive interfaces between Cs oxides and GaAs(110) is directly related to the metallicity of the overlayer. This observation provides strong evidence for the validity of the MIGS model, and conclusively rules out the concept of defect states close to the center of the band gap of GaAs as the origin of the observed band-bending effects in the case of nonreactive interfaces.

The authors acknowledge valuable discussions with Professor Miranda. This work was supported by the Bundesminister für Forshung und Technologie, Project No. 05 313AX B2/TP3.

---

[a]On leave from Departamento de Fisica de la Materia Condensada, Universidad Autonoma de Madrid, Cantoblanco, E-28049 Madrid, Spain.

[1]E. H. Rhoderick, *Metal-Semiconductor Contacts* (Clarendon, Kensington, Australia, 1980).

[2]L. J. Brillson, Surf. Sci. Rep. **2**, 123 (1982).

[3]W. E. Spicer, P. W. Chye, P. R. Skeath, C. Y. Su, and I. Lindau, J. Vac. Sci. Technol. **16**, 1322 (1979).

[4]I. Lindau and T. Kendelewicz, CRC Crit. Rev. Solid State Mater. Sci. **13**, 27 (1986).

[5]V. Heine, Phys. Rev. **138**, A1689 (1965).

[6]E. Louis, F. Yndurain, and F. Flores, Phys. Rev. B **13**, 4408 (1976).

[7]M. L. Cohen, Adv. Electron. Electron Phys. **51**, 1 (1980).

[8]J. Tersoff, Phys. Rev. Lett. **52**, 465 (1984).

[9]F. Flores and C. Tejedor, J. Phys. C **20**, 145 (1987).

[10]M. Prietsch, M.Domke, C. Laubschat, and G. Kaindl, Phys. Rev. Lett. **60**, 436 (1988).

[11]M. Prietsch, C. Laubschat, M. Domke, and G. Kaindl, Europhys. Lett. **6**, 451 (1988).

[12]M. Prietsch, M. Domke, C. Laubschat, T. Mandel, C. Xue, and G. Kaindl, Z. Phys. B (to be published).

[13]G. Landgren, R. Ludeke, Y. Jugnet, J. F. Morar, and F. J. Himpsel, J. Vac. Sci. Technol. B **2**, 351 (1984).

[14]C. Y. Su, I. Lindau, and W. E. Spicer, Chem. Phys. Lett. **87**, 523 (1982).

[15]B. J. Moss and W. A. Goddard III, J. Chem. Phys. **63**, 3523 (1975).

[16]R. W. Grant and J. R. Waltrop, J. Vac. Sci. Technol. B **5**, 1015 (1987).

[17]A. Zur, T. C. McGill, and D. L. Smith, Phys. Rev. B **28**, 2060 (1983).

Metallicity and Gap States in Tunneling to Fe Clusters on GaAs(110)

P. N. First, Joseph A. Stroscio, R. A. Dragoset, D. T. Pierce, and R. J. Celotta
National Institute of Standards and Technology, Gaithersburg, Maryland 20899
(Received 27 February 1989; revised manuscript received 5 July 1989)

We report the characteristics of tunneling to a GaAs(110) substrate with distinct, nanometer-size Fe clusters, as a function of distance from and size of the clusters. We show that Fe clusters of volumes ~150 Å^3, corresponding to ≃13 atoms, are observed to be nonmetallic with a gap at the Fermi level. Larger clusters with > 35 atoms begin to show metallic characteristics. We observe a continuum of cluster-induced gap states in tunneling to the GaAs substrate surrounding the metallic Fe clusters. The decay length of these states has a minimum decay of 3.4 Å at midgap and diverges at the valence- and conduction-band edges.

PACS numbers: 61.16.Di, 73.20.Dx, 73.30.+y

Considerable interest lies in the electronic properties of small metal particles as a probe of the evolution of electronic structure toward that of the solid state. Small particles with $n = 10-100$ atoms span the regime where the electronic properties develop from atomiclike to metallic characteristics with increasing particle size.[1] In conventional electron tunneling experiments, small particles have been used to study the quantization of electron states[2] and single-electron tunneling effects due to the Coulomb interaction.[3] Recent tunneling measurements, emphasizing quantum size effects, have been made on fabricated and point-contact tunnel junctions,[4,5] but not on well-characterized single particles.

The study of metal clusters on semiconductor surfaces has also been of interest in understanding the fundamental electronic properties of metal-semiconductor interfaces.[6,7] Photoelectron measurements have inferred from linewidth trends that bulk electronic structure develops for clusters with ≃100 atoms.[8] Many of these studies, however, determine cluster sizes indirectly and average over a cluster size distribution. The determination of metallic characteristics in cluster systems by examining the Fermi-edge cutoff in photoelectron spectroscopy has been of recent controversy.[9] As shown below, metallic characteristics are quite evident when tunneling into a partially filled band of states at the Fermi level.

In this Letter, we report scanning tunneling microscopy (STM) experiments that determine the electron tunneling characteristics of *single* nanometer-size Fe clusters epitaxially grown on GaAs(110) substrates. We show that clusters with volumes of ~1000 Å^3 (n~85 Fe atoms) display fully metallic characteristics with a finite differential conductance at zero applied bias, similar to thin Fe films. Clusters with volumes ~150 Å^3 (n~13) are found to be nonmetallic, as evidenced by a finite tunneling gap at the Fermi level. We show further that a continuum of cluster-induced gap states is observed in the tunneling spectra obtained from regions of the semiconductor surrounding the metallic Fe clusters. The state density in the gap is found to decay exponentially with distance from the clusters. Specifically, we show that the decay length is dependent on the eigenstate energy within the band gap. The decay length has a minimum value of 3.4 Å at midgap and diverges at both the valence- and conduction-band edges. These observations represent the first measurements reflecting the dispersion of the *complex* solutions of the GaAs electronic band structure, which have been of central importance in theories of metal-semiconductor interfaces.[10-12]

Fe clusters and films were grown by molecular-beam-epitaxy techniques in an ultrahigh-vacuum system, as described in detail elsewhere.[13] Current-versus-voltage (*I-V*) characteristics were obtained by interrupting the STM feedback loop at each point in the topographic image. The tip-sample distance was decreased a fixed amount to increase the dynamic range of the *I-V* measurements in the region of the GaAs band gap.[14]

Figure 1 shows an STM image of the GaAs(110) surface with 0.1-Å coverage of Fe grown at 300 K. The GaAs substrate appears as rows of atoms along the [1$\bar{1}$0] direction, which is at 45° with respect to the $+x$ direction. Only the substrate As atom locations are observed in the bare surface regions, since the image was obtained at negative sample bias. A variety of different size Fe clusters are observed in the image. STM contours show

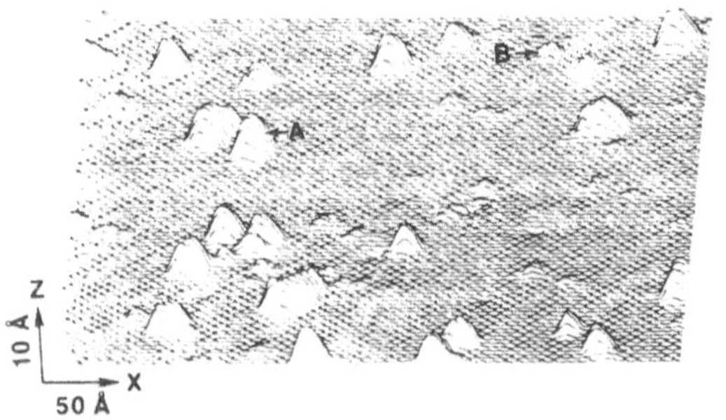

FIG. 1. STM image, 400×384 Å^2, of 0.1-Å-Fe/p-GaAs(110). The sample bias was −2.5 V. Arrow A indicates a cluster of volume 1180 Å^3 and arrow B points to a cluster of volume 160 Å^3.

several different cluster heights indicating 3D cluster growth.[13] LEED observations of coalesced clusters showed the bcc Fe diffraction pattern, consistent with previous work which determined the growth to be epitaxial.[15] Volume estimates of the clusters were achieved by integrating the STM height contours of the clusters, which we feel is a good estimate, since the charge-density contours of metals closely follow the positions of the metal atoms. The volumes of the clusters shown in Fig. 1 range from 100 to 1500 Å^3, corresponding to 9–127 atoms per cluster.

Figure 2(a) shows the tunneling characteristics as a function of cluster size, ranging from a 150-Å^3 cluster to a continuous 17-Å thin film. As a reference, curve *d* in Fig. 2(a) also shows the tunneling characteristics on the bare GaAs surface, 40 Å from the nearest cluster. In

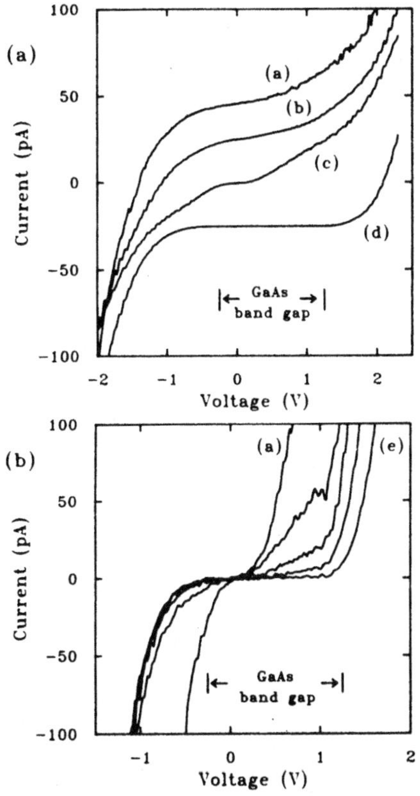

FIG. 2. (a) Top panel: Tunneling current vs voltage characteristics of Fe/*p*-GaAs(110). The four curves correspond to *a*, 17-Å Fe film; *b*, 1150-Å^3 cluster; *c*, 150-Å^3 cluster; and *d*, GaAs(110) surface, 40 Å from the nearest cluster. The curves have been shifted vertically for display. (b) Bottom panel: Tunneling current vs distance from the Fe cluster as indicated in Fig. 3(a). Curve *a* is on the cluster. Curves *b*–*e* correspond to distances from the cluster edge of *b*, 3.7 Å; *c*, 6.7 Å; *d*, 9.6 Å; and *e*, 14.3 Å.

curve *d*, we observe a band gap of 1.5 eV, which is close to the 1.4-eV band gap of GaAs, where no tunneling occurs since cleaved GaAs has no surface states in the band gap. In addition, the *I*-*V* characteristic in curve *d* shows the surface Fermi level pinned at 0.25 eV above the valence band. We found the pinning of the Fermi level to be homogeneous over the surface at the Fe coverages used in Fig. 2.

The *I*-*V* measurement on the thin Fe film in Fig. 2(a), curve *a*, shows current within the band-gap region. In particular, tunneling occurs with a finite differential conductance at zero bias (i.e., there is tunneling both at 0^+ and 0^- V), signifying metallic behavior with a partially filled band of states at the Fermi level. The *I*-*V* characteristic of the 1150-Å^3 cluster (curve *b*) displays metallic behavior similar to that of the thin film, with again a finite differential conductance at zero bias (note also the similarity in the overall characteristics). Deviation from metallic behavior is observed in particles with volumes of ~150–400 Å^3, as shown in Fig. 2(a), curve *c*, where a finite gap in the tunneling current is seen at zero bias. For the small clusters (e.g., curve *c*) gap widths ranging from 0.1 to 0.5 eV have been observed.

The size dependence of the tunneling characteristics in Fig. 2(a) shows that quantum size effects are appreciable for cluster volumes of ~150 Å^3. Such effects are expected when the spacing between energy levels is a few k_BT. The usual simple estimate for the level spacing is given by E_F/N, where E_F is the bulk Fermi energy and N is the number of valence electrons in the particle. This yields a spacing of 0.1 eV for an 11-atom Fe cluster. The cluster size where we observe a band gap of this magnitude corresponds to ~13 atoms, in agreement with the above estimate. The surprising result, perhaps, is not that the smaller particles are nonmetallic, but that metallic behavior begins to be observed in particles with volumes as small as 400 Å^3, corresponding to only ~35 atoms. In photoelectron measurements, bulk metallic behavior has been inferred for particle sizes on the order of 100 atoms.[8] These measurements use bandwidth trends to determine bulk behavior, which may not be a sensitive criterion for the onset of metallicity. Detection of the Fermi edge in photoelectron measurements may be more direct, but is hampered by low intensity and resolution problems.[9] Photoionization measurements, on beams of well-defined clusters, have found indications of band formation for Hg clusters with $n > 12$ atoms.[1] These measurements are consistent with the tunneling results shown here. We also note that detailed calculations for Fe clusters show level spacings of 0.1 eV near the Fermi level for 15-atom clusters,[16] and thus one might expect a transition to room-temperature metallic behavior in the range of 20–30 atoms. A direct comparison with the present results is not possible, since one has to take into consideration the binding of the Fe clusters to the GaAs surface.

The metallic states found on the clusters, which over-

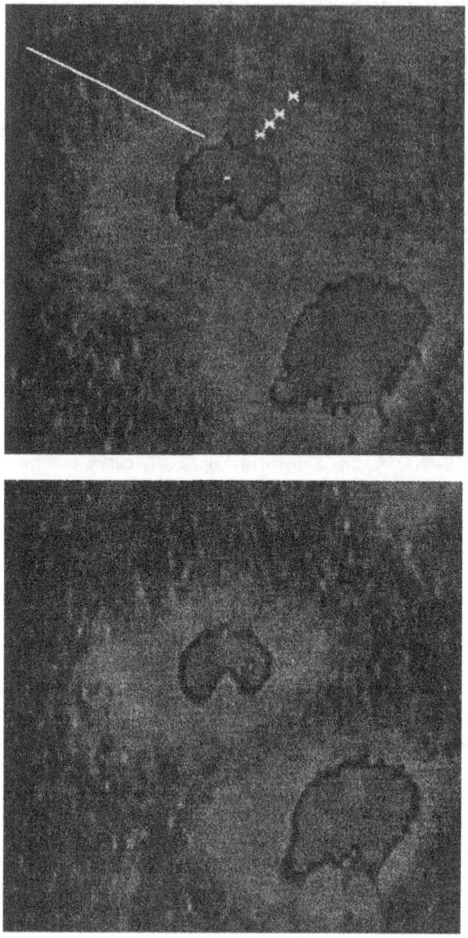

FIG. 3. Images of the logarithm of the differential conductance, $\log(dI/dV)$, obtained from the I-V characteristics of Fe/p-GaAs(110), recorded at each pixel in the normal topograhic image. The images in the top [(a)] and bottom [(b)] panels correspond to voltages 1.1 and 1.0 V, respectively (the observed conduction-band minimum is at 1.25 V). The images are shown with a black-red scale ranging from 0 to 300 pA/V. The blue regions denote areas of large current over the metallic clusters. The positions of the I-V curves, shown in Fig. 2(b), are indicated by the asterisks in (a). The solid line in (a) denotes the position of the [001] line profiles, shown in Fig. 4(a), while the line of asterisks shows the [1$\bar{1}$0] profile direction.

lap in energy the semiconductor band gap, are also observed on the semiconductor surface in the vicinity of the clusters, as shown in Fig. 2(b). Here, curve a displays metallic characteristics from a measurement on one of the Fe clusters shown in the image of Fig. 3(a). Curves b–e still show tunneling current in the GaAs band gap even though they are taken on the GaAs substrate, as indicated in Fig. 3(a). The tunneling current from states in the gap, observed in Fig. 2(b), is found to decrease exponentially with distance from the cluster.

To examine the spatial dependence of the gap states we have recorded I-V characteristics at each point in the topographic image. Images of the differential conductance dI/dV are obtained from the slope of the I-V measurements at voltage V, as shown in Fig. 3. The differential conductance is proportional to the surface density of states;[14] thus, images of dI/dV in the band gap show the spatial distribution of the gap states at an energy $E = eV$ with respect to the Fermi level at 0 V. In Fig. 3(a), we observe a finite conductance surrounding the Fe cluster at a voltage of 1.1 V, which is 0.15 eV below the experimentally observed conduction-band minimum. The contribution from the GaAs gap states is observed to have atomic corrugation with maxima on the Ga sites, as deduced from the phase relationship of the corrugation with that of the topograph. The gap-state amplitude at 1.1 V is found to decay spatially from the cluster with a decay length of 16.8 Å in Fig. 3(a). Figure 3(b) shows the differential conductance image at 1.0 V, just 0.1 eV below that of Fig. 3(a). The region of finite conductance is observed to encompass a smaller region surrounding the Fe cluster with a decay length of 11.4 Å, indicating the gap-state decay to be faster for energies deeper within the gap.

A quantitative examination of the spatial decay of the gap states is made from examining line profiles along the [001] and [1$\bar{1}$0] directions. The line scans along the [001] direction [indicated by the line in Fig. 3(a)] are shown in Fig. 4(a), corresponding to energies with respect to the top of the valence band of 0.95–1.55 eV, in increments of 0.05 eV. The differential conductance is observed to decay exponentially with a slope which increases with decreased energy approaching midgap. The decay length (inverse slope) obtained from a linear fit to these data is shown in Fig. 4(b). The decay length in Fig. 4, obtained on p-type samples from 0.7 to 1.6 eV, is observed to span a range of 3–200 Å, with the minimum corresponding to midgap, and a divergence at the bottom of the conduction band. Similar results are observed for the bottom portion of the gap (−0.15–0.65 eV), obtained on n-type samples, with a minimum decay length at midgap and a divergence at the valence-band edge.

The decay of the gap states in regions of the clean semiconductor, where initially no states exist, arises from matching the cluster wave functions to states in the semiconductor band gap. Within the gap, these wave functions will be evanescent waves determined by the complex solutions of the semiconductor band structure in the band gap.[11,17] At the band edges, the imaginary wave vectors go to zero and hence the decay lengths diverge as the evanescent solutions go over to plane waves. Deeper

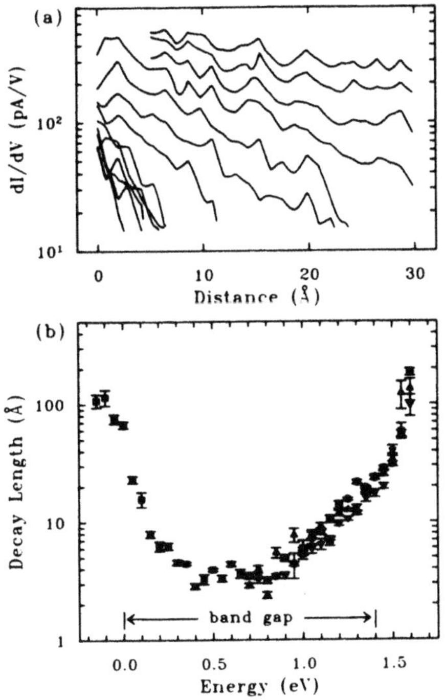

FIG. 4. (a) Line profiles along the [001] direction, indicated by the line in Fig. 3(a), for energies of 0.95–1.55 eV in increments of 0.05 eV. The slower decay corresponds to 1.55 eV. (b) Energy dependence of the decay length of the differential conductance obtained from the inverse slope of the dI/dV line profiles along the [001] (squares and circles) and [1$\bar{1}$0] (up and down triangles) directions. Data from p-type material are shown by the circles and triangles, and from n-type by squares. The tip-sample separation was decreased by 1.0 (circles and up triangles from the data in Fig. 3) and 2.0 Å (squares and down triangles) relative to that used for topographic imaging. The energy scales have been shifted to give zero energy at the valence-band maximum.

in the gap the imaginary wave vectors reach a maximum between the two corresponding bands, usually near midgap, resulting in a loop in complex reciprocal space. The charge-density decay at midgap is ~3 Å for GaAs.[11] The data in Fig. 4 display a decay length at midgap of 3.4 Å and also a divergence at the valence- and conduction-band edges, in agreement with expectations from the complex band structure of GaAs. We attribute the finite decay of the evanescent states in the lower 0.2 eV of the conduction band to the availability of the larger density of states near the L minimum which is at 1.7 eV above the valence-band edge.

In conclusion, we have shown that single nanometer-size Fe clusters, with volumes of 150 Å^3, corresponding to $n \simeq 13$ atoms, are nonmetallic with a gap at the Fermi level. Larger clusters with $n > 35$ atoms show metallic characteristics with a partially filled band at the Fermi level. We attribute this onset of metallicity to the decreasing influence of quantum size effects with increasing cluster size. We have provided a measurement of the spatial characteristics of the semiconductor gap states by showing that the cluster-induced states lying within the semiconductor band gap decay exponentially with distance along the surface. The decay constant is 3.4 Å at midgap and diverges at the valence- and conduction-band edges, reflecting the evanescent character of the complex solutions of the GaAs band structure.

We would like to thank L. Whitman for his expert assistance. We gratefully acknowledge the partial support of the Office of Naval Research.

[1]C. Bréchignac, M. Broyer, Ph. Cahuzac, G. Delecretaz, P. Labastie, J. P. Wolf, and L. Wöste, Phys. Rev. Lett. **60**, 275 (1988).

[2]R. E. Cavicchi and R. H. Silsbee, Phys. Rev. B **37**, 706 (1988).

[3]D. V. Averin and K. K. Likharev, J. Low Temp. Phys. **62**, 345 (1986).

[4]T. A. Fulton and G. J. Dolan, Phys. Rev. Lett. **59**, 109 (1987).

[5]P. J. M. van Bentum, H. van Kempen, L. E. C. van de Leemput, and P. A. A. Teunissen, Phys. Rev. Lett. **60**, 369 (1988).

[6]K. Stiles and A. Kahn, Phys. Rev. Lett. **60**, 440 (1988).

[7]S. Doniach, K. K. Chin, I. Lindau, and W. E. Spicer, Phys. Rev. Lett. **58**, 591 (1987).

[8]G. K. Wertheim, S. B. DiCenzo, and D. N. E. Buchanan, Phys. Rev. B **33**, 5384 (1986).

[9]W. E. Spicer and R. Cao, Phys. Rev. Lett. **62**, 605 (1989).

[10]S. G. Louie, J. R. Chelikowsky, and M. L. Cohen, Phys. Rev. **15**, 2154 (1977).

[11]J. Tersoff, Phys. Rev. Lett. **52**, 465 (1984).

[12]D. Troost, L. Koenders, L.-Y. Fan, and W. Monch, J. Vac. Sci. Technol. B **5**, 1119 (1987).

[13]R. A. Dragoset, P. N. First, J. A. Stroscio, D. T. Pierce, and R. J. Celotta, Mater. Res. Soc. Symp. Proc. (to be published).

[14]R. M. Feenstra, J. A. Stroscio, and A. P. Fein, Surf. Sci. **181**, 295 (1987).

[15]G. A. Prinz and J. J. Krebs, Appl. Phys. Lett. **39**, 397 (1981).

[16]C. Y. Yang, K. H. Johnson, D. R. Salahub, J. Kasper, and R. P. Messmer, Phys. Rev. B **24**, 5673 (1981).

[17]V. Heine, Phys. Rev. **138**, 1689 (1965).

SUBJECT INDEX

AUTHOR INDEX

PERSPECTIVES IN CONDENSED MATTER PHYSICS

Published Volumes

1. G. Margaritondo, *Electronic Structure of Semiconductor Heterojunctions*
2. A.H. MacDonald, *Quantum Hall Effect: a Perspective*
3. M. Ronchetti and G. Jacucci, *Simulation Approach to Solids*
4. W. Mönch, *Electronic Structure of Metal-Semiconductor Contacts*

Forthcoming Volumes

I. Janossy, *Optical Effects in Liquid Crystals*
H. Neddermeyer, *Scanning Tunneling Microscopy*

Executive Editor: L. Miglio

Dipartimento di Fisica dell'Università di Milano
Via Celoria, 16 I-20133 MILANO
Fax + 39/2/2366583; **Telex** 334687 INFNMI; **Tel.** + 39/2/2392.408

finito di stampare nel mese
di settembre 1990
dalla Nuova Timec s.r.l.
Albairate (MI)

Editoriale Jaca Book spa
Via Aurelio Saffi 19, 20123 Milano

spedizione in abbonamento
postale TR editoriale
aut. D/162247/PI/3
direzione PT Milano

GPSR Compliance

The European Union's (EU) General Product Safety Regulation (GPSR) is a set of rules that requires consumer products to be safe and our obligations to ensure this.

If you have any concerns about our products, you can contact us on ProductSafety@springernature.com

In case Publisher is established outside the EU, the EU authorized representative is:

Springer Nature Customer Service Center GmbH
Europaplatz 3
69115 Heidelberg, Germany

Batch number: 10816188

Printed by Printforce, the Netherlands